MW01503297

ELEMENTARY TOPOLOGY

ELEMENTARY TOPOLOGY

DENNIS ROSEMAN

University of Iowa, Iowa City

PRENTICE HALL

Upper Saddle River, New Jersey 07458

Library of Congress Cataloging-in-Publication Data

Roseman, Dennis, date
Elementary Topology / Dennis Roseman
p. cm.
Includes bibliographical references and indexes.
ISBN 0-13-863879-9
1.Topology. I. Title
QA611.R67 1999
514-dc21 97-26005
CIP

Executive editor: George Lobell
Editor-in-chief: Jerome Grant
Executive managing editor: Kathleen Schiaparelli
Managing editor: Linda Mihatov Behrens
Editorial assistants: Nancy Bauer/Gale Epps
Assistant VP production/manufacturing: David W. Riccardi
Manufacturing manager: Trudy Pisciotti
Manufacturing buyer: Alan Fischer
Art director: Jayne Conte
Marketing manager: Melody Marcus
Cover designer: Kiwi Design
Cover image: Dennis Roseman
Text and graphics preparation: Dennis Roseman

© 1999 by Prentice-Hall, Inc.
Upper Saddle River, New Jersey 07458

All rights reserved. No part of this book may be reproduced, in any form or by any means, without permission in writing from the publisher.

10 9 8 7 6 5 4 3 2 1

Printed in the United States of America

ISBN 0-13-863879-9

Prentice-Hall International (UK) Limited, London
Prentice-Hall of Australia Pty. Limited, Sydney
Prentice-Hall Canada Inc., Toronto
Prentice-Hall Hispanoamericana, S.A., Mexico
Prentice-Hall of India, Private Limited, New Delhi
Prentice-Hall of Japan, Inc., Tokyo
Prentice-Hall (Singapore) Pty. Ltd., Singapore
Editora Prentice-Hall do Brazil, Ltda., Rio de Janeiro

TO MY PARENTS

Contents

Preface

Overview

This text provides an introduction to topology. For some students, this is to be their first and last course in topology. For them we wish to present an interesting "stand-alone" text. If this is to be their only exposure to topology, we hope that the student is left with some concrete examples of some interesting objects, some interesting questions involving the plane and three-dimensional space, and the impression that mathematical investigation of these matters is an interesting and useful endeavor.

For those students who continue to a higher level, we wish our introduction to be useful, and yet minimize future duplication of material. We provide a background of interesting examples to hone technical skills that will enable interested students to explore topology at more advanced levels.

Our focus is on examples from Euclidean space, with emphasis on the plane. As much as possible, we maintain an "example-driven" approach, trying to show the need for a concept before formally presenting it. From this text, the student should get some idea of what topology is about, learn more about the plane and space, and gain skills in constructing proofs

The core of this text is the contents of Part I, Basic Topics. In Part II we provide a selection of advanced topics with a focus on using these basic ideas to articulate central questions.

The most common introduction to topology is through the study of general topology, for which several fine texts are available, such as [11]. It is not our intention to replace these approaches, but rather to lead into them, should the student wish to learn more about topology.

Certainly, the more abstract approach is the quickest way, in terms of class time, to introduce the power of topological methods. However, we are not so interested here in matters of speed and power; our focus is on motivation and enjoyment. There is nothing wrong with driving a car, but one should also know how to walk. Sometimes by walking you notice many interesting things along the way. Also, afterwards one may more keenly appreciate a ride for a subsequent journey. So it is with the speed and power of the more abstract views of topology.

Euclidian topology and general topology

The focus of this book is the topology of $\mathbf{R^n}$. However, in optional sections found at the end of each chapter we do cover topics from general topology, with particular emphasis on the fundamentals: topology of sets, continuity and, homeomorphism.

This is done to allow flexibility for the instructor. As "optional" suggests, this material is not required for other sections of this book. Since there is a tradition of introduction of general topology in a first topology course, some instructors may feel that these topics, especially in the first sections, should be part of the syllabus. For example, one might cover this optional material after having finished some, or all, of the non-optional topics in the Basic Topics part of the text. Even if this material is not to be part of the syllabus, the instructor will almost certainly mention that these $\mathbf{R^n}$ topics can be extended, and might be encountered in a future course, should the student continue. By providing these optional sections, we offer a selection of topics so that the instructor can make such points concretely, rather than refer to separate books or future courses.

A technical remark on the topic of quotient spaces: Most topics in general topology correspond to topics in the topology of $\mathbf{R^n}$. One exception is the topic of quotient space. (The problem, of course, is that a quotient space formed from a subset of $\mathbf{R^n}$ may not be embeddable in $\mathbf{R^n}$.)

Here is an overview of each of these optional general topology sections:

***1.8** We define a general notion of topological space and also metric space. Examples are discussed, particularly the function space of continuous real-valued functions and the notion of decomposition space.

***2.7** For the most part, except for the topic of finite sets, definitions, and propositions of Chapter 2 transfer readily to general topology. However, the topic of Cartesian product for infinite products is an important new idea.

***3.8** The basic notions of continuity in Chapter 3 extend to the abstract setting of general topology. We discuss some examples of continuous functions for function spaces, product spaces, and quotient spaces.

***4.6** We discuss a few examples of homeomorphic and non-homeomorphic spaces in general topology, as related to topics of Chapter 4.

***5.6** We discuss the Cantor set as an infinite Cartesian product.

***6.6** We discuss some questions concerning embeddings of topological spaces. When might a topological space embed in $\mathbf{R^n}$, for some n? When might a topological space embed in a metric space?

***7.4** We make a few comments about connectedness as it applies to some of the more abstract examples.

***8.3** We show that the function space of continuous real-valued functions, defined on a interval, is path connected.

***9.5** In Chapter 9 a limit point of a set in $\mathbf{R^n}$ is a limit point of a sequence. In general topology, in non-metric spaces, this is not always the case. We discuss that issue and some ramifications.

***10.5** Compactness is an important property in general topology. We discuss this and give some examples of propositions where compactness is a critical hypothesis.

***11.2** In general topology, local compactness is a significant topic. However, $\mathbf{R^n}$ is locally compact and, as a consequence, all subsets are also locally compact. Local compactness is used as a hypothesis to define an important topology for function spaces.

Text organization

The basic material of this course is contained in the first eleven chapters (plus the appendices); these are the Basic Topics (for further guidance, see page 2). Even without including the optional general topology sections, these basic topics can easily require a semester, if the instructor concentrates on building proof construction skills for the student.

Treatment of advanced topics is brief (for further guidance, see page 268); it is not expected that the instructor will cover all of these.

Our example-driven approach results in an uncommon text organization.

- In Chapter 2, interesting examples are presented which are traditionally found only in later parts of a topology course. For example, in this chapter, we find the topologist's comb and the Hawaiian earing—well-known examples ordinarily used to illustrate relatively advanced topics.

- In Chapter 2, we introduce the basic idea of a homeomorphism in our discussion of rigid motion and similarity—this is even before we define the notion of continuous function in Chapter 3.

- In Chapter 2, in our discussion of subsets which are both open and closed, we prepare the student for a discussion of connectedness in Chapters 4 and 7.

- In Chapter 6, we provide a proof that $\mathbf{R}^1$ is not homeomorphic to $\mathbf{R}^2$, by looking at stable embedding, Proposition 6.5. (We also provide the "usual" proof, using connectedness, in Proposition 8.03 in Chapter 8.)

The end result of this organization is that many of these basic ideas and examples are encountered more than once in the text. This is deliberate. For example, when the student encounters the topologist's comb in the study of local connectivity, it will be a study of a new topological property using a familiar example, rather than a new property and an unfamiliar example.

In the first part of these notes one finds somewhat tedious proofs of results which we later prove in a line or two. We have two motivations for this approach:

- One does not necessarily have to have a neatly ordered body of definitions and theorems to write proofs.

- Having a neatly ordered body of definitions and theorems can make both reading and writing mathematics a lot more pleasant.

The traditional "Definition ⇒ Theorem ⇒ Proof" style of presentation provides a "neater" version of the subject. However, that does not accurately represent the way in which much of mathematics is done. Nor does it adequately prepare the student to write their won proofs. Also, if the student will ever have to answer a topological question outside of a topology class, it will frequently be the case that there will be no readily available definitions and theorems which can be exactly applied.

The usual definition of homotopy type is puzzling to a beginning student. We have chosen, instead, to introduce a notion of "deformation type," Chapter 16. The instructor may recognize this as modeled on the notion of simple homotopy type. For most common topological spaces, it is a notion equivalent to homotopy type.

Appendices

For those students with insufficient mathematical background, we include review background topics in the Appendices. Some of the material in Appendix C on countablility and uncountability may be unfamiliar, at least in the details.

Other than Appendix C, for the well-prepared student, the content of the appendices is for reference only.

Learning how to prove things

Due to the emphasis on proofs, topology may seem very different from other mathematics the student has encountered. This emphasis is largely due to the nature of the subject. In the topics we cover here, there are no calculations to speak of. Also, in elementary topology, there are few "main theorems." Instead, we find a collection of interrelated "smaller theorems," centered around a few basic concepts. Our goal is the understanding of these general concepts. To relate these with each other, we need arguments that are neither numerical calculations nor algorithmic symbolic manipulations. Thus the need for other methods of verification—i.e "proofs".

A few comments and guidance for proofs can be found in Appendix E.

Often in mathematics the final written version of a proof is different from the original argument. In a way, it is much like writing a poem. A finished poem is meant to be read from beginning to end, but it is rarely written that way. The poem, in progress, may differ considerably from the final draft. The original draft may be very sketchy, with an incomplete structure. Often, the initial inspiration does not explicitly appear in the final version.

We are interested in teaching how to prove things as well as how to write proofs. In the beginning of this book we give proofs in considerable detail, providing motivations for many of the steps. To continue that amount of detail for the entire text is unnecessary, and would become repetitive. Thus later proofs tend to have less attention paid to details. The instructor may find it necessary to amplify some of the later proofs depending on the students' needs.

Some technical comments on style

We do not use the hierarchical organization frequently found in mathematics books. Rather than theorems, lemmas, corollaries, propositions, we use a single egalitarian "Proposition."

- ∎ Indicates the end of a proof. At the end of the statement of a Proposition, it indicates the omission of a proof. In most cases, such a proof is left as an exception.

- ◨ Indicates that, not only is the proof omitted, but that the proof is beyond the scope of this text.

- ◆ indicates the end of an Example.

Asking questions

One of our primary goals is to raise interesting questions by means of examining examples.

Everyone likes a good riddle. Riddles are about questions and answers. It would certainly be unsatisfying to give the answer without even stating the riddle. It is only slightly better if we first present the answer, and then the riddle.

Too often, mathematics is presented in the form: first we consider some useful definitions and theorems, later we will see the applications of these. (This is "answer first, riddle last.") Although this is an efficient and logical methodology, it can often be dull to a beginner. Worse still, it does not give insight into the frame of mind that motivated mathematicians to devote their intellectual lives pursuing such investigations.

I begin almost every class I teach with asking "Are there any questions?" Most often the response (if any) begins "What is the answer to ...." or "How do you get the answer to ...." Those are not the questions that we mean. We speak here of the formulation of mathematical questions. The attitude of most students is that mathematics is about the process of finding answers. It is only a slight exaggeration to claim that mathematics is the process of asking questions.

Web Site

Supplementary information including errata, new material, additional exercises, color graphics and interactive graphics is available on the author's Web page:

http://www.math.uiowa.edu/ roseman

Comments and other correspondence concerning this text is welcomed and should be sent to the author at

roseman@math.uiowa.edu

or

Department of Mathematics, University of Iowa
Iowa City, Iowa (52242) U.S.A.

Acknowledgments

The author expresses his gratitude to those that have contributed valuable comments suggestions and corrections to this text.

Colleagues of the Department of Mathematics, University of Iowa, who used earlier versions of this text:
- Oguz Durumeric
- Jon Simon

Reviewers:
- Sue Goodman
  University of North Carolina at Chapel Hill
- Susan Hermiller
  New Mexico State University
- Thomas W. Rishel
  Cornell University
- Dale M. Rohm
  University of Wisconsin at Stevens Point
- Aaron K. Trautwein
  Carthage College

Past students in my Elementary Topology classes.

Special thanks to Brian Treadway, technical typist for the Department of Mathematics, University of Iowa, who cheerfully helped navigate through problems with LaTeX.

Two-dimensional graphics done with Mathematica, and Draw (a NeXTStep application). Three-dimensional graphics is displayed using Geomview, developed by the Geometry Center. In addition San Diego Graphics Utilities package as well as Unix utilities of the author were used in the graphics production.

On a deeper level are contributions of my parents, including my step-parents, to whom this book is dedicated. From my father I learned the spirit of hard work and craftsmanship; from my mother I learned the spirit of enthusiastic curiosity.

Finally, a very special thanks to my wife, Robin, for her support and love, without which this book never would have happened.

Part I

BASIC TOPICS

The basic topics of topology are: open and closed subsets, continuous functions, homeomorphisms, and topological properties. Each of these topics interconnects with all those preceding. In addition, one studies pairs of subsets and the topological properties of such pairs—discussed via the topic of embeddings.

We have organized this material as follows:

- **Open and closed subsets:** In Chapter 1 we introduce these concepts and give some basic examples. In Chapter 2, we give some additional examples, and in Chapter 5 some advanced examples. Further aspects such as limits of sequences and closure of subsets are discussed in Chapter 9.
- **Continuous functions:** In Chapter 3, as well as subsequent chapters, we present examples of continuous functions.
- **Homeomorphisms:** This is the topic of Chapter 4, and subsequent chapters.
- **Topological properties:** First introduced in Chapter 4, the basic topological properties we study are: connectivity in Chapter 7, path connectedness in Chapter 8, compactness in Chapter 10, and local connectedness in Chapter 11.
- **Embeddings:** Covered in Chapter 6.

Throughout this text we use the following notations:

$$\mathbf{N} \text{ denotes the natural numbers: } \mathbf{N} = \{1, 2, \ldots\};$$

$$\mathbf{Z} \text{ denotes the integers: } \mathbf{Z} = \{0, 1, -1, \ldots\}.$$

In all but the optional sections of this book, we assume explicitly that all subsets are subsets of some $\mathbf{R^n}$. A more general viewpoint has several common names: "general topology," "point set topology," or often just "topology." Throughout this text, each chapter has, at the end, an optional section relating topics of that chapter to general topology. These sections are not independent of each other. Topics in one general topology section usually depend on topics in preceding general topology sections. The bulk of the general topology material is the first three optional sections, where basic ideas and examples of general topology are introduced.

To maintain focus on subsets of $\mathbf{R^n}$ we have not included general topology problems in the problem sets at the end of the chapters. However there are implicit problems in these optional sections, namely, filling in some of the details of the exposition.

Most of the definitions generalize in an obvious way and the relationships are also valid in this more abstract setting. The following remark makes precise this notion of putting our topics into a more abstract setting. By "the obvious generalization to general topology" of a definition we mean, using the definition for topological space, Definition 1.44, and Definition 1.51 for the induced topology:

(1) replace each assumption of the form $X \subseteq \mathbf{R^n}$ by the hypothesis $\{X, \mathcal{T}\}$ is a topological space;
(2) replace each assumption of the form $A \subseteq X$ by: A is a subspace of X using the induced topology.

We will call this the "abstract setting" of the definition.

Given a proposition in the text, "the obvious generalization to general topology" is the proposition where we:

(1) replace each hypothesis of the form $X \subseteq \mathbf{R^n}$ by the hypothesis $\{X, \mathcal{T}\}$ is a topological space;
(2) replace each hypothesis of the form $A \subseteq X$ by A is a subspace of X using the induced topology;
(3) use the abstract settings of the definitions used in the statement of the proposition.

We will refer to this as the "abstract setting" of the proposition.

1. OPEN AND CLOSED SUBSETS

OVERVIEW: We begin our exploration of topology by defining and examining the concepts of open subsets, closed subsets, and limit points of subsets of the $\mathbf{R}^n$*. Although the terms here may not be new, our point of view may be different, especially the idea of an open subset of an arbitrary subset of* $\mathbf{R}^n$*.*

We also focus on how to prove some basic propositions. We illustrate how to make the transition from a rough idea to a formal proof.

1.1 What is topology?

A child said, What is the grass ? fetching it to me with full hands;
How could I answer the child ? I do not know what it is, any more than he.
I guess it must be the flag of my disposition, out of hopeful green stuff woven.
Or I guess it is the handkerchief of the Lord,
A scented gift and remembrancer, designedly dropt,
Bearing the owner's name someway in the corners, that we may see and remark, and say, Whose ?
Or I guess the grass is itself a child, the produced babe of the vegetation.
Or I guess it is a uniform hieroglyphic;
And it means, Sprouting alike in broad zones and narrow zones,
Growing among the black folks as among the white;
Kanuck, Tuckahoe, Congressman, Cuff, I give them the same, I receive them the same.
And now it seems to me the beautiful uncut hair of graves.

From Song of Myself, Walt Whitman

As a topologist, I am often asked is "What is topology?" As with the poet asked—"What is grass?", the subject is so familiar I know not one, but several different answers to this, each touching on a different aspect, none providing a good, simple answer to the question.

Here are some possible answers to the question "What is topology?"

- Topology is a study of flexible geometric objects which can bend, shrink, stretch, and twist but are not allowed to be cut, torn, or punctured.

- Topology is an abstraction of geometry where the concept of absolute distance removed, and one looks at geometric subsets independent of size, shape or location.

- Topology is the investigation of the set-theoretic foundations for the concept of continuous function.

- Topology is the study of sets which have some notion of "nearness" of points specified.

These answers are quite different, yet all are correct. The technical terms associated to this list are, in order, deformation type of Chapter 16, homeomorphism type of Chapter 4, continuous functions of Chapter 3, and general topology introduced in Section *1.8.

The poem segment quoted also is about how something as simple and familiar as a blade of grass can be a source of wonderment with seemingly endless associations and implications.

In this text, our mathematical "blades of grass" are familiar objects: a line, a plane, 3-space, and subsets of these. Our seemingly naive mathematical question is "What are these objects?" We will find mathematical wonders in the process of trying to answer this question.

Topology developed from process of examination of certain basic mathematical questions of a general geometric nature. We will learn something of this viewpoint. It will enable us to articulate some interesting questions and to begin to answer some of these questions.

As a graduate student, I had the pleasure of taking the beginning graduate course with R. H. Bing. His introduction, delivered with a thick Texas accent was something like this: "You probably want to know—What is topology ? I could give you an answer but you probably wouldn't really understand it until you finish this course. So we might as well get started ...."

So let us get started.

1.2 Some basic questions

In a sense, basic mathematics is about questions and answers. Often a student's progress is measured in terms of building of problem-solving skills. But question-posing skills are critical for understanding mathematics. As we are now beginning, we will largely measure our progress by the quality of questions we can ask (and, of course, understand).

Here are some mathematical questions and problems that motivate our investigations:

Question 0.1 What is the plane?

Problem 0.2 Classify all subsets of the plane (in some sense).

We will focus on the two items above. In addition, the following questions seem to flow naturally:

Question 0.3 What is space?

Problem 0.4 Classify all subsets of space.

Question 0.5 What is the fourth dimension?

The progression of questions in the above list implicitly contains a powerful mathematical idea—that there *is* a concept of dimension which unites the idea of "line", "plane" and "space". This may seem like an obvious idea, but it is in fact a distillation of thousands of years of mathematical thought.

Question 0.6 What does "dimension" mean?

Finally we add a question which, at first, might seem out of place for this list. However, we will find that, not only does it belong here, but it is fundamental (thus our choice of 0.0 for the number).

Question 0.0 What is the nature of continuity?

To see some connection between the idea of a continuous function and the first six questions on our list, consider the following example.

Example 1.01 Let X be the graph of the function $y = x$, let Y be the graph of the function $y = x + 1$, and let Z be the graph of the function

$$y = \begin{cases} x & \text{if } x < 1 \\ x + 1 & \text{if } 1 \leq x. \end{cases}$$

See Figure 1-1.

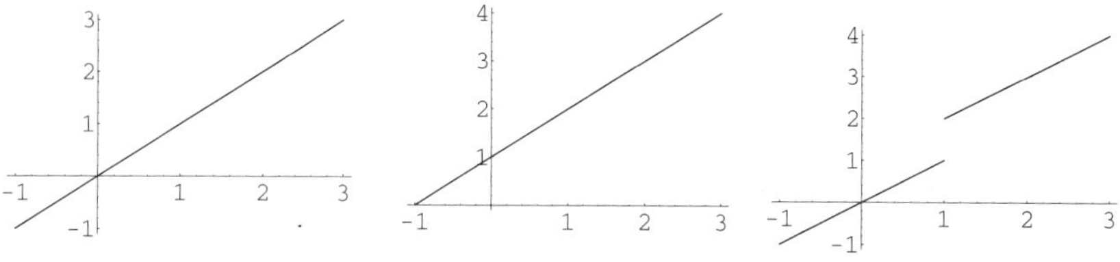

Figure 1-1 Graphs of three functions. On left is X, in the middle is Y, on the right is Z. As subsets of the plane, X and Y are equivalent, X and Z not equivalent; see Example 1.01.

We want to say that X and Y are "the same" as subsets of the plane, but these are quite different from Z. In terms learned from calculus the most obvious way of distinguishing Z from X is that Z is the graph of a function which is not continuous, and X is the graph of a continuous function. That is a clue that the concept of continuity might play a role in answering our questions. ◆

The questions and problems above are not precise. One of our goals is to improve and clarify these questions in mathematical ways. The process of refining such questions and, if possible, answering them is at the core of what mathematics is all about.

To begin thinking about Problem 0.2, we list some subsets of the plane:

(a) points
(b) lines
(c) circles
(d) disks
(e) a perimeter of a triangle
(f) a "filled-in" or "solid"triangle
(g) square and rectangular regions
(h) the whole plane
(i) the half plane
(j) the empty set
(k) solutions of equations in two variables
(l) solutions of inequalities in two variables
(m) any curve
(n) all drawings
(o) black and white reproductions of paintings, famous or obscure
(p) any black and white photograph where we consider the plane as white so that the black points of the photograph define our subset.
(q) any text ever written, in any language, whether written by any

hand or printed—here we think of the text as a subset of a page.

(r) even more generally: take the examples above and take intersections, unions, complements, allowing even infinite unions and intersections.

This list indicates that Problem 0.2 may be difficult. We need some concepts that enable us to sort subsets of the plane into more manageable collections. Topology gives us one good way of achieving this goal.

I have good news and, also, bad news. I don't think it will spoil your enjoyment of our mathematical journey to tell you the bad news: our solution to Problem 0.2 is that we will find no simple classification. This will not be due to ignorance, or lack of success on our part. The good news is that we will know, in a precise way, why there is no such simple classification, reasoning from newly gained knowledge of the plane and our newly gained perspective—the topological viewpoint. In doing so, we will have mathematically demonstrated the richness of our subject. We refer here to Propositions 7.28 and 7.3. A key issue is: what do we want to mean by "simple classification"?

A familiar answer to "what is the plane?", Question 0.1, uses Cartesian coordinates. We consider the plane to be the set of all ordered pairs of real numbers. Distance is measured in the usual way: if $p = (a, b)$ and $q = (c, d)$ are two points, the distance between these points is $\sqrt{(a-c)^2 + (b-d)^2}$. This formula is an analytic formulation of the Pythagorean Theorem. We use the notation $\delta(p, q)$ to denote the distance between p and q. We also use the notation $\mathbf{R}^2$ for the plane. This is short for "real two-dimensional space."

There are other ways to think of the plane—as a two-dimensional vector space or as the set of complex numbers. We consider these later, as needed.

You have certainly been exposed to the following generalization:

Definition 1.02 *We define n-**dimensional space** $\mathbf{R^n}$ to be the set of all ordered n-tuples of real numbers, together with a notion of distance, δ. If $p = (x_1, x_2, \ldots, x_n)$ and $q = (y_1, y_2, \ldots, y_n)$ are two points of $\mathbf{R^n}$, the distance between these points, denoted $\delta(p, q)$, is defined: $\delta(p, q) = \sqrt{\sum_{i=1}^{n}(x_i - y_i)^2}$*

At times it is convenient to use vector notation for points of $\mathbf{R^n}$ and write $\vec{x} = (x_1, x_2, \ldots, x_n)$ with $\vec{0}$ denoting the point $(0, 0, \ldots, 0)$. In this notation, $\delta(\vec{x}, \vec{y}) = |\vec{x} - \vec{y}|$, where the vertical bars refer to length of vector.

For the record, "zero-dimensional Euclidean space" $\mathbf{R^0}$ is defined to be the one-point set consisting of the number zero. In this text we will not need concern ourselves with $\mathbf{R^0}$. So for us, when we speak of $\mathbf{R^n}$, we will always have $1 \leq n$.

At first it might appear that we have clear mathematical answers to our basic questions, Questions 0.1, 0.3, 0.5, and 0.6. The plane is $\mathbf{R}^2$. Space, in the sense of Euclid, is $\mathbf{R}^3$. Four-dimensional space is $\mathbf{R}^4$, and for any $\mathbf{R}^n$, the number n is the dimension.

Before you think that the matter is closed for discussion, consider the following question:

Question 0.7 Are the plane $\mathbf{R}^2$ and space $\mathbf{R}^3$ the same, mathematically?

We feel the answer should be "no," and that is correct. But as we will find out, this not an easy question to answer. In fact, the best we can do completely in this course is to show that the line and the plane are not the same. (See the remark on page 175 and also Proposition 8.03)

1.3 Distance

Properties of distance in $\mathbf{R}^n$ follow from Definition 1.02. Distance is always a positive number. Also, the distance from p to q is the same as the distance from q to p. We note that this formula allows us to speak of the distance from a point to itself, in which case this distance is zero. In fact, this is the only case in which the distance between points could be zero. If we have any third point, r, the distance from p to q is less than the distance from p to r plus the distance from r to q. This inequality is most easily visualized by thinking of the three points as being vertices of a triangle. Geometrically put: any side of a triangle has length no greater than the sum of the lengths of the other two sides. We gather this information into Proposition 1.03, together with the names commonly used for each property. (1.56, 1.57)

Proposition 1.03

(a) *For all* $p, q \in \mathbf{R}^n$, $\delta(p,q) \geq 0$. *Also,* $\delta(p,q) = 0$ *only if* $p = q$ *(positive definite property).*
(b) *For all* $p, q \in \mathbf{R}^n$, $\delta(p,q) = \delta(q,p)$ *(symmetry property).*
(c) *For all* $p, q, r \in \mathbf{R}^n$, $\delta(p,r) \leq \delta(p,q) + \delta(q,r)$ *(triangle inequality).*

Proof: (We supply a proof of this proposition for the plane $\mathbf{R}^2$, and leave, as an exercise, the notational modifications one needs for the case $\mathbf{R}^n$. (Problem 1.1))

To prove the first part of (a), we need only note that, by definition, the square root of a number is always positive.

Next, we consider the second statement of (a). Write $p = (a,b)$, and $q = (c,d)$. We first show: if $p = q$, then $\delta(p,q) = 0$. If $p = q$, then $a = c$ and $b = d$. But then

$$\delta(p,q) = \sqrt{(a-c)^2 + (b-d)^2} = \sqrt{(0)^2 + (0)^2} = 0.$$

Conversely, suppose that $\delta(p,q) = 0$. Then $\sqrt{(a-c)^2 + (b-d)^2} = 0$. This is only possible if $(a-c)^2 + (b-d)^2 = 0$. This implies $a = c$ and $b = d$, which, in turn, implies that $p = q$.

The proof of (b) is simply the algebraic verification of the equation

$$\sqrt{(a-c)^2 + (b-d)^2} = \sqrt{(c-a)^2 + (d-b)^2}.$$

The proof of (c) takes a bit more work. Suppose $r = (e,f)$. Then we need to verify that

$$\sqrt{(a-e)^2 + (b-f)^2} \le \sqrt{(a-c)^2 + (b-d)^2} + \sqrt{(c-e)^2 + (d-f)^2}.$$

We can get rid of some of these variables by using vector notation. Let $\vec{u}$ be the vector from p to r, and $\vec{v}$ the vector from r to q. Then $\vec{u} + \vec{v}$ is the vector from p to q. The distance from p to q is the same as the length $|\vec{u}|$ of the vector $\vec{u}$. In vector notation, our inequality becomes: $|\vec{u} + \vec{v}| \le |\vec{u}| + |\vec{v}|$.

The key algebraic fact we need is known as the Cauchy-Schwarz inequality: $|\vec{u} \cdot \vec{v}| \le |\vec{u}||\vec{v}|$. Here $|\vec{u} \cdot \vec{v}|$ denotes the absolute value of the dot product. (If one accepts as given, the formula $\vec{u} \cdot \vec{v} = \cos(\alpha)|\vec{u}||\vec{v}|$, then the Cauchy-Schwarz inequality is an immediate consequence of this since $|\cos(\alpha)| \le 1$. A proof of the Cauchy-Schwarz inequality is found in Proposition D.26)

Using the Cauchy-Schwarz inequality, we see that

$$\begin{aligned}
|\vec{u} + \vec{v}|^2 &= (\vec{u} + \vec{v}) \cdot (\vec{u} + \vec{v}) \\
&= \vec{u} \cdot \vec{u} + \vec{u} \cdot \vec{v} + \vec{v} \cdot \vec{u} + \vec{v} \cdot \vec{v} \\
&= \vec{u} \cdot \vec{u} + 2\vec{u} \cdot \vec{v} + \vec{v} \cdot \vec{v} \\
&\le |\vec{u}|^2 + 2|\vec{u}||\vec{v}| + |\vec{v}|^2 \\
&= (|\vec{u}| + |\vec{v}|)^2.
\end{aligned}$$

By taking square roots, we obtain our target inequality. ∎

1.4 Open and closed subsets of $\mathbf{R^n}$

The notion of distance in $\mathbf{R^n}$ is fundamental to the mathematical study of $\mathbf{R^n}$. Yet, strange as it may seem, one path to a deeper understanding of $\mathbf{R^n}$ lies in suppressing the concept of distance as much as possible.

The key step in this process is: capture the notion of distance by consideration of special subsets called "open subsets" and "closed subsets". We build up to these ideas in stages: Definitions 1.04, 1.05, and 1.17.

Once we do this, explicit formulas for distance, such as $\sqrt{(a-c)^2 + (b-d)^2}$ for the plane, will rarely be used. We will generally discuss distance *indirectly* by referring, instead, to the properties as listed in Proposition 1.03. This has important implications if one views topology in the more abstract setting of general "metric spaces". This is discussed in the last section of this chapter.

Motivation for these definitions comes from several areas of mathematics. One area is the study of calculus. Question 0.0 relates to the geometric content of "continuity" and "limit." (We directly address continuity in Chapter 3. Limits are treated briefly in this chapter, and in more detail in Chapter 9).

Definition 1.04 *Let $p \in \mathbf{R^n}$ and $\epsilon > 0$ a given number. The* **open n-dimensional ball in $\mathbf{R^n}$ with center p and radius ϵ** *is defined to be the set $N_\epsilon(p) = \{q \in \mathbf{R^n} \mid \delta(p,q) < \epsilon\}$. We also refer to $N_\epsilon(p)$ as an* **ϵ-neighborhood of p in $\mathbf{R^n}$**.

In one-dimensional space $\mathbf{R^1}$, the formula for distance can be expressed in terms of absolute value: $\sqrt{(x_1 - y_1)^2} = |x - y|$. As it applies to $\mathbf{R^1}$, the word "ball" of Definition 1.04 seems inappropriate. If p and ϵ are numbers, $\epsilon > 0$, the open 1-dimensional ball of radius ϵ about p is the set of numbers whose distance from x is less than ϵ. The common term for such a set is "open interval." In interval notation a 1-dimensional ball is $(p - \epsilon, p + \epsilon)$. We can express this algebraically as: all numbers q such that $|p - q| < \epsilon$. (There are two common meanings for (a, b)—a subset of the line or a single point in the plane. The mathematical context should, however, prevent confusion.)

In Definition 1.04, the word "ball" also seems inappropriate for $\mathbf{R^2}$. In this case we sometimes use the term "open disk" instead of "open ball in $\mathbf{R^2}$".

The definition of open ball, Definition 1.04, is a key step in the development of the branch of mathematics called "topology." The definition of open ball, or neighborhood, attempts to capture the idea of nearness in roughly the following way: any open ball with center p must contain all those points in the plane which are "near to p." We are emphasizing the language of sets and suppressing the underlying analytic geometry. Admittedly, at this point, this is a fine distinction, but we eventually reap great benefits from this approach.

The ϵ used in this definition quantifies a notion of "nearness." Points within an ϵ-neighborhood of p can be thought of as "near p," where the standard of "nearness" is: closer to p than ϵ. The underlying problem is, that "nearness" is not an absolute concept but, rather, a relative one.

To an astronomer, the moon is very near; for the rest of us, it is clearly very far away. We continue this discussion in the remark on page 74.

This next definition is central to all that follows in this book. We use ϵ-neighborhoods as building blocks for other subsets of $\mathbf{R^n}$.

Definition 1.05 *A subset U of $\mathbf{R^n}$ is defined to be an* **open subset of $\mathbf{R^n}$** *if, for each $p \in U$, there is an $\epsilon > 0$ such that $N_\epsilon(p) \subseteq U$.*

Definition 1.06 *The collection of all open subsets of $\mathbf{R^n}$ is called the* **topology of $\mathbf{R^n}$**.

Definition 1.07 *A subset C of $\mathbf{R^n}$ is defined to be a* **closed subset of $\mathbf{R^n}$** *if $\mathbf{R^n} - C$ is an open subset of $\mathbf{R^n}$.*

In a more general study of topology, one allows other possible collections to be considered as "open subsets of $\mathbf{R^n}$." In this case, one refers to the topology defined in Definition 1.06 as the "standard $\mathbf{R^n}$" rather than just the "topology for $\mathbf{R^n}$." We will do that in the general topology sections, but in the main portions of the text this will not be necessary.

We will encounter many examples of open or closed subsets of $\mathbf{R^n}$. Most of the figures in this text depict open or closed subsets of $\mathbf{R^2}$ or $\mathbf{R^3}$. For the first two chapters, we will concentrate on a few basic examples and the problems involved with verifying that these subsets are open or closed, as the case may be.

Proposition 1.08 *An open n-dimensional ball is an open subset of $\mathbf{R^n}$.*

Since this is one of our first proofs, we proceed slowly and carefully. At this point, our goal is to show how to construct a proof rather than simply to present a proof. The proof is generated by thinking about some examples and constructing arguments that apply to these examples. With successive refinements, we get closer and closer to a complete proof—one that applies to the general hypotheses.

In Proposition 1.08, the word "open" is used in two different ways. Certainly one would hope that this proposition is true—that the choice of terminology has been well-chosen so that a simple proposition which "sounds like it should be true" is, in fact, true. Mathematicians traditionally use familiar words from everyday language to represent abstract ideas. Ideally, the mathematical usage should be consistent with everyday usage. However, this matching can never be perfect and, inevitably, familiar uses of common words may be misleading.

The first step towards our proof is to draw a picture of a particular disk. Since the proposition is to apply to all disks, our choice of disk should not matter.

(A short digression, first, on graphical conventions for an open disk, say $N_\epsilon(p)$. There are several traditional ways to draw this, and they all begin by drawing the circle of radius ϵ about the given point p. The viewer is to understand that the drawing is to refer to the points *inside* the circle, but not *on* the circle. Sometimes this is indicated by shading in these points inside the circle to call this to the attention of the reader. However, in a complex drawing, such shading may detract from the clarity of the diagram—in that case it is omitted. It is essential to indicate, somehow, that the points on the circle are *not* points of $N_\epsilon(p)$. There are various ways of indicating this: one way is to use a very thin line thickness for the circle. Another is to make use of a dashed line. As with shading, this is often omitted in more complex drawings.)

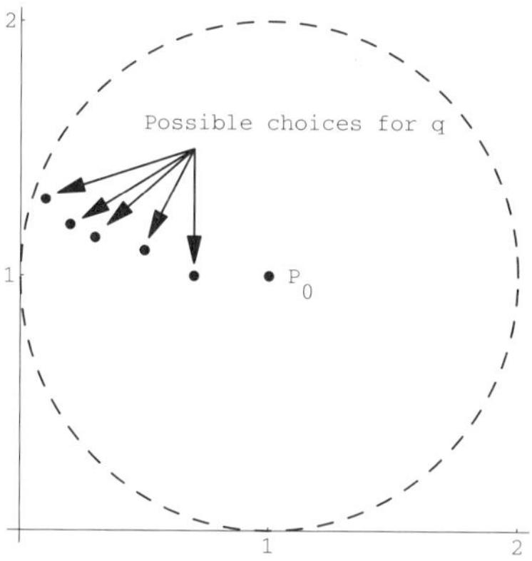

Figure 1-2 Some points of $N = N_1((1,1))$; see proof of Proposition 1.08.

Suppose we draw an open disk N in $\mathbf{R}^2$ with center at (1, 1) and with radius 1 (see Figure 1-2). We need to show: if q is a point of N, we can find a number ϵ_0 so that $N_{\epsilon_0}(q)$ is entirely contained in N. The choice of ϵ_0 depends on the choice of q. For example, if $q = (1, 1/2)$ then a choice of $\epsilon_0 = 1/2$ works; any positive smaller number also works. If we take $q = (1, 1/100)$, then clearly the choice of $\epsilon_0 = 1/2$ is too large. However, the choice of $\epsilon_0 = 1/100$, or even $\epsilon_0 = 1/1000$, will work for this q. For other choices of q, it may be more challenging to calculate an ϵ_0. Consider if q is chosen to be (0.973, 0.17). Calculation will verify that, if we choose ϵ_0 to be $1/1,000,000$, this will work. Such *ad hoc* reasoning may convince us of the correctness of the statement we wish to prove, but this is not a proof for all possible choices of q.

Here is an idea for a proof. Let C be the circle $C = \{(x, y) \in R^2 : (x-1)^2 + (y-1)^2 = 1\}$; then C represents the "edge of N" in our drawing. We note that the closer that q is to C, the smaller we have to choose our ϵ_0. Organizing these thoughts, we proceed to sketch a proof.

First we outline a proof of Proposition 1.08), then fill in the details.

(1) Let $N_\epsilon(p)$ be the given disk; let $q \in N_\epsilon(p)$. We need to show that we can find an $\epsilon_0 > 0$ such that $N_{\epsilon_0}(q) \subseteq N_\epsilon(p)$.
(2) Let ϵ_0 be the distance from q to the circle C.
(3) With this choice we claim that $N_{\epsilon_0}(q) \subseteq N_\epsilon(p)$.

In order to make a more complete proof, we look a bit deeper. We need to make step (2) precise and to justify (3). There is a special way to calculate the distance q to C since C is a circle. Consider a radius of C through q. The distance from q to C is $\epsilon - \delta(p,q)$. So let $\epsilon_0 = \epsilon - \delta(p,q)$. To verify statement (3) we use the triangle inequality for the distance function. Let z be a point of $N_{\epsilon_0}(q)$. Then $\delta(q,z) < \epsilon_0$. Now

$$\delta(p,z) \le \delta(p,q) + \delta(q,z) < \delta(p,q) + \epsilon_0 = \delta(p,q) + (\epsilon - \delta(p,q)) = \epsilon. \quad \blacksquare$$

In step (2) in the proof of Proposition 1.08, we considered the distance between a point inside a circle and that circle—the radius of the circle minus the distance of the point to the center. The general notion of the distance from a point to an arbitrary subset not a simple matter. Suppose $S \subseteq \mathbf{R}^2$ and $p \in \mathbf{R}^2 - S$ and we try to define the distance between p and S as the shortest distance between p and a point of S. The problem is that this might not be a non-zero number.

For example, suppose $S = \{(x,y) \in \mathbf{R}^2 \mid x = \frac{1}{n} \text{ and } y = 0\}$; let $p = (0,0)$. If we choose a point, say $(\frac{1}{N}, 0) \in S$, we can always find a point of S closer to $(0,0)$, namely $(\frac{1}{N+1}, 0)$. So we see that there is no (non-zero) shortest distance between p and points of S. Although we cannot define the notion of distance from a point to *any* subset, we can for *certain kinds* of subsets, as we will see later, in Definition 10.32.

As we have seen, sketches can help visualize relationships. How does one draw a general picture of an open subset of the plane? Experience has shown that for visualization needed for constructing general proofs, thinking of points inside a simple closed curve is useful.

Draw a simple closed curve in the plane. That is, a curve which begins and ends at the same point and does not cross itself. (You have probably seen a definition of this when studying Stokes's Theorem, or perhaps looking at a contour integral in complex variables.) It can be shown that: if U be the "set of points inside C," then U is an open subset of the plane. This generally results in a helpful picture. But this does *not* mean that *all* open subsets of the plane can be depicted in this way. In fact, the you might note, as we proceed, that most of the examples of open subsets in this chapter are not of this type.

We should mention, however, that the precise definition of what "inside" means for an arbitrary curve is surprisingly difficult as it uses the Jordan Curve Theorem (Proposition 4.10) and uses a concept, connectedness, discussed later Chapter 7. Roughly, a connected subset is a single unbroken piece. The theorem states that the complement of a simple closed curve in the plane has two pieces—one "inside" and one "outside."

An important basic closed subset is a closed ball.

Definition 1.09 *If* $p \in \mathbf{R^n}$, $\epsilon > 0$, *the* **closed** n**-dimensional ball with center** p **and radius** ϵ *is defined to be* $\{q \in \mathbf{R^n} \mid \delta(p,q) \leq \epsilon\}$. *We denote this set by* $\overline{N}_\epsilon(p)$ *or, sometimes,* $B_\epsilon(p)$.

The distinction between the open ball and the closed ball lies in the use of the inequality relation— $<$ as opposed to $\leq$.

Proposition 1.10 *A closed* n*-dimensional ball is a closed subset of* $\mathbf{R^n}$.

Proof: Suppose $\overline{N}_\epsilon(p)$ is a given closed ball. We show: $X = \mathbf{R^n} - \overline{N}_\epsilon(p)$ is an open subset of $\mathbf{R^n}$. Let $q \in X$ and let $\epsilon' = \delta(p,q) - \epsilon$. Using the triangle inequality, we can then verify that

$N_{\epsilon'}(q) \subseteq X$ since $N_{\epsilon'}(q) \cap \overline{N}_\epsilon(p) = \varnothing$. ∎ (Problem 1.2)

The term "interval" is found in all calculus and pre-calculus courses. For reference, we list these important subsets of $\mathbf{R^1}$ in the next definition. In Proposition 1.12 we will see this use of the words "open" and "closed" for intervals is consistent with Definitions 1.05 and 1.07.

Definition 1.11 *Let* a *and* b *be numbers with* $a < b$. *The following subsets are all called* **intervals**:

(a) $\{x \in \mathbf{R^1} : a < x < b\}$, *denoted* (a,b), *is called an* **open interval**.
(b) $\{x \in \mathbf{R^1} : a \leq x \leq b\}$, *denoted* $[a,b]$, *is called a* **closed interval**.
(c) $\{x \in \mathbf{R^1} : a < x \leq b\}$ *or* $\{x \in \mathbf{R^1} : a \leq x < b\}$, *denoted* $(a,b]$ *or* $[a,b)$, *and are called* **half-open intervals**.
(d) *The set,* $\{a\}$, *consisting of the single number* a *is sometimes denoted by* $[a,a]$ *and is called a* **degenerate closed interval**.
(e) $\{x \in \mathbf{R^1} : a < x\}$ *and* $\{x \in \mathbf{R^1} : x < a\}$, *denoted* (a,∞) *and* $(-\infty,a)$, *are called* **half-infinite open intervals**.
(f) $\{x \in \mathbf{R^1} : a \leq x\}$ *and* $\{x \in \mathbf{R^1} : x \leq a\}$, *denoted* $[a,\infty)$ *and* $(-\infty,a]$, *are called* **half-infinite closed intervals**.
(g) *Finally, it is sometimes convenient to think of* $\mathbf{R^1}$ *as an interval, using the notation* $(-\infty,\infty)$.

The notational use of ∞ such as appears in $[0,\infty)$ is a convenient use of symbols but does *not* mean that ∞ is a number, even though it appears in a location where one often puts a number.

Proposition 1.12 *In* R^1, *an open interval* (a,b) *is an open subset of* R^1. *Also, a closed interval* $[a,b]$ *is a closed subset of* R^1.

Proof: To prove the first statement, note that $\frac{a+b}{2}$ is the midpoint of the given interval and that $(a,b) = N_{\frac{b-a}{2}}(\frac{a+b}{2})$. Thus (a,b) is an open subset by Proposition 1.08.

Similarly, by writing $[a,b] = \overline{N}_{\frac{b-a}{2}}(\frac{a+b}{2})$, we see that $[a,b]$ is a closed subset of R^1 by Proposition 1.10. ∎

The two most common errors, made by those new to topology, arise from assuming that everyday meanings of the words "open" and "closed" apply to the mathematical definitions. The first mistake is to assume that a set must be *either* open *or* closed. This will be shown to be false in Example 1.15. The second mistake is to assume that a set cannot be *both* open *and* closed. As a trivial example, the empty subset is both open and closed. We will encounter better examples, such as Example 2.22, after we discuss a more general notion of open subset, Definition 1.22.

We have shown that an open disk in the plane is an open subset of the plane, Proposition 1.08. It does not automatically follow that an open disk is *not* a closed subset. However, it is true, and this is our next Proposition.

Proposition 1.13 *If $N_\epsilon(p)$ is an open n-ball in $\mathbf{R}^n$, then $N_\epsilon(p)$ is not a closed subset of $\mathbf{R}^n$.*

Proof: We present a proof for $\mathbf{R}^2$ and leave routine modifications for $\mathbf{R}^n$ as an exercise (Problem 1.3).

First, we present a rough proof. Let $X = R^2 - N_\epsilon(p)$. If $N_\epsilon(p)$ is closed, then X is an open subset. We argue by contradiction. We assume that X is an open subset and find that this is impossible. (If you are unfamiliar with argument by contradiction you might want to consult Appendix A.)

We obtain our contradiction by finding *some* point of X such that there is no open disk, centered at that point, which is contained in X. At this point a sketch of the situation is helpful; see Figure 1-3.

Some points of X will not give us our contradiction. For example, if we take a point q which is a distance d from p where $d > \epsilon$, then we have $N_{(d-\epsilon)/2}(q) \subseteq X$. Thus we focus on the points y in the plane with $\delta(y,x) = \epsilon$. These are the points of the circle C with center p and radius ϵ. Examining a picture, it seems clear that any point q on C, should give us our contradiction. For any positive number ϵ_0, $N_{\epsilon_0}(q)$ contains a point z of $N_\epsilon(p)$. Since $z \notin X$, we have $N_{\epsilon_0}(q) \nsubseteq X$. This shows that X is not open since the definition of open subset fails for the point q.

(Note our introduction of the symbol ϵ_0 above. Our hypothesis refers to a given number ϵ. The symbol ϵ also appears in the definition of "open subset." To avoid confusing these two numbers, we choose something for our "second ϵ" distinct from ϵ, yet similar enough to

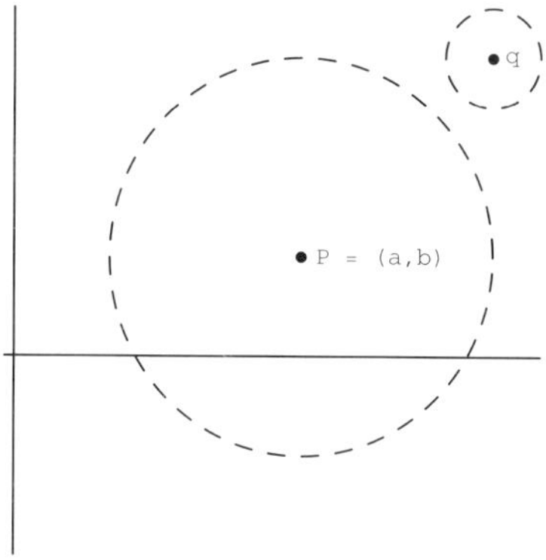

Figure 1-3 Looking at a point q far from $N_\epsilon(p)$; see proof of Proposition 1.13. Points inside large circle C correspond to $N_\epsilon(p)$. Points inside smaller circle correspond to $N_{\frac{d-\epsilon}{2}}(q)$.

remind us of this number's significance. The choice of ϵ_0 was arbitrary, but in a style common in mathematical writing. Other notational choices might have been ϵ', η, ε, etc.)

Our goal is write a version of the proof above which, in the final version, does not depend on reference to drawings. Although any point of C could be used for our argument, we need only find *one* point of X for which the definition of open subset fails. To simplify the proof we specify a particular choice of point. Let us suppose that $p = (a, b)$. Guided by our previous discussion, a good choice for our point of C might be $q = (a + \epsilon, b)$, see Figure 1-4. Then $q \in C$ since $\delta(p, q) = \epsilon$. We next need to show that this choice of q will work, in the sense that $\mathbf{R}^2 - N_\epsilon(p)$ does not contain a ball about q. Specifically, we need to show that given any number $\epsilon_0 > 0$, we can find a point $z \in N_{\epsilon_0}(q)$ with $z \notin X$. Figure 1-4, indicates that $z = (a + \epsilon - \frac{\epsilon_0}{2}, b)$ should work.

At this point, in a small way, our figure is misleading. We need to find a z that will work for *any* given ϵ_0. We have drawn the figure so that ϵ_0 is small relative to ϵ. But what if ϵ_0 is large? Suppose, for example, that $\epsilon_0 = 10\epsilon$; then our choice above for z *will be* in X. This is not a serious problem, but we will have to address it in writing a detailed proof.

There are several ways to deal with this technicality. One way is to refine the choice of z. For example, we could use the following rule: choose z to be $(a+\epsilon-\frac{\epsilon_0}{2}, b)$ if $\epsilon_0 < \epsilon$; choose z to be $(a + \frac{\epsilon}{2}, b)$ if $\epsilon_0 \geq \epsilon$. Another way is to refine our choice of radius for our disk about q. We could let $\epsilon_1 = \min(\epsilon, \frac{\epsilon_0}{2})$. Then if we let $z = ((a + \epsilon) - \frac{\epsilon_1}{2},\ b)$, we get our contradiction. We use this second idea. Putting all this together,

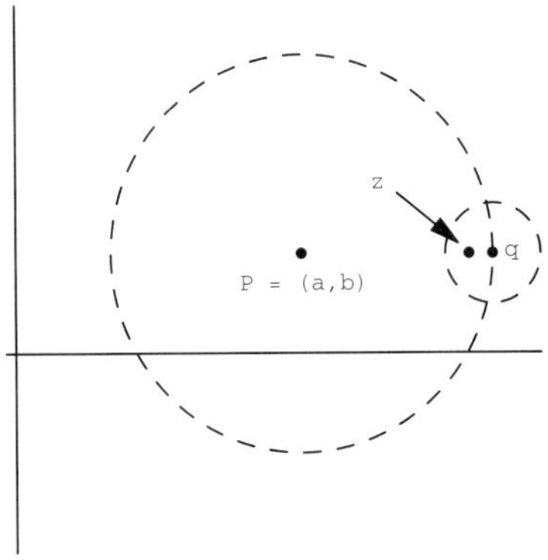

Figure 1-4 Finding a point z which gives a contradiction; see proof of Proposition 1.13. Points inside large circle correspond to $N_\epsilon(p)$. Points inside small circle correspond to $N_{\epsilon_0}(q)$ where $q = (a+\epsilon, b)$. We chose $z = (a + \epsilon - \frac{\epsilon_0}{2}, b)$.

we obtain the following proof:

Proof of Proposition 1.13: Let $X = R^2 - N_\epsilon(p)$. We show $N_\epsilon(p)$ is not closed by showing that X is not open. Let $p = (a,b)$, $q = (a+\epsilon, b)$. Then $q \in X$. If X is open, there is some number ϵ_0 such that $N_{\epsilon_0}(q) \subseteq X$. Let $\epsilon_1 = \min(\epsilon, \frac{\epsilon_0}{2})$ and let $z = (a + \epsilon - \epsilon_1, b)$. Note that $\epsilon_1 < \epsilon_0$. We see that $z \in N_{\epsilon_0}(q)$ since $\delta(z,q) = \epsilon_1 < \epsilon_0$. Also, $z \in N_\epsilon(p)$ since $\delta(p,z) = \sqrt{(\epsilon - \epsilon_1)^2} < \epsilon$. This gives a contradiction since $z \in N_{\epsilon_0}(q)$ and $z \notin X$, contradicting our assumption that $N_{\epsilon_0}(q) \subseteq X$. ∎

A simple, but useful, fact is the following proposition. In Definition 1.11 we defined a single point of $\mathbf{R}^1$ as a degenerate closed interval—this proposition shows that this use of the word "closed" is not inconsistent.

Proposition 1.14 *A point is a closed subset of* $\mathbf{R}^n$. *Also, a point is not an open subset of* $\mathbf{R}^n$. ∎ *(Problem 1.4)*

A technical point on the statement of Proposition 1.14 is that a point of $\mathbf{R}^n$ is not a subset of $\mathbf{R}^n$ but is an *element* of $\mathbf{R}^n$. The distinction here is between an element of a set and a subset consisting of one element. Notationally expressed, if $p \in \mathbf{R}^n$, then $\{p\} \subseteq \mathbf{R}^n$ but $p \nsubseteq \mathbf{R}^n$. A correct way of stating the above proposition would be: "If p is a point of $\mathbf{R}^n$, then $\{p\}$ is a closed subset of $\mathbf{R}^n$."

In general, the purpose of precise mathematical definitions is to make things clear. But sometimes zealous adherence to precision can impede clear communication. We will, in the future, occasionally allow such minor abuse of our language and use the wording as originally stated in Proposition 1.14.

Also, a very careful reader may detect a subtle misuse of terminology in the proofs of some of the propositions above. Note that we have used such phrases as "open subset" or sometimes "open set" to refer to an "open subset of $\mathbf{R}^n$." This is common practice and, in the context of our discussions, is unlikely to lead to confusion. However it is important to realize we are doing this. In Definition 1.17, we generalize the concept to "open subset of a given set." This can lead to a situation in which the word "open" might have two different interpretations. Especially in the beginning of this book, we will be careful in this matter. Later we will sometimes allow a more relaxed vigilance to improve readability.

Example 1.15 Let X be the set of all points in the plane with $0 < x \leq 1$ and $y = 0$. We show that X is neither an open subset nor a closed subset of $\mathbf{R}^2$.

To show that X is not open, consider the point $a = (1/2, 0)$. This is a point of X, but no disk of the form $N_\epsilon((1/2, 0))$ is entirely contained in X. For example, $z = (1/2, \epsilon/2)$ is not in X, but is z in $N_\epsilon((1/2, 0))$, for any choice of ϵ. Thus X is not an open subset.

The choice of a was not critical. It is not much more difficult to show that: for each $x \in X$, and for each $\epsilon > 0$, $N_\epsilon(x)$ is *not* a subset of X. However, for the purposes of our proof, we do not need to show this. We leave proof of this as an exercise (Problem 1.6).

To show X is not closed, we show the complement $R^2 - X$ is not open. Consider the point $b = (0, 0)$. This is a point of $R^2 - X$. Given any $\epsilon > 0$, let $\epsilon_0 = \min(1, \epsilon/2)$. The point $x_0 = (\epsilon_0/2, 0)$ is not in $R^2 - X$ and yet $x_0 \in N_\epsilon((0, 0))$. Thus $R^2 - X$ is not open, and so X is not closed.

(Note: The motivation for the definition of ϵ_0 is to deal with the situation of ϵ being larger than 1. This is similar to the problem of large values of ϵ discussed in the proof of Proposition 1.13.) ◆

Example 1.16 Let Q be the set of rational numbers in $\mathbf{R}^1$; Q is neither open nor closed.

We show Q is not an open subset of $\mathbf{R}^1$. We focus on the rational number 0, and show that there is no $\epsilon > 0$ such that $N_\epsilon(0) \subseteq Q$. We argue by contradiction. Suppose there were such $\epsilon > 0$. By Proposition B.13, there is an irrational number x, $0 < x < \epsilon$. Since $x \notin Q$ and $x \in N_\epsilon(0)$, we have contradicted the assumption that $N_\epsilon(0) \subseteq Q$.

We leave as an exercise to show that Q is not a closed subset of $\mathbf{R}^1$ (Problem 1.5). ◆

1.5 Open and closed subsets of subsets

We next extend the concept of "open" so that we can talk of "an open subset of *any subset* of $\mathbf{R^n}$." Pay careful attention to this deceptively simple yet powerful idea.

Definition 1.17 *Suppose $V \subseteq X \subseteq \mathbf{R^n}$. We say that V is an* **open subset of** *X, if there is an open subset U of $\mathbf{R^n}$ such that $V = X \cap U$.*

Example 1.18 Let X denote the x-axis in the plane: $X = \{(x, y) \in \mathbf{R}^2 : y = 0\}$. Let V be defined by $V = \{(x, y) \in \mathbf{R}^2 : -1 < x < 1 \text{ and } y = 0\}$. Then V is an open subset of X. Let $U = N_1((0, 0))$; then U is an open subset of $\mathbf{R}^2$ and $V = X \cap U$ (see Figure 1-5).

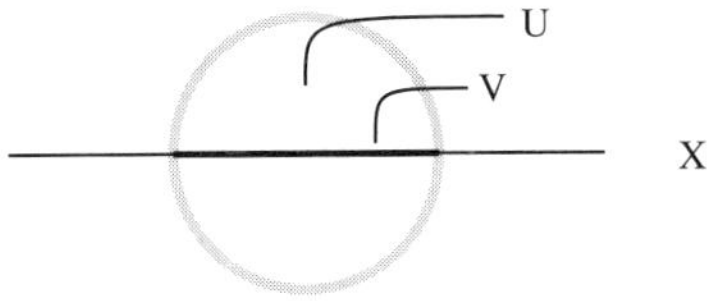

Figure 1-5 An open subset V, shown as a thickened segment, of the x-axis $X \subseteq \mathbf{R}^2$; see Example 1.18. Shown is open subset U of $\mathbf{R}^2$, points inside circle, with $U \cap X = V$.

It is important to note that an open subset of X is not necessarily an open subset of $\mathbf{R^n}$. In our example, V is an open subset of X; however, V is not an open subset of $\mathbf{R}^2$ (the line of reasoning of Example 1.15 may be applied here).

We should be careful when using the word "open." In the literature this is not always done. If the subset X in question is clear from the context, one often finds the terminology "U is an open set" instead of the more accurate "U is an open subset of X." ◆

We now define topology for arbitrary subsets of $\mathbf{R^n}$. If X is a subset of $\mathbf{R^n}$, the topology of X is obtained from the topology of $\mathbf{R^n}$.

Definition 1.19 *Let $X \subseteq \mathbf{R^n}$. The set of all open subsets of X is called the* **topology of** *X.*

Suppose $A \subseteq X \subseteq \mathbf{R^n}$. We want to see how the topology of A relates to the topology of X. To find out if a subset $V \subseteq A$ is an open subset of X, using the above definition, we need to relate it to open subsets of $\mathbf{R^n}$. However we can, alternatively, verify this by relating to open subsets of X, as the next proposition shows.

Proposition 1.20
Suppose $V \subseteq A \subseteq X \subseteq \mathbf{R^n}$. Then V is an open subset of A if and only if there is an open subset U of X such that $V = A \cap U$. ▮ *(Problem 1.7)*

Example 1.21 In this example, X is the x-axis in the plane and A is a unit interval on the x-axis. Let $X = \{(x, y) \in \mathbf{R}^2 \mid y = 0\}$ and $A = \{(x, y) \in \mathbf{R}^2 \mid 0 \le x \le 1 \text{ and } y = 0\}$. We have $A \subseteq X$. Let $V = \{(x, y) \in \mathbf{R}^2 \mid 0 \le x < 1/2 \text{ and } y = 0\}$; V is an open subset of A. For example, $V = N_{\frac{1}{2}}((0,0)) \cap A$.

If we let $U = \{(x, y) \in \mathbf{R}^2 \mid -1/2 < x < 1/2 \text{ and } y = 0\}$, then U is an open subset of X and $V = A \cap U$.

We note, in passing, that V is **not** an open subset of X. We leave this as an exercise to the reader, (Problem 1.9). ◆

Next we have the natural extension of the definition of a closed subset of $\mathbf{R^n}$ to closed subsets of *subsets* of $\mathbf{R^n}$.

Definition 1.22 *Let $X \subseteq \mathbf{R^n}$. A subset C of X is called a* **closed subset of** *X if $X - C$ is a open subset of X.*

If $X \subseteq \mathbf{R^n}$, we sometimes want to describe the subsets of X which are open subsets of X, without direct reference to the open subsets of $\mathbf{R^n}$. The following definition will help to do this.

Definition 1.23 *Let X be a subset of $\mathbf{R^n}$ with $p \in X$. Given a number $\epsilon > 0$ we define an* **ϵ-neighborhood of** *p* **in** *X to be $N_\epsilon(p, X) = N_\epsilon(p) \cap X$.*

So, $N_\epsilon(p, X)$ is all points of X whose distance from p is less than ϵ.

If $X = \mathbf{R^n}$, $N_\epsilon(p) = N_\epsilon(p, \mathbf{R^n})$. Note that $N_\epsilon(p, X)$ is an open subset of X since $N_\epsilon(p)$ is an open subset of $\mathbf{R^n}$. The next proposition uses Definition 1.23 to characterize open subsets.

Proposition 1.24 *If $X \subseteq \mathbf{R^n}$, then U is an open subset of X if and only if, for every $p \in U$, there is an $\epsilon > 0$ such that $N_\epsilon(p, X) \subseteq U$.* ▮ *(Problem 2.14)*

Also, we have the following characterization of the open subsets of X.

Proposition 1.25 *If $X \subseteq \mathbf{R^n}$ and $V \subseteq X$, then V is an open subset of X if and only if V is a union of ϵ-neighborhoods in X of points of V.*

Proof: If $V = \bigcup_{\alpha \in A} N_{\epsilon_\alpha}(p_\alpha, X)$, then V is an open subset of X by Proposition 1.24.

Next, suppose V is an open subset of X. Then there is an open subset, U of $\mathbf{R^n}$ such that $V = U \cap X$. For each $x \in X$, we can find a positive number ϵ_x such that $N_{\epsilon_x}(x) \subseteq U$, and so $N_{\epsilon_x}(x, X) \subseteq U$. We can check that $V = \bigcup_{x \in X} N_{\epsilon_x}(x, X)$. ∎

In Proposition 1.25 above, we allow empty unions. (An empty union is a union indexed by an empty set. An empty union is always an empty set see remark on page 395.)

Note the use of the set X as an *index* for the union found in our proof. This construction is often used in topology. If X is an uncountable set, then such a union is a union of an uncountable collection of open subsets.

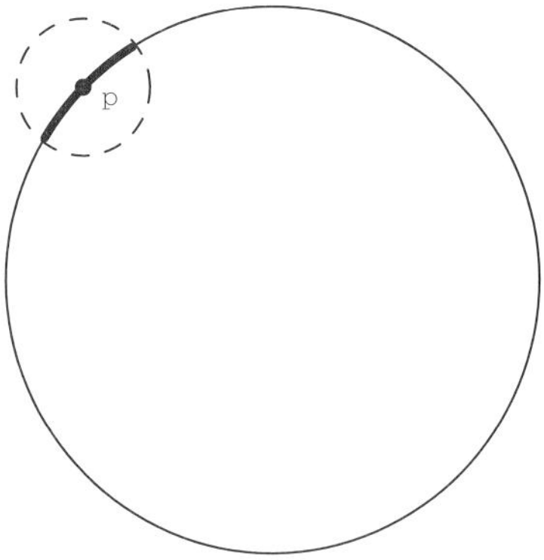

Figure 1-6 An ϵ-neighborhood $N_\epsilon(p, S^1)$ of a point of the unit circle S^1; see Example 1.26. Here $N_\epsilon(p)$ correspond to points inside small circle and $N_\epsilon(p, S^1) = S^1 \cap N_\epsilon(p)$ is shown as thickened arc.

Example 1.26 Consider the possible ϵ-neighborhoods of the unit circle S^1. If $p \in S^1$ and $\epsilon > 2$, then $N_\epsilon(p, X) = S^1$. Otherwise, we can describe $N_\epsilon(p, X)$ as "an open sub-arc of the circle"; see Figure 1-6. In the special case, $\epsilon = 2$, we obtain $S^1 - \{q\}$, where q is the point of S^1 diametrically opposite to p. Thus, using Proposition 1.25, we may say "the open subsets of the unit circle are open sub-arcs together with unions of such sub-arcs." ◆

Example 1.27 Let X be the subset of the plane consisting of the union of the coordinate axes; $X = \{(x, y) \in \mathbf{R}^2 \mid x = 0 \text{ or } y = 0\}$. The ϵ-neighborhoods of points of X are of three types: open intervals on the x-axis, open intervals on the y-axis, or a set which looks like the letter x with the endpoints removed.

More precisely, these three types are obtained as follows. The first type are the neighborhoods where $p = (x, 0)$ and $\epsilon \leq |x|$. The second type occurs when $p = (0, y)$ and $\epsilon \leq |y|$. The third type occurs when

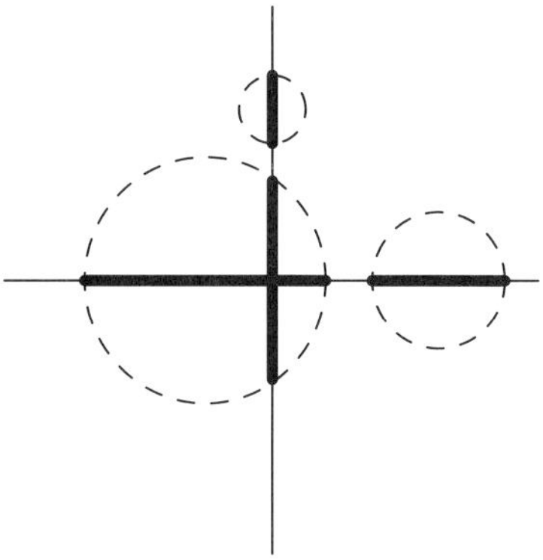

Figure 1-7 Three neighborhoods in union of axes $X \subseteq \mathbf{R}^2$; see Example 1.27. Neighborhoods in X are shown as thickened segments. Points inside circles show corresponding neighborhoods in $\mathbf{R}^2$.

$p = (x, y)$ and $\epsilon > \max(|x|, |y|)$. (See Figure 1-7.) Note that this third type of neighborhood does not "look like" a neighborhood of a point of some Euclidian space. ◆

In $\mathbf{R^n}$, an ϵ-neighborhood $N_\epsilon(p, \mathbf{R^n})$ is never a closed subset (Proposition 1.13). The situation for our new neighborhoods $N_\epsilon(p, X)$ may be different, as this next example shows.

Example 1.28 Let Q be the rational numbers considered as a subset of R^1. We show that some neighborhoods $N_\epsilon(x, Q)$ are closed subsets of Q, and some are not.

We first show that $N_1(0, Q)$ is *not* a closed subset of Q, by showing the complement $X = Q - N_1(0, Q)$ is not an open subset of Q. We need to find some point x_0 of X so that for all $\epsilon > 0$, $N_\epsilon(x_0, Q) \nsubseteq X$.

No open subset of Q which contains x_0 is contained in X. A point of Q which is not in X must be a point of $N_1(0, Q)$. In summary, we need to find some point $x_0 \in X$ so that any $N_\epsilon(x_0, Q)$ must contain a point of $N_1(0, Q)$.

It does not take long to find a candidate for our x_0. Suppose we consider the number $\frac{3}{2} \in X$. We see that $\frac{3}{2}$ is not a suitable choice. We can find an open subset of Q, containing $\frac{3}{2}$, which does *not* contain points of $N_1(0, Q)$, namely, $N_{\frac{1}{2}}(\frac{3}{2}, Q)$. More generally, if we take any rational number, r with $1 < |r|$, we will have $N_{1-r}(r, Q) \cap N_1(0, Q) = \varnothing$. So we are left considering $x_0 = \pm 1$. It turns out that either of these numbers will work. We next will verify that $x_0 = 1$ is a suitable choice.

For any $\epsilon > 0$, we can find a point of $N_1(0, Q)$ in $N_\epsilon(1, Q)$ as follows. Let $\epsilon_0 = \min(\epsilon, 1)$. (As in previous proofs, the reason for this step is to avoid the "large ϵ problem" as in the proof of Proposition 1.13). By Proposition B.11 of Appendix B, we can find a rational number q such that $1 - \epsilon_0 < q < 1$. This completes our proof that $N_1(0, Q)$ is not a closed subset of Q.

On the other hand, $N_{\sqrt{2}}(0, Q)$ *is* a closed subset of Q. To see this we note that $[-\sqrt{2}, \sqrt{2}]$ is a closed subset of R^1, by Proposition 1.12. Let $W = R^1 - [-\sqrt{2}, \sqrt{2}]$; W is an open subset of R^1. Let $U = Q - N_{\sqrt{2}}(0, Q)$; then U is an open subset of Q since $U = Q \cap W$. Thus $N_{\sqrt{2}}(0, Q)$ is a closed subset of Q. ◆

1.6 Properties of open and closed subsets

The next three propositions, Propositions 1.29, 1.30, and 1.31, state some simple, but very important, properties of open and closed subsets.

Proposition 1.29 *Suppose* $X \subseteq \mathbf{R^n}$.

(a) *The union of any number of open subsets of* X *is an open subset of* X.
(b) *The intersection of a finite number of open subsets of* X *is an open subset of* X.

Proof of Proposition 1.29: (Remark: See Definition A.18, and Example A.19, if needed, for background on indexed collection of subsets. If the reader is unfamiliar with working with indexed collections of subsets, we suggest writing these proofs in terms of two subsets at first.)

We first prove statement (a). Suppose that $\{U_\alpha\}_{\alpha \in A}$ is a collection open subset of X where our index set is denoted A. Let $x \in \bigcup_{\alpha \in A} U_\alpha$. By the definition of union, Definition A.20, we know that there is an $\alpha_0 \in A$ such that $x \in U_{\alpha_0}$. Since U_{α_0} is open, there is a positive number, call it ϵ_0, such that $N_{\epsilon_0}(x, X) \subseteq U_{\alpha_0}$. Now $N_{\epsilon_0}(x, X) \subseteq U_{\alpha_0} \subseteq \bigcup_{\alpha \in A} U_\alpha$. Thus $N_{\epsilon_0}(x, X) \subseteq \bigcup_{\alpha \in A} U_\alpha$, and $\bigcup_{\alpha \in A} U_\alpha$ is an open subset of X.

Next we prove part (b). Suppose that $\{U_\alpha\}_{\alpha \in A}$ is a *finite* collection open subsets of X where our index set is denoted A. If A has n elements, we may suppose that our index set, A, consists of the numbers $\{1, \ldots, n\}$. So we can write our collection as $\{U_i\}_{i=1}^{i=n}$.

Suppose $x \in \bigcap_{i=1}^{i=n} U_i$. Since $x \in U_i$, and since each U_i is an open subset of X, there is a number ϵ_i such that $N_{\epsilon_i}(x, X) \subseteq U_i$. Let $\epsilon = \min(\epsilon_1, \ldots \epsilon_n)$.

For $1 \leq i \leq n$, we have $N_\epsilon(x, X) \subseteq N_{\epsilon_i}(x, X) \subseteq U_i$. Therefore if $x \in N_\epsilon(x, X)$, then $x \in \bigcap_{i=1}^{i=n} U_i$. Thus $\bigcap_{i=1}^{i=n} U_i$ is an open subset of X.

We leave, as an exercise, an alternate proof by induction for statement (b). ∎ (Problem 1.10)

The next proposition is really just a version of Proposition 1.29 in terms of closed subsets. The keys to demonstrating this are the DeMorgan rules for subsets, Propositions A.05 and A.07. We have left the proof of Proposition 1.30 as an exercise, but, as a guide, we show a special case: If C and C' are is a closed subsets of X, then $C \cap C'$ is a closed subset of X.

Since C is a closed subset of C, $X - C$ is an open subset of X; similarly, $X - C'$ is an open subset of X. By one of the DeMorgan rules we can write the complement of $C \cap C'$ as a union of two open subsets of X: $X - (C \cap C') = (X - C) \cup (X - C')$. By Proposition 1.29, $X - (C \cap C')$ is an open subset of X. Thus $C \cap C'$ is a closed subset of X.

Proposition 1.30

(a) *The intersection of any number of closed subsets of X is a closed subset of X.*

(b) *The union of a finite number of closed subsets of X is a closed subset of X.* ∎ *(Problem 1.11)*

The proof of the next proposition is an immediate consequence of the definition of an open subset, together with the fact that the empty set is a subset of any set, as noted in the remark on page 393.

Proposition 1.31 *X and the empty subset of X are open subsets of X.* ∎ *(Problem 1.12)*

It is important to carefully compare and contrast Propositions 1.29 and 1.30, paying particular attention to the words "finite" and "infinite." It is not true that any intersection of open subsets is an open subset. Since a finite intersection of open subsets is open, the problem is for the case of infinite intersections. The next two examples demonstrate this and give some interesting ways in which we can express certain sets in terms of infinite union and infinite intersection.

Example 1.32 Consider any point $p \in \mathbf{R}^2$. We may express $\{p\}$ as the (uncountable) intersection of open subsets as follows: $\{p\} = \bigcap_{\epsilon>0} N_\epsilon(p)$ where $0 < \epsilon$. However, as we have noted, Proposition 1.14, $\{p\}$ is not an open subset of $\mathbf{R}^\mathbf{n}$. ◆

Example 1.33 Also, a union of infinitely many closed subsets of $\mathbf{R}^\mathbf{n}$ need not be closed. Take *any* subset S that is not closed. We will later find that there are many such examples. For example, we could take S to be an open disk in the plane. We can express S as the union of closed subsets of $\mathbf{R^n}$ as follows: $S = \bigcup_{p \in S} \{p\}$. By Proposition 1.13, this S not a closed subset of $\mathbf{R^n}$. ◆

Example 1.34 Here is a second example of a union of closed subsets which is not a closed subset. For any $p \in \mathbf{R^n}$, write $\{p\} = \bigcap_{\epsilon > 0} N_\epsilon(p)$. Then $\mathbf{R^n} - \{p\} = \bigcup_{\epsilon \in \mathbf{R}} (\mathbf{R^n} - N_\epsilon(p))$, by the generalized DeMorgan rule for sets, Proposition A.17. Each of the subsets on the right-hand side of this equation is a closed subset of $\mathbf{R^n}$, as follows from Proposition 1.08; yet the subset on the left is not a closed subset by Proposition 1.14. ◆

The basic properties of open subsets, Propositions 1.29 and 1.31, are fundamental for topology. Take note, from time to time, in reading the rest of this text, that most definitions, ultimately, only involve the concept of open subset and closed subset, and most proofs use only these properties and their consequences. In particular, the notion of distance between points will rarely be referred to directly. In the more abstract setting of general topology, Propositions 1.29 and 1.31 are taken as axioms.

We next show how the idea of open subset can be used to articulate other expressions of "nearness." Suppose $A \subseteq X \subseteq \mathbf{R^n}$. We would like to capture the idea that A is dense in X if every point of X is near some point of A. Since we express nearness in terms of open subsets we are lead to the following definition.

Definition 1.35 *Let $A \subseteq X \subseteq \mathbf{R^n}$. We say A* **is dense in** X *if every open subset of X contains a point of A.*

Example 1.36 Let Q denote the rational numbers, and Z denote the irrational numbers. Then Q and Z are dense in $\mathbf{R^1}$. ◆(Problem 1.17)

Any subset of $\mathbf{R^n}$ has a *countable* dense subset. Let $\mathcal{G}$ be the set of all open neighborhoods $N_\epsilon(p)$ in $\mathbf{R^n}$ where ϵ is a rational number and each of the n coordinates of p is a rational number. A countable union of countable sets is a countable set, Proposition C.2. Thus $\mathcal{G}$ will have countably many elements. For each $N_\epsilon(p) \in \mathcal{G}$, choose one point $x_\epsilon^p \in N_\epsilon(p, X)$, (if this set is non-empty). Then $\{x_\epsilon^p\}$ will be a countable subset of X, and one can verify that it is a dense subset of X (Problem 1.18).

1.7 Limit points

We close this chapter by introducing an important definition, "limit point" of a subset, that is especially useful for working with closed subsets. We will soon see, remark on page 28, how one can use this idea to clarify some of the proofs we have presented.

Definition 1.37 *Let $A \subseteq X \subseteq \mathbf{R}^n$, $x \in X$. We say x* **is a limit point of** *A* **in** *X if every open subset U of X, with $x \in U$, contains a point of $A - \{x\}$.*

That is, U contains a point of A *other than* x. Although the definition above mentions only one point of $A - \{x\}$, there may be many points of $A - \{x\}$ in the open set U. Also, differing choices for the open subset U may force different choices for points of $A - \{x\}$.

Generally, we will shorten the phrase "x is a limit point of A in X" to "x is a limit point of A". This is because, if x is a limit point of a subset Z, then x is a limit point of any larger set. Suppose $A \subseteq X \subseteq Z \subseteq \mathbf{R}^n$, $x \in X$, and x is a limit point of A in X. Considering that $x \in Z$, x is a limit point of A in Z. If every open subset of X contains a point of A other than x, then every open subset of Z must contain a point of A other than x.

Example 1.38 Let $X = \mathbf{R}^1$, $A = \{1/n\}$ where $n \in \mathbf{N}$. Let $B = A \cup \{0\}$. Then 0 is a limit point of A ; also, 0 is a limit point of B. ◆

Example 1.38 shows that if x is a limit point of A, it may or may not be true that $x \in A$. The following is a useful characterization of whether a set is closed. Briefly, it says that a set is closed if and only if it contains all its limit points.

Proposition 1.39 *Suppose $X \subseteq \mathbf{R}^n$ and $A \subseteq X$. Then A is a closed subset of X if and only if all points of X, which are limit points of A, are contained in A.*

Proof: We first show: if A is a closed subset, it must contain all of its limit points. We argue by contradiction. Suppose A is a closed subset and there is a limit point x of A with $x \in X - A$. But $X - A$ is an open subset containing x, which does not contain points of A, contradicting the assumption that x is a limit point of A.

Secondly, suppose every limit point of A is contained in A. Let $x \in X - A$. Since x is not a limit point of A, there must be an open subset of X, call it U_x, such that $(U_x - \{x\}) \cap A = \emptyset$. Then $U_x \cap A = \emptyset$. But consider $U = \bigcup_{x \in X-A} U_x$. Since U is a union of open subsets of X, it is an open subset of X. Therefore $A = X - U$ is a closed subset of X. ▮

Suppose $A \subseteq X \subseteq \mathbf{R^n}$. By Proposition 1.39 we can say that A is *not* a closed subset of X if we can find a limit point of A which is not contained in A.

Recall our proof that an open n-ball is not a closed subset of $\mathbf{R^n}$, Proposition 1.13. Using the concept of limit point this proof can now be described as follows:

$N_\epsilon(p)$ cannot be a closed subset of $\mathbf{R^n}$ since, if $p = (a, b)$ and $q = (a + \epsilon, b)$, then one can show that q is a limit point of $N_\epsilon(p)$ which is not contained in $N_\epsilon(p)$. (Problem 1.19)

Also, in Example 1.15 we can more briefly describe our proof that $(0, 1]$ is not a closed subset of the line as "$(0, 1]$ is not a closed subset of the line since 0 is a limit point of $(0, 1]$ which is not a point of $(0, 1]$".

These two modifications do not make these particular proofs any easier since the real work, in each case, is the verification that the given point is, in fact, not a limit point. What we do gain is a general strategy for showing that a subset is closed. Later, as we establish more properties of limit points, this strategy will be more powerful.

Example 1.40 The number 2 is the least upper bound of $X = [0, 2)$; it is also a limit point of X (see Definition B.06 for a definition of a least upper bound). On the other hand, consider the case where X is the union of the interval $[0, 1]$ and the one-point set $\{2\}$. Now 2 is the least upper bound of X but it is not a limit point of X. What makes 2 a special point of X is that it is "isolated." We define this notion below and see that, except for such isolated points, a least upper bound of a subset X of $\mathbf{R^1}$ is a limit point of X. ◆

Definition 1.41 *Suppose $X \subseteq \mathbf{R^n}$ and $x \in X$. We will say that x* **is an isolated point of** X *if there is an ϵ-neighborhood of x $N_\epsilon(x, X)$ such that $N_\epsilon(x, X) = \{x\}$.*

Proposition 1.42 *Suppose $X \subseteq \mathbf{R^1}$ which is bounded above, and x_1 is the least upper bound of X which is not is an isolated point of X. Then x_1 is a limit point of X. In fact, one can say a bit more—for every $\epsilon > 0$ there is a point $c \in X$ such that $x_1 - \epsilon < c < x_1$.*

Proof (of the last statement, by contradiction): Suppose there were an $\epsilon > 0$ for which there were no point $c \in X$ such that $x_1 - \epsilon < c < x_1$. Since x_1 is not an isolated point, it follows that $x_1 \notin X$. Then $x_1 - \epsilon/2$ would be an upper bound of X, contradicting the assumption that x_1 is a *least* upper bound of X. ∎

If $A \subseteq X \subseteq \mathbf{R^n}$ and if every point of x is a limit point of A, then it follows that A is dense in X. The converse is not true if X has isolated points. That is the only problem. Thus, if A is dense in X and $x \in X$, then either x is a limit point of A or an isolated point of A.

Our final example for this section, especially for the irrational angle case, is an interesting non-trivial one. In a future chapter we will restate the problem with a new perspective, Example 3.26.

Example 1.43 Let S^1 denote the unit circle in the plane and let X_θ be the set of points of S^1 corresponding all positive multiples of a given angle θ. Suppose $x_\theta = (\cos(\theta), \sin(\theta))$ is a point of S^1; then

$$X_\theta = \{x_{n\theta}\}_{n\in\mathbf{N}} = \{(\cos(n\theta), \sin(n\theta))\}_{n\in\mathbf{N}}, \ \mathbf{N} \text{ the natural numbers.}$$

We first note that X_θ will be a finite set if and only if θ is a rational multiple of 2π. Suppose θ is a rational multiple of 2π. Write $\theta = \frac{p}{q} 2\pi$ where p and q are positive integers and this fraction is written in reduced form. Then X_θ will consist of q points: $X_\theta = \{x_\theta, x_{2\theta}, \ldots x_{q\theta}\}$. Conversely, if X_θ is finite, there must be a p and a q such that $p\theta = q\theta + N2\pi$ and $\theta = \frac{N}{p-q} 2\pi$.

Thus, if θ is an irrational multiple of 2π, X_θ is infinite. We next claim that, in this case, X_θ is a dense subset of S^1. We first focus on the special case of $x_0 = (1,0)$, and show x_0 is a limit point of X_θ. We show that, for any $\epsilon > 0$, there is a point x_α of X_θ whose distance from x_0 is less than ϵ.

Find an integer N so that if, starting at x_0, we divide S^1 into N equal sub-arcs: $A_1, \ldots A_N$ of arc length less than ϵ.

Since X_θ is infinite, we can find some k such that A_k contains two distinct points, $x_{p\theta}$ and $x_{q\theta}$, of X_θ. So $\delta(x_{p\theta}, x_{q\theta}) < \epsilon$. Write $|p\theta - q\theta| = |p-q|\theta$, and consider the subset of X_θ, $Y_\theta = \{x_{n|p-q|\theta}\}_{n\in\mathbf{N}}$. Consecutive points of Y_θ are a distance less than ϵ apart, so one point x_α of Y_θ must lie in A_1. Thus $\delta(x_\alpha, x_0) < \epsilon$.

We have shown that x_0 is a limit point of X_θ. One can similarly show that any point of S^1 is also a limit point of X_θ. ◆

*1.8 General topology and Chapter 1

In modern mathematics one obtains abstraction by the process of turning a theorem into a definition and examining the logical consequences.

We briefly review the development of the concept of an open subset of $\mathbf{R}^n$. Using the notion of distance in $\mathbf{R}^n$, Definition 1.02 , we define "open set" in terms of ϵ-neighborhoods, Definition 1.04 . Using this, we define open subset, Definition 1.05. Careful analysis of various proofs concerning open subsets reveals that we often rely only on three properties of open sets; see Propositions 1.29 and 1.31 .

(1) The union of any collection of open subsets is an open subset.
(2) The intersection of a finite collection of open subsets is an open subset.
(3) The empty subset and the entire subset are open subsets.

Definition 1.44 takes the notion of "open set" as a basic abstract notion. No notion of distance is used in these definitions. In fact, we will find examples where no sensible notion of distance is possible. In doing this, we begin to develop a new view of topology, called "general topology," or "point set topology."

In this process, we find completely new mathematical examples, which are of interest in their own right. In addition, we get new insight into what "distance" really means, and maybe even more interesting, we get new ideas for what "distance" *could* mean.

General topology begins with the following definition.

Definition 1.44 *Let S be a set, A an index set. Let $\mathcal{T} = \{T_\alpha\}_{a \in A}$ be collection of subsets of S that satisfies the following three properties:*

(a) *Any union of the elements of $\mathcal{T}$ is also an element of $\mathcal{T}$.*
(b) *Any finite intersection of elements of $\mathcal{T}$ is also an element of $\mathcal{T}$.*
(c) *The empty set and S are elements of $\mathcal{T}$.*

We call $\mathcal{T}$ a **topology for** *S and refer to elements of $\mathcal{T}$ as the* **open sets of the topology,** *$\mathcal{T}$. We refer to the pair $\{S, \mathcal{T}\}$ as a* **topological space**.

General topology is the study of all sets, where we investigate collections of subsets which satisfy the properties of Definition 1.44.

Often, in general topology, the word "space" is often used as a synonym for set and "subspace" for subset. Other times one finds "space" used as a shortened version of topological space. A topological space, as we have defined it, is a *pair* of objects: $\{S, \mathcal{T}\}$. But often, we will not have need of reference to the symbol $\mathcal{T}$. In this case we use the simpler language "let S be a topological space, ..."

From the point of view of general topology, $\mathbf{R^n}$ with its familiar notion of open subset, is only one example of a topological space. As a matter of formality, we need to *verify* that this is the case:

Proposition 1.45 *Let $\mathcal{T}$ be the collection of subsets of $\mathbf{R^n}$ defined by: $U \in \mathcal{T}$ if and only if for all $x \in U$ there is a number $\epsilon > 0$ such that if $\delta(x, y) < \epsilon$, then $y \in U$. Then $(\mathbf{R^n}, \mathcal{T})$ is a topological space.*

The proof of consists of the proofs of Propositions 1.29, 1.31. ∎

Example 1.46 A given set can have more than one topology defined for it. Let $\mathcal{H}$ be all subsets of $\mathbf{R^1}$ which are unions of half-open intervals of the form $[a, b)$. One can verify that $\{\mathbf{R^1}, \mathcal{H}\}$ is a topological space. (Note: We get the empty set as an empty union of these intervals.) A common name for this example is the "Sorgenfry line." In $\mathcal{H}$ any half-open interval, say $[0, 1)$, is an open subset. But in the usual topology for $\mathbf{R^1}$, Definition 1.05, $[0, 1)$ is not an open subset. ◆

Once we have verified Proposition 1.45, we are in position to make the following definition:

Definition 1.47 *If $\mathcal{T}$ is the topology for* $\mathbf{R^n}$ *given in Proposition 1.45, we say the topological space* $(\mathbf{R^n}, \mathcal{T})$ *is* $\mathbf{R^n}$ **with the standard topology** *or sometimes* $\mathbf{R^n}$ **with the usual topology**.

Every set (with more than one point) has at least two topologies we can define for it. These are called the "discrete" and the "indiscrete" topologies.

Definition 1.48 *Let S be a set, let $\mathcal{D}$ be the set of all subsets of S. Then $\{S, \mathcal{D}\}$ is called the* **discrete topology** *for S.*

Definition 1.49 *Let S be a set and let $\mathcal{T}$ be the set $\mathcal{T} = \{S, \varnothing\}$. Then $\{S, \mathcal{T}\}$ is called the* **indiscrete topology** *for S.*

So now we have at least three ways to consider the set $\mathbf{R^n}$ as a topological space: the standard topology, the discrete topology, and the indiscrete topology. In fact, there are many more.

The indiscrete topology is not, in general, a useful notion. The discrete topology is of some interest, particularly for finites sets, as we will see in Proposition 1.55. But note that in this topology *all* subsets are open subsets. Thus the *topology* of S is the same as the *set theory* of S.

As in $\mathbf{R^n}$, closed sets in a topological space are defined to be the complements of open subsets; compare with Definition 1.07:

Definition 1.50 *Let $\{S, \mathcal{T}\}$ be a topological space with $C \subseteq S$. We say C* **is a closed subset of** *$\{S, \mathcal{T}\}$ if $S - C \in \mathcal{T}$.*

If $\{S, \mathcal{T}\}$ is a topological space and A is a subset of S, then there is a natural way to define a topology for A. Let $\mathcal{T}'$ be all sets of the form $A \cap O$ where $O \in \mathcal{T}$. Then it is easy to verify that $\{A, \mathcal{T}'\}$ is a topological space. This gives rise to the next definition, which is basically the same as used for subsets of $\mathbf{R^n}$; compare with Definition 1.17.

Definition 1.51 *Suppose $\{S, \mathcal{T}\}$ is a topological space and $A \subseteq S$. Let $\mathcal{T}' = \{A \cap O$ such that $O \in \mathcal{T}\}$. Then $\{A, \mathcal{T}'\}$ is called the* **topology of A derived from** *$\{S, \mathcal{T}\}$. Several common names used for this are the* **derived topology, relative topology, induced topology**, *or* **subset topology** *for A.*

In the standard topology, a point is a closed subset of $\mathbf{R^n}$, Proposition 1.14 . The following example shows that the corresponding statement is not always true when we consider abstract topological spaces.

Example 1.52 Let $X = \{a, b, c\}$ be a set of three points in the plane. Choice of points will not matter here, but we could chose $a = (-1, 0)$, $b = (0, 0)$, and $c = (1, 0)$. Let $\mathcal{T} = \{X, \emptyset, \{a\}, \{b\}, \{a, b\}\}$. One can verify that $\mathcal{T}$ satisfies Definition 1.44; thus $\{X, \mathcal{T}\}$ is a topological space. Yet, since $\{b, c\}$ is not an element of $\mathcal{T}$, the one-point set $\{a\}$ is not a closed subset. ◆

One can invent other topological spaces for finite sets along the lines of Example 1.52. For the most part however these are seldom-used oddities. To avoid such pathological examples, one usually requires that a topological space have additional properties beyond those listed in Definition 1.44. There are several candidates for "good" properties to add. By far, the most common one is the following, which basically says that "points can be distinguished by means of *disjoint* the open subsets of the topology."

Definition 1.53 *Let $\{S, \mathcal{T}\}$ be a topological space. We say $\{S, \mathcal{T}\}$* **satisfies the Hausdorff property** *if, for any distinct points $p, q \in S$, we may find open sets U and V of $\mathcal{T}$ such that $p \in U$, $q \in V$, and $U \cap V = \emptyset$. (Sometimes we express this as: S* **is a Hausdorff space.***)*

In the standard topology for $\mathbf{R^n}$, if $X \subseteq \mathbf{R^n}$, then X is Hausdorff. If $p, q \in X$ and $\epsilon = \delta(p, q)/2$, then $N_\epsilon(p, X) \cap N_\epsilon(q, X) = \emptyset$.

Example 1.52 is not a Hausdorff space. This is easy enough to verify directly, but it also follows from the next proposition.

Proposition 1.54 *If S is a Hausdorff space, and $p \in S$, then $\{p\}$ is a closed subset of the topology of S.* ▮

If every one-point subset of S is a closed subset, then every finite subset, being a finite union of closed subsets of S, is closed. Thus we can conclude:

Proposition 1.55 *If $\{S, \mathcal{T}\}$ is a Hausdorff space, S a finite set, then $\mathcal{T}$ is the discrete topology for S.* ▮

If S has the discrete topology, then S is a Hausdorff space. Also, if S has the indiscrete topology and contains more than one point then S is not a Hausdorff space.

An alternative approach to Definition 1.44 for generalizing the notion of open subset is to generalize the definition of distance. This generalized distance can be used to define neighborhood which, in turn, can be used to define our open sets. The key to this generalization is noting that most uses of distance in $\mathbf{R^n}$ involve only the three properties as listed in Proposition 1.03.

Definition 1.56 *Let S be a set. A* **metric on** *S is a function d which associates to each ordered pair of elements of $(p, q) \in S$ a real number $d(p, q)$ such that:*

(a) *For all $p, q \in S$, $d(p, q) \geq 0$. Also, $d(p, q) = 0$ only if $p = q$.*
(b) *For all $p, q \in S$, $d(p, q) = d(q, p)$.*
(c) *For all $p, q, r \in S$, $d(p, r) \leq d(p, q) + d(q, r)$.*

Definition 1.57 *If d is a metric defined on a set S, then the pair $\{S, d\}$ is called a* **metric space**.

As in the remark on page 30, if we do not intend to use the symbol d in a discussion, we often say "Let S be a metric space ..." rather than "Let $\{S, d\}$ be a metric space ...."

We can now use the notion of a metric to define a topology:

Definition 1.58 *Let $\{S, d\}$ be a metric space, $\epsilon > 0$ and $p \in S$. A* **neighborhood of** *p* **in** *S of radius ϵ, or an ϵ-***neighborhood of** *p consists of all points $x \in S$ such that $d(p, x) < \epsilon$.*

As in $\mathbf{R^n}$, Definition 1.04 , we let $N_\epsilon(p)$ denote the ϵ-neighborhood about p with radius ϵ. Sometimes we use a more elaborate notation, $N_\epsilon^d(p)$, when we wish to notationally emphasize the metric in use.

We can now define the notion of an open subset of a metric space:

Definition 1.59 *Let $\{S, d\}$ be a metric space, $U \subseteq S$. We say U* **is an open set of** *$\{S, d\}$ if for every $p \in U$ there is some $\epsilon > 0$ such that $N_\epsilon^d(p) \subseteq U$.*

We need to justify our use of the term "open subset." We need to verify: if $\{S, d\}$ is a metric space, and if $\mathcal{T}$ is the collection subsets of S defined to be open in Definition 1.59, then $\{S, \mathcal{T}\}$ is a topological space. Once we verify this, a metric gives us a topological space, and we can refer to these open sets as the **topology derived from the metric** d.

Versions of Propositions 1.08 and 1.10 still hold:

Proposition 1.60 *If $\{S, d\}$ is a metric space, $p \in S$ and $\epsilon > 0$, then $N_\epsilon^d(p)$ is an open subset of S.* ∎

Definition 1.61 *If S is a metric space, $p \in S$, $\epsilon > 0$, then the* **closed n-dimensional ball with center** *p* **and radius** *ϵ is defined to be $\{q \in S \mid d(p, q) \leq \epsilon\}$. We denote this set by $\overline{N}_\epsilon^d(p)$, or sometimes also by $B_\epsilon^d(p)$. If d is understood in context, then sometimes this is abbreviated as $\overline{N}_\epsilon(p)$.*

Proposition 1.62 *Let $\{S, d\}$ be a metric space. A closed n-dimensional ball is a closed subset of S.* ∎

Example 1.63 Any set can be considered as a metric space in a somewhat trivial manner. Let S be any set. For any elements of $p, q \in S$, we define $d(p, q)$ to be 0 if $p = q$, and 1 if $p \neq q$. It is easy to check that d is a metric. The reason this is not particularly interesting is that the topology derived from the metric d is the same as the discrete topology for S. For this reason, this metric is called the "discrete metric" for S. ◆

We now have two methods of defining a topology for a set: the first is to somehow describe a set of subsets, $\mathcal{T}$, which satisfy the Definition 1.44, and the second is to define a metric function. It is natural to ask how these methods are related. In particular: Given a topological space, $\{S, \mathcal{T}\}$, is it always possible to find a suitable metric d for S so that the topology derived from this metric is $\mathcal{T}$? The answer to this is "no." This follows from the next proposition.

Proposition 1.64 *If $\{S, d\}$ is a metric space, then, using the topology derived from the metric d, S is a Hausdorff space.*

Proof: If $p, q \in S$, let $\epsilon = \frac{d(p,q)}{2}$. Then $N_\epsilon^d(p)$ and $N_\epsilon^d(q)$ are disjoint open subsets which verify that S is Hausdorff. ∎

Thus the topological space given in Example 1.52 could not have been derived from a metric since this topology does not have the Hausdorff property. It is natural to then ask:

Question 1.65 Given a topological space $\{S, \mathcal{T}\}$, under what conditions can we define a metric d on S so that the topology derived from this metric is the same as $\mathcal{T}$?

This is generally known as the "metrization problem." In Proposition 6.39, we will state one positive answer to this question,

In a metric space any open subset is a union of open balls in that metric. In general topology, it is useful to have definition for this process:

Definition 1.66 *Let $\{S, \mathcal{T}\}$ be a topological space. A collection $\mathcal{B} \subseteq \mathcal{T}$ is called a* **basis for the topology** *$\mathcal{T}$ if every open subset in $\mathcal{T}$ is a union of elements of $\mathcal{B}$.*

Many interesting examples in topology arise from taking a metric space and defining a new metric on the same set, then comparing the two topologies that we get. However, defining a new metric does not always give rise to a new topology, as the next example shows.

Example 1.67 For our set we take the plane. Let p and q be points of the plane with $p = (x_p, y_p)$ and $q = (x_q, y_q)$. We then define

$d(p,q) = |x_p - x_q| + |y_p - y_q|$. For example, in this metric, the distance from $(0,0)$ to $(1,1)$ is 2. One can verify that this is a metric. This metric is in common use and goes by the name of "taxicab metric," or "Manhattan metric." The idea is: suppose a taxicab in New York can travel only along streets that go north-south or east-west. Then the effective distance between two points is the sum of the difference of the north-south coordinates plus the difference of the east-west coordinates.

Given $\epsilon > 0$, one can verify that an ϵ-neighborhood of $p = (x_p, y_p)$, in the taxicab metric consists of the points in the plane inside the square whose corners are $(x_p - \epsilon, y_p)$, $(x_p, y_p + \epsilon)$, $(x_p + \epsilon, y_p)$, and $(x_p, y_p - \epsilon)$. Even though these neighborhoods are very different from the round open-disk neighborhoods of the standard topology as in Definitions 1.04 and 1.05, the topology obtained using the taxicab metric d is the standard topology of the plane. ◆

Example 1.68 More generally one can define a similar metric for $\mathbf{R^n}$. Write $x = (x_1, \ldots, x_n)$, $y = (y_1, \ldots, y_n)$, and define

$$\sigma(x,y) = \max_{i=1}^{n}\{|y_i - x_i|\}.$$

One can verify that this is a metric. ◆

The examples above motivate the following definition

Definition 1.69 *Suppose S is a set and d and d' are two metrics defined on S which have the same derived topology; then we say d* **and** *d'* **are equivalent metrics**.

One of the most important types of topological spaces are function spaces. A **function space** is a topological space in which the points are functions. One familiar example is the following.

Example 1.70 Let I denote the unit interval and let $C(I, \mathbf{R}^1)$ be the set of continuous real-valued functions with domain I. (Here we use Definition D.10 for a definition of continuity.) Given two functions, $f(x)$ and $g(x)$, in $C(I, \mathbf{R}^1)$, define $d(f,g) = \max_{x \in I} |f(x) - g(x)|$. This is a well-defined number since a continuous function has a maximum value on any closed interval; see Proposition D.13, or the more general version, Proposition 10.28.

One can verify that this d *is* a metric.

If $f \in C(I, \mathbf{R}^1)$ and $\epsilon > 0$, then $N_\epsilon^d(f)$ can be described as the set of all functions g such that the graph of g lies between the functions $f(x) + \epsilon$ and $f(x) - \epsilon$. Referring to Figure 1-8, the graph of g must lie between the two dashed graphs. ◆

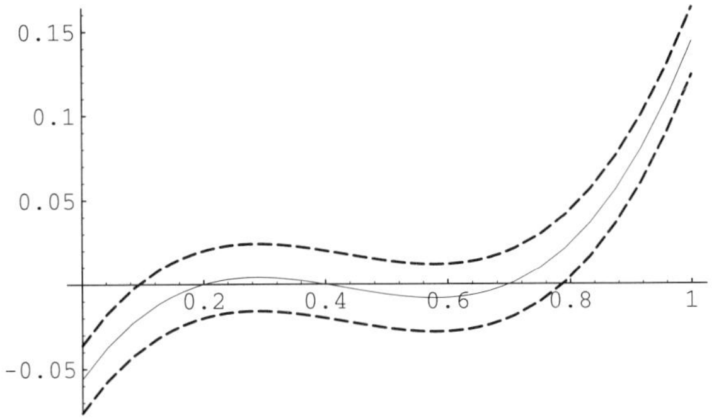

Figure 1-8 An example of a neighborhood of a function in a function space. Here $f(x) = (x - 0.2)(x - 0.4)(x - 0.7)$ and $\epsilon = 0.02$. The neighborhood of f with radius ϵ, using the metric of Example 1.70, is all functions whose graphs lie between $f(x) - \epsilon$ and $f(x) + \epsilon$, shown as dashed curves; see Example 1.70.

Example 1.71 Let $C(I, \mathbf{R}^1)$ be the topological space of continuous real-valued functions defined on the unit interval, Example 1.70. To get a feel for this topology, we consider some closed subsets.

Let $K = \{f \in C(I, \mathbf{R}^1) | f(0) = 0\}$. Then K is a closed subset of $C(I, \mathbf{R}^1)$. Briefly, the idea is to show that the complement of K is an open subset. Let $g \in C(I, \mathbf{R}^1) - K$; then $g(0) \neq 0$. So let $\epsilon = |f(0) - g(0)| = |g(0)|$; $0 < \epsilon$. Then $N^d_\epsilon(g, C(I, \mathbf{R}^1)) \cap K = \emptyset$. ◆

The definitions of dense subset and of limit point, Definitions 1.35 and 1.37, extend without change to the abstract setting of general topology, in the sense defined on page 3.

Example 1.72 If D is a dense subset of a function space $C(X, Y)$ any function in that space can be approximated by a function in D, an important basic idea in analysis.

Let D be the subset of $C(I, \mathbf{R}^1)$ of functions that are piecewise linear with domain I.

We sketch a proof that D is a dense subset of $C(I, \mathbf{R}^1)$. Suppose U is an open subset of $C(I, \mathbf{R}^1)$—we need to find a piecewise-linear function in U. Let $f \in U$. Since U is an open subset, there is an ϵ with $N^d_\epsilon \subseteq U$. We will find a piecewise-linear function g in N^d_ϵ. According to comments in Example 1.70 we need to find a piecewise-linear function p defined on I so that the graph of p lies between the functions $f(x) + \epsilon$ and $f(x) - \epsilon$.

We use the property of continuous functions, defined on an interval known as uniform continuity; see Proposition D.15 (a more general version of this theorem is Proposition 10.30). It assures us that we can choose δ so that if $|x - y| < \delta$, then $|f(x) - f(y)| < \epsilon/2$. Divide I into subintervals of length less than δ, using points $0 = x_0 < x_1 \cdots < x_n = 1$.

One can verify that the piecewise-linear function, g, determined by the points $(x_0, f(x_0)), (x_1, f(x_1)), \ldots, (x_n, f(x_n))$, will be a point of N_ϵ^d. For example, consider the first line segment between $(x_0, f(x_0))$ and $(x_1, f(x_1))$. If $x_0 \leq z \leq x_1$, then since g is monotone for these values and using uniform continuity, $|f(x_0) - g(z)| \leq |f(x_0) - f(x_1)| < \epsilon/2$. Also by uniform continuity $|f(x_0) - f(z)| < \epsilon/2$. By the triangle inequality, for all $x_0 \leq z \leq x_1$,

$$|f(z) - g(z)| \leq |f(z) - f(x_0)| + |f(x_0) - g(z)| \leq \epsilon/2 + \epsilon/2 = \epsilon.$$

Thus this portion of graph of g lies between the graphs of $f(x) - \epsilon$ and $f(x) + \epsilon$. ◆

The definition of limit point, Definition 1.37, extends without change to the setting of general topology, as does the Proposition 1.39 which states a subset $A \subseteq X$ is a closed subset if and only if A contains all its limit points. We illustrate use of these ideas in $C(I, \mathbf{R^1})$.

Example 1.73 Let P be the subspace of $C(I, \mathbf{R^1})$ consisting of polynomial functions restricted to I. We can show that P is not a closed subset of $C(I, \mathbf{R^1})$. If P were a closed subset, it would contain all its limit points. Let $p_n(x) = \sum_0^n \frac{x^n}{n!}$. Let e be the restriction of e^x to I. From calculus we know that e^x is not a polynomial (the only polynomial whose derivative is that polynomial is the zero polynomial). We also know that $e^x = \sum_0^\infty \frac{x^n}{n!}$; this implies that e is a limit point of P. Thus P is not a closed subset. ◆

Here is another example of a topology very different from those considered thus far.

Example 1.74 Let P^n denote the set of lines which pass through the origin of $\mathbf{R^{n+1}}$. In particular, P^2 is the set of lines in $\mathbf{R^3}$. Note that the points of P^n are *not* points on these lines, but the lines themselves.

If $L \in P^n$, then $L \cap S^n$ consists of two points which, we denote by n_L and s_L. (If we are considering P^2, and if L corresponds to the z-axis, then n_L can be thought of as the north pole of S^2, and s_L the south pole.) Such a pair of points is called "an antipodal pair of points of S^n."

If we have two lines, L and L', in $\mathbf{R^{n+1}}$, we define $d(L, L')$ to be the minimum of $\delta(n_L, n_{L'})$ and $\delta(n_L, s_{L'})$ where δ denotes standard distance in $\mathbf{R^{n+1}}$. There is a second way of measuring distance. A "great circle of S^n" is a circle on S^n whose center is $\vec{0} \in \mathbf{R^n}$. We could define $d'(L, L')$ to be the minimum of $\delta'(n_L, n_{L'})$ and $\delta'(n_L, s_{L'})$ where δ' denotes distance as measured along a great circle of S^n.

It can be verified that d and d' are metrics for P^n and that they are equivalent metrics. The resulting topological space is called "n-dimensional projective space." Also, P^2 is called the "projective plane." ◆

Example 1.75 We look at some subsets of projective plane P^2, Example 1.74. We will use the metric d' in this example.

Let L be a line through the origin in $\mathbf{R}^3$ with $L \cap S^2 = n_L \cup s_L$. Given an ϵ with $0 < \epsilon < \pi$, we will describe $N_\epsilon^{d'}(L)$. Let D_n denote the points on S^2 whose distance from n_L, as measured along great circles on S^2, is less than ϵ. Similarly, D_s will denote the points on S^2 whose distance from s_L S^2 is less than ϵ. If n_L and s_L correspond to the north and south poles, then D_n and D_s will be polar caps of radius ϵ, as measured on the surface. One can see that $N_\epsilon^{d'}(L)$ will consist of all lines L' in $\mathbf{R}^3$ such that $L' \cap S^2 \subseteq (D_n \cup D_s)$. The union of lines in $N_\epsilon^{d'}(L)$ will correspond to a solid double cone whose axis is L.

Let A be a plane in $\mathbf{R}^3$ which contains $(0, 0, 0)$ and let K be the set of lines through $(0, 0, 0)$ in $\mathbf{R}^3$ which are contained in A. Claim: K is a closed subset of P^2. The idea is to show that $P^2 - K$ is an open subset of P^2. Let $C = A \cap S^2$; C is a great circle. Let $L \in P^2 - K$, and let $L \cap S^3$ be the antipodal points n_L and s_L. Using some geometry (or calculus if you prefer), one can define the notion of the distance, as measured by d', in S^2 between a point not on a circle and that circle. So let ϵ be the distance between n_L and C (this will be equal to the distance between s_L and C). One then can verify that $N_\epsilon^{d'}(L, P^2) \cap K = \emptyset$. Thus K is a closed subset. ◆

We close this section with one more important way to generate new examples of topological spaces, the quotient space.

Definition 1.76 *Suppose $\{X, \mathcal{T}\}$ is a topological space, and suppose P is a partition of X; see Definition A.35.*

The **quotient space of** X **via** P *is the topological space $(P, \mathcal{T}')$ where $V \in \mathcal{T}'$ if and only if $\bigcup_{p \in V} \{p\}$ is an element of $\mathcal{T}$.*

We may denote this quotient space by X/P.

The function from $q: X \to X/P$ defined by $q(x) = [x]$ is called the **quotient map** *of X to X/P.*

A partition can be viewed as an equivalence relation, as noted in the remark on page 398. If this equivalence relation is denoted by $\sim$, then we denote the quotient space as $X/\sim$

Sometimes, a partition is called **a decomposition**, *in which case, the quotient space is called a* **decomposition space**. *In a quotient space X/P, a subset, $V \subseteq X$, corresponds to a point of X/P. Sometimes this is described as identifying each subset V of the partition to a single point. Using this terminology, X/P is sometimes called the* **identification space**.

The following proposition follows from the Definition 1.76.

Proposition 1.77 *If q is the quotient map for quotient space X/P, then q is continuous.* ▮

Here is another way of looking at the projective spaces, Example 1.74.

Example 1.78 Consider the topological space consisting of the standard n-sphere $S^n \subseteq \mathbf{R}^{n+1}$. The two point subsets of the form $\{\vec{x}, -\vec{x}\}$ define a partition P of S^n. (In the terms of Example 1.74, $\{\vec{x}, -\vec{x}\}$ is a pair of antipodal points of S^n.)

Let Q be the quotient space S^n/P. In Example 4.59 we will show that Q is the same as P^n, where "same as" means homeomorphic. ◆

We introduce a definition for a one-point subset of a partition.

Definition 1.79 *Suppose X is a set and P a partition of X. An element $p \in P$ is said to be a* **degenerate element of** *P if p is a single-point subset.*

An important construction which uses quotient spaces is the mapping cylinder, Definition 1.80. The definition in Chapter 16 of the mapping cylinder, Definition 16.10, as stated is admittedly awkward. In the setting of general topology, it becomes a much simpler definition and is an excellent example of the use of decomposition spaces. In rough terms, the mapping cylinder is obtained from $X \times I$ and Y, identifying points by the rule: "only glue $(x, 1)$ to $f(x)$."

Definition 1.80 *Suppose $f: X \to Y$ is a continuous function from one topological space to another. Consider the disjoint union $(X \times I) \mathring{\bigcup} Y$. Let P be the partition whose only non-degenerate elements are the subsets $(f^{-1}(y) \times \{1\}) \mathring{\bigcup} \{y\}$ where $y \in Im(f)$. The quotient space X/P is called the* **mapping cylinder of** *f. We use the notation M_f.*

The **natural inclusion of** *Y* **into** *M_f is the natural inclusion of Y into $(X \times I) \mathring{\bigcup} Y$, followed by the quotient map of the decomposition.*

A map $\phi: X \to X \times I$, defined by $\phi(x) = x \times 0$ is called the **inclusion of** *X* **into the top of** *$X \times I$. The* **natural inclusion of** *X* **into** *M_f is the inclusion of X into the top of $X \times I$, followed by the natural inclusion of $X \times I$ into $(X \times I) \mathring{\bigcup} Y$, followed by the quotient map of the decomposition.*

Definition 1.81 *Suppose $\{X, \mathcal{T}\}$ is a topological space and $A \subseteq X$. Let P be the partition of X whose only non-degenerate element is the subset A. Then X/P is called* **the topological space obtained from** *X* **by collapsing** *A* **to a point**. *In this case we denote X/P by X/A*

The notion of a quotient space is not an easy one to understand. We need to look at quite a few examples. The quotient space construction can result in some very strange spaces.

Example 1.82 In $\mathbf{R}^2$ with the usual topology, consider $A = N_1(0,0)$, the standard open disk of radius one. Let $X = \mathbf{R}^2/A$.

We show X is not a Hausdorff space. Keep in mind that a point of X corresponds to a subset of $\mathbf{R}^2$. In particular, A is a point of X. Let $y = \{(1,0)\}$; this one-point subset is also a point of X. If X were Hausdorff, then there would be an open subset U of X which contained y and did not contain the point A of X.

Since $A \notin U$, then every point of U is a degenerate (i.e. one-point) subset of $\mathbf{R}^2$. If U is open in X, define W by $W = \bigcup_{x \in U} \{x\}$; then W is an open subset of $\mathbf{R}^2$ such that $W \cap A = \emptyset$, and $(1,0) \in W$. But this is impossible since any open subset of $\mathbf{R}^2$ which contains $(1,0)$ must contain points of A. ◆

On the positive side, for closed subsets, we can show:

Proposition 1.83 *Consider $\mathbf{R^n}$ with the usual topology and let $C \subseteq \mathbf{R^n}$ be a closed subset of $\mathbf{R^n}$; then X/C is a Hausdorff space.* ▮

The key to the proof of Proposition 1.83 is the next proposition. The Hausdorff property for a space says, roughly, that two distinct points can be distinguished by disjoint open subsets. The proposition below (and the definition which follows) says that a closed subset and a point not in that closed subset can be distinguished by disjoint open subsets.

Proposition 1.84 *Consider $\mathbf{R^n}$ with the standard topology. Suppose $C \subseteq \mathbf{R^n}$, C a closed subset of $\mathbf{R^n}$, $p \in \mathbf{R^n}$, $p \notin C$. Then there are disjoint open subsets U and V of $\mathbf{R^n}$ such that $p \in U$ and $C \subseteq V$.* ▮

Proposition 1.84 inspires the next definition.

Definition 1.85 *Let $\{X, \mathcal{T}\}$ be a topological space. We say X is* **regular** *if, for any closed subset C and any point of $p \in X$ with $p \notin C$, there exist disjoint subsets $U, V \in \mathcal{T}$ such that $p \in U$ and $C \subseteq V$.*

We leave as an exercise direct verification that X of Example 1.82 is not a regular space.

1.9 Problems for Chapter 1

1.1 In the text, a proof of Proposition 1.03 was given for the plane (that is, $n = 2$). Prove Proposition 1.03 for the general case of $\mathbf{R^n}$.

1.2 Verify the assertion, made in the proof of Proposition 1.10, that $N_{\epsilon'}(q) \cap \overline{N}_\epsilon(p) = \emptyset$.

1.3 Prove Proposition 1.13 for the $\mathbf{R^n}$ case.

1.4 Prove Proposition 1.14.

1.5 Let Q be the set of rational numbers in $\mathbf{R}^1$; see Example 1.16. Show that Q is not a closed subset of $\mathbf{R}^1$.

1.6 Prove the assertion of Example 1.15:
For each $x \in X$, and for each $\epsilon > 0$, $N_\epsilon(x)$ is *not* a subset of X.

1.7 Prove Proposition 1.20.

1.8 Prove: A one-point subset of the plane is not an open subset of the plane.

1.9 For Example 1.21, verify the assertion that V is not an open subset of X.

1.10 Provide a proof by induction that a finite intersection of open subsets is an open subset, as follows.

1. Let I_n be the statement "an intersection, $\bigcap_{i=1}^{n} U_i$, of n open subsets of X is an open subset of X." The truth of I_1 is clear.
2. Write a proof for the case of the intersection of two open subsets, I_2.
3. Show , for any i, that if I_i is true, then I_{i+1} is true.
   (Hint: Use I_2 and the basic fact that if U, V, and W are any three open subsets of X, then $U \cap V \cap W = (U \cap V) \cap W$.)

1.11 Prove Proposition 1.30. (Hint: See the discussion preceding the statement of Proposition 1.30.)

1.12 Prove Proposition 1.31.

1.13 In reference to the remark on page 14: which of the open subsets given as examples in this chapter can be described as the points inside a simple closed curve?

1.14 Let $X = \{\vec{x} \in \mathbf{R}^2 \mid 1 < |\vec{x}| \leq 2\}$. Show that X is neither an open subset nor a closed subset of the plane.

1.15 Let $A = \{\vec{x} \in \mathbf{R}^2 \mid 1 < |\vec{x}| < 2\}$. Show that A is an open subset of the plane.

1.16 Let X be a finite collection of points in the plane. Show that $\mathbf{R}^2 - X$ is an open subset of $\mathbf{R}^2$.

1.17 Let Q denote the rational numbers and Z denote the irrational numbers. Show Q and Z are dense in $\mathbf{R}^1$.

1.18 Verify the assertion of the remark on page 26, that the subset $\{x_\epsilon^p\}$ is a dense subset of X.

1.19 Write a complete proof of Proposition 1.13 following the outline (see the remark on page 28): $N_\epsilon(p)$ can't be a closed subset of $\mathbf{R^n}$ since, if $p = (a, b)$ and $q = (a + \epsilon, b)$, then one can show that q is a limit point of $N_\epsilon(p)$ which is not contained in $N_\epsilon(p)$.

1.20 Let C denote the unit circle in the plane and let $N_1(0,0)$ denote the open ball about $(0,0)$ in the plane. Show that every point of C is a limit point of $N_1(0,0)$.

1.21 Let Q be the subset of the plane consisting of points for which both coordinates are rational numbers. Which points of the plane are limit points of Q? Give a proof for your answer.

1.22 Let $X = \bigcup_{i=1}^{\infty} C_i$ where C_i is a circle in the plane with center at $(0,0)$ and radius $1/i$. Show that X is not a closed subset of the plane. (Hint: Show that $(0,0)$ is a limit point of X.)

2. BUILDING OPEN AND CLOSED SUBSETS

OVERVIEW: We consider additional examples of open and closed subsets and introduce some techniques for constructing more examples. One technique involves maps such as affine maps, rigid motions, and similarities. Our motivation is threefold. Using linear algebra we can define certain constructions in a very concrete way. Secondly, they will give us a background of examples of homeomorphisms, the main topic of Chapter 4. Also, such maps, introduced as tools here, will become interesting objects of study in later chapters.

2.1 Some basic examples

We open with some simple examples of open subsets and closed subsets of $\mathbf{R}^2$, and invite you to supply verifications based on geometry and the basic material of Chapter 1. We will return to these examples and use less direct methods such as Example 2.33 and the remarks on pages 82, 88, and 88.

Proposition 2.01 *A line in the plane (which we take tc mean the graph of $ax + by = c$ in the x-y-plane where a, b and c are constants, with a and b not both zero) is a closed subset.* ∎ *(Problem 2.1)*

Proposition 2.02 *The unit circle $S^1 = \{(x, y) \in \mathbf{R}^2 \mid x^2 + y^2 = 1\}$ is a closed subset of the plane.* ∎ *(Problem 2.2)*

Roughly, an open rectangle in the plane consists of all points inside a rectangle (but not on the edges). For a closed rectangle, we include the edges. Take a few moments to sketch a direct proof of the next proposition, Proposition 2.05. We soon will look at a more general setting of this (Proposition 2.46 and Problem 2.26) and find some less direct proofs as well, in Example 2.35.

For the next two definitions, we assume a, b, c, and d are real numbers with $a < b$ and $c < d$.

Definition 2.03 *Let* $R = \{(x, y) \in \mathbf{R}^2 \mid a < x < b \text{ and } c < y < d\}$. *We say* R *is an* **open rectangle**. *(See Figure 2-1.)*

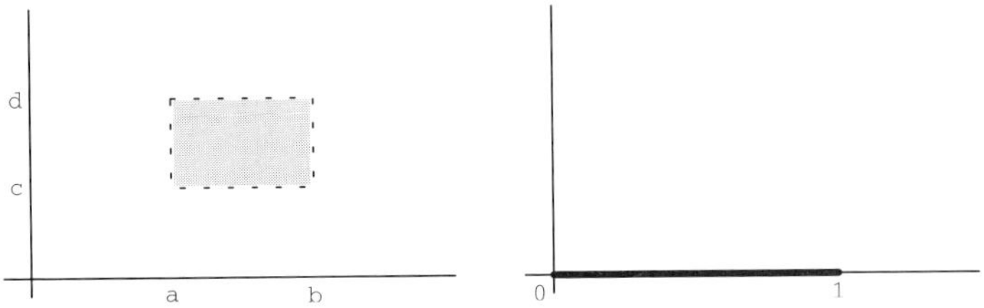

Figure 2-1 On the left, an open rectangle; see Proposition 2.05. Note the edges of the rectangle not contained in the set. On the right, example X from Example 2.06.

Definition 2.04 *Let* $R = \{(x, y) \in \mathbf{R}^2 \mid a \leq x \leq b \text{ and } c \leq y \leq d\}$. *We say* R *is a* **closed rectangle**.

Note the difference between an open and a closed rectangle: the $<$ has been replaced by $\leq$.

Proposition 2.05 *An open rectangle is an open subset of the plane; a closed rectangle is a closed subset of the plane.* ▮ *(Problem 2.5)*

Example 2.06 Let X be the set $X = \{(x, y) \in R^2 : |x| \leq 1 \text{ and } y = 0\}$; X is a closed subset of the plane. (See Figure 2-1.) (Problem 2.3) ◆

Here are some useful subsets of $\mathbf{R^n}$:

Definition 2.07 *We define* n**-dimensional half-space** *to be* $R^n_+ = \{(x_1, x_2, \ldots, x_n) \in \mathbf{R^n} \mid 0 \leq x_1\})$. *(See Figure 2-2.)*

Definition 2.08 *We define* **open** n**-dimensional half-space** *to be* $\mathring{R}^n_+ = \{(x_1, x_2, \ldots, x_n) \in \mathbf{R^n} \mid 0 < x_1\})$. *(See Figure 2-2.)*

Using standard notations for intervals in the line, as in Definition 1.11, $[0, \infty) = R^1_+$, $(0, \infty) = \mathring{R}^1_+$, $(\infty, 0] = R^1 - \mathring{R}^1_+$ and $(\infty, 0) = R^1 - R^1_+$. We refer to R^2_+ as the "standard half-plane"; R^3_+ is called "standard half-space."

Figure 2-2 On the left, a portion of the Euclidian half-plane R^2_+; see Definition 2.07. Points on y-axis shown as thickened. Dotted lines indicates that the subset continues beyond the lines. On the right, the open Euclidian half-plane, $\mathring{R}^2_+$; see Definition 2.08. Note that R^2_+ contains points of the y-axis but $\mathring{R}^2_+$ does not. Dashed lines indicate that $\mathring{R}^2_+$ lies to the right of the y-axis.

For certain subsets of $\mathbf{R^n}$, generalized cylindrical coordinates may be more convenient. In generalized cylindrical coordinates we locate points of $\mathbf{R^n}$, using polar coordinates in the first two coordinates and Cartesian coordinates for the rest.

Definition 2.09 Cylindrical coordinates for $(x_1, x_2, \ldots, x_n) \in \mathbf{R^n}$ *are* $(r, \theta, x_3, \ldots, x_n)$ *where* (r, θ) *are polar coordinates for* (x_1, x_2).

So $r = \sqrt{x_1^2 + x_2^2}$ and $\tan(\theta) = x_2/x_1$ if $x_1 \neq 0$. As with polar coordinates, points have multiple descriptions such as $(1, 0, 2, 3) = (1, 2\pi, 2, 3)$. In addition, the origin can have any real value for θ. For example, $(0, 0, 2, 3) = (0, 1, 2, 3)$.

Definition 2.10 *For an angle,* α, *using generalized cylindrical coordinates, define* $H^{n-1}_\alpha = \{(r, \theta, x_3, \ldots, x_n) \in \mathbf{R^n} \mid \theta = \alpha\}$. *Such a subset is called an* $(n-1)$**-dimensional page of** $\mathbf{R^n}$.

Example 2.11 In Figure 2-3 we indicate eight pages, H^2_0, $H^2_{\pi/4}$, $H^2_{\pi/2}$, $H^2_{3\pi/4}$, H^2_π, $H^2_{5\pi/4}$, $H^2_{3\pi/2}$, and $H^2_{7\pi/4}$ in $\mathbf{R}^3$. Only a square on each page is shown. The term "page" comes from the following image—if we open a book all the way until the front cover meets the back cover the pages of the book will fan out and look something like Figure 2-3. ◆

Definition 2.12 *We define the* **unit** n**-dimensional sphere** *to be the subset* $S^n = \{\vec{x} \in \mathbf{R}^{n+1} \mid \delta(\vec{x}, \vec{0}) = 1\}$ *where* $\vec{0}$ *denotes the origin of* $\mathbf{R}^{n+1}$. *(Sometimes the term* **standard** n**-sphere** *is used.) (See Figure 2-4.)*

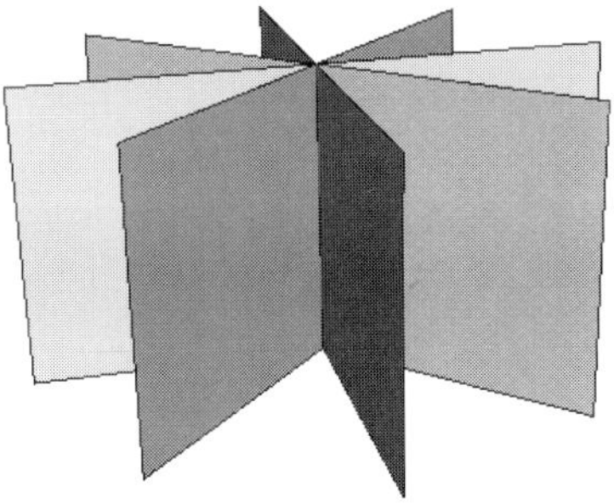

Figure 2-3 Eight pages in $\mathbf{R}^3$; see Example 2.11.

Definition 2.13 *The* **unit n-dimensional ball** *in* $\mathbf{R^n}$ *is the set* $D^n = \{\vec{x} \in \mathbf{R^n} \mid \delta(\vec{x}, \vec{0}) \leq 1\}$. *(Sometimes the term* **standard n-ball** *is used.) (See Figure 2-5.)*

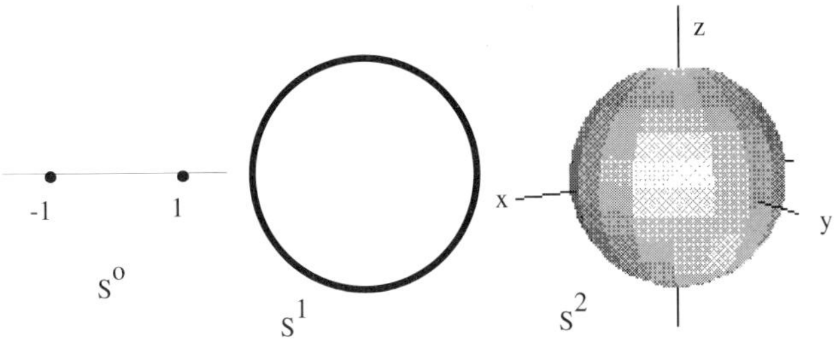

Figure 2-4 On the left, $S^0 \subseteq \mathbf{R^1}$, the two numbers -1 and 1. In the middle, $S^1 \subseteq \mathbf{R^2}$, the unit circle in the plane. On the right, $S^2 \subseteq \mathbf{R^3}$, the unit sphere; see Definition 2.12.

So S^1 is the unit circle in the plane, S^2 is the standard unit sphere in space, and S^0 consists of two points—the numbers -1 and 1 of $\mathbf{R^1}$ since it the set of real numbers such that $|x| = 1$. Also, D^1 is the closed interval [-1, 1] in $\mathbf{R^1}$, D^2 is the unit disk in the plane, D^3 the unit ball in $\mathbf{R^3}$. For any n, $S^n \subseteq D^{n+1}$.

The common usage of the word "dimension" conflicts with the usage in the definition above, as in the definition of n-dimensional sphere. In ordinary usage, we say something is two-dimensional if it has width and height; we say something is three-dimensional if, in addition it has depth. We have defined the 1-dimensional sphere to be a circle. When we think of a circle, we generally have in mind an image, such

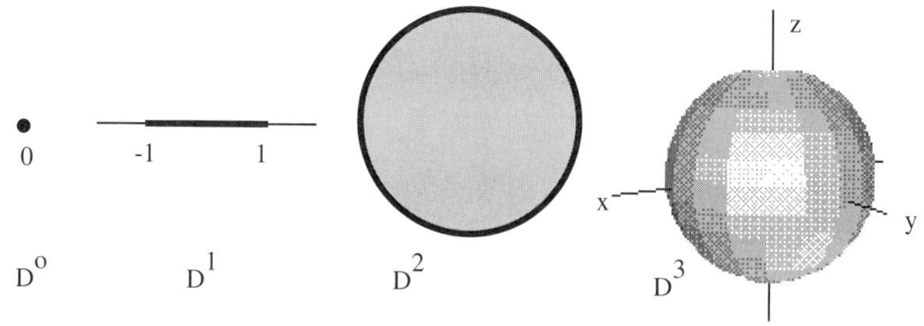

Figure 2-5 On the left, $D^0 \subseteq \mathbf{R^0}$, a single point. In the middle-left, $D^1 \subseteq \mathbf{R^1}$, the interval $[-1, 1]$ in the line. On the middle-right is $D^2 \subseteq \mathbf{R^2}$: a disk whose edge is the unit circle. On the right, $D^3 \subseteq \mathbf{R^3}$, a solid ball consisting of all points on the unit sphere and also those inside it; see Definition 2.13. Unfortunately, there is no simple graphical method to distinguish a drawing of S^2, as in Figure 2-4, from a drawing of D^3.

as the circle in the middle of Figure 2-4. This circle has width and height so it would seem that a case can be made for calling the circle two-dimensional. Motivation for our terminology will become more apparent as we learn more topology—when we discuss the topic of homeomorphism, in Chapter 4, embeddings, in Chapter 6, and manifolds, in Chapter 13. Roughly, here is the idea—we wish to think of the dimension of a circle as a property of the circle and not a property of points outside the circle.

We will say that the circle in Figure 2-4 is a two-dimensional drawing of a 1-dimensional sphere. Similarly, at the right in Figure 2-4, we say we see a three-dimensional drawing of a 2-dimensional sphere. In this text, then, we distinguish between a two-dimensional subset, and 2-dimensional subset.

Using the word "two" for two-dimensional subset means that two dimensions are needed to describe or visualize the subset. Using the number 2 for a 2-dimensional subset has the technical meaning given in the mathematical definition of that subset.

Using the points of $\mathbf{R^{n+1}}$ which correspond to the standard basis vectors of $\mathbf{R^{n+1}}$ as vertices, we construct the some standard objects called "simplexes."

Definition 2.14 *The* **standard** n**-simplex in** $\mathbf{R}^{n+1}$, *denoted* Δ^n, *is defined by* $\Delta^n = \{(x_1, x_2, \ldots, x_{n+1}) \in \mathbf{R}^{n+1} \mid x_1 + \cdots + x_{n+1} = 1$ *with* $0 \leq x_i$ *for all* i, $1 \leq i \leq n+1 \}$.

Example 2.15 The standard 1-simplex in the plane is a line segment with endpoints (1,0) and (0,1). It is the portion of the line with equation

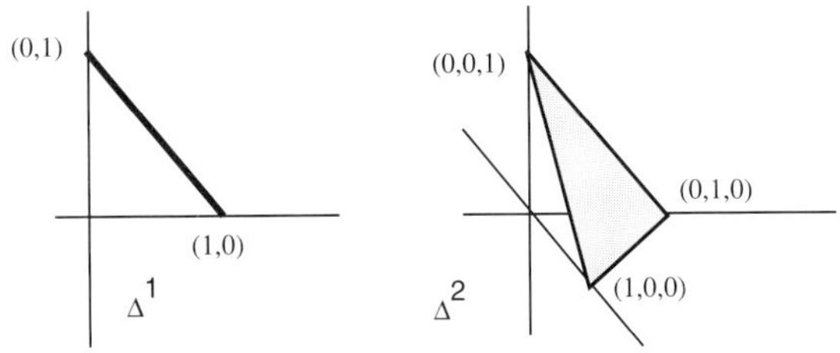

Figure 2-6 On the left, the standard 1-simplex, $\Delta^1 \subseteq \mathbf{R}^2$. On the right, the standard 2-simplex, $\Delta^2 \subseteq \mathbf{R}^3$. See Definition 2.14

$x_1 + x_2 = 1$ which lies in the first quadrant of the plane where $0 \leq x_1$ and $0 \leq x_2$. The standard 2-simplex in space is a triangle with vertices $(1,0,0)$, $(0,1,0)$, and $(0,0,1)$ since it is the portion of the plane $x_1+x_2+x_3 = 1$ contained in the first octant of three-dimensional space; see Figure 2-6.

A more general notion of simplex where coordinates of the vertices correspond to vectors other than the standard basis is discussed in Chapter 17, Definition 17.03. These are used as a basic building block for some important and interesting subsets called "simplicial complexes." ◆

Since $D^n = \overline{N}_1(\vec{0})$, it is a closed subset of $\mathbf{R^n}$, Proposition 1.10. The next proposition gathers similar facts about other subsets.

Proposition 2.16

(a) $\overset{\circ}{R^n_+}$ *is an open subset of* $\mathbf{R^n}$.
(b) R^n_+ *is a closed subset of* $\mathbf{R^n}$.
(c) S^n *is a closed subset of* $\mathbf{R}^{n+1}$.
(d) Δ^n *is a closed subset of* $\mathbf{R}^{n+1}$.

Proof of (a): Let $p \in \overset{\circ}{R^n_+}$; $p = (x_1, x_2, \ldots, x_n)$. Let $\epsilon = x_1$. Then $\epsilon > 0$. We show that $N_\epsilon(p) \subseteq \overset{\circ}{R^n_+}$. Suppose $q \in N_\epsilon(p)$ where $q = (y_1, y_2, \ldots, y_n)$. Then

$$|x_1 - y_1| = \sqrt{(x_1 - y_1)^2} \leq \sqrt{\sum_{i=1}^{n} (x_i - y_i)^2} = \delta(p,q)) < \epsilon = x_1.$$

Since $|x_1 - y_1| < x_1$, it follows that $0 < y_1 < 2x_1$. In particular, since $0 < y_1$, $q \in \overset{\circ}{R^n_+}$; thus $\overset{\circ}{R^n_+}$ is an open subset of $\mathbf{R^n}$.

Proof of (b): Let $X = \mathbf{R^n} - R^n_+$. We need to show that X is an open subset of $\mathbf{R^n}$. Note that $X = \{(x_1, x_2, \ldots, x_n) \in \mathbf{R^n} \mid x_1 < 0\})$. An argument similar to the first part of this proof shows that X is open. Let $p \in X$, $p = (x_1, x_2, \ldots, x_n)$. If $\epsilon = -x_1$, then $N_\epsilon(p) \subseteq X$. So X is an open subset of $\mathbf{R^n}$; thus R^n_+ is a closed subset of $\mathbf{R^n}$.

Proof of (c): Note that $S^n = \overline{N}_1(\vec{0}) \cap (\mathbf{R}^{n+1} - N_1(\vec{0}))$. Now $\overline{N}_1(\vec{0})$ is a closed subset of $\mathbf{R}^{n+1}$ by Proposition 1.10, and $(\mathbf{R}^{n+1} - N_1(\vec{0}))$ is a closed subset since $N_1(\vec{0}))$ is open by Proposition 1.08. Thus S^n is a closed subset, being the intersection of two closed subsets.

We leave the proof of the fourth part, (d), as an exercise (Problem 2.6). ∎

2.2 Some basic properties of open and closed subsets

Example 2.17 Suppose $S \subseteq X \subseteq \mathbf{R^n}$ and S is an open (or a closed) subset of X. Does this imply that S is an open (or a closed) subset of $\mathbf{R^n}$? In general, the answer is no, as these next examples show. Verifications are left as an exercise. (Problem 2.8).

Let $X = R^2_+$ and let $S = N_1((0,0) \cap X$. Then $S \subseteq X \subseteq \mathbf{R}^2$, S is an open subset of X but not an open subset of $\mathbf{R}^2$.

Let $X' = \mathring{R}^2_+$ and let $S' = \overline{N}_1((0,0) \cap X'$, then $S' \subseteq X' \subseteq \mathbf{R}^2$; S' is a closed subset of X', but not a closed subset of $\mathbf{R}^2$. ◆

Despite examples such as Example 1.21 and 2.17, there are some positive things relating open (or closed) subset information to open (or closed) information in the larger set.

Proposition 2.18

(a) *Suppose $U \subseteq X \subseteq \mathbf{R^n}$. If U is an open subset of $\mathbf{R^n}$, then U is an open subset of X.*
(b) *Suppose $C \subseteq X \subseteq \mathbf{R^n}$. If C is a closed subset of $\mathbf{R^n}$, then C is a closed subset of X.* ∎ *(Problem 2.9)*

In Proposition 2.18 we assume that the given subset is open (or closed) in $\mathbf{R^n}$. In Propositions 2.19 and 2.20 our hypothesis deals with open (or closed) in a *subset* of $\mathbf{R^n}$.

In the next two propositions, the (a) part is a special case of the (b) part. We state the redundant (a) part because it is easier to remember, and often it is all one needs.

Proposition 2.19

(a) *If $U \subseteq X \subseteq \mathbf{R^n}$ where U is an open subset of X, and X is an open subset of $\mathbf{R^n}$, then U is an open subset of $\mathbf{R^n}$.*

(b) *More generally, suppose $U \subseteq A \subseteq X \subseteq \mathbf{R^n}$ where A is an open subset of X. If U is an open subset of A, then U is an open subset of X.*

Proof: We prove part (a) of the above and leave the second part as an exercise (Problem 2.10). Suppose U is an open subset of X. By definition, there is an open subset W of $\mathbf{R^n}$ such that $U = X \cap W$. Then U is an open subset of $\mathbf{R^n}$ since it is the intersection of two open subsets of $\mathbf{R^n}$. ∎

Proposition 2.20

(a) *If $A \subseteq X \subseteq \mathbf{R^n}$, A is a closed subset of X, and X is a closed subset of $\mathbf{R^n}$, then A is a closed subset of $\mathbf{R^n}$.*

(b) *More generally, suppose $C \subseteq A \subseteq X \subseteq \mathbf{R^n}$, and A is a closed subset of X. If C is a closed subset of A, then C is a closed subset of X.*

We prove part (a) and leave part (b) as an exercise (Problem 2.12), as well as suggesting an alternative proof for part (a) (Problem 2.11) which uses limit points.

Proof of (a): Since X is a closed subset of $\mathbf{R^n}$, then $U = \mathbf{R^n} - X$ is an open subset of $\mathbf{R^n}$. Since A is a closed subset of X, then $X - A$ is an open subset of X. So there is an open subset $V \subseteq \mathbf{R^n}$ such that $V \cap X = X - A$. Now let $W = U \cup V$; W is an open subset of $\mathbf{R^n}$ since it is the union of open subsets of $\mathbf{R^n}$. And thus $\mathbf{R^n} - W$ is an open subset of $\mathbf{R^n}$

We next claim that $\mathbf{R^n} - W = A$. This will show that A is a closed subset of $\mathbf{R^n}$. A proof for this claim follows from the set equalities:

$$R^n - W = R^n - (U \cup V) = (R^n - U) \cap (R^n - V) = X \cap (R^n - V) = X - V = A.$$

The second equality is by one of the DeMorgan rules for sets, Proposition A.17. The second to last equality comes from the lemma about complements of two subsets, Proposition A.15. The last equality comes from Proposition A.16. ∎

Example 2.21 The x-axis of $\mathbf{R^2}$ is a closed subset of $\mathbf{R^2}$, Proposition 2.01. Using Proposition 2.20 it follows that any closed subset of the x-axis corresponds to a closed subset of the plane. ◆

Propositions 2.19 and 2.20 are sometimes remembered as "an open subset of an open subset is an open subset," and "a closed subset of a closed subset is a closed subset." But keep in mind that we should always be aware, when saying "S is subset," to communicate what S is a subset of. Thus "an open subset of an open subset is an open subset" is really a shortened version of the more precise versions as stated in Proposition 2.18(a) and 2.19. Proposition 2.18 is a special case of Proposition 2.19.

It is possible for a subset to be both open and closed. In particular, for any X, the subset X, and the empty set are both open *and* (therefore) closed subsets of X by Proposition 1.31. There are other examples for subsets which are open and closed, such as Examples 2.22 and 2.23. The nature and significance of this phenomenon will be investigated later, (Proposition 7.18).

Example 2.22 We present a subset X of the plane which has a non-empty subset, $A \neq X$, which is an open *and* a closed subset of X. Let $A = N_1((0,0))$ and $B = N_1((2,0))$. Let $X = A \cup B$. Note that $A \cap B = \emptyset$. Since A is an open subset of the plane and $A = A \cap X$, A is an open subset of X. Since $X - A = B$, and B is an open subset of X, A is also a closed subset. Also, we have B is an open and closed subset of X. ◆

Finite sets in $\mathbf{R^n}$ are very special, as the next proposition shows.

Proposition 2.23 *If $X \subseteq \mathbf{R^n}$, X a finite set, then any subset of X is both an open subset of X and a closed subset of X.*

Proof: Let ϵ_0 be the minimum of the numbers $\delta(p, q)$ where p and q are distinct points of X. The minimum of a finite set of positive numbers is a positive number; thus $\epsilon_0 > 0$. Let A be a subset of X, $A = \{a_1, a_2, \ldots, a_k\}$.

Let $U = \bigcup_{i=1}^{i=k} N_{\epsilon_0}(a_i)$; U is an open subset of $\mathbf{R^n}$ since it is a union of open subsets. We have $U \cap X = A$. Thus A is an open subset of X. The set $X - A$ is finite, and by the argument above is an open subset of X; thus A is a closed subset of X. ▮

If $X \subseteq \mathbf{R^n}$ and x and y are two distinct points of X, then we can find *disjoint* neighborhoods in X of x and y. Let $\epsilon = \delta(x, y)$. By the triangle inequality, $N_{\epsilon/2}(x) \cap N_{\epsilon/2}(y) = \emptyset$. This simple observation becomes an important aspect of subsets when one tries to generalize to more abstract settings (Definition 1.53).

2.3 Linear and affine maps

In geometry we say that two objects such as triangles, are congruent if there is a rigid motion which takes one to the other. This notion can be expressed in terms of affine maps. For the moment we use these notions to construct examples of open and closed subsets, such as in Proposition 2.32. Later, we will broaden this investigation, relating these notions to a fundamental concept called "homeomorphism," Definition 4.01, and the notion of a "space of functions," as in Example 8.19.

For the next few definitions, we use vector notation for $\mathbf{R^n}$. If M is an $m \times n$ matrix, $M\vec{x}$ denotes matrix multiplication where we write the m-dimensional vector $\vec{x}$ as a single column matrix. Also, $|M|$ denotes the determinant of a square matrix..

Definition 2.24 *If M is an $m\times n$ matrix and $\vec{B}$ is an n-dimensional vector, we define a function $F : R^m \to R^n$ by $F(\vec{x}) = M\vec{x} + \vec{B}$. Such a function is called an* **affine transformation,** *or* **affine map.** *If $m = n$ and $|M| \neq 0$, we say that F is a* **non-singular affine map.**

An affine map with $\vec{B} = \vec{0}$ is called a **linear transformation,** *or* **linear map.** *A non-singular affine map with $\vec{B} = \vec{0}$ is called a* **non-singular linear transformation,** *or* **non-singular linear map.**

From standard linear algebra we have:

Proposition 2.25 *Suppose we have two affine maps, $f:\mathbf{R^m} \to \mathbf{R^n}$ with matrix M and $g:\mathbf{R^n} \to \mathbf{R^k}$ with matrix N. Then the composition $g \circ f : \mathbf{R^m} \to \mathbf{R^k}$ is an affine map whose matrix is the matrix product NM.*

An affine map $f:\mathbf{R^1} \to \mathbf{R^1}$ is simply a function of the form $f(x) = mx + b$ where m and b are constants; f is non-singular implies $m \neq 0$.

In the remainder of this chapter we will focus on maps from $\mathbf{R^n}$ to $\mathbf{R^n}$ and thus will only be concerned with the case that $m = n$, and so M is a square matrix. We will focus on two special kinds of non-singular affine maps: rigid motions and similarities.

A critical definition we use here, common in linear algebra, is that of an orthogonal matrix. Here M^T denotes the "transpose" of M. If $M = [a_{ij}]$, then $M^T = [a_{ji}]$.

Definition 2.26 *Let M be an $n \times n$ matrix. We say M is an* **orthogonal matrix** *if $M^T = M^{-1}$.*

The reason "orthogonal" is used for such a matrix is that the column vectors of an $n \times n$ orthogonal matrix M is a set of n mutually orthogonal unit vectors. Note that the i-th row of M^T is the same as the i-th column of M. Thus ij-the entry of M^TM is the same as the dot product of the i-th column and the j-th column. Since M^T is the inverse of M, $M^TM = I$ where I is the $n \times n$ identity matrix. Since the diagonal elements of I are equal to 1 and all other entries of I are 0, the column vectors are unit vectors, and any two distinct column vectors are orthogonal.

We establish a few properties of an orthogonal matrix.

If M is an orthogonal matrix, then M^{-1} is also orthogonal. If M is orthogonal, then $M^T = M^{-1}$. Take the transpose of both sides of this equation, and we get $M = (M^{-1})^T$. Since M is the inverse of M^{-1}, this last equation can be written $(M^{-1})^{-1} = (M^{-1})^T$, which verifies that M^{-1} is orthogonal.

If M is an orthogonal matrix, then the determinant of M is +1 or -1. We have $M^T M = I$. But $\det M^T = \det M$. Take the determinant of both sides of the previous equation, and since the determinant of the product is the product of the determinants, we see $\det M \det M = 1$. Thus $\det M = \pm 1$.

Definition 2.27 *An affine map $F : R^n \to R^n$ defined by $F(\vec{x}) = M\vec{x} + \vec{B}$ is called a* **rigid motion of $\mathbf{R^n}$**, *if M is an orthogonal matrix.*

By the remark on page 52, a rigid motion is a *non-singular* affine map.

Example 2.28 If $M = \begin{pmatrix} \cos(\theta) & \sin(\theta) \\ -\sin(\theta) & \cos(\theta) \end{pmatrix}$, $\vec{x} \in R^2$, function $F(\vec{x}) = M\vec{x}$ corresponds to a clockwise rotation of angle θ about the origin. The inverse of M corresponds to counter-clockwise rotation of angle θ, with matrix

$$M^{-1} = \begin{pmatrix} \cos(-\theta) & \sin(-\theta) \\ -\sin(-\theta) & \cos(-\theta) \end{pmatrix} = \begin{pmatrix} \cos(\theta) & -\sin(\theta) \\ \sin(\theta) & \cos(\theta) \end{pmatrix} = M^T.$$

Thus M is orthogonal.

The matrix $\begin{pmatrix} 1 & 0 \\ 0 & -1 \end{pmatrix}$ corresponds to reflection in the x-axis, while $\begin{pmatrix} -1 & 0 \\ 0 & 1 \end{pmatrix}$ corresponds to reflection in the y-axis.

Next, some examples in $\mathbf{R}^3$. The matrix

$$R_z(\theta) = \begin{pmatrix} \cos(\theta) & \sin(\theta) & 0 \\ -\sin(\theta) & \cos(\theta) & 0 \\ 0 & 0 & 1 \end{pmatrix}$$

corresponds to a rotation about the z-axis of angle θ. Similarly,

$$R_y(\theta) = \begin{pmatrix} \cos(\theta) & 0 & \sin(\theta) \\ 0 & 1 & 0 \\ -\sin(\theta) & 0 & \cos(\theta) \end{pmatrix}$$

corresponds to a rotation of angle θ about the y-axis.

The matrix

$$\begin{pmatrix} 0 & 0 & 1 \\ 0 & 1 & 0 \\ 1 & 0 & 0 \end{pmatrix}$$

might be described as interchanging the x-axis and the z-axis. ◆

In the term "rigid motion," "rigid " refers to the fact is that distances are preserved, as the next proposition shows. The use of the term "motion" is also of interest, a topic we will return to later in the remark on page 312.

Proposition 2.29 *Suppose we have a rigid motion F in $\mathbf{R^n}$. If $\vec{x}$ and $\vec{y}$ are two vectors in $\mathbf{R^n}$, then $\delta(\vec{x},\vec{y}) = \delta(F(\vec{x}),F(\vec{y}))$.*

Proof: Recall the remark on page 8, that $\delta(\vec{x},\vec{y}) = \sqrt{(\vec{x}-\vec{y})\cdot(\vec{x}-\vec{y})}$. Also recall we are writing our vectors as column vectors so relating the dot product and matrix multiplication, we have: $(\vec{x}-\vec{y})\cdot(\vec{x}-\vec{y}) = (\vec{x}-\vec{y})^T(\vec{x}-\vec{y})$.

Write $F(\vec{x}) = M\vec{x}+\vec{B}$. Then,

$$\begin{aligned}
(\delta(F(\vec{x}),F(\vec{y})))^2 &= \\
&= (F(\vec{x})-F(\vec{y}))\cdot(F(\vec{x})-F(\vec{y})) \\
&= ((M\vec{x}+\vec{B})-(M\vec{y}+\vec{B}))\cdot((M\vec{x}+\vec{B})-(M\vec{y}+\vec{B})) \\
&= M(\vec{x}-\vec{y})\cdot M(\vec{x}-\vec{y}) = (M(\vec{x}-\vec{y}))^T M(\vec{x}-\vec{y}) \\
&= ((\vec{x}-\vec{y})^T M^T)M(\vec{x}-\vec{y}) \\
&= (\vec{x}-\vec{y})^T(M^T M)(\vec{x}-\vec{y}) = (\vec{x}-\vec{y})^T I(\vec{x}-\vec{y}) \\
&= = (\vec{x}-\vec{y})\cdot(\vec{x}-\vec{y}) = (\delta(\vec{x},\vec{y}))^2.
\end{aligned}$$

Note that in the third line and the last line we change from dot product to matrix multiplication. Taking square roots, it follows that $\delta(F(\vec{x}),F(\vec{y})) = \delta(\vec{x},\vec{y})$. ∎

One consequence of Proposition 2.29 is that a rigid motion takes an open n-ball to another open n-ball of the same radius:

Proposition 2.30 *If F is a rigid motion in $\mathbf{R^n}$, $p \in U$ and $\epsilon > 0$, then $F(N_\epsilon(p)) = N_\epsilon(F(p))$.* ∎ *(Problem 2.20)*

Proposition 2.31 *Suppose $F(\vec{x}) = M\vec{x}+\vec{B}$ is a rigid motion in $\mathbf{R^n}$. Then F has an inverse F^{-1} which is a rigid motion, and, in fact, $F^{-1}(\vec{y}) = M^{-1}\vec{y} - M^{-1}\vec{B}$.*

Assuming the inverse is as asserted, then the inverse would be a rigid motion by the remark on page 52. One could simply verify that $M^{-1}\vec{y} - M^{-1}\vec{B}$ is an inverse of F, but it is instructive to show how one derives this formula.

If $\vec{y} = M\vec{x}+\vec{B}$, we can obtain the inverse of F by solving for $\vec{x}$ as follows. From linear algebra we know that, since the determinant of M is not zero, the matrix M has an inverse. Multiply both sides of the equation $\vec{y} = M\vec{x}+\vec{B}$ by M^{-1} and obtain $M^{-1}\vec{y} = M^{-1}M\vec{x} + M^{-1}\vec{B}$. Since $M^{-1}M$ is the identity matrix, we have $M^{-1}\vec{y} = \vec{x} + M^{-1}\vec{B}$. From this we solve for $\vec{x}$, and get $\vec{x} = M^{-1}\vec{y} - M^{-1}\vec{B}$. ∎

The composition of affine maps is an affine map (Problem 2.19). If $F(\vec{x}) = M\vec{x} + \vec{B}$ and $F'(\vec{x}) = M'\vec{x} + \vec{B'}$ are rigid motions, then the composition is also a rigid motion since

$$F \circ F'(\vec{x}) = M(M'\vec{x} + \vec{B'}) + \vec{B} = (MM')(\vec{x}) + (M(\vec{B'}) + \vec{B}),$$

and orthogonality of (MM') follows from

$$(MM')^T = (M')^T M^T = (M')^{-1} M^{-1} = (MM')^{-1}. \blacklozenge$$

Here is a proposition relating rigid motions to open and closed subsets of $\mathbf{R^n}$.

Proposition 2.32 *Let F be a rigid motion in $\mathbf{R^n}$.*

(a) *U is an open subset of $\mathbf{R^n}$ if and only if $F(U)$ is an open subset of $\mathbf{R^n}$.*
(b) *C is a closed subset of $\mathbf{R^n}$ if and only if $F(C)$ is a closed subset of $\mathbf{R^n}$.*

Proof: We prove part (a), leaving part (b) as an exercise (Problem 2.21).

Suppose U is an open subset of $\mathbf{R^n}$. We wish to show that $F(U)$ is an open subset of $\mathbf{R^n}$. Let $q \in F(U)$ and let $p = F^{-1}(q)$; then $p \in U$. Since U is an open subset of $\mathbf{R^n}$, we can find an $\epsilon > 0$, so that $N_\epsilon(p) \subseteq U$. By Proposition 2.30, $F(N_\epsilon(p)) = N_\epsilon(F(p)) \subseteq F(U)$ and $F(U)$ is an open subset of $\mathbf{R^2}$.

If $F(U)$ is an open subset of $\mathbf{R^n}$, U is an open subset of $\mathbf{R^n}$. To see this, apply the portion of the proof above to the function F^{-1} which, by Proposition 2.31, is a rigid motion. ∎

For future reference we note that four-dimensional space comes up in a natural way in considering linear functions of the plane. A linear map of $\mathbf{R^2}$ to $\mathbf{R^2}$ is given by a 2×2 matrix, a 4-tuple of numbers, and this in turn corresponds to a point in four-dimensional space.

A general affine map from $\mathbf{R^2}$ to $\mathbf{R^2}$ corresponds to a point of $\mathbf{R^6}$, since it is determined by six real numbers.

2.4 Showing subsets open or closed, using affine maps

Example 2.33 As an illustration of how Proposition 2.32 can be used, we show that any line L in the plane is a closed subset. First consider the x-axis of the plane, X. We give (yet another) proof that X is a closed subset of the plane, using some of our results on rigid motions.

Let $M_1 = \begin{pmatrix} 0 & -1 \\ 1 & 0 \end{pmatrix}$; M_1 corresponds to a rotation of 90 degrees. Let $M_2 = \begin{pmatrix} -1 & 0 \\ 0 & 1 \end{pmatrix}$; M_2 corresponds to a counter-clockwise reflection in the y-axis. We define two rigid motions, $F_1(\vec{x}) = M_1\vec{x}$, and $F_2(\vec{x}) = M_2\vec{x}$.

By Proposition 2.16, R^2_+ is a closed subset of $\mathbf{R}^2$. We express the y-axis, $Y = \{(x, y) \in \mathbf{R}^2 \mid x = 0\}$, as the intersection of closed subsets of $\mathbf{R}^2$ as follows: $Y = R^2_+ \cap F_2(R^2_+)$. Thus Y is a closed subset of $\mathbf{R}^2$.

We note that $X = F_1(Y)$. By Proposition 2.32 b), X is a closed subset of $\mathbf{R}^2$.

For any line L in the plane, there is a rigid motion F such that $F(X) = L$ (Problem 2.22). It then follows that L is a closed subset of $\mathbf{R}^2$. ◆

Granted this is a roundabout method in this case, but soon we will see other examples where this indirect strategy will be useful. A rigid motion is a special kind of function we will study later: a homeomorphism, defined in Definition 4.01.

Example 2.34 Let Q be the first quadrant in the plane; $Q = \{(x, y) \in \mathbf{R}^2 \mid 0 \le x \text{ and } 0 \le y\}$. Then Q is a closed subset of the plane since $Q = F_1(R^2_+) \cap R^2_+$ where F_1 is as defined in Example 2.33. ◆

Example 2.35 Here is one way to prove that the open rectangle

$$R = \{(x, y) \in \mathbf{R}^2 \mid a < x < b \text{ and } c < y < d\}$$

of Proposition 2.05 is an open subset of the plane.

Let F_1 be the rigid motion defined in Example 2.35—a counterclockwise rotation by $\pi/2$. For any two-dimensional vector, $\vec{a}$, we define a rigid motion $T_{\vec{a}}$, a translation by $\vec{a}$, by $T_{\vec{a}}(\vec{x}) = \vec{x} + \vec{a}$.

One can now verify (draw the pictures), that R is an open subset since it is the intersection of four open subsets:

$$R = T_{\vec{a}}(\overset{\circ}{R^2_+}) \cap T_{\vec{d}} \circ F_1(\overset{\circ}{R^2_+}) \cap T_{\vec{b}} \circ F_1 \circ F_1(\overset{\circ}{R^2_+}) \cap T_{\vec{c}} \circ F_1 \circ F_1 \circ F_1(\overset{\circ}{R^2_+})$$

where, using standard basis $\{\vec{i}, \vec{j}\}$,

$$\vec{a} = a\vec{i},\ \vec{b} = b\vec{i},\ \vec{c} = c\vec{j} \text{ and } \vec{d} = d\vec{j}. ◆$$

Example 2.36 Consider the subset $X \subseteq \mathbf{R}^2$ of disjoint closed disks of radius $\frac{1}{3}$ placed along the x-axis at integral points; see Figure 2-7. So, $X = \bigcup_{i \in \mathbf{Z}} \overline{N}_{\frac{1}{3}}((i, 0))$; here $\mathbf{Z}$ denotes the integers. We show that X is a closed subset of $\mathbf{R}^2$, by verifying that $\mathbf{R}^2 - X$ is an open subset of $\mathbf{R}^2$.

Before we present our proof, we consider techniques we have used. We know that each disk, $\overline{N}_{\frac{1}{3}}((i, 0))$, is a closed subset of the plane, and

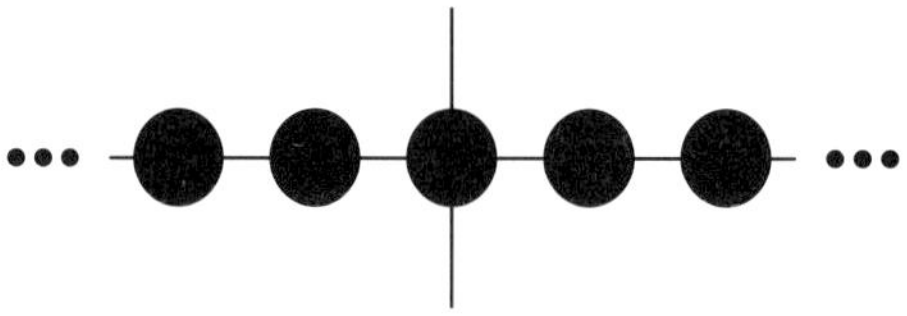

Figure 2-7 The subset X for Example 2.36; dots indicate that disks continue infinitely in both directions.

we would like to take advantage of this fact. However, X is expressed as a union of *infinitely* many closed subsets, and, as we have noted (Examples 1.33 and 1.34), an infinite union of closed subsets is not necessarily a closed subset.

A good approach, modeled after our proof that a single disk is a closed subset (Proposition 1.10), might begin as follows. Let $p \in \mathbf{R}^2 - X$, and let ϵ-δ be the shortest distance from P to a point of X. We could use geometry to calculate this distance. One could continue and complete a proof along these lines. The general calculation of ϵ-δ is only a little more work than it was for just one disk.

Instead, we pursue a different, less geometric technique that is more in the spirit of topology. It focuses on manipulations of open and closed subsets. The point here is the method, not the result. The topological approach will enable us to attack other similar problems where a simple geometric solution is not readily available. The strategy is to express $\mathbf{R}^2 - X$ as a union of open subsets of $\mathbf{R}^2$. Since X is a repeating pattern of disks, it is natural to try to find a related pattern of open subsets of $\mathbf{R}^2$.

Here is one way to do this. Let $H = \{(x, y) \in \mathbf{R}^2 \mid \frac{2}{3} \le |x|\}$. We first show that H is a closed subset of $\mathbf{R}^2$. Define two rigid motions of the plane: $F_1(\vec{x}) = \vec{x} + \vec{b}$, and $F_2(\vec{x}) = -\vec{x} - \vec{b}$ where $\vec{b} = (\frac{2}{3}, 0)$. To fit F_2 into the definition of rigid motion, we view $-\vec{x}$ as $\begin{pmatrix} -1 & 0 \\ 0 & -1 \end{pmatrix} \vec{x}$. Now $H = F_1(R^2_+) \cup F_2(R^2_+)$; thus H is a closed subset of $\mathbf{R}^2$. Let $U_0 = \mathbf{R}^2 - (H \cup \overline{N}_{\frac{1}{3}}((0,0)))$; see Figure 2-8. Then U_0 is an open subset of $\mathbf{R}^2$. Roughly, U_0 is an open, infinite, vertical strip with a closed disk removed.

We now express $\mathbf{R}^2 - X$ as a union of translated copies of U_0. Consider the family of functions in the plane defined by $G_i(\vec{x}) = \vec{x} + \vec{t_i}$ where, for $i \in \mathbf{Z}$, $\vec{t_i} = (i, 0)$. So G_i is a rigid translation in the x-direction by i units. Let $U = \bigcup_{i \in Z} G_i(U_0)$. Then U is an open subset of $\mathbf{R}^2$ with $U = \mathbf{R}^2 - X$. Thus X is a closed subset of $\mathbf{R}^2$. ◆

Proposition 2.37 addresses a more general situation where we have a simple repeating pattern of closed subsets as in Figure 2-9(b).

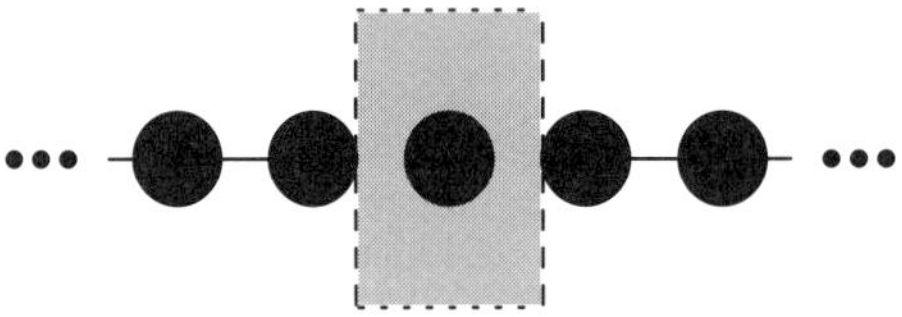

Figure 2-8 The subset, U_0, shown in gray; see Example 2.36. The vertical dashed lines, corresponding to $y = \frac{2}{3}$ and $y = -\frac{2}{3}$, indicate U_0 does not extend beyond these lines. Dotted horizontal line segments indicate U_0 extends above and below these edges. Here $F_1(R^2_+)$ consists of all points to the right of line $y = \frac{2}{3}$, and $F_2(R^2_+)$ consists of all points to the left of line $y = -\frac{2}{3}$.

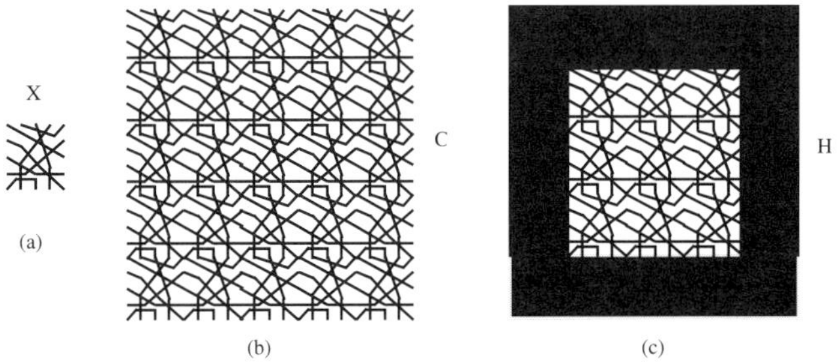

Figure 2-9 (a) The set X; (b) a portion of C; (c) H_2. See Proposition 2.37.

Proposition 2.37 *Consider the standard unit square* $S = \{(x, y) \in \mathbf{R}^2 \mid |x| \leq 1$ *and* $|y| \leq 1\}$. *Let* X *be any closed subset of* S. *For each pair of integers,* i *and* j, *define a rigid motion* F_{ij} *by* $F(\vec{x}) = \vec{x} + \vec{b_{ij}}$ *where* $\vec{b_{ij}}$ *is the vector* $(2i, 2j)$. *Let* $C_{ij} = F_{ij}(X)$ *and let* $C = \bigcup_{i,j \in \mathbf{Z}} C_{ij}$. *Then* C *is a closed subset of* $\mathbf{R}^2$.

Proof: For each natural number i, define an open rectangle by

$$R_i = \{(x, y) \in \mathbf{R}^2 \mid |x| < 2i - 1 \text{ and } |y| < 2i - 1\}.$$

Let $K_i = \mathbf{R}^2 - R_i$. By Proposition 2.05, R_i is an open subset of $\mathbf{R}^2$; thus each K_i is a closed subset of $\mathbf{R}^2$.

Now define $H_i = K_i \cup C$. Roughly, H_i consists of an increasingly large portion of our infinitely repeating pattern together with a frame that goes from the edge of this portion out to infinity in all directions.

We can write H_i as a finite union of closed subsets of $\mathbf{R}^2$:

$$H_i = K_i \cup (\bigcup_{|j|,|k|<i})C_{jk}.$$

For example (see Figure 2-9), $C_{00} = X$ and $H_1 = K_1 \cup C_{00}$. Also,

$$H_2 = K_2 \cup C_{00} \cup C_{10} \cup C_{01} \cup C_{11} \cup C_{-11} \cup C_{1-1} \cup C_{-10} \cup C_{0-1} \cup C_{-1-1}.$$

Finally, $C = \bigcap_i H_i$ and C is a closed subset of $\mathbf{R}^2$ since we can write it as an intersection of closed subsets of $\mathbf{R}^2$. ∎

Another simple non-singular affine map is a scaling. It is, in fact a non-singular linear map. Combined with a rigid motion, we get a similarity, a familiar basic concept of geometry.

Definition 2.38 *Suppose $\vec{x} \in \mathbf{R}^\mathbf{n}$ and k is a positive number. The map of $\mathbf{R}^\mathbf{n}$ defined by $F(\vec{x}) = k\vec{x}$ is called a* **scaling by a factor of** *k.*

To make scaling more in the style of the definition of affine map, write $k\vec{x} = kI_n\vec{x}$ where I_n denote the $n \times n$ identity matrix. An important feature of a scaling $F(\vec{x}) = k\vec{x}$ is that distances are changed by a factor of k. That is, $|F(\vec{x}) - F(\vec{y})| = |k\vec{x} - k\vec{y}| = k|\vec{x} - \vec{x}|$.

Proposition 2.39 *Suppose F is a scaling by a factor of k in $\mathbf{R}^\mathbf{n}$.*

(a) *U is an open subset of $\mathbf{R}^\mathbf{n}$ if and only if $F(U)$ is an open subset of $\mathbf{R}^\mathbf{n}$.*
(b) *C is a closed subset of $\mathbf{R}^\mathbf{n}$ if and only if $F(C)$ is a closed subset of $\mathbf{R}^\mathbf{n}$.* ∎

Definition 2.40 *In $\mathbf{R}^\mathbf{n}$, suppose that $F = F_1 \circ F_2$ where F_1 is a rigid motion of $\mathbf{R}^\mathbf{n}$ and F_2 is a scaling of $\mathbf{R}^\mathbf{n}$. Then we say that F is a* **similarity**. *If $X \subseteq \mathbf{R}^\mathbf{n}, Y \subseteq \mathbf{R}^\mathbf{n}$ and F is a similarity with $F(X) = Y$, we say that X and Y are* **similar.**

For example, all circles in $\mathbf{R}^2$ are similar.

Any similarity can be written as $F(\vec{x}) = kM\vec{x} + \vec{B}$ where M is an orthogonal matrix.

The next proposition follows from Propositions 2.32 and 2.39.

Proposition 2.41 *Suppose F is a similarity in $\mathbf{R}^\mathbf{n}$.*

(a) *U is an open subset of $\mathbf{R}^\mathbf{n}$ if and only if $F(U)$ is an open subset of $\mathbf{R}^\mathbf{n}$.*
(b) *C is a closed subset of $\mathbf{R}^\mathbf{n}$ if and only if $F(C)$ is a closed subset of $\mathbf{R}^\mathbf{n}$.* ∎ *(Problem 2.23)*

Here is an example where we can apply ideas.

Example 2.42 We define a subset H of $\mathbf{R}^2$, frequently called the **Hawaiian earring**. It is an infinite union of smaller and smaller circles all mutually tangent at the point $(0,0)$; see Figure 2-10. For each natural number $n \in \mathbf{N}$ let C_n be the circle in $\mathbf{R}^2$ with center $(\frac{1}{n}, 0)$ and radius $\frac{1}{n}$. Let $H = \bigcup_{n=1}^{\infty} C_n$.

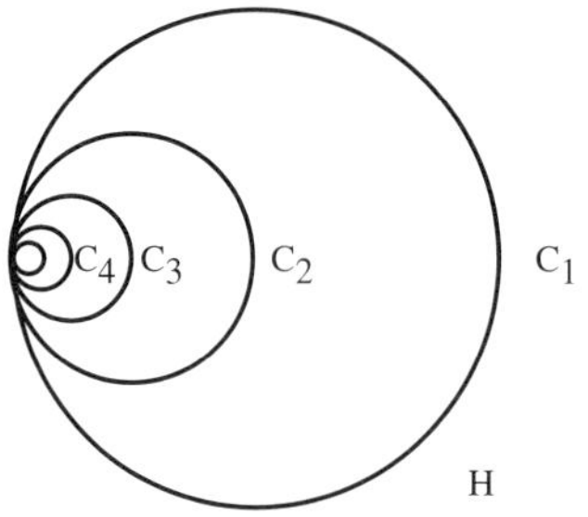

Figure 2-10 The Hawaiian earring. See Example 2.42.

We offer two techniques for showing that H is a closed subset of the plane.

For our first proof, we directly show that $\mathbf{R}^2 - H$ is an open subset of $\mathbf{R}^2$. The idea is to describe $\mathbf{R}^2 - H$ as a union of open subsets, U_n. Here U_0 is the set of points outside the largest circle. We next define $\{U_n\}_{n=1}^{n=\infty}$, an infinite collection of disjoint, similar, crescent-shaped regions.

Let $U_0 = \mathbf{R}^2 - \overline{N}_1((1,0))$. Since $\overline{N}_1((1,0))$ is a closed subset, U_0 is an open subset of $\mathbf{R}^2$. Consider the crescent-shaped region U_1 which lies inside the first circle C_1, and outside the second C_2; see Figure 2-11(a). This is an open subset of $\mathbf{R}^2$ since $U_1 = (\mathbf{R}^2 - \overline{N}_{\frac{1}{2}}(\frac{1}{2}, 0)) \cap N_1(1,0)$. For each natural number $n \in \mathbf{N}$, let $F_n(\vec{x}) = \frac{1}{n}\vec{x}$. Each F_n is a similarity of the plane. For each n, let $U_n = F_n(U_1)$. Each U_n is an open subset of $\mathbf{R}^2$ by Proposition 2.41. Finally, let $U = \bigcup_{n=0}^{\infty} U_n$. This is an open subset of $\mathbf{R}^2$ since it is the union of open subsets. Finally, we note that $U = \mathbf{R}^2 - H$.

Here is a second method to show that H is closed. For each natural number n define $B_n = \overline{N}_{\frac{1}{n}}(\frac{1}{n}, 0)$. Let $H_n = B_n \cup (\bigcup_{i=1}^{n} C_i)$; see Figure 2-11 (b).

Roughly, H_n consists of the first n circles with the n-th one "filled in." For any n, H_n is a closed subset of the plane since it is a union

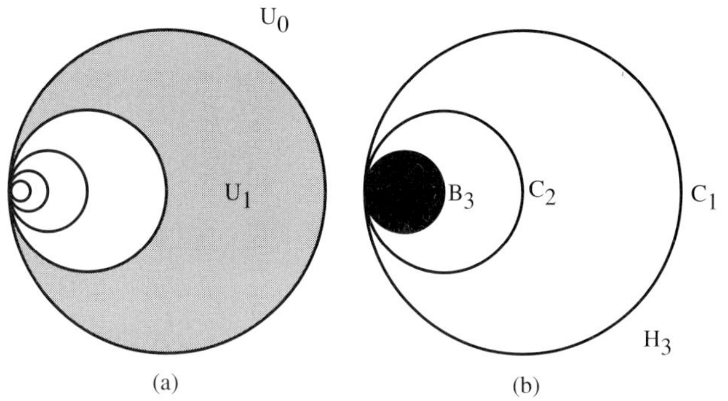

Figure 2-11 (a) A crescent-shaped region. (b) An example of a B_i. See Example 2.42.

of finitely many closed subsets. We can write H as the intersection of closed subsets: $H = \bigcap_i H_i$. Thus H is a closed subset of $\mathbf{R}^2$. ◆

Example 2.43 The next example is called "the topologist's comb." It is a closed subset of $\mathbf{R}^2$ (Problem 2.31). (See Figure 2-12.)

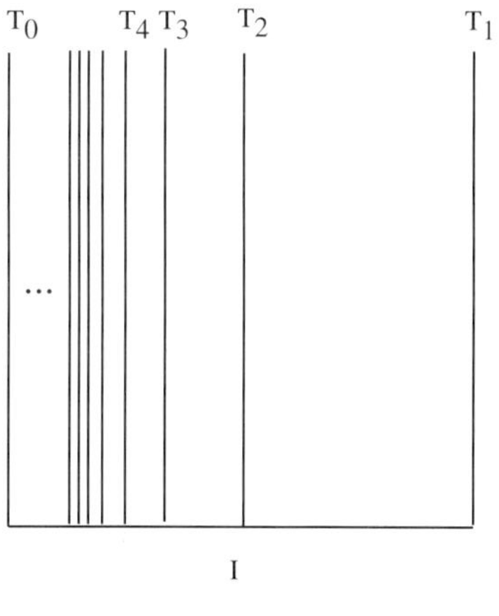

Figure 2-12 The topologist's comb; Example 2.43.

This "comb" has infinitely many "teeth," T_n. For a positive integer n define:

$$T_n = \{(x, y) \in \mathbf{R}^2 \mid x = \frac{1}{n} \text{ and } 0 \leq y \leq 1\},$$

$$T_0 = \{(x,y) \in \mathbf{R}^2 \mid x = 0 \text{ and } 0 \le y \le 1\}, \text{ and}$$
$$I = \{(x,y) \in \mathbf{R}^2 \mid 0 \le x \le 1 \text{ and } y = 0\}.$$

Finally define $X = I \cup T_0 \cup (\bigcup_1^\infty T_n)$. ◆

Example 2.44 A variation of the topologist's comb is called the"shrinking comb"; see Figure 2-13.

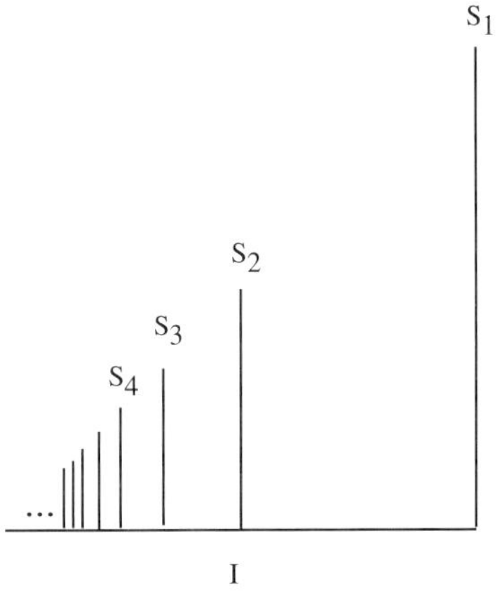

Figure 2-13 The shrinking comb; Example 2.44.

Here the teeth of the comb shrink near the origin. For a positive integer n define:

$$S_n = \{(x,y) \in \mathbf{R}^2 \mid x = \frac{1}{n} \text{ and } 0 \le y \le \frac{1}{n}\},$$
$$I = \{(x,y) \in \mathbf{R}^2 \mid 0 \le x \le 1 \text{ and } y = 0\}.$$

Finally, define $Y = I \cup (\bigcup_1^\infty T_n)$. This is a closed subset of $\mathbf{R}^2$ (Problem 2.32). ◆

2.5 Cartesian products of open and closed subsets

The next propositions concern open and closed subsets of Cartesian products. For basic facts and definitions for Cartesian products; see

Appendix A, especially, the remark on page 395.

The problem is how to relate open subsets of a Cartesian product with open subsets of the factors. Here is the crux of the problem. As sets, $\mathbf{R}^2 = \mathbf{R}^1 \times \mathbf{R}^1$. How does this relate to open subsets, in particular to open balls? In $\mathbf{R}^1$ an open ball is an open interval. The product of two of these is an open rectangle in the plane. But an open ball of the plane is a round disk.

Proposition 2.45 brings to mind the saying "You can't fit a square peg in a round hole." True, but you can at least put a very small square peg into a round hole. In the plane this proposition says that inside any disk, no matter how small, there is an open rectangle. (We have already established that any open rectangle, no matter how small, contains an open disk, Proposition 2.05.)

Proposition 2.45 *Suppose* $p = (a, b) \in \mathbf{R^n} \times \mathbf{R^m} = \mathbf{R^{n+m}}$, *and* U *is an open subset of* $\mathbf{R^{n+m}}$ *with* $p \in U$. *Then there are numbers* ϵ *and* ϵ', *so that* $N_\epsilon(a) \times N_{\epsilon'}(b) \subseteq U$. ▮ *(Problem 2.24)*

One also has a version of Proposition 2.45 for *subsets* of $\mathbf{R^{n+m}}$. Suppose $X \subseteq \mathbf{R^n}$ and $Y \subseteq \mathbf{R^m}$, $p = (a, b) \in X \times Y$, and U is an open subset of $X \times Y$ with $p \in U$. By intersecting the open rectangles such as obtained in Proposition 2.45 with $X \times Y$ we can show (Problem 2.25): there are numbers ϵ and ϵ', so that $N_\epsilon(a, X) \times N_{\epsilon'}(b, Y) \subseteq U$.

Proposition 2.05 showed that the rectangle $R = \{(x, y) \in \mathbf{R}^2 \mid a < x < b \text{ and } c < y < d\}$ is an open subset. We can consider R to be a product of open balls of $\mathbf{R}^1$:

$$R = N_{\frac{b-a}{2}}(a + \frac{b-a}{2}) \times N_{\frac{d-c}{2}}(a + \frac{d-c}{2}).$$

More generally, it is true that if $(a, b) \in \mathbf{R^n} \times \mathbf{R^m}$, $N_\epsilon(a) \times N_{\epsilon'}(b)$ is an open subset of $\mathbf{R^{n+m}}$. This together with the proposition above imply the product of open subsets is an open subset.

Proposition 2.46

(a) *Suppose* $A \subseteq \mathbf{R^n}$ *and* $B \subseteq \mathbf{R}^m$ *and* U *is an open subset of* A *and* V *is an open subset of* B; *then* $U \times V$ *is an open subset of* $A \times B$.

(b) *Suppose* $A \subseteq \mathbf{R^n}$ *and* $B \subseteq \mathbf{R^m}$ *and* C *is a closed subset of* A, K *a closed subset of* B; *then* $C \times K$ *is a closed subset of* $A \times B$. ▮ *(Problem 2.26)*

Since a point in any Euclidian space is a closed subset, Proposition 1.14, and since any subset is a closed subset of itself, Proposition 1.31, it follows that for any point $a \in A$ and for any point $b \in B$, $\{a\} \times B$ and $A \times \{b\}$ are closed subsets of $A \times B$.

From this we get (yet another) proof that the x-axis and the y-axis are closed subsets of the plane since these are $\mathbf{R}^1 \times \{0\}$ and $\{0\} \times \mathbf{R}^1$, respectively, products of closed subsets of $\mathbf{R}^1 \times \mathbf{R}^1 = \mathbf{R}^2$.

2.6 Cones and suspensions

We close this chapter with two standard constructions we will use several times in this book. Recall we consider $\mathbf{R^n} \subseteq \mathbf{R^{n+1}}$ by identifying $\mathbf{R^n}$ with $\mathbf{R^n} \times \{0\} \subseteq \mathbf{R^n} \times \mathbf{R^1}$.

Definition 2.47 *Suppose $X \subseteq \mathbf{R^n} \subseteq \mathbf{R^{n+1}}$. Let $p \in \mathbf{R^{n+1}}$ be a point whose last coordinate is non-zero. The* **cone on** X **from** p*, denoted pX, is the union of all line segments in $\mathbf{R^{n+1}}$ from p to a point of X.*

Definition 2.48 *Suppose $X \subseteq \mathbf{R^n} \subseteq \mathbf{R^{n+1}}$ $p, q \in \mathbf{R^{n+1}}$ such that $p = (p_1, \ldots, p_{n+1})$ and $q = (q_1, \ldots, q_{n+1})$ where $q_{n+1} < 0 < p_{n+1}$. (Roughly, p and q lie on opposite sides of $\mathbf{R^n}$ in $\mathbf{R^{n+1}}$.) The* **suspension of** X **from** p **and** q *is the union of all line segments in $\mathbf{R^{n+1}}$ from p to a point of X and from q to a point of X.*

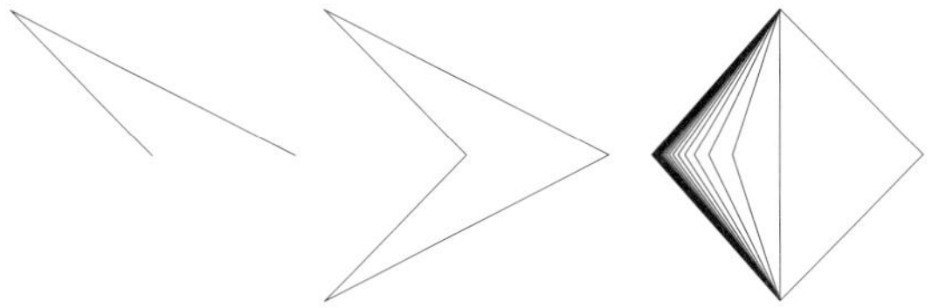

Figure 2-14 From left to right: the cone on $A = \{1/2, 1\}$ from p, the suspension of A from p and q, and the suspension of $B = \{1/n\}_{n \in \mathbf{N}}$ from p' and q'; see Example 2.49.

Example 2.49 Suppose $p = (0, 1)$, $q = (0, -1)$, $p' = (1/2, 1)$, and $q' = (1/2, -1)$. We consider two subsets of $\mathbf{R^1}$: $A = \{1/2, 1\}$ and $B = \{1/n\}_{n \in \mathbf{N}}$. The cone on A from p is the union of two line segments, the suspension of A from p and q is the perimeter of a quadrilateral, and the suspension of B from p' and q' is the union of infinitely many line segments, as shown in Figure 2-14.

If we consider $S^1 \subseteq \mathbf{R^2} \subseteq \mathbf{R^3}$, then the cone from any point p, as in Definition 2.47, is a cone in the sense commonly used in geometry. The suspension is sometimes called a "double cone." See Figure 2-15.

Let X be the subset of $\mathbf{R^2}$ which is the union of two line segments: one from $(-1, 0)$ to $(1, 0)$, the other from $(0, -1)$ to $(0, 1)$. Let $p = (0, 0, 1)$, $q = (0, 0, -1)$. The cone on X from p is the union of two triangles. The suspension of X from p and q is the union of two squares.

Also note that if Σ is the suspension of X from p and q, we can write Σ as a union of two cones pX and qX with $pX \cap qX = X$. ◆

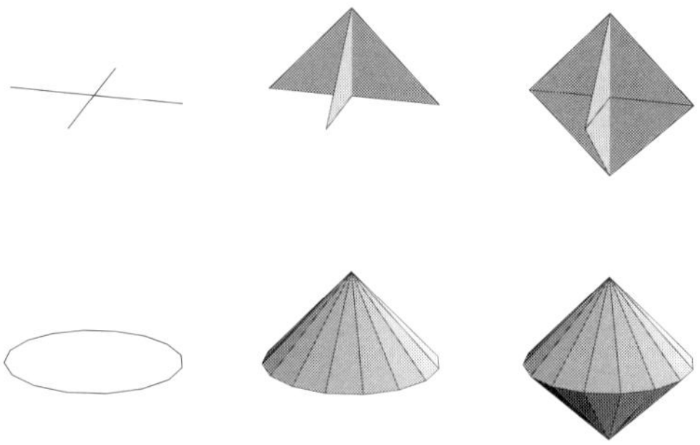

Figure 2-15 A cone from $p = (0,0,1)$ and suspension using $q = (0,0,-1)$. Top row: X of Example 2.49 (two line segments in $\mathbf{R}^2$), the cone on X from p and suspension of X from p and q. Bottom row: $S^1 \subseteq \mathbf{R}^2$, the cone on S^1 from p, and suspension of S^1 from p and q.

*2.7 General topology and Chapter 2

Most basic propositions about open and closed subsets of $\mathbf{R}^n$ remain true in their abstract settings.

In $\mathbf{R}^n$ one-point subsets are closed, Proposition 1.14. However, in general topological spaces we have seen that a one-point subset might not be closed, Example 1.52. More generally finite subsets in general topology are not necessarily all open and closed subsets as in $\mathbf{R}^n$; see Proposition 2.23.

However, finite *Hausdorff* spaces, are not so strange; compare Proposition 1.54 and the following:

Proposition 2.50 *Suppose S is a Hausdorff space, $X \subseteq S$ where X is a finite set. Then any subset of X is both an open subset of X and closed subset of X.* ∎

The notions of rigid motion and of scaling, Definitions 2.27 and 2.38, were defined using matrix operations, which have no meaning in a general topological space. We find suitable generalizations for metric spaces by noting the effect these functions have on distance, such as Proposition 2.29, and defining similar concepts.

Let $\{X, d\}$ be a metric space and k a number, $0 < k$. Define $d'(x, y) = kd(x, y)$. One can verify that d' is a metric for X, and that the topology for X corresponding to d is the same as the topology for X corresponding to d'. In other words, d and d' are equivalent metrics.

Definition 2.51 *Suppose $\{S, d\}$ is a metric space. An* **isometry** *is a function $F : S \to S$ such that for all x and $y \in S$ then $\delta(x, y) = d(F(x), F(y))$.*

A more general notion found in the literature is the idea of similarity in a metric space. This generalizes our affine notion of similarity.

Definition 2.52 *Suppose $\{S, d\}$ is a metric space. A* **similarity** *is a function $F : S \to S$ such that for some positive number ρ and, for all x and $y \in S$, $\delta(x, y) = \rho d(F(x), F(y))$.*

So a similarity with $\rho = 1$ is an isometry.

Using these definitions one can show that a similarity of a metric space is a one-to-one function that maps open subsets to open subsets and closed subsets to closed subsets, as in Proposition 2.30.

Given topological spaces $\{S, \mathcal{T}\}$ and $\{S', \mathcal{T}'\}$, we define a topology for $S \times S'$; compare with Proposition 2.46(a). This is just a "warm-up" for Definition 2.55.

Definition 2.53 *Suppose $\{S, \mathcal{T}\}$ and $\{S', \mathcal{T}'\}$ are topological spaces, and $U \subseteq S \times S'$. We say is W is an* **open subset of** *$S \times S'$ if W is a union of sets of the form $U \times U'$ where U is an open subset of S and U' is an open subset of S'*

So, products of open subsets provide a basis for the topology W of $S \times S'$.

The disjoint union of topological spaces $\{S, \mathcal{T}\}$ and $\{S', \mathcal{T}'\}$ is a topological space, in the "obvious way" as:

$$\{S \underset{\circ}{\bigcup} S' \ , \ \mathcal{T} \underset{\circ}{\bigcup} \mathcal{T}'\}.$$

This is called the "disjoint union topology."

As an exercise in these basic concepts we consider a way of looking at the disjoint union topology, using Cartesian product. Consider the two-point set $X = \{0, 1\}$. We can identify the disjoint union of A and B as a subset of $A \times B \times X$ corresponding to $(A \times \{b\} \times \{0\}) \cup (\{a\} \times B \times \{1\})$. Using the discrete topology for X, we get a product topology for $A \times B \times X$.

With the induced topology, open subsets of $(A \times \{b\} \times \{0\})$ are exactly the sets of the form $(U \times \{b\} \times \{0\})$ where U is an open subset of A. Similarly, with the induced topology, open subsets of $(\{a\} \times B \times \{1\})$ are exactly the sets of the form $(\{a\} \times V \times \{1\})$ where V is an open subset of B.

The definition of product topology, Definition 2.53, extends to a finite product of sets, but it is mathematically important to consider *infinite* products. There are two competing notions for the "best" way to proceed, giving rise to two topologies that are, in general, distinct: the box topology and the product topology. At first glance, the box topology might seem to be the obvious choice for "best" generalization. In fact, as we will indicate, there are problems with it, and it is seldom used. We leave as an exercise the verification that the box topology *is* a topology.

Definition 2.54 *Suppose* $\{S_\alpha, \mathcal{T}_\alpha\}_{\alpha \in A}$ *is a collection of topological spaces. We say* W *is an* **open subset of the box topology for** $\prod_{\alpha \in A} S_\alpha$, *if* W *has, as basis, sets of the form* $\prod_{\alpha \in A} U_\alpha$ *where, for each* $\alpha \in A$, U_α *is an open subset of* S_α.

The definition of the product topology is similar, adding an important finiteness condition.

Definition 2.55 *Suppose* $\{S_\alpha, \mathcal{T}_\alpha\}_{\alpha \in A}$ *is a collection of topological spaces. We say* W *is an* **open subset of the product topology for** $\prod_{\alpha \in A} S_\alpha$ *if* W *has, as basis, sets of the form* $\prod_{\alpha \in A} U_\alpha$ *where, for each* $\alpha \in A$, U_α *is an open subset of* S_α *and for all but a finite number of* α, $U_\alpha = S_\alpha$.

One of the sets studied in general topology is the set of all functions from one set to another, Y^X; this set is a Cartesian product, Definition A.41. This accounts for our great interest in topologies for infinite products. If X and Y are topological spaces, our product topology, Definition 2.55, gives us a topology for Y^X.

It turns out that this is not always the most useful topology to consider. One basic reason is that, since Y^X is a product of a number of copies of Y (see Example A.41), this topology is entirely derived from that of Y and has nothing to do with the topology of X. This does not cause us to abandon use of this topology, but to look for ways to improve on it such as we will eventually do in Definition 10.50.

Example 2.56 It is helpful in understanding the product topology to look at the set of real-valued functions. Using our notations, we would write this as $\mathbf{R^1}^{\mathbf{R^1}}$. This looks a bit strange, so we will use the simpler notation, $\mathbf{R}^\mathbf{R}$. Take a finite collection of numbers $\{x_i, \ldots, x_n\}$ and a corresponding number of open intervals $\{(a_1, b_1), \ldots (a_n, b_n)\}$. Consider the open, vertical, line segments in the plane

$$U_i = \{(x, y) : x = x_i \text{ and } a_i < y < b_i\}.$$

Let $H = \cup_i U_i$. A basic open subset in the topology of $\mathbf{R}^\mathbf{R}$ consists of all functions $f : \mathbf{R}^1 \to \mathbf{R}^1$ whose graphs intersect U. (You might think

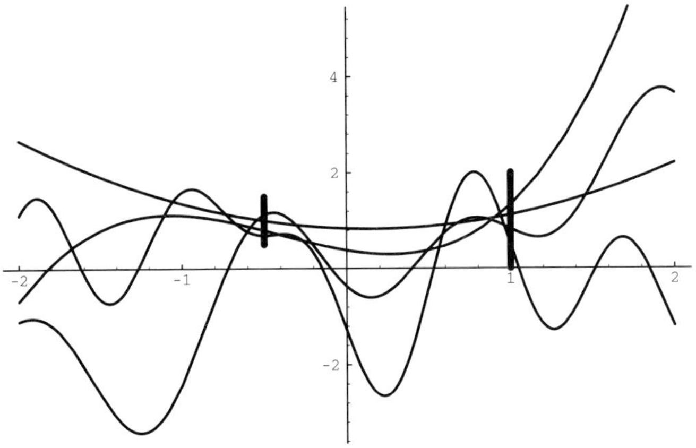

Figure 2-16 Some points in an open subset of $\mathbf{R}^{\mathbf{R}}$; see Example 2.56.

of H as being a collection of "hoops" that the function, f, must jump through; see Figure 2-16. In this figure, the open subset U corresponds to $\prod_{\alpha \in \mathbf{R}^1} U_\alpha$ where $U_\alpha = \mathbf{R}^1$ for all $\alpha \in \mathbf{R}^1$, except that $U_{-0.5} = (0.5, 1.5)$ and $U_{0.5} = (0, 2)$, indicated as thickened vertical line segments in the figure. Any function whose graph meets these vertical segments is a point of the open subset the subset $U \subseteq \mathbf{R}^{\mathbf{R}}$.) ◆

The Cartesian product of two metric spaces is a metric space. In fact, many metrics are possible. Here is one standard definition:

Definition 2.57 *Suppose $\{S, d\}$ and $\{S', d'\}$ are two metric spaces. The* **product metric** *d''* **for** *$S \times S'$, denoted $d'' = d \times d'$, is defined by $d''((a, b), (A, B)) = \sqrt{d(a, A)^2 + d'(b, B)^2}$.*

Proposition 2.58 *Suppose $\{S, d\}$ and $\{S', d'\}$ are two metric spaces; then $d \times d'$ is a metric for $S \times S'$.* ∎

Countably infinite products of metric spaces are take a bit more work:

Proposition 2.59 *Suppose $\{S_i, d_i\}_{i \in N}$ is a countable collection of metric spaces; then $\prod_{i \in N} S_i$ is a metric space.*

Proof: Suppose p and $q \in \prod_{i \in N} S_i$ with $p = (p_1, p_2, \ldots)$ and $q = (q_1, q_2, \ldots)$. Let $\rho_i(x, y) = \min(1, d_i(x, y))$. (This definition will be encountered again in our discussion of bounded metrics; see page 125.) Define

$$\rho(p, q) = \sum_{i=1}^{\infty} \frac{\rho_i(p_i, q_i)}{2^i}.$$

One can show that ρ is a metric for $\prod_{i \in N} S_i$. ∎

One of the more important examples of infinite product is the Hilbert cube.

Definition 2.60 *The* **Hilbert cube** *is the topological space which is the Cartesian product of I with itself a countable number of times. Here I is the unit interval with the standard topology, and we use the product topology for the Cartesian product.*

By Proposition 2.59, the Hilbert cube is a metric space. More directly, we could use the proof of Proposition 2.59, and view the Hilbert cube as $\prod_{n \in \mathbf{N}} I_n$ where I_n is the interval $[\frac{-1}{2^n}, \frac{1}{2^n}]$. In this case we use the following definition for the metric. If $x = (x_1, x_2, \ldots)$, $y = (y_1, y_2, \ldots)$ are points in $\prod_{n \in \mathbf{N}} I_n$, the distance from x to y is given by $\sqrt{\sum_1^\infty (y_i - x_i)^2}$.

Uncountable products of metric spaces might not have a metric definable. To see this, we take note of a special property of metric spaces.

Definition 2.61 *Let $\{X, \mathcal{T}\}$ be a topological space. We say that $\{X, \mathcal{T}\}$ is* **first countable** *if, for every $x \in X$, there is a countable collection of open subsets $\{B_i\}$, each containing x, such that, if U is any open subset of $\mathcal{T}$ with $x \in U$, then there is an N such that $B_N \subseteq U$.*

Clearly any metric space is first countable—for any X, let $B_i = N^d_{1/i}(x)$.

Example 2.62 On the other hand, $\mathbf{R}^\mathbf{R}$ with the product topology is not first countable. In fact we will show that the countability condition fails at every point of $\mathbf{R}^\mathbf{R}$. This implies that $\mathbf{R}^\mathbf{R}$, with the product topology, is not a metric space. The proof we present is a good exercise in understanding the product topology.

Consider the function $f(x) = x$. Suppose we had a countable collection of open subsets $\{B_i\}$ of $\mathbf{R}^\mathbf{R}$ which contain f, a potential candidate for showing first countability. We will find a contradiction. We may assume without loss of generality that the sets $\{B_i\}$ are basis elements of the product topology. Recall that the basic open subsets of $\mathbf{R}^\mathbf{R}$, used to define the product topology, are products of open subsets which are equal to $\mathbf{R}^1$, in all but a finite number of factors. Consider a B_i. There is a finite collection of points $b_{ij} \in \mathbf{R}^1$ such that $B_i \cap (\{b_{ij}\} \times \mathbf{R}^1) \neq \{b_{ij}\} \times \mathbf{R}^1$. The collection of all such points for all i and j is again a countable collection. Denote this countable collection of points by $\mathcal{B} = \{b_{ij}\}$. Let z be a number not in $\mathcal{B}$.

Let U be the open subset of $\mathbf{R}^\mathbf{R}$ which is the product of copies of all of $\mathbf{R}^1$ except for the factor corresponding to z, in which case we use the subset $N_1(z)$. We can check that $f \in U$ but for all i, $B_i \not\subseteq U$. To verify $B_i \not\subseteq U$, let

$$g(x) = \begin{cases} x & \text{if } x \neq z \\ z+1 & \text{if } x = z. \end{cases}$$

(Recall that $\mathbf{R}^{\mathbf{R}}$ is the set of all functions $f: \mathbf{R}^1 \to \mathbf{R}^1$, not just the continuous ones). Note that, for all i, $g(x) \in B_i$, but $g(x) \notin U$. ◆

Proposition 2.63 clarifies the (meager) possibilities of metrics for with uncountable products.

Proposition 2.63 *A non-empty product space is only metrizable if all but a countable number of factors are single-point spaces.* ∎

2.8 Problems for Chapter 2

2.1 Prove Proposition 2.01. (Hint: Let Q be the intersection of the line L with the unique line containing P which is perpendicular to L. You may assume that the shortest distance between a line L and a point P not on the line is the distance from P to Q).

2.2 Prove Proposition 2.02.

2.3 Prove that X of Example 2.06 is a closed subset of the plane, using the definition of closed subset and some geometry.

2.4 Prove that S^2 is a closed subset of D^3.

2.5 Prove, using the definition of closed subset, open subset, and some geometry, that a closed rectangle is a closed subset of the plane, and an open rectangle is an open subset of the plane. (This is Proposition 2.05.)

2.6 Prove Proposition 2.16(d).

2.7 Let C be the line segment in $\mathbf{R}^3$ with endpoints $(1,0,0)$ and $(0,1,0)$. Show that C is a closed subset of the standard 2-simplex.

2.8 In Example 2.17, verify the assertions: $D \cap \mathring{R}^2_+$ is a closed subset of $\mathring{R}^2_+$ but it is not a closed subset of $\mathbf{R}^2$. Also, verify that $U \cap R^2_+$ is an open subset of R^2_+ but not an open subset of $\mathbf{R}^2$.

2.9 Prove Proposition 2.18.

2.10 Prove the second part of Proposition 2.19.

2.11 Provide an alternative proof of part (a) of Proposition 2.20, using Proposition 1.39 and the remark on page 27.

2.12 Prove the second part of Proposition 2.20.

2.13 In reference to Proposition 2.20, find an example of subsets A, C, and X of the plane such that C is a closed subset of A and yet C is not a closed subset of X. (Remarks: We want an example where $X \neq \mathbf{R}^2$. Of course, A is *not* a closed subset of X. Once you find your example, you are expected to prove that, for your example, the set C is not a closed subset of X.)

2.14 Prove Proposition 1.24.

2.15 Using the space in Example 1.27, let $A = \{(x, y) \in X \mid x = 0\}$. Show that A is a closed subset of X, and A is not an open subset of X.

2.16 Let Q be the set of rational numbers in $\mathbf{R}^1$. Show that if U is a non-empty subset of Q, then U is not an open subset of $\mathbf{R}^1$. (Hint: You might want to make use of Proposition 1.28.)

2.17 Let Q denote the rational numbers. We consider the subset $Q \times Q$ of $\mathbf{R}^2$; in other words, points such that both coordinates are rational numbers. We wish to see if the discussion of Example 1.28 generalizes to $\mathbf{R}^2$. Is there an $\epsilon > 0$ such that $N_\epsilon((0,0))$ is a closed subset of $Q \times Q$? (Justify your answer.)

2.18 Describe, geometrically, as in Example 2.28, the rigid motion of $\mathbf{R}^3$ which uses the matrix:

$$\begin{pmatrix} 0 & 1 & 0 \\ 1 & 0 & 0 \\ 0 & 0 & 1 \end{pmatrix}.$$

2.19 Prove the assertion of the remark on page 55, that a composition of affine maps is an affine map.

2.20 Prove Proposition 2.30.

2.21 Prove part (b) of Proposition 2.32.

2.22 Verify the claim made in Example 2.33 that any line in the plane is the image, under a rigid motion, of the x-axis.

2.23 Prove Proposition 2.41.

2.24 Prove Proposition 2.45.

2.25 Verify the assertion made in the remark on page 63, that by intersecting the open rectangles such as obtained in Proposition 2.45 with $X \times Y$. we can show:

Suppose $X \subseteq \mathbf{R}^n$ and $Y \subseteq \mathbf{R}^m$, $p = (a, b) \in X \times Y$, and U is an open subset of $X \times Y$ with $p \in U$.

2.26 Prove Proposition 2.46. Note that the discussion preceding the statement of Proposition 2.46 outlines a proof of part (a).

2.27 Let Δ^2 denote the standard 2-simplex in $\mathbf{R}^3$. Let $B = \{(x,y,z) \in \mathbf{R}^3 \mid x = 0 \text{ or } y = 0 \text{ or } z = 0\}$. Roughly, B is the edges of Δ^2. Show that B is a closed subset of Δ^2.

2.28 Prove that a page H^2_α is a closed subset of $\mathbf{R}^3$.

2.29 Let $G = \{(x,y) \in \mathbf{R}^2 \mid \text{ either } x \text{ or } y \text{ are integers}\}$. We may describe G as union of all the integer grid lines in the plane. Show that G is a closed subset of the plane.

2.30 Viewing the plane as $\mathbf{R}^1 \times \mathbf{R}^1$, let S_i be the subset of the plane corresponding to the product of closed intervals: $S_i = [i, i+1] \times [i, i+1]$. Let $S = \bigcup_{i=-\infty}^{i=\infty} S_i$. We can describe this as "S consists of unit squares placed along the diagonal line $x = y$." Show that S is a closed subset of $\mathbf{R}^2$.

2.31 Verify that the topologist's comb, Example 2.43, is a closed subset of the plane.

2.32 Verify that the shrinking comb, Example 2.44, is a closed subset of the plane.

2.33 Let M be the subset of the topologist's comb, defined $M = I \cup (\bigcup_1^\infty T_n)$. (See Example 2.43 for notations; also refer to Figure 2-17.) Is M a closed subset of T? Explain.

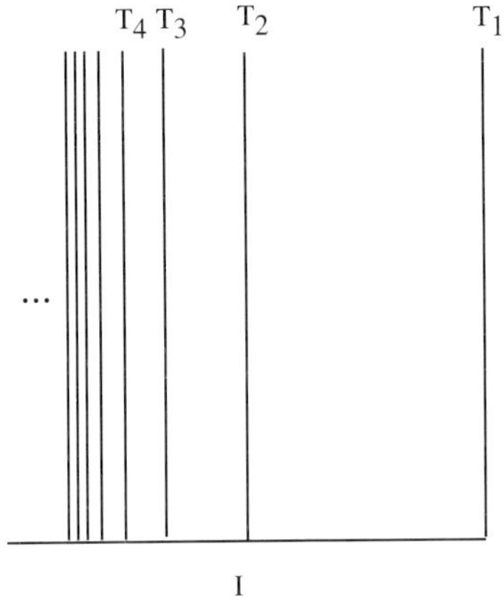

Figure 2-17 The broken topologist's comb, Problem 2.33.

2.34 (Nested rectangles) For each natural number i, $1 < i$, let P_i be the perimeter of the rectangle with vertices at $(i, 1 - \frac{1}{i}), (-i, 1 - \frac{1}{i}), (-i, -(1 -$

$\frac{1}{i}$)), and $(i, -(1-\frac{1}{i}))$. So each P_i is the union of four line segments. Let L_1 be the line $y = 1$, and L_{-1} be the line $y = -1$. Finally, let $Z = L_1 \cup L_{-1} \cup (\bigcup_i P_i)$; see Figure 2-18. Show that Z is a closed subset of $\mathbf{R}^2$.

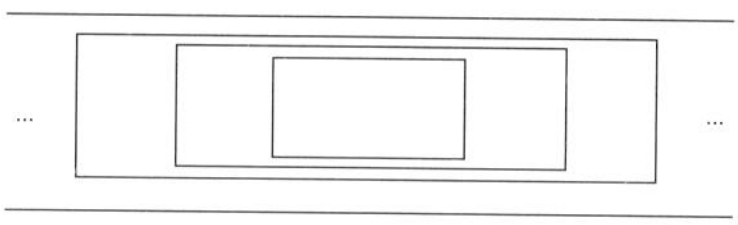

Figure 2-18 Nested rectangles, Problem 2.34.

2.35 Let X be the x-axis in the $\mathbf{R}^2$ and consider $X \subseteq \mathbf{R}^2 \subseteq \mathbf{R}^3$. Let $p = (0,0,1)$. Show that the cone on X from p is neither an open subset nor a closed subset of $\mathbf{R}^3$.

2.36 Let X be a closed subset of D^2, and consider $X \subseteq \mathbf{R}^2 \subseteq \mathbf{R}^3$. Let $p = (0,0,1)$, $q = (0,0,-1)$. Show that the cone on X from p is a closed subset of $\mathbf{R}^3$. Also, show that the suspension of X from p and q is a closed subset of $\mathbf{R}^3$.

2.37 View $\mathbf{R}^k \subseteq \mathbf{R}^n$ as the first k coordinates of $\mathbf{R}^n$. Show that $\mathbf{R}^k$ is a closed subset of $\mathbf{R}^n$ and *not* an open subset of $\mathbf{R}^n$. (Hints: To show not open, follow the proof found in Example 1.15. One way to show $\mathbf{R}^k$ is a closed subset is to show it is the intersection of $n-k$ $(n-1)$-dimensional hyperplanes of $\mathbf{R}^n$.)

3. CONTINUITY

OVERVIEW: We define a pivotal notion: continuous function from one subset of a Euclidean space to another. We show how this relates to, and extends, the more familiar notions of continuity such as those frequently used in calculus. Then we discuss methods of generating continuous functions.

3.1 Definition of continuous function

We turn to the problem of understanding the basic nature of continuity, Question 0.0 of Chapter 1. In calculus one usually studies functions from $\mathbf{R^n}$ to $\mathbf{R^m}$, for some n and m, or perhaps functions where the domain is a simple subset of $\mathbf{R^n}$. For functions of a single variable, the domain is typically an interval of some type. In the plane the domain is frequently a rectangle or a disk. What might be new to you, is that we consider functions whose domain is an *arbitrary subset* of $\mathbf{R^n}$.

For a real-valued function $f(x)$, we say $f(x)$ is continuous if, at each point x_0 in the domain of f, $\lim_{x \to x_0} f(x) = f(x_0)$ (for example, Definition D.06). The concept of the limit of $f(x)$ at a point x_0 is frequently encountered in a first semester calculus text, such as Definition D.01. This is sometimes referred to as the ϵ-δ-form for continuity.

The statement $\lim_{x \to x_0} f(x) = f(x_0)$ roughly says: if x is near x_0, then $f(x)$ is near $f(x_0)$. There are two problems with such a statement. The first is that the word "near" is not well-defined. Does "near" mean within a distance of $1/10$? or within $1/1,000,000$? We need to have some criterion to test whether or not something is "near." Secondly, the two occurrences of the word "near" refer to two different, yet related, situations. We clarify this by considering *two* criteria for nearness; one measured by a number ϵ-δ, and one measured by a number δ. Recalling our discussion of ϵ-near in the remark on page 11, we can write our statement of continuity as:

Definition of $\lim_{x\to x_0} f(x) = f(x_0)$ (rough draft): "If we have a criterion of nearness to $f(x_0)$, call it ϵ-δ, then we may find a criterion of nearness to x_0, call it δ, so that if x is δ-near to x_0, then $f(x)$ is ϵ-near to $f(x_0)$."

To better articulate the term δ-near, we say that x is δ-near to x_0 means that the distance between x and x_0 is less than δ. In the line, the distance between x and x_0 is expressed, algebraically, by the number $|x - x_0|$. Putting this all together, we arrive at:

Definition of limit (final algebraic version): $\lim_{x\to x_0} f(x) = f(x_0)$ means that given any number $\epsilon > 0$, there exists a number $\delta > 0$ so that if $|x - x_0| < \delta$, then $|f(x) - f(x_0)| < \epsilon$.

The advantage of this last formulation is that it is precise, and it is stated entirely in terms of algebra. It is free from direct mention of geometric concepts. This makes the definition (sometimes) technically easier to use. The disadvantage is that the basic, simple, geometric idea seems to have gotten lost.

We restore a geometric spirit into the definition, by expressing this statement in terms of open subsets and functions. Let U be the open interval $(f(x_0)-\epsilon, f(x_0)+\epsilon)$ and let V be the open interval $(x-\delta, x+\delta)$. Then the statement "if $|x - x_0| < \delta$, then $|f(x) - f(x_0)| < \epsilon$" translates to the statement "$V \subseteq f^{-1}(U)$." This motivates the next definition. We will see other equivalent formulations of continuity in Propositions 3.43, and 9.39. (Definition 3.53).

Definition 3.01 *Suppose $X \subseteq \mathbf{R^n}$, $Y \subseteq \mathbf{R}^m$, and $f: X \to Y$. We say f* **is continuous** *if, for each open subset U of Y, $f^{-1}(U)$ is an open subset of X.*

In Definition 3.01, we refer to $f^{-1}(U)$. Although this *set* is well-defined it does not necessarily mean that there is a *function* f^{-1}. For clarification on this important point, if necessary; see the remark on page 396.

We have defined continuity for functions whose domain is an arbitrary subset of $\mathbf{R^n}$. To begin to appreciate the scope of this definition, note that we can even apply it to finite sets. Finite sets have special properties in terms of open subsets and continuous functions. The proof of the proposition below follows immediately from the definition of continuity and Proposition 2.23.

Proposition 3.02 *If X is a finite set, $X \subseteq \mathbf{R^n}$, $Y \subseteq \mathbf{R^m}$, and f is any function $f: X \to Y$, then f is a continuous function.* ▮ *(Problem 3.1)*

Example 3.03 Any rigid motion F of $\mathbf{R^n}$ is a continuous function. Rigid motions have inverses, and the inverse F^{-1} of a rigid motion is a rigid motion (Proposition 2.31). By Proposition 2.30 F^{-1} takes open subsets to open subsets. Similarly, one can show that any similarity of $\mathbf{R^n}$ is a continuous function. ◆

It is useful to have a local definition of continuity. One finds similar definitions in calculus, such as (Definition D.06).

Definition 3.04 *Suppose $X \subseteq \mathbf{R^n}$, $Y \subseteq \mathbf{R^m}$, $f: X \to Y$, and $x \in X$. We say f is* **continuous at** *x if, for each open subset U of Y which contains $f(x)$, there is an open subset V of X such that $x \in V$ and $f(V) \subseteq U$.*

In Proposition 3.05 we connect continuity at a point, with continuity.

Proposition 3.05 *Suppose $X \subseteq \mathbf{R^n}$, $Y \subseteq \mathbf{R^m}$, and $f: X \to Y$. Then f is continuous if and only if f is continuous at every point of X.* ▮

Using Proposition 3.05, we get a slightly different way of testing continuity at a point, in terms of neighborhoods of a point.

Proposition 3.06 *Suppose $X \subseteq \mathbf{R^n}$, $Y \subseteq \mathbf{R^m}$, $f: X \to Y$, and $x \in X$. Then f is continuous at x if and only if for $\epsilon > 0$ there is a $\delta > 0$ such that $f(N_\delta(x, X)) \subseteq N_\epsilon(f(x), Y)$.* ▮

Example 3.07 It might seem that the statement of Definition 3.04 is strangely put.

You may, perhaps, expect this definition to be stated: "f is continuous at x if for any open subset U of Y, which contains $f(x)$, $f^{-1}(U)$ is an open subset of X." This implies our definition since, if we let $V = f^{-1}(U)$ then $f(V) \subseteq U$. (If f is surjective, then $f(V) = U$.)

However, we present an example where we have continuity at a point, yet the quoted statement above does not apply. Define a function (see Figure 3-1) $f: \mathbf{R^1} \to \mathbf{R^1}$ by :

$$f(x) = \begin{cases} \frac{1}{n} & \text{if } \frac{1}{n+1} < x \le \frac{1}{n}, \text{ for some natural number } n \\ 2 & \text{if } 1 < x \\ 0 & \text{if } x \le 0. \end{cases}$$

We use Proposition 3.06 to show that f is continuous at 0. Note that $f(0) = 0$. Let U be an open subset containing 0. By the definition of open subset, there is an ϵ-δ with $N_\epsilon(0) \subseteq U$. Now $N_\epsilon(0)$ is an open subset, by Proposition 1.08. We show that $N_\epsilon(0)$ will serve as our "V" as in Definition 3.04.

We first note that if $2 \le \epsilon$ there is no problem finding our δ. In this case $f^{-1}(N_\epsilon(0) = \mathbf{R^1}$. Thus for any value of δ, $f(N_\delta(0))) \subseteq N_\epsilon(0)$.

So, now suppose that $\epsilon < 2$. Let N be an integer such that $\frac{1}{N} < \epsilon$. Then one can see that $f^{-1}(N_\epsilon(0)) = (-\infty, \frac{1}{N+1}]$, which is *not* an open subset of $\mathbf{R^1}$. (We could prove this directly along the lines of Example 1.15.) Nevertheless, we may find our δ. Choose $\delta = \frac{1}{N+1}$; then verify that $f(N_\delta(0)) \subseteq N_\epsilon(0)$. ◆

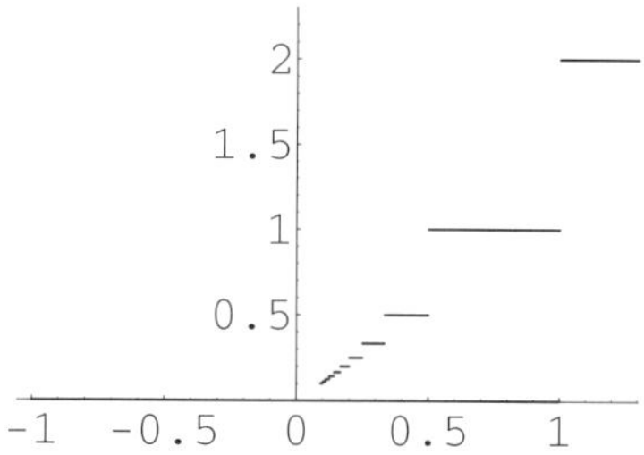

Figure 3-1 A function that is continuous at 0, but inverse images of open subsets of the line, containing $f(0)$, are not open subsets of line. See Example 3.07.

A version of Proposition 3.06 is frequently found in calculus texts as a definition of continuity, such as the remark on page 416. We next formulate such a definition, using the notion of distance directly, rather than referring to open neighborhoods. (Be careful to distinguish the two uses of the symbol δ in the statement: the *distance function* $\delta(x, y)$ and the *number* δ.)

Definition 3.08 *If* $F: \mathbf{R^n} \to \mathbf{R^m}$ *is a function and* $x_0 \in \mathbf{R^n}$, *we say that* F **is continuous at** x_0 **in the** ϵ-δ **sense,** *if for any number* $\epsilon > 0$ *there exists a* $\delta > 0$ *such that if* $\delta(x, x_0) < \delta$, *then* $\delta(f(x), f(x_0)) < \epsilon$. *We say* F **is continuous in the** ϵ-δ **sense** *if it is continuous at every point of* $\mathbf{R^n}$.

This is equivalent to our definition of continuity, Definition 3.01:

Proposition 3.09 $F: \mathbf{R^n} \to \mathbf{R^m}$ *is continuous in the* ϵ-δ *sense if and only if* F *is continuous.* ▮ *(Problem 3.3)*

Although it is hard to make general statements on how one proves things, it is often the case that the ϵ-δ form of continuity is more convenient for examining specific examples, and our definition in terms of open subsets more convenient for general proofs.

Consider the situation, familiar from the first semester of calculus where we wish to formulate the definition of a continuous function from a closed interval, $[a, b]$, to the real numbers. This is an example of a function defined on a *subset* of $\mathbf{R^1}$.

An approach frequently taken is to define right-hand limits, and left-hand limits, and use these to define continuity from the left and continuity from the right; see Definitions D.02, D.03, D.08, and D.09. Then one defines a function to be continuous on $[a, b]$, if it is continuous at each point of (a, b), and if the function is continuous from the right at a and continuous from the left at b.

Definition 3.01 gives us a way of expressing the ideas of left-handed or right-handed continuity in terms of open subsets. For example, the way we consider continuity from the right at a is that the ϵ-δ-neighborhoods of a in the set $[a,b]$ are half-open intervals of the form $[a,c)$ where $c < b$ (except for one "large" ϵ-δ-neighborhood, which is all of $[a,b]$).

3.2 Some basic constructions of continuous functions

Any function $f\colon \mathbf{R^n} \to \mathbf{R^m}$ can be described in terms of functions from $\mathbf{R^n}$ to $\mathbf{R^1}$ in the following simple way.

Definition 3.10 *For each i, $1 \leq i \leq m$, we define a map, $\pi_i\colon \mathbf{R^m} \to \mathbf{R^1}$, that gives the i-th coordinate of each point by $\pi_i(x_1, \ldots, x_i, \ldots x_m) = x_i$. We call π_i the* **projection of $\mathbf{R^m}$ onto the i-th coordinate**.

Suppose we are given $f\colon \mathbf{R^n} \to \mathbf{R^m}$. If we define $f_i(x_1,\ldots,x_n) = \pi_i \circ f$, then we can write:

$$F(x_1,\ldots,x_n) = (f_1(x_1,\ldots,x_n), f_2(x_1,\ldots,x_n),\ldots, f_m(x_1,\ldots,x_n)).$$

We call f_i the **i-th coordinate function of** *F.*

We can further generalize Definition 3.10 for Cartesian products.

Definition 3.11 *Suppose $A \subseteq \mathbf{R^n}$, $B \subseteq \mathbf{R^m}$; $X \subseteq \mathbf{R}^k$. Define $\pi\colon \mathbf{R}^{n+m} \to \mathbf{R^n}$ by $\pi(x_1, \ldots, x_n, x_{n+1}, \ldots x_{n+m}) = (x_1, \ldots x_n)$, and $\pi'\colon \mathbf{R}^{n+m} \to \mathbf{R^m}$ by $\pi'(x_1, \ldots, x_n, x_{n+1}, \ldots x_{n+m}) = (x_{n+1}, \ldots x_{n+m})$. Given any function $f\colon X \to A \times B$, we can define $f_A = \pi \circ f$, and $f_B = \pi' \circ f$. Note that f_A is a function from X to A, and f_B is a function from X to B. We call F_A the* **projection of** *F* **on** *A, F_B the* **projection of** *F* **on** *B.*

A function $F(t) = (x(t), y(t))$ is a parametric curve. The following is a standard theorem from calculus on parametric curves.

Proposition 3.12 *If $F\colon \mathbf{R^1} \to \mathbf{R^2}$ is given by $F(t) = (x(t), y(t))$, then $F(t)$ is continuous if and only if $x(t)$ and $y(t)$ are continuous.* ▮ *(Problem 3.4)*

Proposition 3.12 is about continuous functions from $\mathbf{R^1} \to \mathbf{R^1} \times \mathbf{R^1}$. The next proposition generalizes this in two ways. In part (a), we replace $\mathbf{R^1}$ by $\mathbf{R^n}$, and replace $\mathbf{R^1} \times \mathbf{R^1}$ with the Cartesian product of m copies of $\mathbf{R^1}$ with itself. In part (b), we replace $\mathbf{R^1}$ by an arbitrary subset of $\mathbf{R^1}$, and replace $\mathbf{R^1} \times \mathbf{R^1}$ with $A \times B$, the cartesian product of two arbitrary subsets of $\mathbf{R^n} \times \mathbf{R^m}$.

Proposition 3.13

(a) *Suppose* $F: \mathbf{R^n} \to \mathbf{R^m}$ *with coordinate functions* f_i*:*

$$F(x_1, \ldots, x_n) = (f_1(x_1, \ldots, x_n), f_2(x_1, \ldots, x_n), \ldots, f_m(x_1, \ldots, x_n)).$$

Then F *is continuous if and only if all of the functions* f_1*,* f_2*, ...,* f_m *are continuous.*

(b) *Suppose that* $A \subseteq \mathbf{R^n}$*,* $B \subseteq \mathbf{R^m}$*,* $X \subseteq \mathbf{R}^k$*, and* $f: X \to A \times B$*. Define* $f_A = \pi \circ f$*, and* $f_B = \pi' \circ f$*. Then* f *is continuous if and only if both* f_A *and* f_B *are continuous.* ▮ *(Problem 3.5)*

Another way to express Proposition 3.13(b) is the as follows:

Let $X \subseteq \mathbf{R^n}$, $A \subseteq \mathbf{R^m}$ and $B \subseteq \mathbf{R}^k$. Suppose also we are given functions $f: X \to A$ and $g: X \to B$. The function $f \times g : X \to A \times B$ is defined by $(f \times g)(x) = (f(x), g(x))$. Then $f \times g$ is continuous if and only if f and g are continuous. We say that $f \times g$ is the "product of f and g."

Having now established a connection between our topological definition of continuity and that from calculus, we may use Proposition 3.13 together with continuity, as established in calculus, to verify continuity of many other functions.

In Appendix D we list some background material about continuity of functions of a single variable and real-valued functions of several variables; see Proposition D.11. From this point on, we use these without explicit reference. Also you may use these results, without need of proof, in the exercises.

Restricting a continuous function is one simple, but useful, way to get examples of continuous functions whose domain is a subset.

Definition 3.14 *Suppose* $A \subseteq X \subseteq \mathbf{R^n}$*,* $Y \subseteq \mathbf{R^n}$*, and* $f: X \to Y$*. The function from* A *to* Y*, defined* $f|_A(x) = f(x)$*, for all* $x \in A$*, is called the* **restriction of** f **to** A .

Proposition 3.15 *Suppose* $A \subseteq X \subseteq \mathbf{R^n}$*,* $Y \subseteq \mathbf{R^n}$*, and* $f: X \to Y$*. If* f *is a continuous function, then so is* $f|_A$*.* ▮ *(Problem 3.6)*

Example 3.16 For example, suppose we define $F: \mathbf{R}^2 \to \mathbf{R}^2$ by $F(x, y) = (x^2 + y^2, x + y)$ and let R be the rectangle defined by $R = \{(x, y) \in \mathbf{R}^2 \mid 0 \leq x \leq 1 \text{ and } 0 \leq y \leq 1\}$ Then, we could consider $F|_R$ to be a map from R to R^2_+. The proposition above would imply that since the map F is continuous, then $F|_R$ is also continuous. ◆

It is not in general true that if $A \subseteq \mathbf{R}^1$ and $f: A \to \mathbf{R}^1$ is a continuous function real-valued function, then f is the restriction of some continuous function defined for all numbers.

Here is a familiar example. Suppose that $A = \mathbf{R}^1 - \{0\}$. Define $f(x) = x/|x|$. Note that $f(x) = -1$ is $x < 0$ and $f(x) = +1$ is $0 < x$. As a continuous function from A to $\mathbf{R}^1$, f is continuous. However, we cannot extend f to a map of all of $\mathbf{R}^1$, (Problem 3.7). That is, there is no continuous function $F: \mathbf{R}^1 \to \mathbf{R}^1$ such that $F|_A = f$.

Example 3.17 Here is another example. Suppose $A = (0, \infty)$. Define $f: A \to \mathbf{R}^1$ by $f(x) = 1/x$. Again, f is continuous, but there is no continuous map $F: \mathbf{R}^1 \to \mathbf{R}^1$ such that $F|_A = f$. The problem with defining F is intuitively clear from the graph of f near the y-axis; details are left as an exercise, (Problem 3.8). ◆

Allowing for arbitrary subsets of $\mathbf{R}^n$ as domains for our functions opens up a new world, mathematically. Although open subsets of the subset A are obtained by restricting open subsets of $\mathbf{R}^n$ (here by "restricting" we refer to taking the intersection), the resulting continuous functions are not all obtained by restriction.

There are common situations in which a continuous function, defined on a subset, can be extended. The next example is often used in calculus. Note that the nature of interval for the domain is a critical factor.

Example 3.18 Suppose $[a, b]$ is a closed interval in $\mathbf{R}^1$, and $f: [a, b] \to \mathbf{R}^1$ is any continuous function. Then we can define a continuous function $F: \mathbf{R}^1 \to \mathbf{R}^1$ by

$$F(x) = \begin{cases} f(a) & \text{if } x \leq a \\ f(x) & \text{if } a \leq x \leq b \\ f(b) & \text{if } b \leq x. \end{cases}$$

Make a sketch of the graph of an example of such an F. Continuity of F is readily verified; it also follows from Proposition 3.45.

In contrast, for the *open* interval $J = (-\pi/2, \pi/2)$, define $g: J \to \mathbf{R}^1$ by $g(x) = \tan(x)$. Then g cannot be extended to a continuous map of $\mathbf{R}^1$ to itself. ◆

The next two definitions define simple, but useful, functions.

Definition 3.19 *Given any $X \subseteq \mathbf{R}^n$, define a map from X to itself by $f(x) = x$, for all $x \in X$. This map is called the* **identity map of** *X and is denoted by $Id|_X$.*

Definition 3.20 *If $A \subseteq X \subseteq \mathbf{R}^n$, the map from A to X, defined by $i_A(x) = x$ for all $x \in A$, is called the* **inclusion map of** *A* **into** *X.*

Clearly, the inclusion map is the restriction to A of the identity map of X. That is: $i_A = (Id|_X)_A$. Later, in discussing continuous maps, it will be useful to note the following.

Proposition 3.21

(a) *If $X \subseteq \mathbf{R}^n$, then $Id|_X$ is continuous.*

(b) *If $A \subseteq X$, then i_A is continuous.* ▮ *(Problem 3.9)*

Definition 3.22 *If $n < m$, the map $i{:}\mathbf{R^n} \to \mathbf{R^m}$ defined by $i(x_1, \ldots, x_n) = (x_1, \ldots, x_n, 0, \ldots, 0)$ is called the* **standard inclusion of $\mathbf{R^n}$ into $\mathbf{R^m}$**.

A standard way of building complicated functions from simpler functions is by taking compositions of functions.

Proposition 3.23 *Suppose $X \subseteq \mathbf{R^n}$, $Y \subseteq \mathbf{R}^m$, and $Z \subseteq \mathbf{R}^k$. Suppose f is a function, $f{:}X \to Y$, and g is a function $g{:}Y \to Z$. If f and g are continuous, then composition, $g \circ f$ is continuous.* ▮ *(Problem 3.10)*

Suppose $X \subseteq \mathbf{R^n}$ and function $f{:}X \to X$ is continuous. Using Proposition 3.23 we can derive an infinite set of continuous functions from X to itself by taking compositions of f with itself: $\{f,\ f\circ f,\ f\circ f\circ f,\ \ldots\}$. Later in the text we will see examples where iterations of a function are used to define subsets, Examples 5.31 and 11.12.

Definition 3.24 *The composition $f \circ f$ is called the* **second iterate of** *f; the composition of n copies of f is called the* **n-th iterate of** *f, denoted $f^{(n)}$. We identify $f^{(1)}$ with f.*

Definition 3.25 *Suppose $X \subseteq \mathbf{R^n}$, and also have a continuous function $f{:}X \to X$, $x \in X$. The set of all iterates of x under f is call* **orbit of** *x* **under** *f.*

Example 3.26 Let $f{:}S^1 \to S^1$ be rotation by an angle θ. We can restate the results of Example 1.43 as follows, (Problem 3.11).

Suppose $x \in S^1$. If θ is a rational multiple of 2π, then the orbit of x under f is a finite subset. If θ is a irrational multiple of 2π, the orbit of x is dense in S^1. ◆

3.3 Graphs of functions

A graph of a function is not simply a drawing of a function; it is also a special kind of subset.

Given $f{:}\mathbf{R}^1 \to \mathbf{R}^1$, define $F{:}\mathbf{R}^1 \to \mathbf{R}^2$ by $F(x) = (x, f(x))$. Continuity of F follows since it is continuous in each coordinate. The image of F is called "the graph of F." If $g{:}\mathbf{R}^2 \to \mathbf{R}^1$ define a continuous map $G{:}\mathbf{R}^2 \to \mathbf{R}^3$ by $G(x,y) = (x, y, g(x,y))$. Again, the image of G is called the "graph of G."

More generally, we make the following definition.

Definition 3.27 *Suppose that $f: X \to Y$ is a function where $X \subseteq \mathbf{R^n}$ and $Y \subseteq \mathbf{R^m}$. Write*

$$f(x_1,\ldots,x_n) = (f_1(x_1,\ldots,x_n), f_2(x_1,\ldots,x_n), \ldots, f_m(x_1,\ldots,x_n)).$$

The **graph of** f *is the set*

$$\{(x_1,\ldots,x_n,y_1,\ldots,y_m) \in \mathbf{R}^{n+m}: (x_1,\ldots,x_n) \in X \text{ and } y_i = f_i(x_1\ldots,x_n)\}, 1 \le i \le m\}.$$

The definition above is the most common use of the word "graph" found in other parts of mathematics. However, there is a second common usage of "graph" found in topology and combinatorics. In this text, for example, we will define topological graph, Definition 17.17. For this reason we should be careful and speak of "graph of a function" rather than just "graph."

The graph of f is a subset of $X \times Y$. In other notation, it is all points of the form $x \times f(x)$. The graph of a function from $\mathbf{R}^2$ to $\mathbf{R}^2$ is a subset of four-dimensional space. The same is true if we are considering a function from $\mathbf{R}^3$ to $\mathbf{R}^1$, or from $\mathbf{R}^1$ to $\mathbf{R}^3$. Thus the topic of subsets of four-dimensional space arises in a natural way from consideration of such lower-dimensional problems.

As an application of the graph of a function as a subset, we present yet another proof that a line is a closed subset of the plane, Proposition 2.01.

Consider a line with equation $y = mx + b$. Define a function $F: \mathbf{R}^2 \to \mathbf{R}^1$ by $F(x, y) = y - (mx + b)$. The function is continuous, by standard results of calculus. Consider the one-point subset $\{0\} \subseteq \mathbf{R}^1$. This is a closed subset of $\mathbf{R}^1$ by Proposition 1.14; thus $\mathbf{R}^1 - \{0\}$ is an open subset of the line. Since F is continuous, $F^{-1}(\mathbf{R}^1 - \{0\})$ is an open subset of $\mathbf{R}^2$. Thus $\mathbf{R}^2 - F^{-1}(\mathbf{R}^1 - \{0\})$ is a closed subset of $\mathbf{R}^2$. But it is easy to check that $F^{-1}(\{0\}) = \mathbf{R}^2 - F^{-1}(\mathbf{R}^1 - \{0\})$ and that this subset is the given line. This completes the proof. We will revisit this proof shortly and shorten it considerably in the remark on page 88.

Graphs of continuous functions are, properly considered, closed subsets; this topic is the focus of the next two propositions. Proposition 3.28 is a special case of Proposition 3.29.

Proposition 3.28 *Suppose $f: \mathbf{R^n} \to \mathbf{R^m}$ is a continuous function. Then the graph of f is a closed subset of $\mathbf{R^{n+m}}$.*

The idea of the proof is to follow the proof above for a linear function and write

$$f(x_1,\ldots,x_n) = (f_1(x_1,\ldots,x_n),\ldots,f_m(x_1,\ldots,x_n)).$$

Consider the function $D:\mathbf{R^{n+m}} \to \mathbf{R^m}$ defined by

$$D(x_1, \ldots x_n, \ldots, x_{n+m}) = \\ (f_1(x_1, \ldots, x_n) - x_{n+1}, \ldots, f_m(x_1, \ldots, x_n) - x_{n+m}).$$

One can verify that the graph of f is $D^{-1}((0, \ldots, 0))$. ▮

Take care in using Proposition 3.28. Note that this is for functions defined on *all of* $\mathbf{R^n}$. If we have a function whose domain is a *subset* of $\mathbf{R^n}$, the graph of this function is still a subset of $\mathbf{R^n} \times \mathbf{R^m} \approx \mathbf{R^{n+m}}$, but it is *not* always true that the graph of a function is a closed subset of $\mathbf{R^{n+m}}$. Suppose that $X = \{x \in \mathbf{R^1} \mid 0 < x < 1\}$. Consider the graph G of f where we define $f(x) = x$. Then the graph of f is $G = \{(x, x) \in \mathbf{R^2} \mid 0 < x < 1\}$. Now G is *not* a closed subset of $\mathbf{R^2}$. (For example $(1, 1)$ is a limit point of G but not a point of G.)

However, recall the remark on page 82. Note that G is a closed subset of $X \times \mathbf{R^1}$. Since $f(X) = X$, $G \subseteq X \times X$. In fact, G is a closed subset of $X \times X$. The discussion in the remark above is prelude to:

Proposition 3.29 *Suppose $f: X \to Y$ is a continuous function where $X \subseteq \mathbf{R^n}$ and $Y \subseteq \mathbf{R^m}$. Let $Z = f(X)$. Then the graph of F is a closed subset of each of the following three sets:*

(a) $X \times \mathbf{R^m}$
(b) $X \times Y$
(c) $X \times Z$. ▮ *(Problem 3.12)*

Next, we consider a more complicated closed subset of $\mathbf{R^2}$.

Example 3.30 Define a function $f: (\mathbf{R^1} - \{0\}) \to \mathbf{R^1}$ by $f(x) = \sin(\frac{1}{x})$, if $x \neq 0$. Let Γ be the graph of f. This is pictured in Figure 3-2(a). Let $L = \{(x, y) \in \mathbf{R^2} \mid x = 0 \text{ and } -1 \leq y \leq 1\}$; let $G = \Gamma \cup L$. (See Figure 3-2 (b).)

Then G is a closed subset of the plane. One way to see this is to verify that L is the set of all limit points of Γ that are not points of Γ, (Problem 3.13). This would also show that Γ is not a closed subset of the plane.

As an exercise in the use of the propositions we have developed, we offer an alternative proof which expresses $\mathbf{R^2} - G$ as a union of three open subsets of $\mathbf{R^2}$. Let $V = \mathbf{R^1} - \{0\}$; V is an open subset of $\mathbf{R^1}$. By Proposition 3.29, we can view Γ as a closed subset of $V \times \mathbf{R^1}$. Let $U_0 = V \times \mathbf{R^1} - \Gamma$. Then U_0 is an open subset of $V \times \mathbf{R^1}$. But by Proposition 2.46, the set $V \times \mathbf{R^1}$ is an open subset of $\mathbf{R^2}$, so we may apply Proposition 2.19 to show that U_0 is an open subset of $\mathbf{R^2}$. Now $\mathbf{R^2} - U_0$ is not the same as G since $\mathbf{R^2} - U_0$ contains all points on the y-axis, while G has only the points of L. Our plan is to get some open

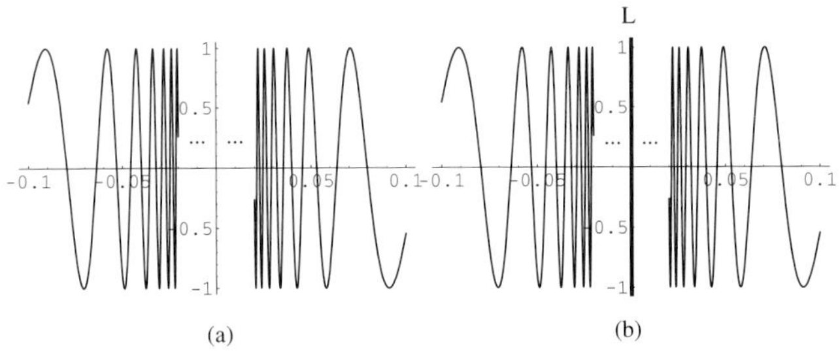

Figure 3-2 (a) A portion of Γ, (b) the subset G; see Example 3.30.

subsets that will correct this by "covering up" the points of the y-axis not in L while not "covering up" points of Γ.

One idea is to consider the parabolas $f_1(x) = x^2 + 1$ and $f_2(x) = -x^2 - 1$. Let U_1 be the open subset of $\mathbf{R}^2$ consisting of points above the graph of f_1, U_2 the set of points below f_2. To verify that U_1 is an open subset of $\mathbf{R}^2$, define a map from $\mathbf{R}^2$ to $\mathbf{R}^1$ by $F_1(x, y) = y - f_1(x)$. Then $U_1 = F_1^{-1}((0, \infty))$, so U_1 is an open subset of the plane, being the inverse image of an open subset under a continuous function. Clearly, openness of U_2 follows similarly. The union $U = U_0 \cup U_1 \cup U_2$ is an open subset of $\mathbf{R}^2$ since it is the union of three open subsets. Finally, we check that $U = \mathbf{R}^2 - G$. ◆

Example 3.31 Let Γ denote the graph of $\ln(x)$. We show that Γ is a closed subset of the plane. Proposition 3.28 does not apply since $\ln(x)$ is only defined if $0 < x$. If we apply Proposition 3.29, this only proves that Γ is a closed subset of $\overset{\circ}{R}{}^2_+$. However, let Γ' denote the graph of $y = e^x$. We *can* use Proposition 3.28 to show that Γ' is a closed subset of $\mathbf{R}^2$. Let $M = \left(\begin{smallmatrix} 0 & 1 \\ 1 & 0 \end{smallmatrix}\right)$ and $F(\vec{x}) = M\vec{x}$. Then F is a rigid motion and $F(\Gamma') = \Gamma$; thus Γ' is a closed subset of $\mathbf{R}^2$ by Proposition 2.32. ◆

3.4 More examples of continuous functions

We continue with more examples of continuous functions.

Example 3.32 Consider the continuous map from the line to the plane defined by $e(\theta) = (\cos(\theta), \sin(\theta))$. The image of e is the unit circle.

Note that the inverse of any point on the circle is an infinite set of points on the line.

This is an important function. In trigonometry this is called the "wrapping function." Sometimes it is called the "exponential map" since it describes the complex exponential function, $w = e^z$, restricted to the imaginary axis. As complex numbers, points on the y-axis are of the form yi and we have $e^{yi} = \cos(y) + i\sin(y)$. In topology, this map is critical in the analysis of the circle, found in Chapter 15. ◆

Example 3.33 We can get continuous functions from complex functions. Here is one example. Let $c = a + bi$ be a complex number and let $z = x + yi$ be a complex variable. Then the function $f(z) = z^2 + c$. can be viewed as a function from the plane to the plane. We can express this without using complex numbers. Write

$$z^2 + c = (x + yi)(x + yi) + (a + bi) = (x^2 - y^2 + a) + (2xy + b)i.$$

We can thus view the given function $f(z)$ as being equivalent to $f(x, y) = (x^2 - y^2 + a,\ 2xy + b)$.

This function, fully considered, is a surprisingly interesting and mathematically rich topic. A fuller investigation of the iterates of such functions is developed in a branch of mathematics called "complex dynamical systems," [13]. ◆

Continuity is defined in terms of inverse images of subsets. The next two definitions involve images rather than inverse images. These ideas are useful, but not as fundamental as the concept of continuity.

Definition 3.34 *Suppose $X \subseteq \mathbf{R^n}$, $Y \subseteq \mathbf{R^m}$ and $f: X \to Y$. We say f is an* **open map** *if, for any open subset U of X, $f(U)$ is an open subset of Y.*

Definition 3.35 *Suppose $X \subseteq \mathbf{R^n}$, $Y \subseteq \mathbf{R^m}$ and $f: X \to Y$. We say f is a* **closed map** *if, for any closed subset C of X, $f(C)$ is a closed subset of Y.*

A projection of a Cartesian product to a factor gives an open map:

Proposition 3.36 *Suppose $A \subseteq \mathbf{R^n}$, $B \subseteq \mathbf{R^m}$. Define $\pi_A: A \times B \to A$ by $\pi_A(a, b) = a$; also, $\pi_B: A \times B \to B$ by $\pi_B(a, b) = b$. Then π_A and π_B are continuous and open maps.* ▮ *(Problem 3.15)*

We now have three concepts: f is continuous, f is an open map, and f is a closed map. These three are independent concepts, in that none of these properties implies the other, nor do any two imply the third. We show some of this in Examples 3.37 and 3.38. Some of the other possible implications are left as an exercise. (Problem 3.16).

Example 3.37 We give an example of an closed map which is not open. Let X be the union of the x-axis and y-axis; $X = \{(x, y) \in \mathbf{R^2} \mid x =$

0 or $y = 0\}$. Let $Y = \{(x, y) \in \mathbf{R}^2 \mid x = 0\}$; Y is the y-axis. Define $f: X \to Y$ by $f(x, y) = y$.

We will see that f is continuous and closed but not open. Certainly f is continuous since it is a restriction of a continuous function. To see f is not an open function, consider an open interval on the x-axis, $U = \{(x, y) \in \mathbf{R}^2 \mid 1 < x < 3 \text{ and } y = 0\}$. This is an open subset of X since $U = N_1((2, 0), X)$. But $f(U) = \{(0, 0)\}$ is a one-point subset of Y and is not an open subset of Y. It is not difficult to verify that f is a closed map. (Problem 3.17). ◆

Example 3.38 We give an example of continuous function whichû is an open map and not a closed map. Let X be the open rectangle in the plane $X = \{(x, y) \in \mathbf{R}^2 \mid 0 < x < 1 \text{ and } 0 < y < 1\}$. As in the example above, let $Y = \{(x, y) \in \mathbf{R}^2 \mid x = 0\}$, and define $f: X \to Y$ by $f(x, y) = y$. Again, f is continuous and, by Proposition 3.36, an open map.

Let $C = \{(x, y) \in X \mid y = \frac{x}{2}\}$. Then C is a closed subset of X. To see this, let L be the line in the plane with equation $y = \frac{x}{2}$. The line L is a closed subset of the plane, Proposition 2.01. The open rectangle X is an open subset of the plane, Proposition 2.05. Since $X - C = (\mathbf{R}^2 - L) \cap X$, $X - C$ is an open subset of X. Thus C is a closed subset of X. But $f(C) = \{(x, y) \in \mathbf{R}^2 \mid x = 0 \text{ and } 0 < y < \frac{1}{2}\}$; this is not a closed subset of Y.

Also, X is a closed subset of itself, and $f(X)$, which corresponds to an open subinterval of the y-axis, is not a closed subset of Y. ◆

When we write affine maps (see Definition 2.24) in terms of coordinates, all the coordinate functions are linear and therefore continuous. Thus we have:

Proposition 3.39 *If F is an affine map, then F is a continuous function.* ∎

Example 3.40 Let us examine an example of an affine map. Let $M = \begin{pmatrix} 1 & 2 \\ 1 & 4 \end{pmatrix}$, $\vec{B} = \begin{pmatrix} 5 \\ -7 \end{pmatrix}$ and $\vec{X} = \begin{pmatrix} x \\ y \end{pmatrix}$. If $\vec{Y} = M\vec{X} + \vec{B}$, then $\vec{Y} = \begin{pmatrix} x+2y+5 \\ x+4y-7 \end{pmatrix}$. This is equivalent to the function from the plane to the plane given by $f(x, y) = (x + 2y + 5, 3x + 4y - 7)$.

Conversely, if $f(x, y) = (f_1(x), f_2(x))$ where $f_1(x)$ and $f_2(x)$ are linear functions, then f is an affine function, (Problem 3.18). ◆

3.5 More properties of continuous functions

Non-singular affine maps are continuous open and closed. The following proposition implies captures this (and also provides a proof of Propositions 2.32 and 2.41):

Proposition 3.41 *Suppose $F: \mathbf{R^n} \to \mathbf{R^n}$ is a non-singular affine map; then*

(a) *U is an open subset of $\mathbf{R^n}$ if and only if $F(U)$ is an open subset of $\mathbf{R^n}$.*
(b) *U is a closed subset of $\mathbf{R^n}$ if and only if $F(U)$ is a closed subset of $\mathbf{R^n}$.*

Proof: The essential observation is that a non-singular affine map has an inverse, and this inverse is also a non-singular affine map. In fact, the argument given in the remark on page 54 shows that if $F(\vec{x}) = M\vec{x} + \vec{B}$, then $F^{-1} = \vec{x} = M^{-1}\vec{y} - M^{-1}\vec{B}$. We note $|M^{-1}| = \frac{1}{|M|}$. Since $|M| \neq 0$, $|M^{-1}| \neq 0$. We note in passing that since we have found an inverse of F and since the inverse of F^{-1} is F, it follows that F (and F^{-1}) are one-to-one correspondences.

Suppose U is an open subset of $\mathbf{R^n}$. Let $W = F^{-1}(U)$. Since F^{-1} is continuous we know that the inverse of F^{-1} takes open subsets of $\mathbf{R^n}$ to open subsets $\mathbf{R^n}$. But the inverse of F^{-1} is F. Thus we have established part (a).

The argument for part (b) follows along similar lines. ∎

Example 3.42 Let E be an ellipse in the plane with equation $\frac{x^2}{a^2} + \frac{y^2}{b^2} = 1$, and let U be the points inside E. That is, $U = \{(x, y) \in \mathbf{R^2} \mid \frac{x^2}{a^2} + \frac{y^2}{b^2} < 1\}$.

We can show that U is an open subset of the plane since it is the image, under a non-singular affine map, of an open subset of $\mathbf{R^2}$. Let $M_1 = \begin{pmatrix} a & 0 \\ 0 & b \end{pmatrix}$, and define $F(\vec{x}) = M\vec{x}$; then $U = F(N_1(0,0))$.

Here is another way to prove U is an open subset. Consider the function $G: \mathbf{R^2} \to \mathbf{R^1}$ defined by $G(x, y) = \frac{x^2}{a^2} + \frac{y^2}{b^2}$. Then let $V = \{x \in \mathbf{R^1} \mid x < 1\}$. Then V is an open subset of $\mathbf{R^1}$, G is continuous, and $U = G^{-1}(V)$. It follows that U is an open subset of $\mathbf{R^2}$. ◆

Our definition of continuity is expressed in terms of open subsets. Sometimes it is more convenient to consider closed subsets:

Proposition 3.43 *Suppose $X \subseteq \mathbf{R^n}$, $Y \subseteq \mathbf{R}^m$, $f: X \to Y$. Then f is a continuous function if and only if the inverse of any closed subset of Y is a closed subset of X.*

Proof: Suppose f is continuous and C is a closed subset of Y. We show $f^{-1}(C)$ is a closed subset of X. The set $Y - C$ is an open subset of Y, and thus $f^{-1}(Y - C)$ is an open subset of X. But $f^{-1}(Y - C) = f^{-1}(Y) - f^{-1}(C) = X - f^{-1}(C)$. Since $X - f^{-1}(C)$ is an open subset of X, $f^{-1}(C)$ is a closed subset of X.

Next suppose, for any closed subset C of Y, $f^{-1}(C)$ is a closed subset of X. Let U be any open subset of Y. We show that $f^{-1}(U)$ is an open subset of X. Let $C = Y - U$; then C is a closed subset of Y, and thus $f^{-1}(C)$ is a closed subset of X. But $f^{-1}(C) = f^{-1}(Y - U) = f^{-1}(Y) - f^{-1}(U) = X - f^{-1}(U)$. Since $X - f^{-1}(U)$ is a closed subset of X, $f^{-1}(U)$ is an open subset of X. ∎

We can use Proposition 3.43 to provide a different proof that $\overset{\circ}{R^n_+}$ is an open subset of $\mathbf{R^n}$, and that R^n_+ and S^n are closed subsets, Proposition 2.16. Consider π_i, the projection onto the i-th coordinate; $\pi_i(x_1, \ldots, x_n) = x_i$. This function is continuous, by Proposition 3.13. Now $\overset{\circ}{R^n_+} = p^{-1}((0, \infty))$. Thus $\overset{\circ}{R^n_+}$ is open since it is the inverse image of an open subset of $\mathbf{R^1}$. Similarly, R^n_+ is a closed subset of $\mathbf{R^n}$ since it is the inverse image of a closed subset of $\mathbf{R^1}$ (namely, $R^n_+ = p^{-1}([0, \infty))$).

Also, we could prove that the sphere S^n is a closed subset of $\mathbf{R}^{n+1}$, by considering the continuous function $f: \mathbf{R^n} \to \mathbf{R^1}$ defined by $f(x_1, \ldots, x_{n+1}) = \sum_1^n x_i^2$. Then $\{1\}$ is a closed subset of $\mathbf{R^1}$, and so $S^n = f^{-1}(\{1\})$ is a closed subset of $\mathbf{R}^{n+1}$.

We can use Proposition 3.13 to "improve" the argument of the remark on page 82 and provide a shorter proof that a line is a closed subset of the plane. As in the remark on page 82, define a continuous function $F: \mathbf{R^2} \to \mathbf{R^1}$ by $F(x, y) = y - (mx + b)$. Then since $\{0\}$ is a closed subset of $\mathbf{R^1}$, $F^{-1}(\{0\})$ is a closed subset of X, and $F^{-1}(\{0\})$ is the same subset as the graph of the line $y = mx + b$. We note that the reason this proof seems shorter is that we have encapsulated the argument of the proof for the remark on page 82 in our proof of Proposition 3.13.

By examining limit points, we can decide if a subset is closed, Proposition 1.39. By Proposition 3.43, continuity of a function can be decided by looking at closed subsets. The following Proposition puts these observations together and makes a direct connection between continuity and limit points.

Proposition 3.44 *Suppose $X \subseteq \mathbf{R^n}$, $Y \subseteq \mathbf{R^m}$, and $f: X \to Y$. Then f is a continuous function, if and only for all subsets $A \subseteq X$, if a is a limit point of A, then $f(a)$ is either a point of $f(A)$ or a limit point of $f(A)$.*

Proof: Suppose f is continuous, $A \subseteq X$, with a a limit point of A, and yet $f(a)$ is neither a limit point of $f(A)$ nor a point of $f(A)$. Then there is an open subset U of Y with $f(a) \in U$ and $f(A) \cap U = \emptyset$. But then

$f^{-1}(U)$ is an open subset of X which contains a with $f^{-1}(U) \cap A = \emptyset$, contradicting the assumption that a is a limit point of A.

Next, suppose, for all subsets $A \subseteq X$: if a is a limit point of A, then $f(a)$ is a limit point of $f(A)$ or a point of $f(A)$. Suppose C is a closed subset of Y and let $A = f^{-1}(C)$. We show that A is a closed subset of X. Continuity of f will then follow from Proposition 3.43.

We will verify the contrapositive: if A is not a closed subset of X, then there is a limit point a of A with $a \notin A$. Let $c = f(a)$. Since $A = f^{-1}(C)$, $c \notin C$. By our hypothesis, c is a limit point of C, and so C is not closed. ∎

3.6 The gluing lemma

The following proposition is a useful one and is sometimes called the "gluing lemma" since it tells how to "glue" two continuous functions together.

Proposition 3.45 *(gluing lemma) Suppose that $X \subseteq \mathbf{R}^n$, $Y \subseteq \mathbf{R}^m$, $X = A \cup B$, and $f: X \to Y$, with $f|_A$ and $f|_B$ both continuous.*

(a) *If A and B are both open subsets of X, then f is a continuous function.*
(b) *If A and B are both closed subsets of X, then f is a continuous function.*

Proof: We begin with a proof of part (a). Let U be an open subset of Y, we need to show that $f^{-1}(U)$ is an open subset of X. Let $U_A = f^{-1}|_A(U)$, and $U_B = f^{-1}|_B(U)$. By our continuity hypotheses, U_A and U_B are open subsets A and B, respectively. By Proposition 2.19, U_A and U_B are open subsets of X. Now $f^{-1}(U)$ is an open subset of X since $f^{-1}(U) = U_A \cup U_B$. This completes the proof of part (a).

It seems natural to want to prove part (b) using the proof of part (a) as a model. A few drawings of examples (see Figure 3-3), together with thinking of how to use the hypotheses that A and B are closed subsets of X, leads to the following proof.

Proof of part (b): Let $U_A = f^{-1}|_A(U)$, $U_B = f^{-1}|_B(U)$ and $W_A = U_A \cup (X - A)$. We show W_A is an open subset of X. The set $A - U_A$ is closed in A, and A is closed in X. Now we apply Proposition 2.20, and conclude that $A - U_A$ is closed in X. Therefore $W_A = X - (A - U_A)$ is open in X. We similarly define $W_B = U_B \cup (X - B)$, and show W_B is an open subset of X. We check that $f^{-1}(U) = W_A \cap W_B$. Then $f^{-1}(U)$ is

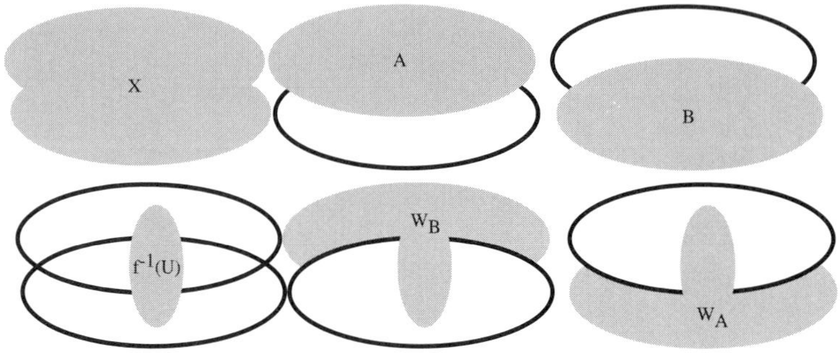

Figure 3-3 Figures for proof of gluing lemma for closed subsets, Proposition 3.45(*b*).

open in X since it is the intersection of two open subsets of X. Details of these set equalities are left as an exercise, (Problem 3.20). ∎

The proof above of part (*b*) leaves something to be desired. It is straightforward in its approach, but complicated in its detail. Even though we have a proof, it is a healthy attitude to wonder—isn't there a simpler way? There is, and we will outline this.

Let's try and analyze what makes the proof so awkward. The basic problem is that we are using the definition of continuity directly, and so we were looking to describe everything in terms of open subsets. Yet our hypothesis is about closed subsets. But, we can express continuity in terms of *closed* subsets, Proposition 3.43. You are invited to write a short alternative proof of part (*b*) of the proposition above (Problem 3.21), which uses Proposition 3.43.

Example 3.46 We must have some restrictions on the subsets A and B in Proposition 3.45. For example, let X, A, and B be subsets of the line; $X = [0, 2]$ $A = [0, 1)$, and $B = [1, 2]$. Define

$$f(x) = \begin{cases} 1 & \text{if } x \in A \\ 2 & \text{if } x \in B. \end{cases}$$

We see that $f|_A$ and $f|_B$ are continuous, but f is not.

The only problem is at the point $x = 1$. Removing this point will remove the problem in the following sense. If we let $X' = X - \{1\}$, $A' = [0, 1)$, and $B' = (1, 2]$, and define $f'(x)$ by

$$f'(x) = \begin{cases} 1 & \text{if } x \in A' \\ 2 & \text{if } x \in B'. \end{cases}$$

f' is a continuous function. Of course, A' and B' are open subsets of X'.

Here is another example. It shows that we have to be careful extending the gluing lemma to an infinite collection of subsets. Let f be a function from $\mathbf{R^n}$ to $\mathbf{R^m}$ which is *not* continuous. For any $x \in \mathbf{R^n}$, $\{x\}$ is a closed subset. We write $\mathbf{R^n}$ as a union of closed subsets: $\mathbf{R^n} = \bigcup_{x \in \mathbf{R^n}} \{x\}$. For all x and y in $\mathbf{R^n}$, $\{x\} \cap \{y\}$ is a closed subset, and $f|_{\{x\}}$ is continuous. ◆

Definition 3.47 *A continuous function $f: I \to \mathbf{R^n}$ is* **piecewise linear** *if there is a finite collection of numbers t_i with $0 = t_0 < t_1 < \cdots < t_n = 1$ and, for each interval $[t_i, t_{i+1}]$, $f|_{[t_i, t_{i+1}]}$ is a linear map.*

More generally, suppose there is collection of numbers, t_i, i is an integer, such that $\{t_i\}_{i \in \mathbf{Z}}$ has no limit points in $\mathbf{R^1}$ and such that if $i < j$, $t_i < t_j$. We say a continuous function, $f: \mathbf{R^1} \to \mathbf{R^n}$ is **piecewise linear** *if for each interval $[t_i, t_{i+1}]$, $f|_{[t_i, t_{i+1}]}$ is a linear map.*

It might be tempting to simplify the definition above in the case $f: \mathbf{R^1} \to \mathbf{R^n}$ by saying: "we can write $\mathbf{R^1}$ as a union of closed intervals and f is linear on each interval." However, we don't want to allow a function such as shown in Figure 3-4.

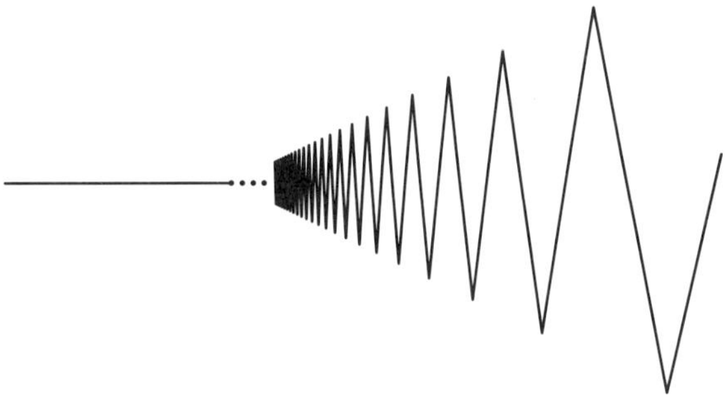

Figure 3-4 The function $f(x)$ shown is not a piecewise-linear function; see the remark on page 91 and Definition 3.47.

Here, $f(x) = 0$ if $x \leq 0$; otherwise, the graph of the restriction of f to $[1/(n+1), 1/n]$ is the line segment with endpoints $(1/(n+1), (-1)^{(n+1)}/(n+1))$ and $(1/n, (-1)^n/n)$.

Example 3.48 The gluing lemma, Proposition 3.45, implies that any piecewise-linear function is continuous, (Problem 3.22). ◆

3.7 Two more examples of continuous functions

As an illustration of use of the gluing lemma, consider a problem of defining a continuous function in polar coordinates. Suppose we have a continuous function $F\colon [0,\infty)\times[0,2\pi] \to \mathbf{R}^1\times[0,2\pi]$. We describe conditions under which the function F can be considered to be a continuous function in polar coordinates. For example, $F(r,\theta) = (r\cos(\theta),\theta)$ is certainly a continuous function of the Cartesian products $[0,\infty) \times [0,2\pi] \to \mathbf{R}^1 \times [0,2\pi]$. But, considered as a function in polar coordinates, it gives a map from $\mathbf{R}^2$ to $\mathbf{R}^2$.

In the next proposition we investigate the continuity of such a function. It is not as easy to prove as might first appear.

Proposition 3.49 *Suppose $F\colon [0,\infty) \times [0,2\pi] \to \mathbf{R}^1 \times [0,2\pi]$ is a continuous function such that*

(a) *$F(r,0) = F(r,2\pi)$, for all $r \in [0,\infty)$.*
(b) *$F(0,\theta) = 0$, for all $\theta \in [0,2\pi]$.*

Then F, considered as a map in polar coordinates, corresponds to a function $F'(r,\theta)$, which is a well-defined continuous map from $\mathbf{R}^2$ to $\mathbf{R}^2$.

Proof (Sketch): Conditions (a) and (b) are exactly what is needed to show that F' gives a well-defined map in polar coordinates. We will use the gluing lemma to show that $F'|_{(\mathbf{R}^2-\vec{0})}$ is continuous. What is left, then, is to show continuity of F' at the origin. This is done in Chapter 10, Example 10.16.

Consider the map $\gamma\colon [0,\infty) \times \mathbf{R}^1 \to \mathbf{R}^2$ defined by $\gamma(r,\theta) = (r,\theta)$ where we are using polar coordinates for $\mathbf{R}^2$. Next, avoiding the case $r = 0$, consider the restrictions of γ: $\alpha = \gamma|_{(0,\infty)\times[0,2\pi]}$ and $\beta = \gamma|_{(\mathbf{R}^1-\{0\})\times[0,2\pi]}$. Then we have the following diagram, which commutes in the sense that $\beta \circ F = F' \circ \alpha$.

$$\begin{array}{ccc}
(0,\infty)\times[0,2\pi] & \xrightarrow{F} & (\mathbf{R}^1-\{0\})\times[0,2\pi] \\
\big\downarrow \alpha & & \big\downarrow \beta \\
\mathbf{R}^2-\vec{0} & \xrightarrow{F'} & \mathbf{R}^2-\vec{0}.
\end{array}$$

It would seem that continuity of F should imply continuity of F'. It is tempting to say something like "$F' = \beta \circ F \circ \alpha^{-1}$ is a composition of continuous functions; thus F is continuous." The problem is that α is *not* one-to-one and so there is no function α^{-1}.

One way to resolve this problem is to write $\mathbf{R}^2 - \vec{0}$ as a union of two closed subsets and use the gluing lemma. Using polar coordinates, write

$$\begin{aligned} C &= \{(r,\theta) \in (\mathbf{R}^2 - \vec{0}) \mid 0 \le \theta \le \pi\} \text{ and} \\ K &= \{(r,\theta) \in (\mathbf{R}^2 - \vec{0}) \mid \pi \le \theta \le 2\pi\}. \end{aligned}$$

Each of C and K are homeomorphic to $R^2_+ - \vec{0}$. In Cartesian coordinates, C is the upper half of $\mathbf{R}^2 - \vec{0}$, and K the lower half. Clearly, C and K are closed subsets of $\mathbf{R}^2 - \vec{0}$ whose union is $\mathbf{R}^2 - \vec{0}$. Write $X_C = (0,\infty) \times [0,\pi]$ and $X_K = (0,\infty) \times [\pi, 2\pi]$. Then $\alpha|_{X_C}: X_C \to C$ and $\alpha|_{X_K}: X_K \to K$ are one-to-one functions; these functions and their inverses are continuous, (Problem 3.36). We now have the following diagram:

$$\begin{array}{ccc} (0,\infty) \times [0,2\pi] & \xrightarrow{F} & (\mathbf{R}^1 - \{0\}) \times [0,2\pi] \\ \uparrow (\alpha|_{X_C})^{-1} & & \downarrow \beta \\ C & \xrightarrow{F'|_C} & \mathbf{R}^2 - \vec{0}. \end{array}$$

We can check that $F'|_C = \beta \circ F \circ (\alpha|_{X_C})^{-1}$. Since $F'|_C$ is a composition of continuous functions, it is continuous. Similarly, we see that $F'|_K = \beta \circ F \circ (\alpha|_{X_K})^{-1}$ is continuous. ∎

We close this chapter with one more example of a continuous function.

Definition 3.50 *Let $X \subseteq \mathbf{R}^n$ and $p \in \mathbf{R}^n$. We define the* **distance from** p **to** X *to be $g.l.b._{x \in X}\delta(p,x)$. We denote this by $\delta(p,X)$.*

It is important to note that this notion of "distance" does not satisfy the basic properties such as those listed in Proposition 1.03: positive definite property, symmetry, and the triangle inequality.

Certainly we have $\delta(X,p) \ge 0$, but we have a problem with the second statement of the positive definite property. We generally expect that, if the distance between two things is zero, then they should be the same. If $p \in X$, then $\delta(X,p) = 0$. So $\delta(X,p) = 0$ does not imply that $p = X$ or even $\{p\} = X$. For example, in $\mathbf{R}^1$, if $X = (-1,1)$, then $0 \in X$ and so $\delta(0,X) = 0$. We note that $\delta(p,X) = 0$ does not imply that $p \in X$ since in our example $\delta(1,X) = 0$. In fact, it is not hard to see that if x is a limit point of X, then $\delta(x,X) = 0$, (Problem 3.29).

This notion is not symmetric: we do not define $\delta(X,p)$, only $\delta(p,X)$.

However a version of the triangle inequality does hold—for any three points of $\mathbf{R}^n$ p, q, and r, (Problem 3.30):

$$\delta(p,X) \le \delta(q,X) + \delta(r,X).$$

We will return to this topic in the remark on page 248.

Proposition 3.51 *Let X be a subset of $\mathbf{R^n}$. The function $f: \mathbf{R^n} \to \mathbf{R^1}$ defined by $f(p) = \delta(p, X)$ is continuous.*

Proof: We verify continuity of F, using the ϵ-δ formulation, Proposition 3.09.

Choose $\delta = \epsilon$. We show that, given an $\epsilon > 0$, if $\delta(p, q) < \epsilon$, then $|\delta(p, X) - \delta(q, X)| < \epsilon$.

We do this by showing that $|\delta(p, X) - \delta(q, X)| \leq \delta(p, q)$.

For all $x \in X$, we have $\delta(p, x) \leq \delta(p, q) + \delta(q, x)$. Using this we can argue that $\delta(p, X) \leq \delta(p, q) + \delta(q, X)$, (Problem 3.31). Similarly, we can argue that $\delta(q, X) \leq \delta(p, q) + \delta(p, X)$. Putting these together we conclude that $|\delta(p, X) - \delta(q, X)| \leq \delta(p, q)$. ∎

Example 3.52 We can use Proposition 3.51 to get some interesting open and closed subsets.

Let $X \subseteq \mathbf{R^n}$ and a a number with $0 < a$. Let $D_a(X) = \{p \in \mathbf{R^n} \mid \delta(p, X) = a\}$. Since, by Proposition 3.51, $\delta(p, X)$ is a continuous function, then $D_a(X)$ is a closed subset of $\mathbf{R^n}$.

Let X be the circle in $\mathbf{R^3}$ given by equations $x^2 + y^2 = 4$ and $z = 0$. The set of points $D_1 = \{p \in \mathbf{R^3}$ such that $\delta(p, X) = 1\}$ is a donut-shaped surface, called a "torus."Ł We give a definition of a torus later, Definition 13.24.

Let L be the subset of $\mathbf{R^3}$ consisting of all lines parallel to a coordinate axis which pass through a point with all integer coordinates; L is sometimes called the three-dimensional integer grid. In vector notation, L consists of all lines with vector equation $\vec{P} + t\vec{Q}$ where the coordinates of $\vec{P}$ are all integers and $\vec{Q}$ is one of the three vectors $\vec{i}$, $\vec{j}$, or $\vec{k}$.

The set $D_{1/4}(L)$ is an interesting infinite surface in $\mathbf{R^3}$. We leave it as an exercise to draw it, (Problem 3.32). ◆

*3.8 General topology and Chapter 3

We have considered two ways to generalize the notions of open subset for $\mathbf{R^n}$. The first is the setting of a general topological space, Definition 1.44; the second is the setting of a metric space 1.59. In each of these settings we generalize the notion of continuous function, Definition 3.01.

For general topological spaces, we define:

Definition 3.53 *If X and Y are topological spaces and if $f: X \to Y$ is a function, we say f is* **continuous** *if, for any open subset U of Y, $f^{-1}(U)$ is an open subset of X.*

A topological space is really a pair consisting of a set and a collection of subsets. If we have a function between two topological spaces and it is important to emphasize the topologies, we sometimes use notation such as $f:\{X,\mathcal{T}\} \to \{Y,\mathcal{T}'\}$. If the topologies are clear from context, we write $f:X \to Y$.

Example 3.54 In calculus, the derivative and integral can be considered as continuous functions. For example, consider the set of polynomials, P, restricted to the unit interval I. Let $\mathcal{T}$ denote the topology for P induced from $P \subseteq C(I,\mathbf{R}^1)$; see Example 1.73. The function $D:\{P,\mathcal{T}\} \to \{P,\mathcal{T}\}$ is defined by $D(p) = p'$ where p' denotes the derivative of p. From standard results of calculus, we can show that D is a continuous function.

For polynomial p, define $I(p)$ to be the polynomial $\int p\,dx$ whose value at 0 is 0. Then $I:\{P,\mathcal{T}\} \to \{P,\mathcal{T}\}$. One can check that I is a continuous function. Also, using the standard topology S for $\mathbf{R}^1$, one can define a function $E:\{P,\mathcal{T}\} \to \{\mathbf{R}^1,S\}$ by $E(p) = \int_0^1 p\,dx$ where $\int_0^1 p\,dx$ is the definite integral. One can show that this E is a continuous function. ◆

In the setting of metric spaces, one has a version of the ϵ-δ definition as in Definition 3.08.

Definition 3.55 *Suppose (X,d) and (Y,d') are metric spaces, $f:X \to Y$ is a function, and $x_0 \in X$. We say f is* **continuous at x_0 in the ϵ-δ sense,** *if for any number $\epsilon > 0$ there exists a $\delta > 0$ such that if $d(x,x_0) < \delta$, then $d'(f(x),f(x_0)) < \epsilon$. We say f is* **continuous in the ϵ-δ sense** *if it is continuous at every point of X.*

These definitions are compatible. If f is a continuous map of metric spaces in the ϵ-δ sense, then f is also a continuous map between the corresponding topological spaces obtained from the two metrics.

The basic properties of continuous functions $\mathbf{R}^n$, discussed in this text, are generally valid in the abstract setting.

If $\{S,\mathcal{T}\}$ is a topological space, then the identity map of S is a continuous function from $\{S,\mathcal{T}\}$ to $\{S,\mathcal{T}\}$. In general topology we have to be careful. The identity set function is not always continuous since we might have more than one notion of topology for a given set.

Example 3.56 Consider $\mathbf{R}^1$ with the standard topology S and the half-open topology, $\mathcal{H}$; see Example 1.46. The identity map $i:(\mathbf{R}^1,S) \to (\mathbf{R}^1,\mathcal{H})$ is not continuous, but the identity map $i:(\mathbf{R}^1,\mathcal{H}) \to (\mathbf{R}^1,S)$ is continuous. ◆

The general notion of continuity has the expected properties with regard to the general notion of product of topological spaces, using the product topology. The proofs need to be modified to take into account this product topology when one has an infinite product.

One can define the graph of a function $f: X \to Y$ as a subset of $X \times Y$; see the remark on page 82. We need to make some modification for the theorem that a graph of a function is a closed subset, Proposition 3.29, as the next example shows.

Example 3.57 Let $\{X, \mathcal{T}\}$ be the three-point non-Hausdorff space of Example 1.52 and let $\mathcal{D}$ be the discrete topology for X; see Definition 1.48. Let f be the identity map on the set X. This gives rise to a map $f: \{X, \mathcal{D}\} \to \{X, \mathcal{T}\}$. Since every subset of $\{X, \mathcal{D}\}$ is an open subset, f is continuous. Let Γ denote the graph of f; $\Gamma = \{(a, a), (b, b), (c, c)\}$. Then Γ is not a closed subset of $X \times X$ where we use the topology $\mathcal{D} \times \mathcal{T}$. One can verify this by listing the open subsets of $\mathcal{D} \times \mathcal{T}$—there are not that many. But the crux of the problem is that the only element of $\mathcal{T}$ which contains c is X. Thus, in $\mathcal{D} \times \mathcal{T}$, any open subset containing (a, c), which is not in Γ, must contain the point (a, a) which is in Γ. ◆

However, the problem above is relatively minor:

Proposition 3.58 *If $f: \{X, \mathcal{T}\} \to \{Y, \mathcal{T}'\}$ is a continuous map of topological spaces with Y a Hausdorff space, then the graph of f is a closed subset of $X \times Y$.* ▮

Concerning the quotient space construction, Definition 1.76, we should mention:

Definition 3.59 *Given a partition, P of X, the function from X to X/P that assigns to x the subset of P which contains x, is called the* **quotient map of** X/P

Proposition 3.60 *Suppose $\{X, \mathcal{T}\}$ is a topological space, and suppose P is a partition of X. Let $\mathcal{Q}$ be the quotient topology for X/P. The map $q: \{X, \mathcal{T}\} \to \{X/P, \mathcal{Q}\}$ defined by $y = q(x)$, if $x \in y$, is continuous.* ▮

3.9 Problems for Chapter 3

In Appendix D we list some results on continuity of real-valued functions such as Proposition D.11. You are expected to make use these results where appropriate, without need of proof, in subsequent exercises.

3.1 Prove Proposition 3.02. (Follow hint given in remark preceding statement of the proposition.)

3.2 Suppose that $f: X \to Y$. Show that f is continuous if and only if for all $N_\epsilon(y, Y)$, $f^{-1}(N_\epsilon(y, Y))$ is an open subset of X.

3.3 Prove Proposition 3.09.

3.4 Prove Proposition 3.12.

3.5 Prove Proposition 3.13. (Hint: You might want to use Proposition 2.45 in part of your proof since it relates open subsets in a product to open subsets in each factor.)

3.6 Prove Proposition 3.15.

3.7 Verify the claims made in the remark on page 79. (Hint: Suppose there were such an F; consider $F(0)$.)

3.8 Verify statements made in Example 3.17.

3.9 Prove Proposition 3.21.

3.10 Prove Proposition 3.23.

3.11 Verify the assertion of Example 3.26 that the results of Example 1.43 implies the statements about orbits.

3.12 Prove Proposition 3.29.

3.13 Prove the set G of Example 3.30 is a closed subset of the plane by showing that G contains all its limit points.

3.14 Suppose $f(z) = \frac{az+b}{cz+d}$ where $z, a, b, c,$ and d are complex numbers with $ad \neq bc$. Suppose $c \neq 0$ and let $z_0 = -\frac{d}{c}$. Viewing complex numbers as points of $\mathbf{R}^2$, show that $f: \left(\mathbf{R}^2 - z_0\right) \to \mathbf{R}^2$ is a continuous function, as in Example 3.33.

3.15 Prove Proposition 3.36.

3.16 Find an example of a function which is an open map and is not continuous. Find an example of a function which is a closed map and is not continuous.

3.17 Verify that f of Example 3.37 is a closed map.

3.18 Show, as claimed in Example 3.40: if $f(x, y) = (f_1(x), f_2(x))$, where $f_1(x)$ and $f_2(x)$ are linear functions, then f is an affine function.

3.19 Let T be the topologist's comb, Example 2.43. Define a map $p: T \to \mathbf{R}^1$ by $p(x, y) = x$. Is p an open map ? Explain.

3.20 Verify the claims in the proof of Proposition 3.45 that $U_A \cup (X - A) = X - (A - U_A)$ and that $f^{-1}(U) = W_A \cap W_B$.

3.21 Provide an alternative proof for Proposition 3.45 by verifying the continuity of f, using Proposition 3.43.

3.22 Prove the assertion of Example 3.48.

3.23 Suppose that $X \subseteq \mathbf{R^m}$, $Y \subseteq \mathbf{R^n}$ and that $f: X \to Y$. Reformulate the definition of continuity of f (Definition 3.08) in terms of neighborhoods of points in X and in Y, and prove that your reformulation is equivalent to the given definition.

3.24 Define a function $f: \mathbf{R}^1 \to \mathbf{R}^1$ by

$$f(x) = \begin{cases} x^2 & \text{if } x \text{ is rational} \\ x^3 & \text{if } x \text{ is irrational.} \end{cases}$$

Show that f is continuous only at 0.

3.25 Show that the graph of $y = \frac{1}{x}$ is a closed subset of $\mathbf{R}^2$.

3.26 Show that the graph of $y = \tan(x)$ is a closed subset of the plane. (Hint: You might want to start by considering $G^{-1}((-\infty, 0])$ where $G(x, y) = y - \arctan(x)$, and $H^{-1}([0, \infty))$ where $H(x, y) = y - \pi - \arctan(x)$; see Figure 3-5.)

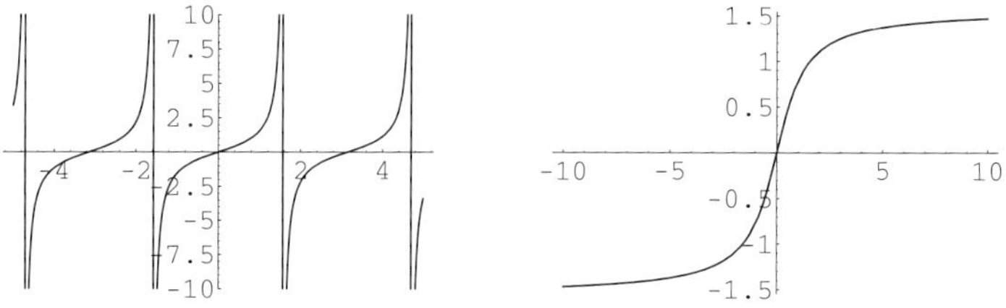

Figure 3-5 A portion of the graph of $\tan(x)$ and $\arctan(x)$; see Problem 3.26.

3.27 Show that the graph of $y = \sec(x)$ is a closed subset of the plane. (See the hint for Problem 3.26.)

3.28 Let $P(x, y) = 0$ be a polynomial equation in two variables. Show that the graph of this equation is a closed subset of the plane.

3.29 Verify the assertion that if x is a limit point of X, then $\delta(c, X) = 0$, as asserted in the remark on page 93.

3.30 Prove the triangle inequality mentioned in the remark on page 93.

3.31 Verify the assertion in the proof of Proposition 3.51 that $\delta(p, X) \leq \delta(p, q) + \delta(q, X)$.

3.32 Make a sketch of the set $D_{1/4}(L)$ of Example 3.52.

3.33 Suppose that $A \subseteq X \subseteq \mathbf{R^n}$, $Y \subseteq \mathbf{R^m}$, and $f: X \to Y$ is continuous. Show that if A is dense in X, then $f(A)$ is dense in $f(X)$.

3.34 Suppose that $A \subseteq X \subseteq \mathbf{R^n}$, with A dense in X, and $Y \subseteq \mathbf{R^m}$. Suppose $f: X \to Y$ and $g: X \to Y$ are two continuous functions, with $f|A = g|A$. Show that $f = g$.

3.35 A "contraction" is a function $f: \mathbf{R^n} \to \mathbf{R^n}$ such that there is a number ρ, $0 < \rho < 1$ so that, for all x and y, $|f(x) - f(y)| \leq \rho|x - y|$. Prove: If f is a contraction, then f is continuous.

3.36 Verify the assertions, in the proof of Proposition 3.49, that $\alpha|_{X_C}: X_C \to C$ and $\alpha|_{X_K}: X_K \to K$ and that their inverses are continuous.

3.37 Suppose we want to consider a function such as $F(r, \theta) = (r\cos(\theta), 2\theta)$ as giving a continuous map from of $\mathbf{R}^2$ in polar coordinates. Revise the statement of Proposition 3.49 so as to apply, and give a proof for continuity away from the origin.

3.38 Recall the Hawaiian earring, H, Example 2.42. Define a continuous function from the unit interval, I onto H, and prove that your function is continuous.

(Hint: Intuitively, the idea is to wrap a part of I about each of the circles that make up H. One of the technical difficulties is that the sum of the circumferences of these is infinite.)

3.39 Show that there is a continuous function f which maps the topologist's comb, X, Example 2.43, onto the Hawaiian earring H, Example 2.42, such that, for $i = 1, 2, \ldots$, $f(T_i) = C_i$.

4. HOMEOMORPHISM

OVERVIEW: The focus of this chapter is the concept of homeomorphism. It is our first and most basic answer to the question: When are two subsets of a Euclidean space the same? Our answer is: two subsets are the same if they are homeomorphic.

4.1 Homeomorphism and homeomorphism type

Suppose $X \subseteq \mathbf{R^n}$ and $Y \subseteq \mathbf{R^m}$. The concept of homeomorphism will capture a basic idea of "sameness." This is of fundamental importance, and we will devote much of our subsequent efforts trying to understand and apply this definition.

Consider the question: given any two sets A and B, when are they the same size, as sets? The answer is: when there is a one-to-one correspondence between A and B. (Definitions and discussions of the concepts of cardinality of sets appear in Appendix C.)

So, if two subsets of the plane are to be "the same," they must, *at least*, have the same size as sets. Our key additional requirement uses the concept of continuity as discussed in Chapter 3.

Definition 4.01 *Suppose $X \subseteq \mathbf{R^n}$ and $Y \subseteq \mathbf{R^m}$. A function $f: X \to Y$ is called a* **homeomorphism** *if*

(a) *f is a one-to-one correspondence between X and Y,*
(b) *f is continuous, and*
(c) *the inverse of f is continuous.*

Definition 4.02 *Suppose $X \subseteq \mathbf{R^n}$ and $Y \subseteq \mathbf{R^m}$. We say X is* **homeomorphic to** *Y, if there exists a homeomorphism from X to Y.*

Two subsets are "homeomorphic" if and only if there is a homeomorphism from one to the other. Among all subsets of Euclidean space, the relation X is homeomorphic to Y is an equivalence relation:

Proposition 4.03 *If $X \subseteq \mathbf{R^n}$, $Y \subseteq \mathbf{R^m}$ and $Z \subseteq \mathbf{R^k}$, then*

(a) *If X is homeomorphic to Y and Y is homeomorphic to Z, then X is homeomorphic to Z.*
(b) *If X is homeomorphic to Y, then Y is homeomorphic to X.*
(c) *X is homeomorphic to X.* ▮ *(Problem 4.1)*

Definition 4.04 *Consider the set $\mathcal{G}$ of all sets T such that T is a subset of some $\mathbf{R^n}$. Then X is homeomorphic to Y is an equivalence relation on $\mathcal{G}$. If X and Y are equivalent, we say that X and Y have the same* **homeomorphism type**. *Sometimes we use the phrase* X **and** Y **are topologically equivalent.**

In Definition 4.01(c) we refer to the *function* which is the inverse of f. Only one-to-one correspondences have inverse functions (see the remark on page 75), but condition (a) assures us that f is one-to-one.

Condition 4.01(c) is equivalent to: for any open subset U of X, $f(U)$ is an open subset of Y. As a consequence, we may say: a homeomorphism is a one-to-one correspondence which induces a one-to-one correspondence between open subsets.

By Proposition 3(b), we can replace the phrase "X is homeomorphic to Y" by the more symmetric "X and Y are homeomorphic."

Recall the notion of an open map, Definition 3.34. Condition (c) of Definition 4.01 is equivalent to f is an open map, assuming that f is a one-to-one-correspondence. Thus, *in the context of 4.01(a) and (b)*, we can replace 4.01(c) by "f is an open map." (But this does *not* mean we can generally replace continuity of f^{-1} by requiring that f be open since there are continuous maps which are open and not homeomorphisms; for example, projections as in Proposition 3.36).

If F is a non-singular affine transformation of $\mathbf{R^n}$, then F is a homeomorphism. The fact that F is a one-to-one correspondence, continuity of F, and its inverse have all been addressed in the proof of Proposition 3.41. Thus the special cases of rigid motion and similarity are also homeomorphisms.

Example 4.05 Consider the function $f: \mathbf{R}^2 \to \mathbf{R}^2$, defined by $f(x,y) = (x, x^2 + y)$. This is not a function of particular importance, but we use it as an illustration of use of the definition of homeomorphism. We show that f is a homeomorphism.

We need to verify that the three conditions (a), (b), and (c) of Definition 4.01 are satisfied. The easiest is condition (b) since the functions x and $x^2 + y$ are continuous functions by standard results from Appendix D, and Proposition 3.12.

We next verify f is a one-to-one correspondence. There are two parts to this: f is onto, and f is one-to-one. Let (a, b) be a point in the plane. To show f is onto, we need to find a point (x, y) such that $f(x, y) = (a, b)$. Thus we must have $(x, x^2 + y) = (a, b)$. This gives us two equations to be satisfied: $x = a$, and $x^2 + y = b$. We solve these equations for x and y : $x = a$ and $y = b - a^2$. We see that $f(x, y) = (a, b)$ since

$$f(x, y) = (x, x^2 + y) = (a, a^2 + (b - a^2)) = (a, b).$$

Next, we show that f is one-to-one by showing that $f(x_0, y_0) = f(x_1, y_1)$ implies $x_0 = x_1$ and $y_0 = y_1$. If $f(x_0, y_0) = f(x_1, y_1)$, then $(x_0, x_0^2 + y_0) = (x_1, x_1^2 + y_1)$. Thus $x_0 = x_1$ and $x_0^2 + y_0 = x_1^2 + y_1$. From these two equations we see that, indeed, we must have $x_0 = x_1$ and $y_0 = y_1$.

Finally, we must verify the third condition, that f^{-1} is a continuous function. We have basically done the work for this when demonstrating that f was onto. We showed that (a, b) corresponded to the point $(a, b - a^2)$. Thus we showed that $f^{-1}(a, b) = (a, b - a^2)$. We verify that this is the inverse of f by showing that $f^{-1} \circ f(x, y) = (x, y)$:

$$f^{-1} \circ f(x, y) = f^{-1}(x, x^2 + y) = (x, (x^2 + y) - x^2) = (x, y).$$

Since the functions a and $b - a^2$ are continuous, it follows that f^{-1} is continuous.

Thus, we have shown that f is a homeomorphism. A more general version of this example is found in (Problem 4.05).◆

We next give an example of a homeomorphism from an n-sphere minus a point and $\mathbf{R^n}$. See Figure 4-1 for the case of a 2-sphere minus a point. In describing the homeomorphism we identify $\mathbf{R^n}$ as a coordinate hyperplane of $\mathbf{R^{n+1}}$, corresponding to the first n coordinates. That is, we identify $\mathbf{R^n}$ with its image, under the standard inclusion map.

Proposition 4.06 *Let $S^n \subseteq \mathbf{R^{n+1}}$ be the standard n-sphere and let $P \in S^n$ be the point $P = (0, 0, \ldots, 0, -1)$.*

For $x \in S^n - \{P\}$ let l_x be the line in $\mathbf{R^{n+1}}$ containing x and P. Then $h(x)$ is defined to be the intersection of the line l_x and the hyperplane $\mathbf{R^n}$. The map $h\colon (S^n - \{P\}) \to \mathbf{R^n}$ is a homeomorphism called **stereographic projection**.

Proof: Suppose $x \in S^n$, $x = (x_1, x_2, \ldots, x_n, x_{n+1})$. In vector notation we can write l_x as $l_x = \vec{P} + t(\vec{x} - \vec{P})$. In term of coordinates this is

$$(tx_1, tx_2, \ldots, tx_n, -1 + t(x_{n+1} + 1)).$$

The intersection we want is the point on l_x where the last coordinate is zero. This happens if $t = \frac{1}{x_{n+1}+1}$.

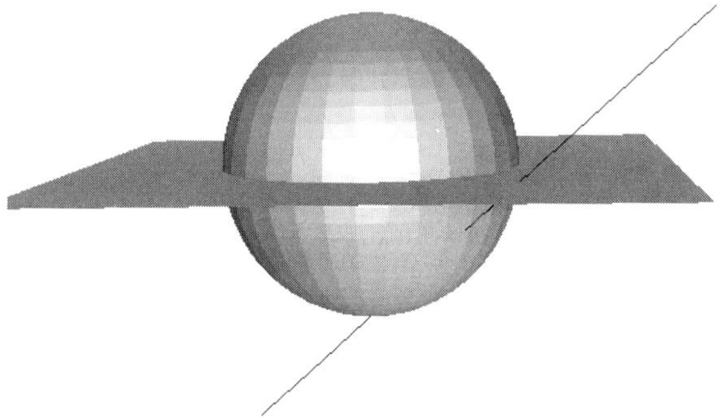

Figure 4-1 Figure for stereographic projection of a 2-sphere mapping minus the "south pole," homeomorphically, to the xy-coordinate plane of $\mathbf{R}^3$; see Proposition 4.06.

Thus we see that

$$h((x_1, x_2, \ldots, x_n, x_{n+1})) = (\frac{x_1}{x_{n+1}+1}, \frac{x_2}{x_{n+1}+1}, \ldots, \frac{x_n}{x_{n+1}+1}, \ 0).$$

Using this formulation, we can now verify that h is, in fact, a homeomorphism. ▮ (Problem 4.4)

Example 4.07 Let $Q \subseteq \mathbf{R}^1$ denote the set of rational numbers, and let $Q' = \{q \in Q \mid |q| < 1\}$. We will show that Q' and Q are homeomorphic.

Both Q' and Q are countably infinite subsets and so there exist one-to-one correspondences. However, we need a *continuous* one-to-one correspondence with continuous inverse. There are many homeomorphisms from $(-1, 1)$ to $\mathbf{R}^1$ but we would like one that sends Q' to Q. For example, $f(x) = \tan(\frac{\pi x}{2})$ is a homeomorphism from $(-1, 1)$ to $\mathbf{R}^1$ but this won't do since, for example, $f(1/3) = \tan(\frac{\pi}{6}) = \frac{\sqrt{3}}{3}$.

We will give two descriptions of a homeomorphism. In the first, we will describe a function using the binary decimal notation. We will also give a second simpler description of this function. Our motivation for use of the binary decimal notation at this point is as a "warm-up" for definitions used in the Chapter 5, as well as later in this text.

We will first focus on the positive rational numbers in Q' and Q; negative numbers will be handled in a similar fashion. Written in decimal form, rational numbers can be recognized as those which have a repeating pattern. The strategy is to define a map $\phi: Q' \to Q$ that maps:

- rational numbers in the interval [0, 1/2] to rational numbers in [0,1]

- rational numbers in the interval [1/2, 3/4] to rational numbers in [1,2]
- rational numbers in the interval [3/4, 7/8] to rational numbers in [2,3]
- in general, rational numbers in the interval $[1 - \frac{1}{2^n}, 1 - \frac{1}{2^{n+1}}]$ to rational numbers in $[n, n+1]$.

In binary decimal notation, these numbers $1 - \frac{1}{2^n}$ are written:

Fraction		Binary decimal
$\frac{1}{2}$	$\longleftrightarrow$	0.1
$\frac{3}{4}$	$\longleftrightarrow$	0.11
$\frac{7}{8}$	$\longleftrightarrow$	0.111

Take a number α in [0,1], and write it as a binary decimal with no terminal infinite string of 1's. Then the expression is of the form "a sequence of N 1's to the right of the decimal, with a zero in the $(N+1)$-st position, followed by a binary string, x." We map α to the number which has binary expression $X.x$ where X is the binary expression for N.

For example, $\phi(0.11111010101010\ldots) = 101.1010101010\ldots$. For negative numbers we use the obvious definition so that, for example, $\phi(-0.11111010101010\ldots) = -101.1010101010\ldots$.

Here is another way of writing ϕ. Let Φ be the piecewise-linear function defined by $\Phi(1 - \frac{1}{2^n}) = n$, $\Phi(-(1 - \frac{1}{2^n})) = -n$ and $\Phi(0) = 0$; see Figure 4-2. We claim that $\Phi|_Q = \phi$. These functions agree on points of the form $\pm(1 - \frac{1}{2^n})$, thus we need only verify that ϕ is linear between these values.

For example, consider the number x which, written in binary form, is $0.11111010101010\ldots = 0.11111 + 0.00000010101010\ldots$. The number x, written in decimal form, is a number between $1 - \frac{1}{32} = \frac{31}{32}$ and $1 - \frac{1}{64} = \frac{63}{64}$. In the decimal system, $\Phi(\frac{31}{32}) = 5$, $\Phi(\frac{63}{64}) = 6$, and Φ is linear on the interval $[\frac{31}{32}, \frac{63}{64}]$ with slope 64. As we have defined it,

$$\phi(0.11111010101010\ldots) = \phi(0.11111) + \phi(0.00000010101010\ldots).$$

So we consider $\phi(0.00000010101010\ldots) = 0.10101010\ldots$; the effect of ϕ can be described as shifting the decimal point six places to the right. In the binary system, the process of shifting the decimal point one place to the right corresponds to multiplication by 2; shifting 6 places to the right corresponds to multiplication by $2^6 = 64$. Thus we see that ϕ is indeed piecewise linear.

We leave it as an exercise to verify that ϕ is a homeomorphism from Q' to Q (Problem 4.8). ◆

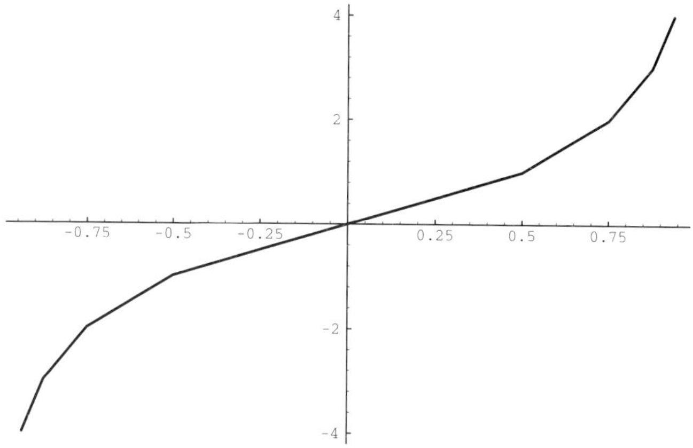

Figure 4-2 A portion of the graph of the piecewise-linear function Φ; see Example 4.07.

In calculus (for example, in Green's Theorem) one studies circular paths in the plane, usually defined as simple closed curves. Here is a definition frequently found in calculus books:

Definition 4.08 *Let $X \subseteq \mathbf{R^n}$ and $[a, b]$ a closed interval. A simple closed curve in X is a continuous function $f: [a, b] \to X$ such that $f(x) \neq f(y)$, except that $f(a) = f(b)$.*

Example 4.09 Using the idea of homeomorphism, we can connect the definition of a simple closed curve with the notion of a circle. Let $X \subseteq \mathbf{R^n}$, and suppose f is a simple closed curve in X. Then the image of $f(I)$ is homeomorphic to a circle (Problem 4.9). Also, if $C \subseteq \mathbf{R}^2$ and C is homeomorphic to a circle, then there is a simple closed curve whose image is C. Suppose $h: S^1 \to C$ is a homeomorphism. Define: $f: I \to S^1$ by $f(t) = (\cos 2\pi t, \sin 2\pi t)$ where $I = [0, 1]$. Then $h \circ f$ will be a closed simple path in the plane, with $C = h \circ f(I)$. ◆

The following theorem, although intuitively obvious, is very difficult to prove—so difficult that we will not attempt to do so in this text. (There are two basic methods used to prove the Jordan Curve Theorem—a basic topological approach can be found in [2, 10, 35, 43]; approaches using methods of algebraic topology can be found in [32, 33, 36] or almost any text in algebraic topology. Yet it is such a basic theorem about the topology of the plane that we feel you should know and understand the statement. We begin with a rough statement of the theorem, which we will later refine. (The "roughness" lies in the current lack of definition of "inside" or "outside.")

Proposition 4.10 Jordan Curve Theorem: Rough Version 1 *Let C be a simple closed curve in the plane. Then* $\mathbf{R}^2 - C$ *is the union of two disjoint connected subsets— namely, the points "inside C" and the points "outside C."* ∎

As a consequence of this we can restate the Jordan Curve Theorem, Proposition 4.10, replacing the term "simple closed curve" by "a subset homeomorphic to a circle":

Proposition 4.11 Jordan Curve Theorem: Rough Version 2 *Let C be a subset of the plane homeomorphic to a circle. Then* $\mathbf{R}^2 - C$ *is the union of two, disjoint, connected subsets— namely the points "inside C" and the points "outside C."* ∎

We will give a complete version of the Jordan Curve Theorem, as Proposition 7.20.

4.2 Refining basic questions about $\mathbf{R}^n$

One of our stated basic goals is to understand subsets of the plane (Problem 0.2). One simple question along these lines is:

Question 4.12 How many subsets of the plane are there?

As stated, this is not a difficult question. The answer is: there are uncountably many. There are uncountably many points in the plane; the collection of all one-point subsets of the plane is uncountable, so the collection of all subsets of the plane is an uncountable.

However, this is an unsatisfactory answer since all one-point sets are, topologically, the same. More precisely, all one-point subsets are homeomorphic to each other. In fact, we can, more generally, show that:

Proposition 4.13 *Suppose* $X \subseteq \mathbf{R}^n$ *and* $Y \subseteq \mathbf{R}^m$ *where X and Y are finite sets. Then X and Y are homeomorphic if and only if they have the same number of points.*

Proof: Certainly if they are homeomorphic, they must have the same number of points—this follows from the "one-to-one correspondence" condition in the definition of homeomorphism, Definition 4.01.

Next, assume they have the same number of points—then there must be a one-to-one correspondence $f: X \to Y$. This f is a homeomorphism. Continuity of f, and its inverse, follow from the fact that all subsets of X and all subsets of Y since X and Y are finite, are open subsets, Proposition 2.23. ∎

The fact that Question 4.12 has such a simple answer does not mean we should abandon the line of inquiry, but rather that we need to refine the question to make the answer more interesting. But first we explore the concept of homeomorphism.

Example 4.14 From Proposition 4.13 we can begin to get a sense of how topology differs from classical geometry. For example, let $X = \{(-1,1),(0,0),(1,1)\}$ and $Y = \{(-1,0),(0,0),(1,0)\}$. These subsets are equivalent in the sense of homeomorphism type, yet the points of X determine a triangle and those of Y lie on a straight line. In geometry, the concept of colinearity of points is fundamental. In topology, it is not significant. ◆

Finite subsets are important, but they are very special. For example, one way to reformulate Proposition 4.13 is to say that any one-to-one correspondence between finite sets is a homeomorphism. However, it is not generally the case that any one-to-one correspondence is a homeomorphism, as the following example shows.

Example 4.15 Let N and G be the subsets of the plane: $N = \{(1,0),\ (2,0),\ldots\}$ and $G = \{(0,0),(1/1,0),(1/2,0),(1/3,0),\ldots\}$. Since N and G are infinite and countable, there is a one-to-one correspondence between these sets. We show G and N are not homeomorphic.

Before we begin a proof, let us give some thought to strategy. We need to show that a homeomorphism between G and N does not exist. As a general rule, the strategy used to show that something does not exist entails an argument by contradiction. So we assume that a homeomorphism from G to N exists, and we hope to contradict something. That "something" is likely to be one of the three properties that make up the definition of homeomorphism: one-to-one correspondence, continuity, continuity of the inverse function.

There *does* exist a one-to-one correspondence between these sets. So the contradiction must lie in the continuity of such a function, or its inverse.

To investigate continuity, we need to know about the open subsets of G and the open subsets of N. At this point one should draw pictures of these two sets and find examples of open subsets. Notice that the one-point subset $\{(1,0)\}$ is an open subset of N since $N_1((1,0)) \cap N = \{(1,0)\}$. Similarly *any* one-point subset of N is an open subset of N. Since any subset is a union of its one-point subsets, and since the union of open subsets is an open subset, it follows that *any* subset of N is an open subset. This implies that *any* function from N to G is continuous! Thus our contradiction must somehow be obtained by showing that a one-to-one correspondence from G to N could not be continuous.

Let's look at the open subsets of G. Note that $\{(1,0)\}$ is an open subset of G since $N_{1/2}((1,0)) \cap G = \{(1,0)\}$. One similarly can argue that any one-point subset, $\{(1/i,0)\}$, is an open subset of G since $N_{1/i(i+1)}((1/i,0)) \cap G = \{(1/i,0)\}$. All that remains are open subsets of G which contain $(0,0)$. Here things are different: $\{(0,0)\}$ is not an open subset of G. Now that we have found the ingredients of the proof, we are ready to write:

Proof of Example 4.15: Suppose there is a homeomorphism, $f: G \to N$. Consider the point of $n_0 \in N$ where $n_0 = f((0,0))$. Such a point must exist since f is an onto function. Note that $N_1(n_0, N) = \{n_0\}$. Thus $\{n_0\}$ is an open subset of N. Since f is one-to-one, $f^{-1}(\{n_0\})$ is the single point $\{(0,0)\}$. Since f is continuous, $f^{-1}(\{n_0\})$ is an open subset of G. But the set $\{(0,0)\}$ is not an open subset of G.

We prove this last statement by showing that: if U is an open subset of X which contains (0,0), then U must also contain other points of X. Since $(0,0) \in U$, there is an $\epsilon > 0$ with $N_\epsilon(0,0) \subseteq U$. By Proposition B.09, we can find an integer N such that $\frac{1}{N} < \epsilon$. But then $(\frac{1}{N}, 0) \in U$.

Therefore, G and N are not homeomorphic. ◆

Recall that Proposition 3.44 allowed us to express continuity in terms of limit points. Putting this together we obtain the next proposition, which says that a homeomorphism sends limit points to limit points.

Proposition 4.16 *Suppose $X \subseteq \mathbf{R^n}$, $Y \subseteq \mathbf{R^m}$ and $f: X \to Y$ is a homeomorphism. If $x \in X$, x is a limit point of X if and only if $f(x)$ is a limit point of Y.* ▮ *(Problem 4.10)*

For example, consider the proof of Example 4.15. The key step is to focus on the point $(0,0)$ of G. This point is a special point of G; it is the only limit point of G. On the other hand, no point of N is a limit point of N. Thus by Proposition 4.16 there could not be a homeomorphism between N and G.

Another useful fact is that a restriction of a homeomorphism gives a homeomorphism:

Proposition 4.17 *Suppose $X \subseteq \mathbf{R^n}$, $Y \subseteq \mathbf{R^m}$ and $f: X \to Y$ is a homeomorphism with $A \subseteq X$. Then $f|_A$ is a homeomorphism from A to $f(A)$.* ▮ *(Problem 4.11)*

We return to the question of how many subsets of the plane there are, Question 4.12. We want to consider homeomorphic subsets as "the same." Thus we refine Question 4.12 as:

Question 4.18 How may different subsets of the plane are there, if we consider homeomorphic subsets to be the same?

For *any* natural number $n \in \mathbf{N}$ let S_n be the set of all subsets of the plane with n points. The collection of subsets $\{S_n\}_{n\in\mathbf{N}}$ is an infinite collection.

We now have an answer to Question 4.18: there is an infinite collection of subsets of the plane, no two of which are homeomorphic.

There is something bothersome about this answer. It is based on examination of very simple subsets, and does not consider the variety of possibilities that lie in the plane. In fact, the same analysis shows that there is an infinite collection of subsets of the line, no two of which are homeomorphic.

Perhaps we are not asking the right question. By Proposition 4.13, $\{S_n\}_{n\in N}$ is a *countable* collection. Thus we might ask:

Question 4.19 Is there an *uncountable* collection of subsets of the plane, no two of which are homeomorphic ?

Now we have an interesting but tough question. The posing of this question is a high point of our discussion thus far. Another high point will come when we answer this question; see Proposition 7.28. We first need to look at many more examples and also lay the groundwork for the proof of this proposition.

4.3 Intervals and homeomorphisms

Intervals are important subsets of $\mathbf{R}^1$. The next proposition says that any two intervals of the same type are homeomorphic. Since different intervals may have different lengths, we once again see that this notion of sameness (homeomorphism) contrasts with basic notions of sameness (congruence) of geometry.

Proposition 4.20

(a) *Any two open intervals in* $\mathbf{R}^1$ *are homeomorphic.*
(b) *Any two non-degenerate closed intervals in* $\mathbf{R}^1$ *are homeomorphic.*
(c) *Any two half-open intervals in* $\mathbf{R}^1$ *are homeomorphic.*
(d) *An open interval is homeomorphic to* $\mathbf{R}^1$.
(e) *A half-open interval is homeomorphic to* R^1_+.

Proof: We first prove part (a). Suppose (a, b) and (c, d) are two open intervals. Define a map from $\mathbf{R}^1$ to $\mathbf{R}^1$ by

$$f(x) = \frac{d-c}{b-a}x + c - \frac{d-c}{b-a}a, \text{ for all } x \in \mathbf{R}^1\text{; see Figure 4-3.}$$

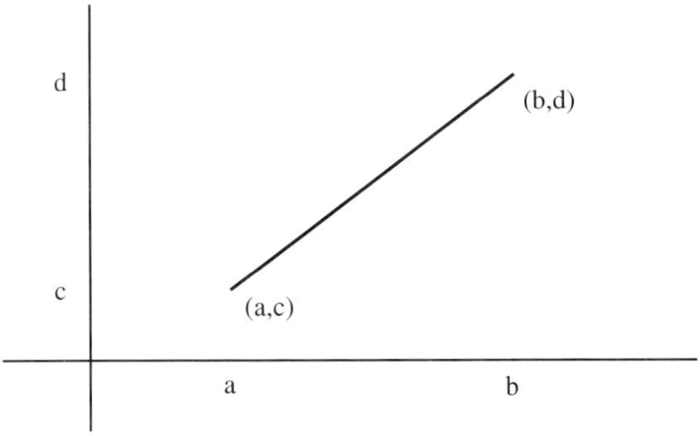

Figure 4-3 The graph of function f, used in the proof of Proposition 4.20, that any two open intervals are homeomorphic.

(To derive the formula for f, calculate the equation of the line that passes through the points (a, c) and (b, d).) Note that $f(a) = c$ and $f(b) = d$. This is a homeomorphism of $\mathbf{R^1}$. It is easy enough to verify this directly, or we might note that f is a non-singular affine map and use Proposition 4.1. The restriction of f to (a, b) is a homeomorphism from (a, b) to (c, d) by Proposition 4.17. Proof of part (b) follows similarly by considering $f|_{[a,b]}$.

For part (c), if the given intervals are of the form $[a, b)$ and $[c, d)$, or if they are of the form $(a, b]$ and $(c, d]$, the proof follows by restricting f. However in the remaining two cases where the intervals are $[a, b)$ and $(c, d]$, or $(a, b]$ and $[c, d)$, we need to use the function $F = \frac{c-d}{b-a}x + d - \frac{c-d}{b-a}a$ in place of the function f. Note that $F(a) = d$ and $F(b) = c$. (Here F is an equation for the line which passes through the points (a, d) and (b, c).)

Part (d) can be shown by using the function $\tan(x)$ for the homeomorphism. This maps $(-\pi/2, \pi/2)$, in a homeomorphic way, onto the whole line. From part (b) it follows that any other open interval is homeomorphic to $\mathbf{R^1}$.

Proof of part (e) is obtained by restricting the domain of $\tan(x)$ to $[0, \pi/2)$. ∎

Proposition 4.20 (d) has the following generalization:

Proposition 4.21 *If B is an open n-ball in $\mathbf{R^n}$, it is homeomorphic to* $\mathbf{R^n}$.

Proof: All open n-balls are homeomorphic (since they are similar), so it is sufficient to show that $N_1(\vec{0})$ is homeomorphic to $\mathbf{R^n}$. The map,

$f: N_1(\vec{0}) \to \mathbf{R^n}$ defined by

$$f(\vec{x}) = \left(\frac{1}{1 - |\vec{x}|}\right) \vec{x},$$

is such a homeomorphism (Problem 4.21). ∎

Propositions 4.23 and 4.24 reveal some special properties of continuous maps and homeomorphisms of the line and of closed intervals. First, we need a definition.

Definition 4.22 *Suppose* $X \subseteq \mathbf{R}^1$ *and* $f: X \to \mathbf{R}^1$. *We say* f *is a* **strictly increasing function** *if, for all* $a, b \in X$, *if* $a < b$, *then* $f(a) < f(b)$. *We say* f *is a* **strictly decreasing function** *if for all* $a, b \in X$, *if* $a < b$, *then* $f(a) > f(b)$. *If a function is either strictly increasing or strictly decreasing, we say* f *is a* **strictly monotone function**.

The definitions above are special because they are based on an order of points. Subsets of the plane or higher-dimensional spaces do not have a similar natural concept of order.

In calculus one frequently finds the notion of an increasing function: "f is increasing" if $a < b$, then $f(a) \leq f(b)$. We will not need this idea and thus use the word "strictly" to distinguish our usage, emphasizing our use of "$<$" as opposed to "$\leq$."

We will use this definition, and the two propositions which follow, in our examination of examples. Useful as these are for the examples of this book, we do not want to leave the impression that monotone functions are an important part of topology. They relate only to real-valued functions of a single real variable.

If $f: X \to Y$ is strictly monotone and $A \subseteq X$, then $f|_A$ is also strictly monotone.

In the next proposition, for parts (d) and (e), recall that when we say "X is an interval" we mean to include the possibilities of infinite intervals, as in Definition 1.11.

Proposition 4.23

(a) *If* $f: \mathbf{R}^1 \to \mathbf{R}^1$ *is continuous and one-to-one, then* f *is a strictly monotone function.*
(b) *If* f *is a continuous, strictly monotone function from a closed interval* $[a, b]$ *to* $\mathbf{R}^1$, *then* $f([a, b])$ *is the closed interval with endpoints* $f(a)$ *and* $f(b)$.
(c) *If* $f: \mathbf{R}^1 \to \mathbf{R}^1$ *is continuous, one-to-one and onto, then* f *is a homeomorphism.*
(d) *If* X *is an interval in* $\mathbf{R}^1$ *and* $f: X \to \mathbf{R}^1$ *is continuous and one-to-one, then* f *is a strictly monotone function.*
(e) *If* X *is an interval in* $\mathbf{R}^1$ *and* $f: X \to X$ *is continuous, one-to-one and onto, then* f *is a homeomorphism.*

Proof of part (a): We argue by contradiction. Suppose there is a one-to-one continuous map f of $\mathbf{R}^1$ which is not strictly monotone. If f is one-to-one and is not monotone, then there are two possibilities. In the first case, the function "goes up, then goes down." In the second, the function "goes down, then goes up." More precisely, in first case there are numbers A, B, and C, with $A < B < C$ and $f(A) < f(B)$, $f(B) > f(C)$. In the second case there are numbers A, B, and C, with $A < B < C$ and $f(A) > f(B)$, $f(B) < f(C)$ (Problem 4.22).

Consider the first case: $f(A) < f(B)$, $f(B) > f(C)$. Let $M_0 = \max(f(A), f(C))$. Note that $M_0 < f(B)$. Let M be the number midway between M_0 and $f(B)$; that is, $M = M_0 + \frac{f(B)-M_0}{2}$. We now have $f(A) < M < f(B))$ and $f(C) < M < f(B))$. Now consider the restriction $f|_{[A,B]}$. We apply the Intermediate Value Theorem, Proposition D.14, to $f|_{[A,B]}$ with M as intermediate between $f(A)$ and $f(B)$. We find a number x with $A \leq x \leq B$ such that $f(x) = M$. Since $M \neq f(B)$, we know that $x \neq B$, and thus $x \in [A, B)$. We may similarly apply the Intermediate Value Theorem, to $f|_{(B,C]}$ and the value M, and conclude there exists a number y with $y \in [B, C]$ such that $f(y) = M$. However, now we have two numbers, x and y, with $x \neq y$, and $f(x) = f(y)$; see Figure 4-4. This contradicts the assumption that f was one-to-one. Clearly the argument for the second case is similar (Problem 4.23). This finishes the proof of part (a).

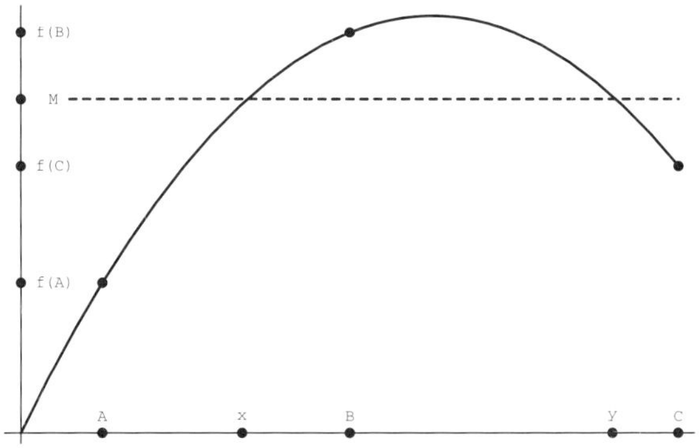

Figure 4-4 Figure for proof of Proposition 4.23(a) for the first case, $f(A) < f(B)$, $f(B) > f(C)$.

Before continuing in this proof, we remark that we used the Intermediate Value Theorem in an essential way. We needed ,somehow, to show the existence of a point, $x \in \mathbf{R}^1$ which gives us our contradiction. Since the hypotheses are very general ones, we find ourselves in need of a very general result which shows the existence of our point. The Intermediate Value Theorem is one of the few such results at our disposal. In this sense we say that our use of this theorem is "natural." We have not yet given a proof of the Intermediate Value Theorem; a proof can be found in most calulus texts. However, a proof of Intermediate Value Theorem is forthcoming; see the remark on page 185.

We leave part (b) as an exercise, as well as parts (d) and (e) (Problem 4.24).

To prove part (c) we first note that since f is a one-to-one correspondence, it has an inverse function defined. We need to establish the continuity of f^{-1}. By part (a), f is a strictly monotone function. By part (b), if $[a, b]$ is any closed interval in the line, $f([a, b])$ is a closed interval with endpoints $f(a)$ and $f(b)$. Considering the corresponding open intervals, it follows, $f((a, b))$ is the open interval with endpoints $f(a)$ and $f(b)$. This implies that f is an open map. To see this, let U be an open subset of $\mathbf{R}^1$. We can then write U as the union of intervals (using, say, Proposition 1.25 since ϵ-neighborhoods in $\mathbf{R}^1$ are open intervals.). Then $f(U)$ is the union of the images of these intervals which, being the union of open subsets, is an open subset. ▮

In addition, we can make the following conclusions regarding what happens at the endpoints of intervals.

Proposition 4.24

(a) *If $[a, b]$ and $[c, d]$ are closed intervals in $\mathbf{R}^1$ and if f is a homeomorphism from $[a, b]$ to $[c, d]$, then $f(a)$ is either c or d. Furthermore if $f(a) = c$, then $f(b) = d$; if $f(a) = d$, then $f(b) = c$.*

(b) *If $[a, b)$ and $[c, d)$ are half-open intervals in $\mathbf{R}^1$ and if f is a homeomorphism from $[a, b)$ to $[c, d)$, then $f(a) = c$.* ▮ *(Problem 4.25)*

4.4 The circle and the half-open interval

We have seen, in Example 4.15, a continuous one-to-one function whose inverse is not continuous. Example 4.27 provides another example. First we note the following fact from trigonometry.

Proposition 4.25 *Let S^1 denote the unit circle in the plane; let $X \in \mathbf{R}^1$ be the half-open interval $[0, 2\pi)$. The map $f: X \to S^1$, defined by $f(x) = (\cos(x), \sin(x))$, is a one-to-one correspondence.* ▮

Example 4.26 Let S^1 denote the unit circle in the plane and $X = [0, 2\pi)$. Consider $f: X \to S^1$ defined by $f(x) = (\cos(x), \sin(x))$;

this is continuous, by Proposition 3.12, and since $\sin(x)$ and $\cos(x)$ are continuous. Also, f is a one-to-one correspondence by Proposition 4.25. However, we show that f^{-1} is not continuous.

We argue by contradiction. Let $U = \{x \in X: 0 \leq x < \pi\}$; U is an open subset since $U = X \cap N_{\pi}(\{0,0\})$. If f^{-1} is continuous then $f(U)$ is an open subset of S^1. However, this is not the case.

To see that $f(U)$ is not an open subset of S^1, consider the point $p = (1,0)$. Then $p \in S^1 \subseteq f(U)$. If $f(U)$ is an open subset, there is an ϵ such that $N_{\epsilon}(p, S^1)$ is entirely be contained in S^1. Let $L = S^1 - f(U)$. Since f is a one-to-one correspondence, $S^1 - f(U) = f(X - U)$. From what we know about the geometry of the circle (the circumference of the circle is 2π) and the definitions of the sine and cosine functions, we see that L is the lower "half" of the circle; that is, $L = \{(x,y) \in S^1: y < 0\} \cup \{(-1,0)\}$. However, any $N_{\epsilon}(p, S^1)$ must contain some points $(x,y) \in S^1$ with y negative. Such points are points of L. Thus we see that f^{-1} is not continuous. ◆

Example 4.27 In Example 4.26, we have only shown that the given function is not a homeomorphism. We would like to show something stronger: that X and S^1 are not homeomorphic. That is, not only does the "obvious" candidate function fail to be a homeomorphism, but that all possible attempts to construct a homeomorphism must fail.

We will, in fact, offer several ways to show this. Our first attempt is try to adapt the reasoning found in the example above. This straightforward approach, although not difficult, does not lead to a particularly "nice" proof. However, in the analysis, we examine examples of open subsets and interesting functions from X to S^1 and that gives a better understanding of the problems involved.

Suppose there were some homeomorphism $f: X \to S^1$. As before, we center our attention on the point $f(0)$; let $c_0 = f(0)$. We wish to choose an open subset of X similar to the U above, whose image lies within a semicircle. Then one could use the argument to show that $f(U)$ cannot be an open subset of S^1.

Unlike the situation in Example 4.26, we are not working with a specified function. There we knew what $f(U)$ was since we were given f. In the present case, we do not know exactly what $f(U)$ is. In particular, $f(U)$ might be a large set in that it consists, in some sense, of most points of the circle. It is helpful at this point to look at some examples of functions from X to S^1, which are continuous one-to-one correspondences, to see some possibilities for f.

Here is a collection of functions from X to S^1. Consider any homeomorphism β where $\beta: [0, 2\pi) \to [0, 2\pi)$. Then $f'(x) = (\cos(\beta(x)), \sin(\beta(x)))$, is a continuous one-to-one correspondence. There are many such functions. For example, we could choose $\beta(x) = (2\pi x)^{1/2}$ or, more generally, let $\beta_n(x) = ((2\pi)^n x)^{\frac{1}{n+1}}$; see Figure 4-5.

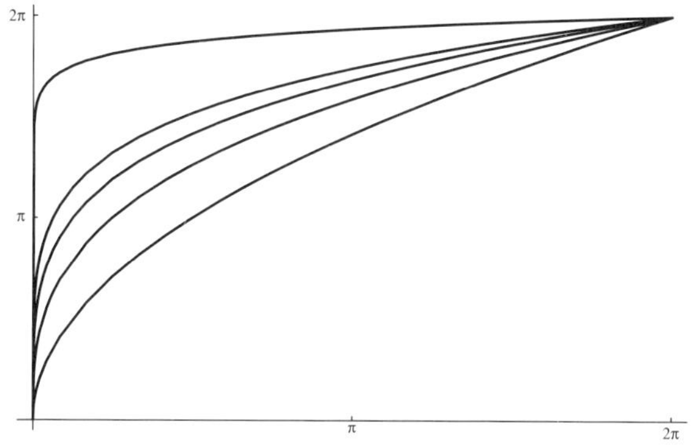

Figure 4-5 Graphs of some functions $\beta_n(x)$: (from lower to upper) of $\beta_1 = (2\pi x)^(1/2)$, $\beta_2 = ((2\pi)^2 x)^{1/3}$, $\beta_3 = ((2\pi)^3 x)^{1/4}$, $\beta_4 = ((2\pi)^4 x)^{1/5}$, and $\beta_{20} = ((2\pi)^{20} x)^{1/21}$; see proof for Example 4.27.

For large values of n we see that $f'([0,\pi))$ becomes much larger than a semicircle (Problem 4.27). There are several ways to resolve this problem. The one we try here is to replace the interval $U = [0,\pi)$ of Example 4.26 by another open interval.

Recall $c_0 = f(0)$. Write $c_0 = (\cos(\theta_0), \sin(\theta_0))$; the point diametric to c_0 is $c_0' = (\cos(\theta_0 + \pi), \sin(\theta_0 + \pi))$. Let $x_0' = f^{-1}(c_0')$. Our plan is to let $U = [0, x_0')$ and show that $f([0, x_0'])$ is contained in a semicircle S with endpoints c_0 and c_0'. (In fact, we could prove that $f([0, x_0']) = S$, but we don't need this for our proof.)

There are two semicircles with endpoints c_0 and c_0', so we need to deal with this uncertainty. Let z be a number with $x_0' < z < 2\pi$. For example, we could choose the midpoint $z = 2\pi - \frac{2\pi - x_0'}{2}$. Let $a = f(z)$ since $a \notin f(U)$. The strategy is to choose the semicircle which does *not* contain a. Also, if we remove a from S^1, we can analyze the situation more easily.

Suppose $a = (\cos(\theta_a), \sin(\theta_a))$; then every point of S^1 can be written as $(\cos(\theta_a + \phi), \sin(\theta_a + \phi))$ where $0 < x < 2\pi$. Thus we have a homeomorphism $h\colon (S^1 - \{a\}) \to (0, 2\pi)$ where $h(\cos(\theta_a + \phi), \sin(\theta_a + \phi)) = \phi$.

Consider $H = h \circ f|_{[0,x_0']}$; this is continuous and one-to-one since each of the functions is. Let $\phi_0 = h(c_0)$ and $\phi_0' = h(c_0')$.

Then H is a monotone function $H(0) = \phi_0$ and $H(x_0') = \phi_0'$. It follows that $H([0, x_0'])$ is contained in the interval J, with endpoints ϕ_0 and ϕ_0'. If not, then there would be a w with $0 < w < x_0'$ with either $H(w) < \min(\phi_0, \phi_0')$ or $H(w) > \max(\phi_0, \phi_0')$, either of which would contradict the monotonicity of H. The image $h^{-1}(J)$ is a semicircle which contains $f(U)$; and we may finish our proof modeled on the proof in Example 4.26, by showing that $f(U)$ is not an open subset of S^1 (Problem 4.28). ◆

Example 4.28 Here is another proof that the interval $[0, 2\pi)$ is not homeomorphic to the circle S^1, Example 4.27.

Let U be the upper semicircle and L be the lower semicircle. That is, $U = \{(x, y) \in S^1 : 0 \le y\}$, and let $L = \{(x, y) \in S^1 : y \le 0\}$. Of course, $S^1 = U \cup L$.

Consider two maps, g_1 and g_2, from the interval [-1, 1] to $\mathbf{R^1}$ defined by $g_1(x) = (x, \sqrt{1 - x^2})$ and $g_2(x) = (x, -\sqrt{1 - x^2})$. Note that $Im(g_1) = U$ and $Im(g_2) = L$. Suppose there is homeomorphism $g: X \to S^1$. Then the maps G_1 and G_2 defined by $G_1 = g^{-1} \circ g_1$ and $G_2 = g^{-1} \circ g_2$ are continuous maps from a closed interval to $[0, 2\pi)$. Thus each has a maximum, Proposition D.13. Suppose the maxima of G_1 and G_2 are a_1 and a_2, respectively. Let $a = \max(a_1, a_2)$. Consider $w = a + \frac{2\pi - a}{2}$; this is the point midway between a and 2π, so $0 < a < w < 2\pi$. We claim $g(w)$ is not a point of S^1 or, equivalently, $w \notin g^{-1}(S^1)$. This would contradict the assumption that g has domain $[0, 2\pi)$. To verify this claim, note that the images of G_1 and G_2 are contained in the interval $[0, a]$, and so

$$\begin{aligned} g^{-1}(S^1) &= g^{-1}(U \cup L) = g^{-1}(U) \cup g^{-1}(L) \\ &= g^{-1}(g_1(I)) \cup g^{-1}(g_2(I)) = G_1(I) \cup G_2(I) \subseteq [0, a]. \end{aligned}$$

This second proof is shorter than the first, but it is much less direct. What is the utility of having two proofs of the same proposition? It is not because this particular proposition is so central to topology. The answer is that our basic interest is in the methods (tools) rather than the results (products). Soon we will have other proofs that the circle and the half-open interval are not homeomorphic, as shown in Example 4.36, and, also, in the remark on page 121). Also, we will revisit this proof in the remark on page 246.

One often finds in mathematics that the key step in finding a proof of an unsolved problem is to first find a new proof of a solved problem.◆

4.5 Topological properties

Proposition 4.24 involves homeomorphisms from a subset of $\mathbf{R^n}$ (in this case $\mathbf{R^1}$) to itself. This prompts the following definition.

Definition 4.29 *If $X \subseteq \mathbf{R^n}$, a homeomorphism from X onto itself is called a* **self-homeomorphism**.

In geometry, the maps of an object to itself are the congruences. Self-congruences are important in geometry. These are the symmetries of the object. The study of symmetries of a geometric object leads to deep understandings of the geometric properties of the object. This is a fundamental point of view of modern geometry. Similarly, we can get important understanding of objects from the topological viewpoint by studying self-homeomorphisms of a subset.

For example, $\mathbf{R}^n$ has the property that all points look alike. The following definition uses the concept of self-homeomorphism to make this notion precise.

Definition 4.30 *Suppose $X \subseteq \mathbf{R}^n$. We say X is* **homogeneous** *if, for every two points $x, y \in X$, there is a self-homeomorphism $f: X \to X$ such that $f(x) = y$.*

Example 4.31 We show $\mathbf{R}^1$ is homogeneous by using the homeomorphism $f(t) = t + (y - x)$ since $f(x) = x + (y - x) = y$. To show $\mathbf{R}^n$ is homogeneous we use an rigid motion that takes x to y. Using vector notations, our self-homeomorphism is defined: $F(\vec{t}) = \vec{t} + (\vec{y} - \vec{x})$. ◆

Example 4.32 Using polar coordinates, write $x = (\rho_x, \theta_x)$ and $y = (\rho_y, \theta_y)$. Rotation by an angle of θ where $\theta = \theta_y - \theta_x$, sends x to y. Let R be the matrix $R = \begin{pmatrix} \cos(\theta) & -\sin(\theta) \\ \sin(\theta) & \cos(\theta) \end{pmatrix}$. The restriction of $F(\vec{t}) = R\vec{t}$ to the circle is a homeomorphism which verifies homogeneity. ◆

Example 4.33 By Proposition 4.24 (a), the closed interval $[a, b]$, $a \neq b$, is not homogeneous since a homeomorphism of $[a, b]$ must send a to either a or b, and cannot send it to, say, the interval midpoint $(b - a)/2$. ◆

The concept of a homogeneous space gives us our first interesting example of a topological property.

Definition 4.34 *Let P be a property which makes sense for subsets of a Euclidean space. That is, P is a statement which is either true or false for subsets of Euclidean spaces. We say that P is a* **topological property** *if, whenever $X \subseteq \mathbf{R}^n$ and $Y \subseteq \mathbf{R}^m$ with X and Y homeomorphic, then property P is true for X if and only if property P is true for Y.*

An equivalent way of stating Definition 4.34 is: P is a topological property if, whenever $X \subseteq \mathbf{R}^n$ and $Y \subseteq \mathbf{R}^m$ where X is homeomorphic to Y, then if P is true for X then property P is true for Y. Note that the "if and only if" has been replaced by "if ...then." This follows from the fact that homeomorphism is an equivalence relation. Bye the symmetric property "X is homeomorphic to Y" implies "Y is homeomorphic to X." Suppose we have established that for *any* subsets X and Y of some Euclidean spaces, if there is a homeomorphism f from X to Y, and property P is true for X, then property P holds for Y. We could then apply this result for the subsets Y, X, and the homeomorphism f^{-1} from Y to X, and conclude that if property P holds for Y then it must hold for X.

As a rule, when proving something is a topological property we use this reformulation since the burden of proof is (slightly) less.

Another expression of this notion often encountered is "P is a topological property if it is invariant (or preserved) under homeomorphisms."

Topological properties are fundamental since we can use them to distinguish subsets. Suppose P is a topological property, $X \subseteq \mathbf{R^n}$ and $Y \subseteq \mathbf{R}^m$. If we find that property P is true for X and false for Y, then X and Y are not homeomorphic.

Proposition 4.35 *The property "X is homogeneous" is a topological property.*

Proof: Suppose $f: X \to Y$ is a homeomorphism and X is homogeneous. We wish to show that Y is homogeneous.

Let y_1 and y_2 be points of Y; we need find a self-homeomorphism, H, of Y such that $H(y_1) = y_2$. Let $x_1 = f^{-1}(y_1)$ and $x_2 = f^{-1}(y_2)$. Since X is homogeneous, there is a self-homeomorphism of X, call it h, such that $h(x_1) = x_2$. We need to define a self-homeomorphism of Y. It is helpful to draw a diagram of some of the information to date, using a question mark to indicate the map we seek:

$$\begin{array}{ccc} X & \xrightarrow{h} & X \\ \downarrow f & & \downarrow f \\ Y & \xrightarrow{?} & Y \end{array}$$

One way to get a self-homeomorphism of Y is to let $H = f \circ h \circ f^{-1}$. We will see that this works. We can diagram this as

$$\begin{array}{ccc} X & \xrightarrow{h} & X \\ \uparrow f^{-1} & & \downarrow f \\ Y & \xrightarrow{H} & Y \end{array}$$

Now H is a homeomorphism since a composition of homeomorphisms is a homeomorphism. We have $y_1 = f(x_1)$ and $y_2 = f(x_2)$. We verify that $H(y_1) = y_2$ as follows:

$$\begin{aligned} H(y_1) &= (f \circ h \circ f^{-1})(y_1) = (f \circ h \circ f^{-1})(f(x_1)) \\ &= (f \circ h) \circ (f^{-1} \circ f)(x_1) = (f \circ h) \circ Id_X(x_1) \\ &= (f \circ h)(x_1) = f(x_2) = y_2. \end{aligned}$$

By the remark on page 117 we have now completed the proof that we have a topological property. ∎

The use of diagrams, like those in the proof of Proposition 4.35, is a useful technique, ubiquitous in much of modern mathematics.

We will often find ourselves in the situation where we want to define a map from Y to itself, and are given a homeomorphism f from X to Y and another continuous function h whose domain is X and whose range is X. (In some future problems we find applications where h might not be a homeomorphism, but only continuous.) It is often helpful to consider the continuous map $f \circ h \circ f^{-1}$. This map is called the **conjugate of h by f**. (This terminology comes from abstract group theory.)

Example 4.36 Here is a typical example of the use of a topological property. The interval $[0, 2\pi)$ is not homogeneous since no homeomorphism of $[0, 2\pi)$ can send 0 to anything except 0, by Proposition 4.24(b). On the other hand we have noted that the circle is homogeneous, Example 4.32. Since homogeneity is a topological property, it follows that $[0, 2\pi)$ is not homeomorphic to the circle. ◆

One important example of a topological property is connectedness. This will be the focus of Chapter 7. We introduce this now as an example of a topological property.

Definition 4.37 *Suppose $X \subseteq \mathbf{R^n}$. X is connected if and only if it* cannot *be written as the union of two non-empty disjoint open subsets.*

We have encountered subsets which are not connected in our discussion of subsets which are both open and closed subsets. If X is not connected, then we can write union of two non-empty disjoint open subsets, say $X = U \cup V$. Note that since $X - U = V$ is an open subset, U must be closed as well as open. Similarly, V must be an open and a closed subset of X.

Proposition 4.38 *The property "X is connected" is a topological property.* ∎ *(Problem 4.30)*

The proof of the next proposition uses the Intermediate Value Theorem. In Proposition 7.1 we give a more elementary proof of this which does not rely on the Intermediate Value Theorem.

Proposition 4.39 *The unit interval I is connected. Also, the real line $\mathbf{R}^1$ is connected.* ∎

Proof (for I): The argument is by contradiction. Suppose I is a union of two disjoint open sets, U and V. Let $a \in U$ and $b \in V$. There are such points since U and V are non-empty. Without loss of generality we may assume that $a < b$ (if not, then we could relabel U and V). Also note that $[a, b] \subseteq I$; this observation is critical to our proof. Define a function $f: [a, b] \to \mathbf{R}^1$ by: $f(x) = 0$ if $x \in U$; $f(x) = 1$ if $x \in V$. One can check that f is a continuous function, using the gluing lemma. But this contradicts the Intermediate Value Theorem since $1/2$ is an intermediate value between $f(a) = 0$ and $f(b) = 1$, yet there is no z with $a \leq z \leq b$ such that $f(z) = 1/2$ since f takes on only the values 0 and 1. We have a contradiction; thus we conclude that I is connected.

The proof above can also be modified for the case of $\mathbf{R}^1$. ∎ (Problem 4.31)

We discuss some other topological properties. Definitions 4.45, 4.48 are non-standard and are used for discussion and exercises only. On the other hand, Definitions 4.41 and 4.37 are found elsewhere in the literature, and are important ideas.

Definition 4.40 *Let $X \subseteq \mathbf{R}^n$. Suppose f is a function from X to itself. Let $x \in X$. We say x is a* **fixed point of** f *if $f(x) = x$.*

The following definition represents a significant concept in mathematics, both pure and applied.

Definition 4.41 *Let $X \subseteq \mathbf{R}^n$; we say X has the* **fixed-point property** *if every continuous map $f: X \to X$ has a fixed point.*

Proposition 4.42 *The property "X has the fixed-point property" is a topological property.* ∎ *(Problem 4.36)*

Proposition 4.43 *The unit interval I has the fixed-point property.*

Proof: Let $f: I \to I$ be a continuous function. If either $f(0) = 0$ or $f(1) = 1$, we are done since we have found a fixed point.

Suppose $f(0) \neq 0$ and $f(1) \neq 1$; then we must have $0 < f(0)$ and $f(1) < 1$.

Consider the graph of $f(x)$ and the graph of $y = x$. The fixed points of f correspond to the values of x where the two graphs meet; see Figure 4-6.

To articulate the proof, we use the Intermediate Value Theorem. It is useful to introduce $g(x) = f(x) - x$. Then $0 < g(0)$ and $g(1) < 0$. By the Intermediate Value Theorem, there is an $x_0 \in I$ with $g(x_0) = 0$. But then x_0 is a fixed point of f since $g(x_0) = f(x_0) - x_0 = 0$. ∎

By this time, use of the Intermediate Value Theorem should seem "expected" in that we need to come up with the existence of a point without having sufficient information for directly finding it.

Here is an example illustrating the utility of these ideas.

Example 4.44 The standard circle of the plane, S^1, is not homeomorphic to the unit interval I of the line. By proposition 4.43, I has the

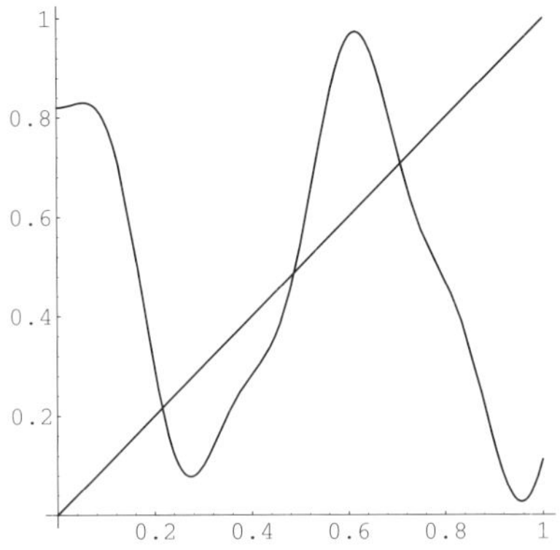

Figure 4-6 A graph of $y = x$ and another function $f: [0,1] \to [0,1]$. In this particular case the graphs intersect at three points, showing that f has three fixed points.

fixed-point property; but S^1 does not since any rotation of the circle (by an angle which is not a multiple of 2π) has no fixed points. ◆

A common technique of mathematics is to turn a theorem into a definition. Useful as this can be, it often results in complex and convoluted definitions. We can view the statement of Proposition 4.24(b) as giving us another topological property. We make the following definition:

Definition 4.45 *Let $X \subseteq \mathbf{R}^n$. We say X has the* **self-homeomorphism fixed-point property** *if every self-homeomorphism of X has a fixed point.*

Proposition 4.46 *The self-homeomorphism fixed-point property is a topological property.*

Proof: Suppose X has the self-homeomorphism fixed-point property and f is a homeomorphism from X to Y. We show Y has the self-homeomorphism fixed-point property.

Let H be a self-homeomorphism of Y. Let $h = f^{-1} \circ H \circ f$. This h is a self-homeomorphism of X since it is a composition of homeomorphisms. (It is helpful here to draw a diagram similar to that used in the proof of Proposition 4.35.) Since X has the self-homeomorphism fixed-point property, there exists an $x \in X$ such that $h(x) = x$. Let $y = f(x)$. We next verify that $H(y) = y$. We know that $h(x) = x$; that is, $(f^{-1} \circ H \circ f)(x) = x$. We can write this as $(f^{-1} \circ H)(y) = x$. Apply f to both sides of this last equation, and we get $(f \circ f^{-1} \circ H)(y) = f(x)$; thus $H(y) = f(x)$ or $H(y) = y$. ∎

We can use Proposition 4.46 to get (yet) another proof that $[0, 2\pi)$ is not homeomorphic to a circle. By Proposition 4.24(b), $[0, 2\pi)$ has the self-homeomorphism fixed-point property. But the circle has self-homeomorphisms that do not have fixed points–any rotation by an angle which is not a multiple of 2π will do. Thus the circle does not have the self-homeomorphism fixed-point property. Since this property is a topological property, it follows that $[0, 2\pi)$ and the circle cannot be homeomorphic.

Seeing how useful this technique can be, we examine Proposition 4.24(a) and try to somehow define a topological property. Although it does not say that self-homeomorphisms of a closed interval must have a fixed point, it does say that any self-homeomorphism must send endpoints to endpoints. So let E be the set of endpoints of $[a, b]$; that is, E is the two-point subset $\{a, b\}$. Proposition 4.24(a) then says that if h is a self-homeomorphism of $[a, b]$, that $h(E) = E$. This motivates the next definitions.

Definition 4.47 *Let $X \subseteq \mathbf{R^n}$ and suppose $f: X \to X$ is a continuous function and $A \subseteq X$. We say that* **A is invariant under** f *if $f(A) = A$.*

Definition 4.48 *Let $X \subseteq \mathbf{R^n}$, we say that X has the* **two-point invariant subset self-homeomorphism property** *if, for every self-homeomorphism h of X, there is a two-point subset of X which is invariant under h.*

The proof of the next proposition is very similar to the proof of Proposition 4.35.

Proposition 4.49 *The two-point invariant self-homeomorphism property is a topological property.* ∎ *(Problem 4.37)*

By Proposition 4.24(a), any closed interval has the two-point invariant subset self-homeomorphism property. On the other hand, a half-open interval does not have this property. For example, consider the map $f: [0, 1) \to [0, 1)$ defined by $f(x) = x^2$. One can verify that this is a self-homeomorphism that does not have the two-point invariant subset (Problem 4.38). Thus it follows that a closed interval cannot be homeomorphic to a half-open interval.

Most of the examples discussed so far in this chapter have been circles or subsets of the line. We next discuss some other examples of homeomorphic spaces.

Example 4.50 Let S be the square in the plane with vertices at $(1, 0)$, $(0, 1)$, $(-1, 0)$, and $(0, -1)$. Note that $S = \{(x, y) \in \mathbf{R^2} \mid |x| + |y| \leq 1\}$. Let D^2 be the unit disk; that is, $D^2 = \{(x, y) \in \mathbf{R^2} \mid x^2 + y^2 \leq 1\}$. One can show that D^2 and S are homeomorphic. The idea is to use polar coordinates and map a line segment from the origin to a point on the square $|x| + |y| = 1$ to the corresponding radius for the disk. Details are left as exercise (Problem 4.32). ◆

An elementary fact about homeomorphisms is

Proposition 4.51 *Suppose $X \subseteq \mathbf{R^n}$, $Y \subseteq \mathbf{R^n}$, $P \subseteq \mathbf{R^p}$, $Q \subseteq \mathbf{R^q}$, and there are homeomorphisms $f: X \to P$, and $g: Y \to Q$. Then $f \times g$ is a homeomorphism from $X \times Y$ to $P \times Q$.* ▮ *(Problem 4.12)*

We illustrate the use of this proposition in the next example:

Example 4.52 Roughly, out example X is a closed rectangle with an open rectangle removed. Consider numbers $a < b < c < d$ and $A < B < C < D$. Let

$$X = ([a,d] \times [A,D]) - ((b,c) \times (B,C)).$$

We claim that X is homeomorphic to an annulus $S^1 \times I$. We verify this using several steps; see Figure 4-7.

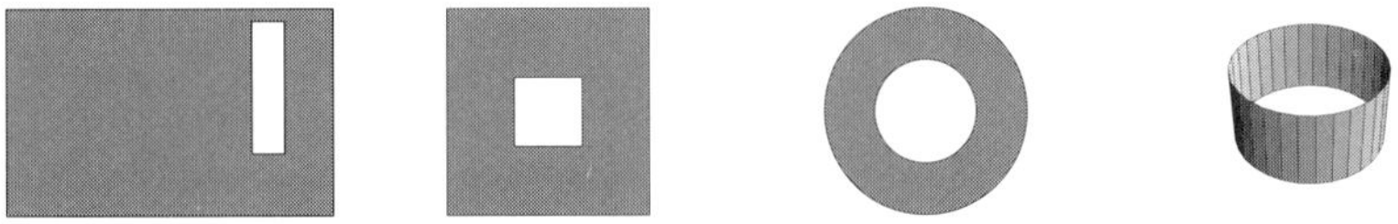

Figure 4-7 Shown on left is the closed rectangle with an open rectangle removed; this is X of Example 4.52. It is homeomorphic to $S^1 \times I$, shown on the right. Progression is shown from left to right. First, X is homeomorphic to X', a standard square with a half-sized sub-square removed; then X' is homeomorphic to the A, a region between two concentric circles; and, finally, A is homeomorphic to $S^1 \times I$.

Let

$$X' = ([-1,1] \times [-1,1]) - ((-0.5,0.5) \times (-0.5,0.5)), \text{ and let}$$

$$A = \{x, y \in \mathbf{R}^2 \mid 0.5 \le \sqrt{x^2 + y^2} \le 1\}.$$

The strategy is to find homeomorphisms $h_1: X \to X'$, $h_2: X' \to A$, $h_3: A \to S^1 \times I$, and then to define our homeomorphism as $h_3 \circ h_2 \circ h_1$.

There is a unique, piecewise-linear function $\phi: [a,d] \to [-1,1]$ such that $\phi(c) = -0.5$ and $\phi(d) = 0.5$. Similarly, there is a unique piecewise-linear function $\psi: [a,d] \to [-1,1]$ such that $\psi(C) = -0.5$ and $\psi(D) = 0.5$. We leave it as an exercise to verify that ϕ and ψ are homeomorphisms. (Hint: Show they are monotone.) By Proposition 4.51,

$$\phi \times \psi : [a,d] \times [A,D] \to [-1,1] \times [-1,1]$$

is a homeomorphism. Note that

$$(\phi \times \psi)((b,c) \times (B,C)) = ((-0.5,0.5) \times (-0.5,0.5)).$$

It then follows that $h_1 = \phi \times \psi|_X$ is a homeomorphism from X to X' since a restriction of a homeomorphism is a homeomorphism, Proposition 4.17.

In Example 4.50, we gave hints for the construction of a homeomorphism, call it H, from the square $S = [-1,1] \times [-1,1]$ to the unit disk. If this has been done in a simple way, you should find that $H(X') = A$. Thus we can let $h_2 = H|_{X'}$.

In polar coordinates, $A = \{(r, \theta \in \mathbf{R}^2 \mid 0.5 \le r \le 1\}$; in cylindrical coordinates, $S^1 \times I = \{(r, \theta, z) \in \mathbf{R}^3 \mid r = 1 \text{ and } 0 \le z \le 1\}$. We define h_3 by $h_3(r, \theta) = (1, \theta, 2r - 1)$, and check that h_3 is a homeomorphism. ◆

It is proofs, such as the proof for Example 4.52, that give rise to the rough description of topology as "the study of objects as if they were made of rubber."

By now it should be clear that showing two spaces not homeomorphic, even in the "most obvious" cases, may prove rather difficult.

Question 4.53 Are $\mathbf{R}^1$ and $\mathbf{R}^2$ homeomorphic?

The answer to this is no, but to prove this is not easy. In Appendix C.12, we see that there is a one-to-one-correspondence between these sets! So the problems must lie in issues of continuity.

We will have more to say about Question 4.53 later in the remark on page 175 and Proposition 8.03.

Of course, one might naturally also ask

Question 4.54 Is $\mathbf{R}^2$ homeomorphic to $\mathbf{R}^3$?

The answer is no. We will give an idea of the proof; see Proposition 15.31, but a complete proof involves concepts more advanced than we will consider in this course.

More generally, it can be shown that $\mathbf{R}^n$ is homeomorphic to $\mathbf{R}^m$ if and only if $n = m$.

To dispel the possible impression that all we do in topology is show that things that are "obviously different" really *are* different, here is just one (challenging) example where it is not immediately clear whether two particular given spaces are homeomorphic or not.

Question 4.55 Let $\mathbf{Q}$ be the set of rational numbers in the line and let Q' be the set of points of the plane, both of whose coordinates are rational. Are $\mathbf{Q}$ and Q' homeomorphic ?

*4.6 General topology and Chapter 4

The concept of homeomorphism, in its abstract setting, is the basic equivalence relation for topological spaces. Two topological spaces are homeomorphic if there is a one-to-one correspondence between the sets which is continuous with continuous inverse. Because the spaces we study in general topology are more abstract, the problems defining homeomorphisms or showing spaces not homeomorphic are more diverse (and challenging). We mention only a few aspects here.

In general topology, as we have seen, there are many ways to define a topology for a given set. Often in general topology one is interested in relating several topologies defined on the same set.

Suppose we have a set S and two metrics d and d'; then we have two metric spaces $\{S, d\}$ and $\{S, d'\}$. In Definition 1.69, we introduced the notion of equivalent metrics. If d and d' are equivalent metrics for S, then the identity map is a homeomorphism.

Here is one general application of equivalent metrics. Suppose $\{S, d\}$ is a metric space; one can define a new metric for S as follows. For $x, y \in S$, define $d'(x, y) = \min(1, d(x, y))$. One can verify that d' is also a metric for S, and d and d' are equivalent metrics. Since, for all $x, y \in S$ we clearly have $d'(x, y) \leq 1$, such a metric is called a "bounded metric."

In $\mathbf{R^n}$, finite subsets with the same number of elements are homeomorphic, Proposition 4.13. However, in general topology, this is not necessarily the case, as in Example 1.52.

Recall the definition of quotient map, Definition 1.76. We investigate the question: Given a continuous function, when could it be considered to be a quotient map of some sort?

Definition 4.56 *Suppose X and Y are topological spaces and $f: X \to Y$ is a continuous surjection. We say f is an* **identification map** *if V is open in Y whenever $f^{-1}(V)$ is an open subset of X.*

Identification maps can be described using closed subsets: Suppose X and Y are topological spaces and f is a continuous surjective function, $f: X \to Y$. Then f is an identification map if C is closed in Y whenever $f^{-1}(C)$ is a closed subset of X.

Proposition 4.57 *Suppose X and Y are topological spaces and $f: X \to Y$ is a continuous surjection. If f is a closed map, or if f is an open map, then f is an identification map.* ∎

Example 4.58 One can have an identification map which is neither open nor closed.

For example, define $X \subseteq \mathbf{R}^2$ by

$$X = N_1((0,0)) \cup N_{\frac{3}{2}}((3,0)) \cup \overline{N}_1((6,0)).$$

Define $f: X \to (-\frac{3}{2}, \frac{3}{2}))$ by $f((x,y)) = y$ if $(x,y) \in N_1((0,0)) \cup N_{\frac{3}{2}}((3,0))$; $f((x,y)) = 0$ if $(x,y) \in \overline{N}_1((6,0))$. One can check that f is an identification map. However, f is not a closed map since $f(N_1((0,0)))$ is not a closed subset. Also, f is not an open map since $f(\overline{N}_1((6,0)))$ is not a open subset. We leave verification of details as an exercise. ◆

We use the idea of homeomorphism to express the idea that, in general topology, an identification map is a quotient map, in a sense.

Suppose that $f: X \to Y$ is an identification map, Definition 4.56. For $y \in Y$ let $P_y = f^{-1}(y)$ and $P = \{P_y\}_{y \in Y}$; P is a partition of X into inverses of points of Y. Define a map $g: X/P \to Y$ by $g(P_y) = y$. Then g can be seen to be a homeomorphism from X/P to Y and, if q is the quotient map of X/P, one has $f = g \circ q$.

Example 4.59 As defined in Example 1.74, the projective plane P^2 is defined using the set of lines in $\mathbf{R}^3$ which pass through the origin.

In Example 1.78, we defined a decomposition space S^2/P where we partition S^2 into two-point antipodal subsets. We show S^2/P is homeomorphic to P^2. (The same proof works to show S^n/P is homeomorphic to P^n.)

Define a function $h: P^2 \to S^2/P$ as follows. Let $L \in P^2$; L is a line in $\mathbf{R}^3$ which passes through the origin. Then let $A_L = L \cap S^2$; A_L is a pair of antipodal points. Define $h(L) = A_L$. Verification that h is a homeomorphism is not difficult. The problem is to verify continuity of h and h^{-1} since the topology for P^2 is defined from a metric and the topology for S^2/P is the quotient topology. ◆

Here is a proposition concerning infinite products. It may seem mysterious, but will seem less so (in the case of a countably infinite product) if one considers Proposition 5.34.

Proposition 4.60 *Let $\{Y_{\alpha \in A}\}$ be an infinite collection of topological spaces where each Y_α is a finite set with at least two elements, and the topology for Y_α is the discrete topology. For all $\alpha \in A$, let X_α be the two-point set $\{0,1\}$ with the discrete topology. Then $\prod_{\alpha \in A} X_\alpha$ is homeomorphic to $\prod_{\alpha \in A} Y_\alpha$.* ∎

Here is a simple example of a homeomorphism that involves a function space. We can consider the set of orthogonal transformations of $\mathbf{R}^2$ as a subset of $\mathbf{R}^4$, as mentioned in remark on page 55. This makes it a topological space, with topology induced from the standard topology of $\mathbf{R}^4$.

The key to defining a homeomorphism needed for this next Proposition is to show that an orthogonal transformation of $\mathbf{R}^2$ with determinant +1 is a rotation by some angle, θ, which can then be associated to a point of the unit circle.

Proposition 4.61 *Let R_2 be the topological space of orthogonal transformations of $\mathbf{R}^2$ which have determinant +1; R_2 is homeomorphic to the circle, S^1.* ∎

In $\mathbf{R}^3$ one has a similar result, where we use one of the quotient spaces we have defined. This gives more of a flavor for the sorts of things one can do with our more general notions.

Proposition 4.62 *Let R_3 be the topological space of orthogonal transformations of $\mathbf{R}^3$ which have determinant +1; R_3 is homeomorphic to the three-dimensional projective space, P^3.* ∎

4.7 Problems for Chapter 4

4.1 Prove Proposition 3.

4.2 Show that for any two points, p and q of the plane, and any two numbers, ϵ and η, that $N_\epsilon(p)$ is homeomorphic to $N_\eta(q)$.

4.3 Show that if $y = g(x)$ is a continuous function, $g: \mathbf{R}^1 \to \mathbf{R}^1$, then the function $f(x, y) = (x, y + g(x))$ is a homeomorphism of $\mathbf{R}^2$ to itself. (Hint: Follow the proof in Example 4.05.)

4.4 Verify that stereographic projection (see Proposition 4.06) is a homeomorphism.

4.5 Prove that $\overset{\circ}{R}{}^2_+$ is homeomorphic to $\overset{\circ}{D}{}^2$. (Hint: as a first step think of stereographic projection.)

4.6 Let $P = \{(x, y) \in \mathbf{R}^2 \mid 0 \leq x \text{ and } y = 0\}$; P corresponds to the non-negative real numbers on the x-axis. Prove that $\mathbf{R}^2 - P$ is homeomorphic to the open half-plane $\overset{\circ}{R}{}^2_+$. (Hint: Consider the complex function $w = \sqrt{z}$.)

4.7 In $\mathbf{R}^2$, let $X = N_1((-1, 0)) \cup N_1((1, 0))$; X is a union of two disjoint open disks. Let L be a line in $\mathbf{R}^2$. Prove that $\mathbf{R}^2 - L$ is homeomorphic to X. (Hint: Use the result stated in Problem 4.5. You do not need to prove Problem 4.5„ only use the fact that such a homeomorphism exists.)

4.8 Prove that the map ϕ defined in Example 4.07 is a homeomorphism.

4.9 Verify the claim of Example 4.09 that a simple closed curve is homeomorphic to a circle (Problem 4.9).

4.10 Prove Proposition 4.16.

4.11 Prove Proposition 4.17

4.12 Prove Proposition 4.51.

4.13 Let $\mathbf{N}$ be the set of natural numbers and $\mathbf{Z}$ the set of integers. Show that $\mathbf{R}^1 - \mathbf{N}$ is homeomorphic to $\mathbf{R}^1 - \mathbf{Z}$.

4.14 Consider the subsets of $\mathbf{R}^2$ defined by $N = \{(1,0),(2,0),\ldots\}$ and $Z = \{(0,0),(1,0),(-1,0),\ldots\}$. Show that N and Z are homeomorphic.

4.15 Let $F \subseteq \mathbf{R}^2$ be the set, $F = \{(1/1,0),(1/2,0),(1/3,0),\ldots\}$, and let N be as in Problem 4.15. Show that F and N are homeomorphic.

4.16 For $i \in \mathbf{N}$, let $S_i = \{(1/i,0),(0,1/i),(-1/i,0),(0,-1/i)\}$. You can think of points of S_i as corners of a square centered about the origin with sides of length $2/i$. Let $S = \bigcup_{i\in N} S_i$. Show that S is homeomorphic to N where N is as in Problem 4.14

4.17 For $i \in \mathbf{N}$, let $B_i = \{(\frac{1}{i},0),(\frac{1}{i},1)\}$ and let $B = \bigcup_{i\in N} B_i$; let F be as in Problem 4.14. Is B homeomorphic to F?

4.18 Let N be as in Problem 4.14; let $Q_0 = \{(x,y) \in \mathbf{R}^2 : x \in \mathbf{Q} \text{ and } y = 0\}$; here $\mathbf{Q}$ denotes rational numbers. Show that Q_0 is not homeomorphic to N.

4.19 Let S^1 be the unit circle in the plane; let c_0 be a point, $c_0 \in S^1$. Show $S^1 - c_0$ is homeomorphic to $\mathbf{R}^1$.

4.20 Suppose $X \subseteq \mathbf{R}^n$ and $Y \subseteq \mathbf{R}^n$. Show that $X\times Y$ is homeomorphic to $Y\times X$.

4.21 Show that $f(\vec{x}) = \frac{1}{1-|\vec{x}|}\vec{x}$ is a homeomorphism from $N_1(\vec{0})$ to $\mathbf{R}^n$, as claimed in the proof of Proposition 4.21.

4.22 Verify the statement in the proof of Proposition 4.23 that if f is one-to-one and is not monotone, then there are numbers A, B, and C, with $A < B < C$, and either $f(A) < f(B)$, $f(B) > f(C)$, or there are numbers A, B, and C, with $A < B < C$ and $f(A) > f(B)$, $f(B) < f(C)$,

4.23 Write the proof for the second case of Proposition 4.23.

4.24 Prove parts (b), (d), and (e) of Proposition 4.23.

4.25 Prove Proposition 4.24.

4.26 Let $\mathbf{Q} \subseteq \mathbf{R}^1$ be the subset of rational numbers. Does every map $F: \mathbf{Q} \to \mathbf{Q}$, which is a homeomorphism, have to be a strictly monotone function ?

4.27 Sketch $f'([0,\pi))$ using $\beta_2(x)$, $\beta_3(x)$, and $\beta_4(x)$ as in Example 4.27.

4.28 Finish the proof of Example 4.27.

4.29 Let $\mathbf{Q}$ be the set of rational numbers. Show $\mathbf{Q}$ is homogeneous.

4.30 Prove Proposition 4.38.

4.31 Fill in details from the proof of Proposition 4.39. Verify that the map f is continuous.

4.32 Let S be the square in the plane with vertices at $(1,0)$, $(0,1)$, $(-1,0)$, and $(0,-1)$. Note that $S = \{(x,y) \in \mathbf{R}^2 \mid |x| + |y| \leq 1\}$. Let D^2 be the unit disk; that is, $D^2 = \{(x,y) \in \mathbf{R}^2 \mid x^2 + y^2 \leq 1\}$. Show that D^2 and S are homeomorphic.

4.33 Let T be a solid triangle in $\mathbf{R}^2$. That is, T consists of all points inside a triangle of the plane together with all points on its perimeter. Show that T is homeomorphic to D^2. (Hint: Consider the centroid of the triangle, c. A ray from c will intersect the ray in a subset homeomorphic to a closed interval, and we can write the triangle as the union of these subsets. We can write D^2 in a similar way: D^2 is the union of radial line segments. This gives a guide for how to define a homeomorphism from T to D^2.)

4.34 Assuming Problem 4.33, show the standard 2-simplex Δ^2 is homeomorphic to D^2.

4.35 Let H^2_α be a page in $\mathbf{R}^3$; see Definition 2.10. Show that H^2_α is homeomorphic to R^2_+.

4.36 Prove Proposition 4.46.

4.37 Prove Proposition 4.49.

4.38 Verify that the map $f(x)$, of the remark on page 122, is a self-homeomorphism that does not leave invariant any two-point subset.

4.39 Let X be one of the crescent-shaped regions in the complement in $\mathbf{R}^2$ of the Hawaiian earring; see Example 2.42. Show that X is homeomorphic to the standard open unit disk $\overset{\circ}{D}{}^2$. (Hint: This might take several steps. First consider the inversion in the unit circle of the plane given by $f(\vec{x}) = \frac{\vec{x}}{|\vec{x}|}$. You will find the image $f(X)$, although not an open disk, but it is easier to relate to an open disk.)

4.40 Suppose $X \subseteq \mathbf{R}^n$, $p \in (\mathbf{R}^{n+1} - \mathbf{R}^n)$, $Y \subseteq \mathbf{R}^m$, and $q \in (\mathbf{R}^{m+1} - \mathbf{R}^m)$. Show that the cones pX and qY (recall Definition 2.47) are homeomorphic.

5. CANTOR SETS AND ALLIED TOPICS

OVERVIEW: Using general mathematical methods, such as infinite intersections, we can define subsets of $\mathbf{R}^n$ *with strange and sometimes wonderful properties. In this chapter we look at several of these subsets. These examples provide a first test of usefulness of the concepts introduced thus far, especially homeomorphism, as well as provide motivation for further mathematical development.*

Our first example is a subset of the line, the standard Cantor set, Definition 5.01. We then attempt several variations of this example. Some of the attempts will result in interesting failures to produce something "new." The successes will show the need for additional mathematical tools.

5.1 The standard Cantor set

Constructions that involve infinite processes pervade topology. We can define a subset in a simple manner, yet have some difficulty understanding what it is. An example is the standard Cantor set. In this chapter, we introduce this example and some similar subsets. In later chapters, after additional tools have been developed, we reexamine these subsets.

Definition 5.01 *The* **standard ternary Cantor set K** *is the subset of all numbers* $x \in \mathbf{R}^1$ *such that* $x = \sum_{n=1}^{\infty} \frac{a_n}{3^n}$, *where, for each* n, a_n *is either 0 or 2. Sometimes we will refer to this simply as the* **standard Cantor set**.

Recall the formula for the sum of a geometric series:

$$\sum_{n=0}^{\infty} x^n = \frac{1}{1-x} \text{ if } |x| < 1.$$

The requirement for $x \in K$ is that it *can be* written in a certain form. For example, $1 \in K$ since we can write $1 = \frac{2}{3} + \frac{2}{9} + \frac{2}{27} + \cdots$. To see this, begin with the formula above with $x = \frac{1}{3}$: $1 + \frac{1}{3} + \frac{1}{9} + \frac{1}{27} + \cdots = \frac{3}{2}$. Thus $\frac{1}{3} + \frac{1}{9} + \frac{1}{27} + \cdots = \frac{1}{2}$. Multiplying both sides by 2, we get $1 = \frac{2}{3} + \frac{2}{9} + \frac{2}{27} + \cdots$. Thus we see $1 \in K$ and $\frac{1}{2} \notin K$.

An alternative way of expressing Definition 5.01 is to say that the standard Cantor set consists of all numbers in the unit interval whose decimal expansion, using base 3, contains no ones. The base 3 system is also called the **ternary system**. Thus $1 = \frac{2}{3} + \frac{2}{9} + \frac{2}{27} + \cdots$, in the base 3 decimal system, is just the statement that $1 = 0.\overline{2}$; here the bar notation indicates infinite repetition, 0.2222222.... So, $0.\overline{1}$, in the ternary system, is the number $\frac{1}{2}$.

Proposition 5.02 *The standard Cantor set is an uncountable set.*

Proof: Consider a point in the standard Cantor set $x = 0.02020022202\ldots$ where we are writing in the ternary decimal system. A binary string is an expression consisting of a sequence of the symbols 0 or 1. We can associate to x a binary string by replacing the twos by ones. In our example, we would get 01010011101.... By Propositions C.11 and C.13, the set of such binary strings is uncountable. Thus the standard Cantor set is uncountable. ∎

For many purposes, it will be more convenient, and mathematically revealing, to have an alternative, and more geometric, description of the standard Cantor set as an infinite intersection.

We begin with a set C_1, obtained by removing the middle third of the unit interval. So, $C_1 = [0, 1/3] \cup [2/3, 1]$. The set C_2 is obtained by removing the middle third of each of the two intervals making up C_1:

$$C_2 = [0, 1/9] \cup [2/9, 1/3] \cup [2/3, 7/9] \cup [8/9, 1].$$

Inductively, we define C_i to be the set obtained from C_{i-1} by removing the middle third of each of the sub-intervals of C_{i-1}. (See Figure 5-1.)

The standard Cantor set is the same as $\bigcap_{i=1}^{\infty} C_i$. We leave details of the verification of this as an exercise (Problem 5.1). The following analysis will be helpful for this verification.

Every point of the standard Cantor set can be written as an intersection of smaller and smaller closed intervals—these being closed subintervals of the sets $\{C_i\}$. We will elaborate and provide notations for later use.

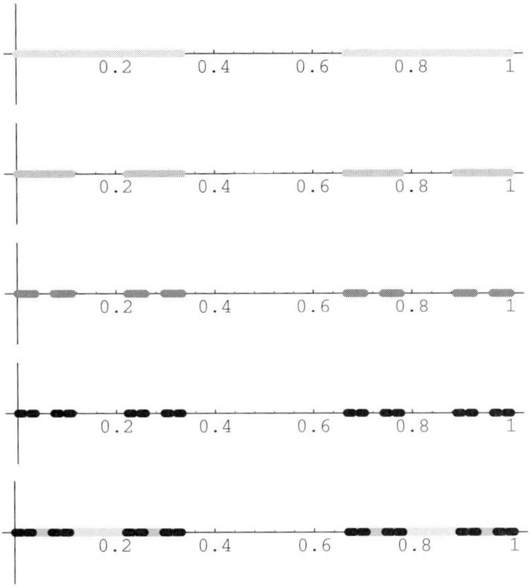

Figure 5-1 Subsets used to define the standard Cantor set; see remark, page 133. At top, points of C_1 indicated in light gray. Below, in increasingly darker gray are C_2, then C_3 and C_4. At the bottom we show all four of these sets simultaneously.

Write C_1 as the union of two disjoint closed intervals, I_0 and I_2, where I_0 lies to the left of I_2. Next, let $I_1 = I - C_1$ where I is the unit interval. Note that I_1 is an *open* interval. The subscripts have the following interpretation: I_0 consists of all numbers in I, which, when written in the ternary decimal system, can be expressed with a zero in the first position to the right of the decimal point. This includes the decimal number 0.1 since this number can be written as $0.0\overline{2}$. Similarly, I_2 consists of all numbers in I which, in the ternary decimal system, can be written with a two immediately to the right of the decimal point. For example, $0.1\overline{2}$ is in I_2.

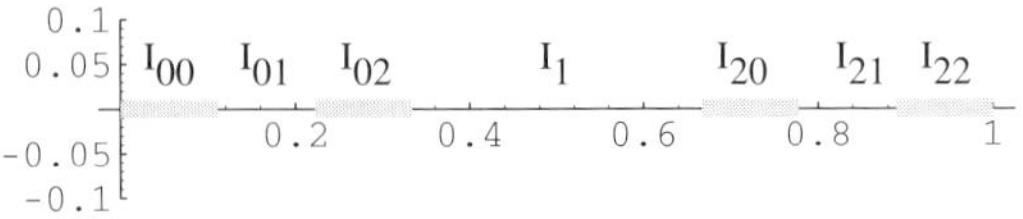

Figure 5-2 Labeling of intervals for C_2; see discussion in remark, page 133

Next, express C_2 as the union of four disjoint closed intervals, I_{00}, I_{02}, I_{20} and I_{22}, listed in order from left to right; see Figure 5-2. That is, I_{00} lies to the left of I_{02}, which in turn lies to the left of I_{20}, which lies to the left of I_{22}.

As before, one can give an interpretation of these subscripts in terms of the decimal expansion of the elements. Numbers in I_{00} are all numbers which can be written in the base 3 decimal system so as to begin 0.00.... Numbers in I_{02}, when written in the base 3 system, can be written so as to begin 0.02.... Let $I_{01} = I_0 - C_2$. Note that I_{01} is an open interval and $I_{01} = I_0 - (I_{00} \cup I_{02})$. Numbers in I_{01} all can be written in base 3 decimal system as beginning 0.01....Similarly, define $I_{21} = I_2 - C_2$; see Figure 5-2. We invite you to supply an interpretation of the subscripts of I_{01} and I_{21}. In the next stage, we define the sets I_{000}, I_{001}, I_{002}, I_{020}, I_{021}, I_{022}, I_{200}, I_{201}, I_{202}, I_{220}, I_{221}, and I_{222}.

Proposition 5.03 *The standard Cantor set K is a closed subset of* $\mathbf{R}^1$.

Proof: By remark, page 133, the standard Cantor set is the same as $\bigcap_{i=1}^{\infty} C_i$. Each C_i is a closed subset of the line since it is the union of a finite number of closed intervals. Since it is the intersection of closed subsets, K is closed. ∎

The standard Cantor set does not have any isolated points and, in fact, every point of K is a limit point of K (Problem 5.4).

The standard Cantor set has some remarkable properties. One property is that it is a disjoint union of two homeomorphic copies of itself. In fact, $K \cap I_0$ is homeomorphic to K since the restriction of $f(x) = 3\,x$ to the subset $K \cap I_0$ is a homeomorphism (Problem 5.5). Similarly, using a restriction to K_2 of $f(x) - 3x - 2$,, one sees $K \cap I_2$ is homeomorphic to K. Thus, K is the disjoint union of $K \cap I_0$ and $K \cap I_2$, each of which is homeomorphic to K.

There are many subsets of K homeomorphic to K. Suppose $I_\alpha = [a, b]$ is one of the sub-intervals of C_i, for some i, and $I_\beta = [c, d]$ is one of the sub-intervals of C_j. Then there is a homeomorphism $f_{(\alpha,\beta)} : K \cap I_\alpha \to K \cap I_\beta$. It is not difficult to verify that the restriction of $f(x) = \frac{d-c}{b-a}x + c - \frac{d-c}{b-a}$ gives a homeomorphism. (See Proposition 4.20 for a reference to this formula).

The standard Cantor set is not as familiar as other subsets we have considered up to this point. It certainly seems different, but we need to prove this. For example, consider the following proposition.

Proposition 5.04 *The standard Cantor set K is not homeomorphic to a closed interval in the line, I.*

Proof: The idea is to use the fixed-point property of I, Proposition 4.43. Define $f : K \to K$ by $f(x) = 1 - x$. This is a continuous map from K to itself (in fact, it is a homeomorphism). But it has no fixed points since the only real number such that $x = 1 - x$ is $\frac{1}{2}$, but $\frac{1}{2} \notin K$, as shown in remark, page 131. By Proposition 4.42, the fixed-point property is a topological property. We conclude that K and I are not homeomorphic. ∎

An alternative proof of the proposition above can be obtained using the fact that I is connected, Proposition 4.39. The standard Cantor set is not connected. Let $U = (-\infty, 1/2)$ and $V = (1/2, \infty)$; $U' = K \cap U$, and $V' = K \cap V$. Then U' and V' are non-empty, disjoint open subsets of K whose union is K. Since connectedness is a topological property, Proposition 4.38, K and I are not homeomorphic.

5.2 Variations of definition of Cantor set in $\mathbf{R}^1$

We consider a variation of Definition 5.01.

Example 5.05 The **half Cantor set** $K_{\frac{1}{2}}$ is the subset of all numbers $x \in \mathbf{R}^1$ such that $x = \sum_{n=1}^{\infty} \frac{a_n}{3^n}$ where, for each n, a_n is either 0 or 1. The largest number in $K_{\frac{1}{2}}$ is the number $\frac{1}{2}$, which corresponds to all of the a_n being equal to 1.

It is not difficult to see that K is homeomorphic to $K_{\frac{1}{2}}$. If we define $F : K \to K_{\frac{1}{2}}$ by $F(x) = x/2$, then F is a homeomorphism. Because of this homeomorphism, we have chosen the name "half Cantor set." ◆

Example 5.06 Let T be the set of numbers in the unit interval which can be written as a decimal string in the base seven system by using only the digits 0 and 6, terminal sequences of 6's being allowed. We call T the **thin Cantor set**; see Figure 5-3.

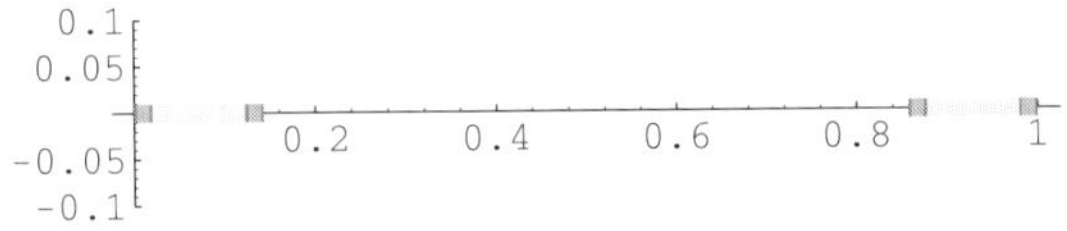

Figure 5-3 The first two sets (light and dark gray) used in defining the thin Cantor set; see Example 5.06. The third set not shown since it is too small.

Using an alternative description, as in the proof of remark, page 133, we could describe T as an infinite intersection, as follows. Instead of removing the middle third of the unit interval, divide I into seven equal pieces; remove all but the first and the last (or, in other words, remove the five interior sevenths). Let T_1 be the union of these two intervals. Next, divide each of these two intervals into seven equal pieces; in each, remove all but the first and the last. Thus we are left with four small intervals. Let T_2 be the union of these four intervals. Repeating this process we get a sequence of subsets. We can see that $T = \bigcap_{i=1}^{\infty} T_i$. ◆

There is a natural one-to-one correspondence between T and K that is easy to define. Points of T are expressible as decimals (in the base seven system) which use only 0's and 6's. We obtain a point of K by formally replacing the 6's by 2's in the string. For example, 0.06006606006...corresponds to 0.02002202002.... One can verify that this function is a homeomorphism.

Later, we present a different proof, Proposition 5.13.

Example 5.07 The fat Cantor set. The fat Cantor set is most easily described in terms of removing intervals. First, choose a number ϵ where $0 < \epsilon < 1/3$. For the first stage, divide the unit interval into three pieces. The middle piece is an open interval with length $\epsilon/2$, the other two pieces are of equal length. Remove this middle piece, leaving two closed intervals. Let F_1 be the union of these two intervals, $F_1 = [0, \frac{1}{2} - \frac{\epsilon}{4}] \cup [\frac{1}{2} + \frac{\epsilon}{4}, 1]$; see Figure 5-4.

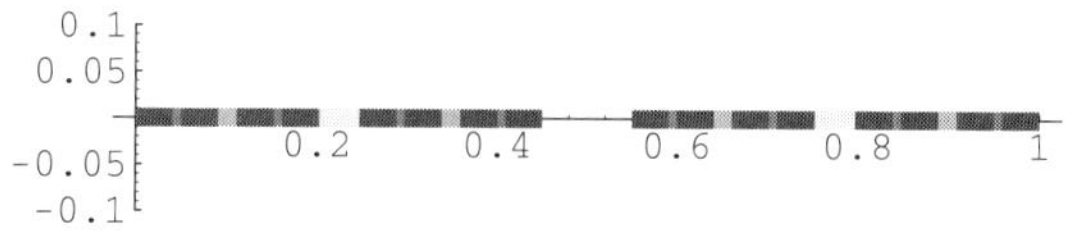

Figure 5-4 The first four sets, $F_1, \ldots F_4$, used to define a fat Cantor set, with $\epsilon = 0.2$, Example 5.07. Because of the large size of these sets, most of what is seen is F_4, in darkest gray.

In the next stage, divide each of these intervals into three pieces. The middle piece is an open interval of length $\epsilon/8$; the other intervals being of equal length. Remove these two open intervals, leaving four closed intervals altogether. Let F_2 denote the union of these four closed intervals. (Note that the sum of the lengths of the intervals removed at this second stage is $\epsilon/4$.) At the n-th stage we have 2^n intervals, whose union is denoted by F_n. Remove, from each, a small, open sub-interval so that the sum of the lengths of the intervals removed is $\epsilon/2^n$. The fat Cantor set is the intersection, $\bigcap_{i=1}^{\infty} F_i$.

We obtain F_1 from the unit interval by removing an interval of length $\epsilon/2$. Then F_2 is obtained by further removing a set of total length $\epsilon/4$, etc. Thus F is obtained by removing sets of total length ϵ from the unit interval since $\epsilon = \epsilon/2 + \epsilon/4 + \epsilon/8 \ldots$. ◆

At first consideration it might seem that the three types of Cantor sets defined above are different. Yet, from the point of view of topology, we will see that they are, in fact, the same. In all three examples, a key observation is that a point in any one of these Cantor sets is expressible as the intersection of smaller and smaller closed intervals. First, we need to articulate what we mean by the of the size of a set.

Definition 5.08 *Let $X \subseteq \mathbf{R}^n$; let $\delta(X) = l.u.b.(\delta(x,y))$ where $x, y \in X$, if this number exists. The number $\delta(X)$ is called the* **diameter of** X. *If $X = \emptyset$, define the diameter of X to be zero. If $X \neq \emptyset$ and $\delta(X)$ does not exist, then X has* **infinite diameter.**

The diameter of a set is not a topological property. The diameter of an open interval (a, b) in the line is $b - a$. Then $(0, 1)$ with diameter 1 and $(0, 2)$ with diameter 2 are homeomorphic intervals, but have different diameters. A line in the plane has infinite diameter, but is homeomorphic to an open interval which has finite diameter.

The next four propositions are used as tools in showing that our examples, called Cantor sets, are homeomorphic, Proposition 5.13.

Proposition 5.09 *Suppose $\{A_i\}$ is a collection of subsets of $\mathbf{R}^n$, and $z \in \bigcap_{i=1}^{\infty} A_i$. If $\lim_{i\to\infty} \delta(A_i) = 0$, then $\{z\} = \bigcap_{i=1}^{\infty} A_i$.* ▮ *(Problem5.7)*

Proposition 5.10 *Suppose $\{I_j\}_{j\in\mathbf{N}}$ is a collection of closed intervals such that, for all $j \in \mathbf{N}$, $I_{j+1} \subseteq I_j$. Then $\bigcap_{j=1}^{\infty} I_j \neq \emptyset$. In fact, if $I_j = [a_j, b_j]$, then $l.u.b.\{a_j\}_{j\in\mathbf{N}} \in \bigcap_{j=1}^{\infty} I_j$.*

Proof: The set $A = \{a_j\}$ is bounded above by b_1. By the least upper bound property, there is a least upper bound B for A. We claim that $B \in \bigcap_{j=1}^{\infty} I_j$. If not, then there is an N with $B \notin [a_N, b_N]$. We cannot have $B < a_N$ since B is an upper bound of A. Thus $b_N < B$. So now we have

$$a_1 \leq a_2 \leq \ldots b_{N+1} \leq b_N < B.$$

Define B' to be the point midway between b_N and B by $B' = B - \frac{B-b_N}{2}$; then

$$a_1 \leq a_2 \leq \ldots b_{N+1} \leq b_N < B' < B,$$

and we see that B' is an upper bound of A less than B. Having arrived at a contradiction, we conclude that $B \in \bigcap_{j=1}^{\infty} I_j$. ∎

Our goal is to show that the subsets we have defined, the Cantor set, fat Cantor set, and thin Cantor set are all homeomorphic, Proposition 5.13. For this we need to look at the topology for these examples. These examples can all be described as an infinite intersection of closed subsets. Thus, it is more convenient to describe topology in terms of the closed subsets rather than the open subsets. Propositions 5.11 and 5.12 lay the groundwork for this investigation.

Proposition 5.11 *Suppose A is a closed subset of the standard Cantor set K. Write $K = \bigcap_{i=1}^{\infty} C_i$, (see remark, page 133). For each i let A_i denote the union of all sub-intervals of C_i which contain a point of A. Then $A = \bigcap_{i=1}^{\infty} A_i$.*

Proof: From the definition of intersection, $A \subseteq \bigcap_{i=1}^{\infty} A_i$. We next show that if $x \notin A$, then $x \notin \bigcap_{i=1}^{\infty} A_i$. We argue by contradiction.

Suppose $x \in \bigcap_{i=1}^{\infty} A_i$ and $x \notin A$. Note that A is a closed subset of $\mathbf{R}^1$ since A is a closed subset of K, and K is a closed subset of $\mathbf{R}^1$, Proposition 2.20. So we can find an $\epsilon > 0$ such that $N_\epsilon(x) \cap A = \phi$. Let N be an integer such that $1/3^N < \epsilon$. Each of the sub-intervals of C_i has length $1/3^N$. We have $x \in A_N$ and there is a sub-interval J of A_N, which contains a point of A. But since J is of length $1/3^N$ it must be entirely contained in $N_\epsilon(x)$, which contains no points of A. This gives us our contradiction. ∎

In this discussion, we need to be careful about the possibility that a closed subset is empty. For example, note in the proposition above that if $A = \varnothing$, then $\bigcap_{i=1}^{\infty} C_i$ is an empty subset.

A key ingredient in the proof of Proposition 5.11 is that points of the set could be nicely described as the intersection of smaller and smaller subsets. The fact that these subsets are closed or that they are intervals is not used. We needed A to be closed, together with the fact that K is the intersection of sets whose "pieces" were of increasingly small diameter. Proposition 5.12 generalizes Propositions 5.09 and 5.11:

Proposition 5.12 *Suppose we have a collection of subsets $\{C_i\}$ of $\mathbf{R}^\mathbf{n}$; let $X = \bigcap_{i=1}^{\infty} C_i$. Suppose that each C_i can be written as a union of sets $\{P_\alpha^i\}$*

such that $\lim_{i\to\infty}(\max_\alpha \delta(P^i_\alpha)) = 0$. *Let A be a closed subset of X, and let* A_i *be the union of all the sets,* P^i_α, *whose intersection with A is non-empty. Then* $A = \bigcap_{i=1}^{\infty} A_i$. ∎ *(Problem 5.8)*

Proposition 5.13 *The standard Cantor set K, the thin Cantor set T, and the fat Cantor set F are all homeomorphic.*

Proof: We show that F and K are homeomorphic; the same technique shows K and T are homeomorphic.

We need to define a function $h : F \to K$ that is a homeomorphism. The basic problem is that points of F have not been explicitly defined, so it is not easy to directly define a function from F to K. So we are proceed indirectly. The strategy is to use Proposition 5.09 to code points of F; then to use this code to define our function.

We may write F_1 as the union of two disjoint closed intervals, J_0 and J_2, where J_0 lies to the left of J_2, in the same manner as in remark, page 133. We also write F_2 as the union of four disjoint closed intervals, J_{2_1}, J_{2_2}, J_{2_3}, and J_{2_4}, where J_{2_1} lies to the left of J_{2_2} which in turn lies to the left of J_{2_3} which lies to the left of J_{2_4}. Continuing, write F_3 as the union of intervals of the form F_{3_i}, etc. (We are using a different style for the subscripts for these sets since they do not correspond to decimal expansions of numbers in those intervals, as was the case for the other subsets as in remark, page 133.)

Let $x \in F$. We can find a sequence of intervals J_{i_j} such that $J_{i_j} \subseteq F_i$ for all n, and such that $\{x\} = \bigcap_{i=1}^{\infty} J_{i_j}$. Define $h(x)$ to be the point of K $\bigcap_{i=1}^{\infty} I_{i_j}$. In this definition we are using Propositions 5.09 and 5.10 which assert that the sets $\bigcap_{i=1}^{\infty} J_{i_j}$ and $\bigcap_{i=1}^{\infty} I_{i_j}$ are each, in fact, one point subsets.

We claim that h is a homeomorphism. It is not difficult to verify that h is a one-to-one correspondence. We leave this as an exercise,(Problem 5.9). Hint: Define the inverse of h.

To see that h is continuous, we show that the inverse of a closed subset of K is a closed subset of F. We write K as the intersection of subsets C_i where C_i is the union of 2^i intervals C_{i_j}; here the index j has values $1 \le j \le 2^i$. Let $J(i)$ be this set of values: $J(i) = \{1, 2, \ldots, 2^i\}$. Let A be a closed subset of K. By Proposition 5.12, we may write $A = \bigcap_{i=1}^{\infty} A_i$. Write each A_i as the union of sub-intervals of C_i; $A_i = \bigcup_{j\in A(i)} I_{i_j}$ where $A(i) \subseteq J(i)$. Define $B_i = \bigcup_{j\in A(i)} J_{i_j}$. So, for example, if $A_3 = A_{3_2} \cup A_{3_7}$, then $B_3 = B_{3_2} \cup B_{3_7}$. Now let $B = \bigcap_{i=1}^{\infty} B_i$. One then verifies that $h^{-1}(A) =$

B. We see that B is closed since it is the intersection of closed subsets. Thus h is continuous. We may similarly see that the inverse of h is continuous. Thus h is a homeomorphism. ∎

Example 5.14 The middle sevenths Cantor set. Define the **middle sevenths Cantor set** to be all those numbers which can be written as a decimal in the base seven system (allowing terminal sequences of 6's) using only the digits 0, 2, 4, or 6. ◆

Proposition 5.15 *The middle sevenths Cantor set is homeomorphic to the standard Cantor set.*

Proof (Outline): Given a point in the middle sevenths Cantor set, write it as a decimal in the base seven system (allowing terminal sequences of 6's) using only the digits 0, 2, 4, or 6. In this string, replace each 0 by the string 00, 2 by 02, 4 by 20, 6 by 22. We then obtain a string in the base 3 system which corresponds to a point of the standard Cantor set. This process defines a function from the middle sevenths Cantor set to the standard Cantor set. One can then show that this function is a homeomorphism. ∎

The following challenging question shows that there is much of interest to explore in this example:

Question 5.16 Is the standard Cantor set homogeneous?

On first analysis, it seems unclear what the answer should be. Each of the C_i is a union of intervals and these endpoints are special points of C_i. No C_i is homogeneous since there is no homeomorphism of C_i which takes an endpoint to a non-endpoint. Furthermore, any of these endpoints is a point of K. We will refer to such points as endpoints of K, but keep in mind that these are not endpoints of intervals of K, since in fact, K contains no intervals. It seems plausible that these endpoints of K should be special points of K and that there would be no homeomorphism of K that takes an endpoint of K to a non-endpoint.

On the other hand, one can find a lot of self-homeomorphisms of K. In fact, the answer to Question 5.16 is: yes, K is homogeneous.

Example 5.17 Here is one idea of how to show that K is homogeneous. Suppose, in base 3 notation, we take $x = 0.220202222\ldots$ and $y = 0.220022020\ldots$. We first describe a function f such that $f(x) = y$. We can describe $f(x) = y$ in the following way: change some of the 0's to 2's and some of the 2's to 0's. The idea is to articulate this change and apply it to other points of K.

For our x and y, the base 3 decimals agree in the first three decimal places, disagree in the 4th and 5th, agree in the 6th, disagree in the 7th, agree in the 8th, disagree in the 9th, etc: If, for example, $p =$

0.020202020... we change the 0 to a 2 if x and y agree, or change the 2 to a 0, if x and y disagree. Thus $f(p) = 0.020022222\ldots$:

$$\begin{array}{rcllllll} x & = & 0.220 & 20 & 2 & 2 & 2 & 2\ldots \\ y & = & 0.220 & 02 & 2 & 0 & 2 & 0\ldots \\ p & = & 0.020 & 20 & 2 & 0 & 2 & 0\ldots \\ f(p) & = & 0.020 & 02 & 2 & 2 & 2 & 2\ldots. \end{array}$$

Here are some details on how one can define f. For i and j, where i and j can have the values 0 or 2, define four functions α_{ij}:

$$\begin{array}{ll} \alpha_{00}(0) = \alpha_{22}(0) = 0 & \alpha_{00}(2) = \alpha_{22}(2) = 2 \\ \alpha_{02}(0) = \alpha_{20}(0) = 2 & \alpha_{02}(2) = \alpha_{20}(2) = 0. \end{array}$$

Note that α_{ij} is the identity map on the two element set $\{0, 2\}$ if $i = j$; otherwise, it switches these elements. Note that

$$a_n(y) = \alpha_{a_n(x)a_n(y)}(a_n(x)).$$

For any point of K, let a_n denote the number in the n-th position when this number is written in base 3 decimal notation; this is the same a_n as in Definition 5.01. Consider two points x and y in K written in base 3 decimal notation, and suppose that p denotes an arbitrary point of K. Define $f: K \to K$ by $q = f_{xy}(p)$ where

$$a_n(q) = \alpha_{a_n(x)a_n(y)}(a_n(p)).$$

Certainly we have $f(K) \subseteq K$. We leave it as an exercise to show that f is a homeomorphism (Problem 5.10). As noted, $f(x) = y$; thus K is homogeneous. ◆

5.3 Cantor sets in the plane

Our attempts to alter the definition of a Cantor set and obtain a new subset of the line, have not resulted to anything really new, where by new we mean a subset not homeomorphic to the standard Cantor set. Our next exploration will generalize this infinite construction to the plane in search of something new.

Example 5.18 Disks in the plane, (version 1). For $i \in \mathbf{N}$ we define a set D_i which a union of 2^i closed disjoint disks in the plane, as shown in Figure 5-5.

Note that $D_1 \supseteq D_2 \supseteq \cdots$. Let $X = \bigcap_{i=1}^{\infty} D_i$. It might at first seem that X might be different than the standard Cantor set K. In remark, page 133,

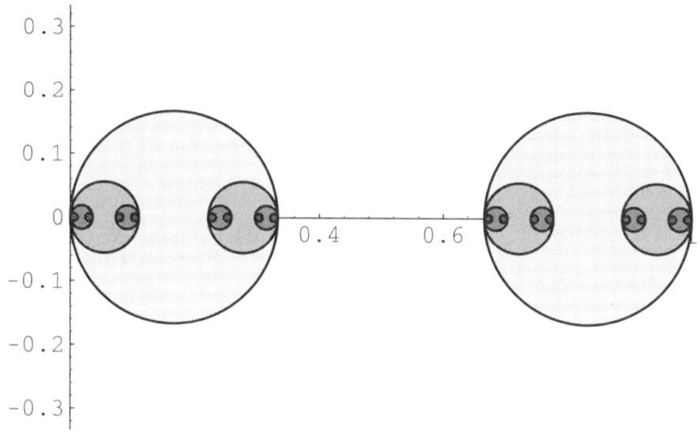

Figure 5-5 The first four sets, $D_1, \ldots D_4$, used in the definition of the subset of Example 5.18, shown with increasingly darker shades of gray.

we showed that the standard Cantor set K could be written $K = \bigcap_{i=1}^{\infty} C_i$ where each C_i is the union of 2^n of intervals. It is tempting to think that, since C_i and D_i are not homeomorphic, this would somehow imply that K and X were different. However, K and X *are* homeomorphic.

Think of the ternary Cantor set as a subset of the plane by using the standard inclusion map. Let $K' = i(K)$ where $i : \mathbf{R}^1 \to \mathbf{R}^2$ is the standard inclusion; then K' is homeomorphic to K. (We refer to K' as the **standard Cantor set in the plane**). Note that $X = K'$ thus K and X are homeomorphic. ◆

Next, we try a second construction so that the resulting subset does not lie on a line.

Example 5.19 Disks in the plane, (version 2—alternate vertical and horizontal placement). As before, we define a sequence of subsets. For each $i \in \mathbf{N}$ we define a set D'_i, consisting of a union of 2^i closed disjoint disks in the plane, as shown in Figure 5-6.

The difference between this and the previous example is that, for even values of i, the disks are placed above one another, and for i odd, they are placed horizontally. Again we have $D'_1 \supseteq D'_2 \supseteq \cdots$. Let $X' = \bigcap_{i=1}^{\infty} D'_i$.

Certainly, $X' \neq K'$, so the proof used for Example 5.18 won't work.

However, X' and K *are* homeomorphic. We sketch the proof; details are left as an exercise (Problem 5.12). Denote the sub-disks of D'_1 by J_0 and J_2. Then label the sub-disks of D'_2 by J_{00}, J_{02}, J_{20}, and J_{22}. In the discussion of remark, page 133, we introduced similar notation.

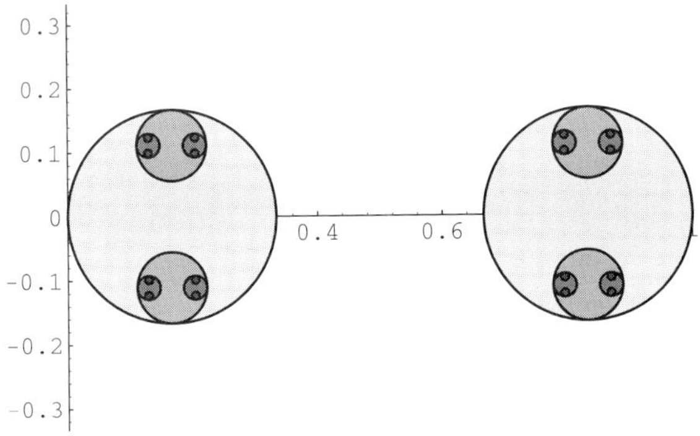

Figure 5-6 The first four sets, $D'_1, \ldots D'_4$, used in definition of set of Example 5.19, shown with increasingly darker shades of gray.

Next, label the eight sub-disks of D'_3 as J_{000}, J_{002}, J_{020},.... If $x \in X'$, then $x = \bigcap_{i=1}^{\infty} J_{n_i}$ where J_{n_i} is one of the disks mentioned above with $J_{n_i} \subseteq D'_i$ (here each n_i is a string of zeros or twos of length i). Define $h(x) = \bigcap_{i=1}^{\infty} I_{n_i}$ where the I_{n_i} are as in remark, page 133. It can be shown that h is a homeomorphism from X' to K. ◆

Here is yet another variation:

Example 5.20 Begin with the square region in the plane,

$$Q = \{(x, y) \in \mathbf{R}^2 : 0 \le x \le 1 \text{ and } 0 \le y \le 1\}.$$

Let $Q_1 = \{(x, y) \in Q \mid$ *both* x and y can be written as decimals in the ternary system, with either a 0 or a 2 in the first place to the right of the decimal point $\}$. A moment's reflection should convince the you that Q_1 is the union of four sub-squares of Q, one in each corner. In a similar way, one then defines sets Q_i for any integer i, so that Q_i is the union of 4^i squares. For example, Q_2 is all points $(x, y) \in Q$ such that x and y both can be written as decimals in the ternary system, using either a 0 or a 2 in the first two places to the right of the decimal point; see Figure 5-7. We then define $K^2 = \bigcap_{i=1}^{\infty} Q_i$.

One can show that K^2 is homeomorphic to K,(Problem 5.14). (Hint: One possible proof begins like this. It is enough to show that K^2 is homeomorphic to the middle sevenths Cantor set of Example 5.14. Call this middle sevenths Cantor set S. Our motivation for considering this is that one description of S involves an iterated dividing of intervals

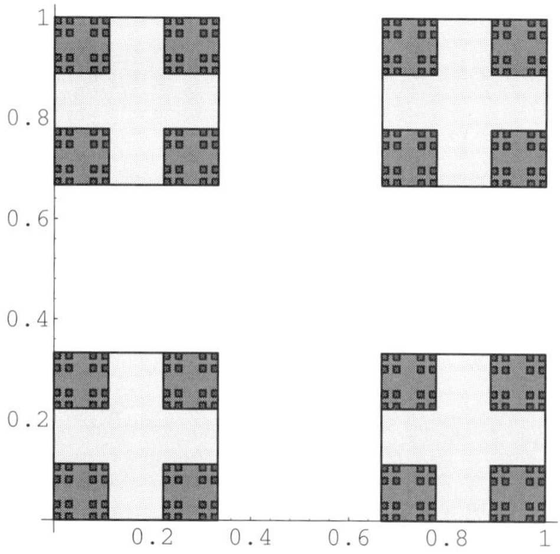

Figure 5-7 The first four sets, $Q_1', \ldots Q_4'$, used in the definition of the subset set of Example 5.20, shown with increasingly darker shades of gray.

into four parts, obtaining a sequence of sets S_i and writing $S = \bigcap_{i=1}^{\infty} S_i$. For example,

$$S_1 = [0, 1/7] \cup [2/7, 3/7] \cup [4/7, 5/7] \cup [6/7, 1].$$

Similarly, Q_1 has four sub-squares.) ◆

Actually, K^2 is the Cartesian product of the standard Cantor set with itself: $K^2 = K \times K$ (Problem 5.15). Note this is equality of *sets.*

In particular, the standard Cantor set K has the very unusual property that K is *homeomorphic* to $K \times K$!

It may seem that the reason that Examples 5.18, 5.19, and 5.20 end up being homeomorphic to the standard Cantor set is that, in many ways, a closed disk, or square, is a two-dimensional generalization of a closed interval. Perhaps we could use some other kinds of two-dimensional subsets for our basic sets. The plane contains some interesting subsets such as an annulus.

Definition 5.21 *Any set homeomorphic to $S^1 \times I$ is called an* **annulus**.

We leave as an exercise (Problem 5.18), to show that the region consisting of two concentric circles in the plane and all points between these circles is an annulus. Recall that, as we have defined Cartesian product, $S^1 \times I \subseteq \mathbf{R}^3$.

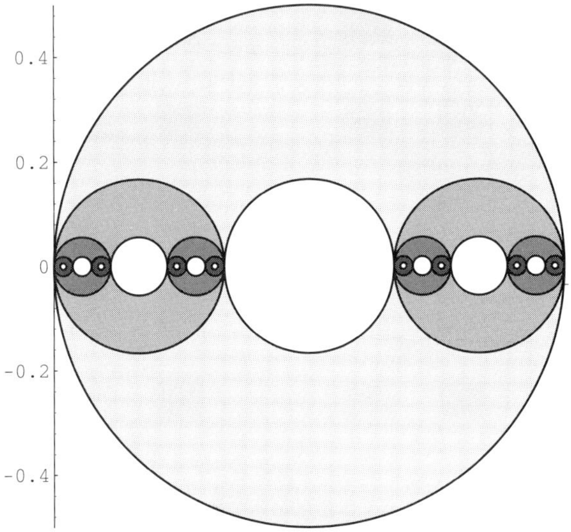

Figure 5-8 The first four sets, $A_1, \ldots A_4$, used in definition of set of Example 5.22, shown with increasingly darker shades of gray.

Here is a variation of the definition of a Cantor set, which uses annuli.

Example 5.22 Let A_1 be the annulus corresponding to two concentric circles with center $(1/2, 1)$ and with radii $1/2$ and $1/6$. For each $i \in \mathbf{N}$ we define a set A_i, consisting of a union of 2^i closed disjoint annuli in the plane, as shown in Figure 5-8. Note that we have $A_1 \supseteq A_2 \supseteq \ldots$. Let $A = \bigcap_{i=1}^{\infty} A_i$.

However, as in Example 5.18, a little reflection shows that $A = K'$. Thus A is homeomorphic to the standard Cantor set. ◆

Perhaps, in the example above, we didn't take advantage of the fact that the annulus has a "hole." We try one more example in the plane:

Example 5.23 Intersection of annuli, version 2. Let A_1 be the annulus, as in Example 5.22. For each $i \in \mathbf{N}$ define a set A'_i, consisting of a union of 2^i closed disjoint annuli in the plane as shown in Figure 5-9.

The difference here is that all of the annuli "go around" A_1. As before, $A'_1 \supseteq A'_2 \supseteq \ldots$. Let $A' = \bigcap_{i=1}^{\infty} A'_i$.

The intersection of A' and the x-axis is K' and so is homeomorphic to a Cantor set. However A' is not homeomorphic to a Cantor set. Later

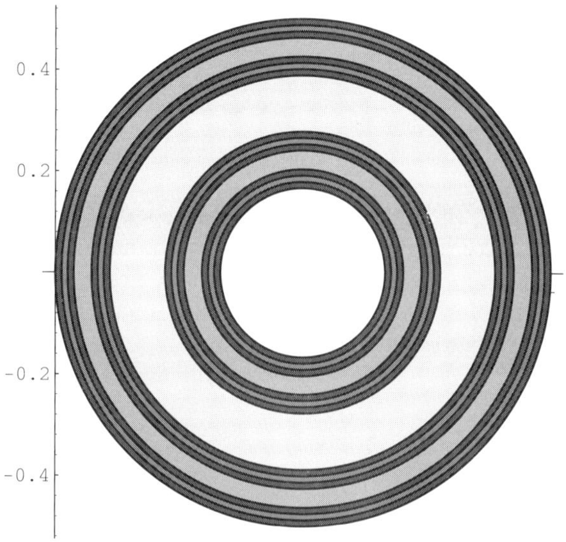

Figure 5-9 The first four sets, $A'_1, \ldots, A'_4$, used in the definition of the subset of Example 5.23, shown with increasingly darker shades of gray.

in this text, in Problem 7.24, we will show how to prove this. Roughly, the difference is that A' contains subsets which are homeomorphic to a circle, yet the standard Cantor set does not contain such sets.

We note that one thing that makes this example different from the others is that the "pieces" of the sets, A_i, do not have diameters that are getting smaller and smaller, even though the annuli are getting thinner and thinner.

In fact one can show (Problem 5.19) that A' is homeomorphic to $K \times S^1$. However, we have not yet established that $K \times S^1$ is not homeomorphic to K. ◆

5.4 Other infinite intersections

Example 5.24 This next example is frequently called **Sierpiński's gasket**.

Begin with an equilateral triangle, G_0. The subset $G_1 \subseteq G_0$ consists of the three equilateral triangles formed by the vertices of G_0 and the midpoints of the edges of G_0. In effect, we have removed an open sub-triangle. Continue inductively to divide each triangle of G_i into three smaller equilateral sub-triangles to obtain G_{i+1}; see Figure 5-10.

Figure 5-10 Four subsets, shown separately, G_0, G_1, G_2, and G_3, used to define Sierpiński's gasket; see Example 5.24.

Sierpiński's gasket, $\mathcal{G}$, is defined to be $\bigcap_i G_i$; see Figure 5-11. ◆

Figure 5-11 Five subsets, G_0, G_1, G_2, G_3, and G_4, shown in increasingly dark shades of gray, used to define Sierpiński's gasket; see Example 5.24.

Example 5.25 The next example, S, is known as **Sierpiński's carpet**. The simplest description of S uses the standard Cantor set K. Define $S = K \times I \cup I \times K$.

We can also view $S = \bigcap_{i=i}^{\infty} S_i$ where each S_i is obtained from the square $Q = \{(x, y) \in \mathbf{R}^2 : 0 \le x \le 1 \text{ and } 0 \le y \le 1\}$ by removing additional smaller and smaller sub-squares, in a process similar to the definition of the standard Cantor set. We describe the first two of these subsets.

Let $S_1 = \{(x, y) \in Q \mid$ *at least one* of x and y can be written as decimals in the ternary system, with either a 0 or a 2 in the first place to the right of the decimal point $\}$. An alternative description is that S_1 is Q with an open sub-square removed: $S_1 = Q - ((\frac{1}{3}, \frac{2}{3}) \times (\frac{1}{3}, \frac{2}{3}))$. Next, let $S_2 = \{(x, y) \in Q :$ *at least one* of x and y can be written as decimals in the ternary system, with either a 0 or a 2 in the first two places to the right of the decimal point $\}$. Then S_2 can be described as S_1 with 8 small open sub-squares removed; see Figure 5-12.

We leave the verification that $K \times I \cup I \times K = \bigcap_{i=i}^{\infty} S_i$ as an exercise (Problem 5.16). The key is to recall that we can write $K = \bigcap_{i=i}^{\infty} C_i$, and also note that $S_i = (C_i \times I) \cup (I \times C_i)$.

Although we cannot yet readily prove it with the tools we have, neither Sierpiński's gasket nor Sierpiński's carpet is homeomorphic to the standard Cantor set. One idea for a proof is to note that $\mathcal{G}$ and S contain subsets homeomorphic to a closed interval. For example, S contains the

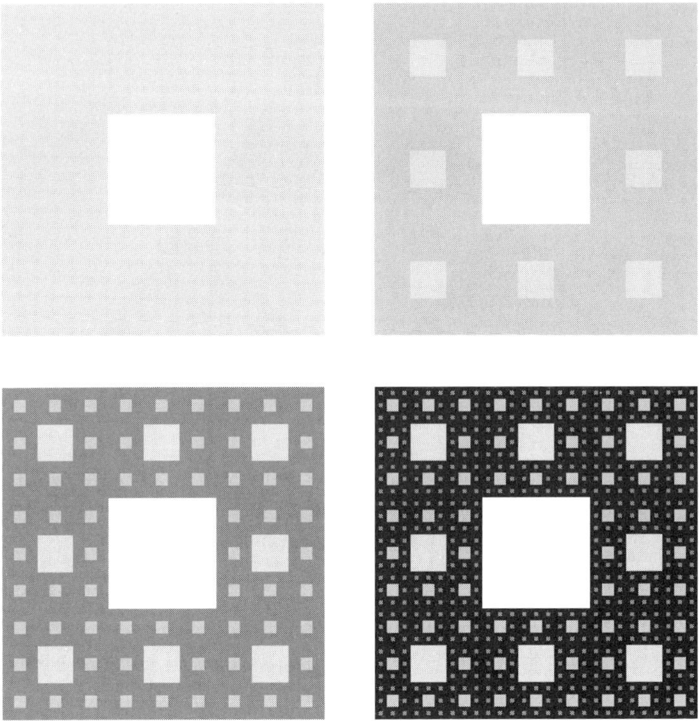

Figure 5-12 Four stages, S_1, S_2, S_3, and S_4, in the construction of Sierpiński's carpet, Example 5.25.

subset $A = \{(x, y) \in \mathbf{R}^2 : y = 0 \text{ and } 0 \leq x \leq 1\}$. We will later see, in Problem 7.18, that the standard Cantor set does not contain such a subset.

It might seem that Sierpiński's gasket and Sierpiński's carpet might be homeomorphic since they are defined by similar processes. Later, in Example 7.30, we will show that they are not homeomorphic. ◆

5.5 Cantor subsets in $\mathbf{R}^3$

We next consider subsets of $\mathbf{R}^3$. One obvious way to define a sort of three-dimensional Cantor set might be the following.

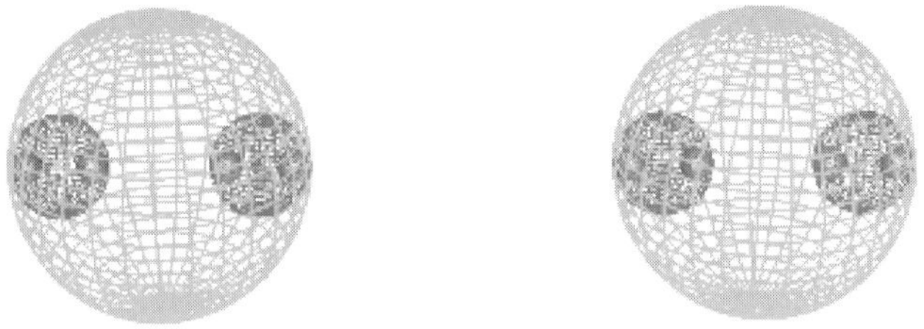

Figure 5-13 The first three sets, B_1, B_2, B_3, used in definition of set of Example 5.26, with spheres shown as wire-frames shown with increasingly darker shades of gray.

Example 5.26 For each $i \in \mathbf{N}$ define a set B_i to be of a union of 2^i closed, disjoint, solid, three-dimensional balls in space, with centers on the x-axis, as shown in Figures 5-13 and 5-14.

Again, $B_1 \supseteq B_2 \supseteq \ldots$. Let $B = \bigcap_{i=1}^{\infty} B_i$. However, one can see that B is homeomorphic to the standard Cantor set (Problem 5.20). (Hint: Note that B is a subset of the x-axis.) ◆

Example 5.27 There is a subset of $\mathbf{R}^3$, known as "Menger's sponge", which is a three-dimensional generalization of Sierpiński's carpet. We define $\mathcal{M}$ to be the union of three subsets of $\mathbf{R}^3$: $\mathcal{M} = X_1 \cup X_2 \cup X_3$. Here $X_1 = S \times I$ where S denotes Sierpiński's carpet. Let f be the rigid motion in $\mathbf{R}^3$ which interchanges the x-axis and z-axis; let g be the rigid motion which interchanges the y-axis and z-axis. To be specific, f and g can be defined, respectively, using matrix multiplication, by

$$\begin{pmatrix} 0 & 0 & 1 \\ 0 & 1 & 0 \\ 1 & 0 & 0 \end{pmatrix} \text{ and } \begin{pmatrix} 1 & 0 & 0 \\ 0 & 0 & 1 \\ 0 & 1 & 0 \end{pmatrix}.$$

Then $X_2 = f(X_1)$, and $X_3 = g(X_1)$. We might note that $X_2 = I \times S$.

Recalling the description in Example 5.25 of Sierpiński's carpet as an infinite intersection: $S = \bigcap_i S_i$; let $M_i = (S_i \times I) \cup f(S_i \times I) \cup g(S_i \times I)$. You should draw a sketch of M_1. In Figure 5-15 we show M_4. Then $\mathcal{M} = \bigcap_{i=1}^{\infty} M_i$.

Here is a third description of Menger's sponge, the style of iterated removal of "middles" of sets. Divide each side of the unit cube into

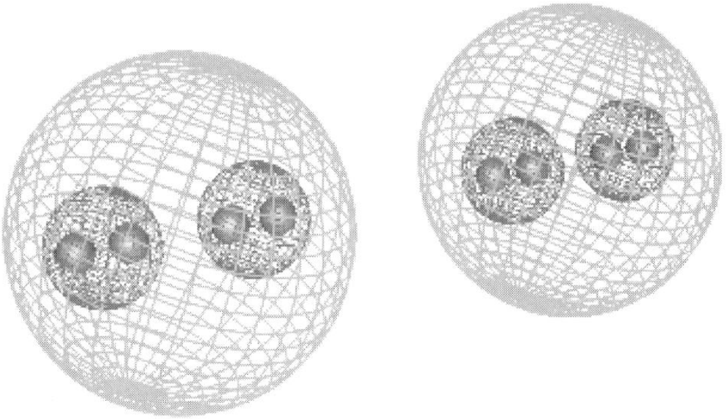

Figure 5-14 Another view of the subsets shown in Figure 5-13.

three, thus dividing the cube into 27 sub-cubes. From this, remove the middle seven cubes pictured in Figure 5-16.

The remaining 20 cubes is denoted by M_{1j}, where $j = 1, \ldots, 20$. We can verify that $M_1 = \bigcup_{j=1}^{20} M_{1j}$. Next, divide each of the cubes M_{1j} into 27 sub-cubes, remove middle seven cubes from each of these, leaving $20 \times 20 = 400$ of these smaller cubes. We can check that the union of these is M_2. (Figure 5-15 is the union of $400 \times 400 = 160,000$ cubes!) ◆

Example 5.28 Here is another famous subset of $\mathbf{R}^3$ known as "Antoine's necklace".

Let L_0 be a solid (closed) tubular region in space. Inside L_0 we put four smaller solid tubular regions, so that they are linked to form a chain, as shown in Figure 5-17. Let L_1 denote the union of these four smaller tubes. Next, L_2 consists of sixteen small tubes: four in each of the tubes of L_1, as shown in Figure 5-18. Figure 5-18 is complex—we offer Figures 5-19 and 5-20 to clarify that image.

Continuing, for each $i \in \mathbf{N}$ define L_i so that each L_i is the union of 4^n small solid tubes. Let $L = \bigcap_{i=1}^{\infty} L_i$. The set L is called "Antoine's necklace." This seems to be an interesting subset. However, the techniques we have established in the previous examples of this chapter can be put together to show that L is homeomorphic to the standard Cantor set. We sketch this proof.

We make a correspondence between points of L and points of the middle sevenths Cantor set. Each subset L_i, used to construct Antoine's necklace is a union of 4^i disjoint solid tubes, L'_j. We label each of these

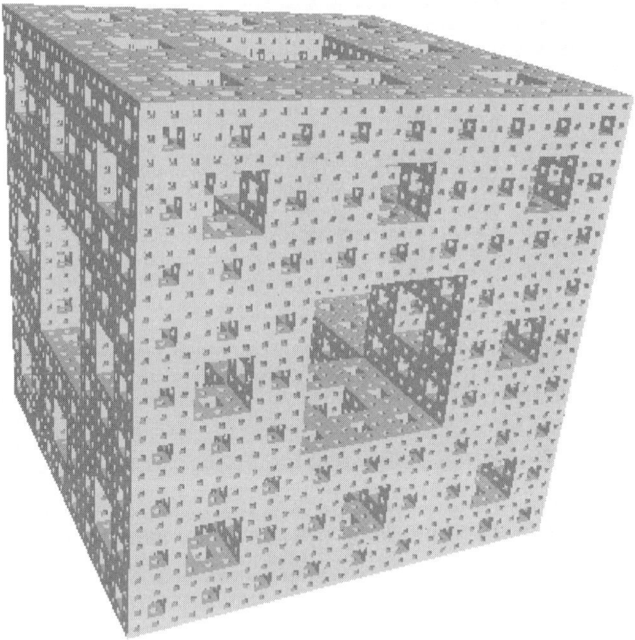

Figure 5-15 Menger's sponge can be written as $\mathcal{M} = \bigcap_{i=1}^{\infty} M_i$. Shown here is the set M_4; see Example 5.27.

tubes with a subscript consisting of a string of length j formed using the set $\{0, 2, 4, 6\}$. For example, we write

$$L_2 = L'_{00} \cup L'_{02} \cup L'_{04} \cup L'_{06} \cup L'_{20} \cup L'_{22} \cup L'_{24} \cup L'_{26} \cup \\ L'_{40} \cup L'_{42} \cup L'_{44} \cup L'_{46} \cup L'_{60} \cup L'_{62} \cup L'_{64} \cup L'_{66}.$$

Furthermore, we label so that if $n \le m$ and we have two substrings $s = a_1 a_2 \ldots a_n$ and $t = a_1 a_2 \ldots a_n \ldots a_m$, then $L'_s \subseteq L'_t$. So, for example, $L'_{426} \subseteq L'_{42} \subseteq L'_4$.

Next, we verify that for every point x in Antoine's necklace there is a unique infinite string, $\sigma = a_1 a_2 \ldots$, where $a_i \in \{0, 2, 4, 6\}$ such that $x = L'_{a_1} \cap L'_{a_1 a_2} \cap L'_{a_1 a_2 a_3} \ldots$.

The map ϕ which assigns to each $x \in L$ the real number written $\sum \frac{a_i}{4^i}$ gives a one-to-one correspondence between L and the middle sevenths Cantor set. Finally, one checks that ϕ is a homeomorphism. ◆

The example above is a bit unsettling. It really seems that this time, in some sense, we have come up with something "new," yet the set is homeomorphic to something "old". In the next chapter we introduce

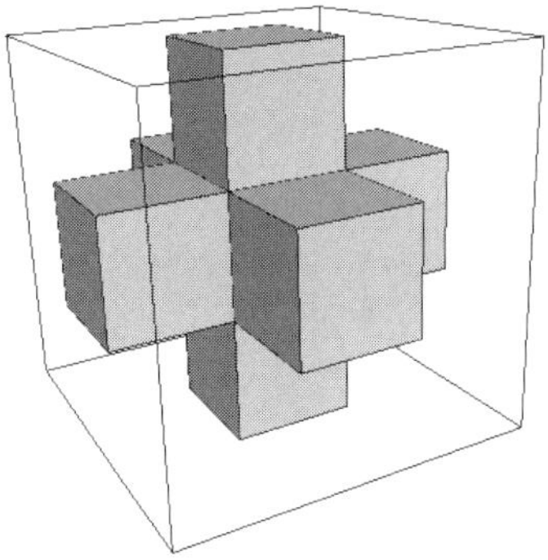

Figure 5-16 A subset that can be used to define Menger's sponge. If we divide a cube into 27 equal sub-cubes, the refer to the 7 sub-cubes shown as the" 7 middle cubes," see Example 5.27.

concepts that let us distinguish Antoine's necklace from the standard Cantor set. Roughly, what we will find is that these two homeomorphic subsets are distinguished by the way they are situated in $\mathbf{R}^3$. That is, we have two very different ways of placing the standard Cantor set into $\mathbf{R}^3$.

The standard Cantor set is of interest in branches of mathematics other than topology. Much of this interest has to do with the concept of measure. It is not our purpose to discuss measure theory in detail, but we can give a brief idea of what is involved for subsets of $\mathbf{R}^1$.

The concept of the measure gives us one way of describing the size of a subset. For an interval $J = [a, b]$, size is taken to be its length, $b - a$. We denote this by $\mu(J)$. (In the $\mathbf{R}^2$, the measure of a subset is related to the concept of area; in $\mathbf{R}^3$ it is related to volume.)

Measure theory in $\mathbf{R}^1$ extends this idea of measurement to non-interval subsets.

Definition 5.29 *Let* $X \subseteq \mathbf{R}^1$. *Suppose* $\{J_i\}_{i \in N}$ *is a collection of closed intervals that* $X \subseteq \bigcup_{i=1}^{\infty} J_i$ *such that* $\sum_1^\infty \mu(J_i)$ *converges. Define the* **measure** *of* X *to be the infimum of all possible numbers* $\sum_1^\infty \mu(J_i)$, *for all such collections of intervals.*

One mathematical way of saying that a set is small is to require that it be a set of measure zero. Any finite subset of the line has measure zero.

The standard Cantor set has measure zero. Recall that we write the Cantor set as $K = \bigcap_{i=i}^{\infty} C_i$ where C_i is the union of 2^i closed intervals of length $\frac{1}{3^i}$. Thus $\mu(C_i) = 2^i(\frac{1}{3^i}) = (\frac{2}{3})^i$. Since the infimum of the set

Figure 5-17 A chain of four small linked tubes, the second step in the construction of Antoine's necklace, Example 5.28. Note: Links have be drawn, alternating light and dark, as a visual aid.

Figure 5-18 A chain of sixteen small linked tubes, the third step in the construction of Antoine's necklace, Example 5.28. Note: Links have be drawn, alternating light and dark, as a visual aid.

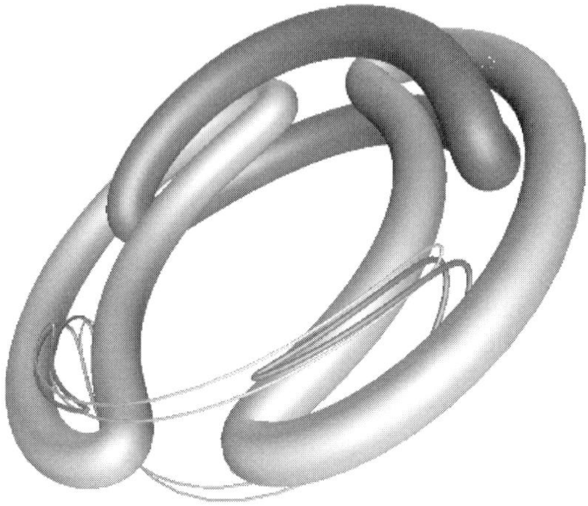

Figure 5-19 Some portions of the sets used to define Antoine's necklace, Example 5.28. Shown are three of the tubes of L_1 and the four tubes as shown in Figure 5-17.

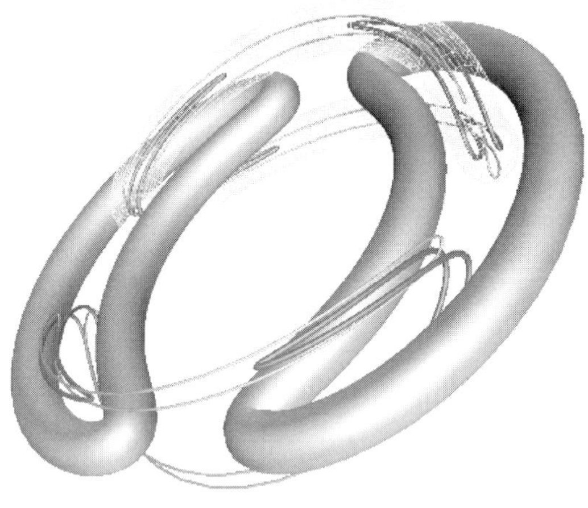

Figure 5-20 A variation of Figure 5-19. Here we also show one tube of L_1 in wire-frame mode, so that we can see how the links of l_2 fit inside.

of numbers $\{(\frac{2}{3})^i\}_{i\in\mathbf{N}}$ is zero, it follows that the standard Cantor set K has measure zero.

The standard Cantor set is of mathematical interest since it is small (measure zero) yet large (uncountable).

As one would expect, the measure of the interval [a, b] is $b - a$, but the proof of this is not easy and we do not present it here. Since all closed intervals are homeomorphic, the measure of a subset is not a topological property. The fat Cantor set, F, is homeomorphic to the standard Cantor set, but the measure of the fat Cantor set is $1 - \epsilon$. A basic fact about measure is that if $A \subseteq X$ then $\mu(X) = \mu(A) + \mu(X - A)$. In particular $\mu(F) = \mu(I) - \mu(I - F) = 1 - \mu(I - F)$. Recall $F = \bigcap_{i=1}^{\infty} F_i$. From the definition of F_1 we see that $I - F_1$ is a single open interval of length $\frac{\epsilon}{2}$, so $\mu(I - F_1) = \frac{\epsilon}{2}$. Next, $I - (F_1 \cap F_2)$ consists of the open interval $I - F_1$ plus two open intervals, each of length $\frac{\epsilon}{4}$; $I - (F_1 \cap F_2 \cap F_3)$ will consist of all three of these intervals, and in addition four intervals, each of length $\frac{\epsilon}{32}$. The total measure of all seven of these intervals is $\frac{\epsilon}{2} + 2\frac{\epsilon}{8} + 4\frac{\epsilon}{32} = \frac{\epsilon}{2} + \frac{\epsilon}{4} + \frac{\epsilon}{8}$. Continuing one can show that

$$\mu(F) = 1 - \mu(I - F) = 1 - \sum_{j=1}^{\infty} \frac{\epsilon}{2^j} = 1 - \epsilon.$$

We conclude that measure of the fat Cantor set is $1 - \epsilon$.

This shows that even "X has measure zero" is not a topological property.

One last topic concerning measure is the remarkable Cantor function.

Example 5.30 It follows from the proof of Proposition 5.02 that there is a function from the standard Cantor set K to the unit interval I which is a one-to-one correspondence between these two sets. This map is not continuous (Problem 5.3). We cannot find a function from K to I which is both one-to-one and continuous since, as we will be later able to show using Proposition 10.38, such a map would be a homeomorphism.

However, we *can* show that there is a continuous onto function $f : K \to I$. We will define a continuous function $F : I \to I$, then define $f = F|K$. We follow the construction used in Proposition 5.02. If $x \in K$, then x can be written as a string in the ternary decimal system, using only 0's and 2's. Define $F(x)$ to be the decimal in the binary decimal system obtained by replacing each 2 by a 1 in this expression for x. If $x \notin K$, then x lies in a unique open interval I_s where s is a string in the symbols 0 and 1; see remark, page 133. In this case we define $F(x)$ to be the number written as $0.s$ in the binary decimal system. It is helpful to look at a particular interval. Consider, for example the interval $[1/3, 2/3]$. In the base 3 system the endpoints of this interval are written as $0.0\overline{2}$ and $0.2\overline{0}$, while other points are of the form $0.1x$

where x is a string using 0, 1 and 2.

Note that $F(0.0\overline{2}) = 0.0\overline{1} = 0.1$ The map f is certainly not a one-to-one map since $F(0.1x) = F(0.0\overline{2}) = F(0.2\overline{0}) = 0.1$ where we use base 3 for domain and base two for range of F. Note that f maps K onto I. We leave it as an exercise to verify that F is, in fact, continuous (Problem 5.21). The map F is called the "Cantor function."

If $x \notin K$, then it is contained in an open interval (one of the I_j's) such that the restriction of F to this interval is a constant. At each of these points, one can define the notion of a derivative, and that derivative is zero. So F has the remarkable property that it is continuous, has zero derivative almost everywhere (that is, for all points except a set of measure zero), and yet it is not a constant function. In fact, it takes on all values in I!

An alternate formulation of the Cantor function is found in (Problem 5.22) ◆

Example 5.31 The Cantor set occurs naturally in many problems in topology. Here is one such example.

One way to study a function is to look at the iterates of the function. Let $f : X \to Y$ be a continuous function. In Definition 3.24 we introduced the notation $f^{(n)}$ for the composition of f with itself $n - 1$ times.

Consider a function $f\colon I \to \mathbf{R}^1$ where I is the unit interval. Take one value of x and consider all points $\{f^{(j)}(x) : \text{for all } j\}$. This is the orbit of x under f. Of course, $f^{(j)}(x)$ would not be defined if $f^{(j-1)}(x)$ is less than zero, or greater than 1. Define $Z_1 = f^{-1}(I)$. Since f is continuous, Z_1 is a closed (possibly empty) subset of I. We can see that Z_1 is the set of points such that $f^{(2)} = f \circ f$ is defined. More generally, we can define, for any i, $Z_i = (f^{(i)})^{-1}(I)$. It is not hard to verify that $Z_1 \supseteq Z_2 \supseteq Z_3 \supseteq \ldots$. Now, let $Z_\infty = \bigcap_{i=1}^{\infty} Z_i$. Then Z_∞ is a (possibly empty) set on which we could study $f^{(n)}$ for any n.

Consider the particular function $f(x) = -3|x - 1/2| + 3/2$. (Figure 5-21 shows the graphs of f, $f^{(2)}$, $f^{(3)}$, and $f^{(4)}$). Here Z_∞ is the ternary Cantor set (Problem 5.23). That a Cantor set should appear is not an isolated example. For similar examples, consult a text dealing with one-dimensional dynamical systems, such as [13]. ◆

We next discuss some interesting subsets of the plane, which arise from variations of our definition of the half Cantor set.

Definition 5.32 *Let b be a fixed number. Define the* **pseudo-binary numbers with base** *b, denoted $N(b)$, to be the set of all numbers of the form $\sum_1^\infty a_n b^n$ where, for all n, either a_n is 0 or 1. We refer to b as the* **base** *of $N(b)$.*

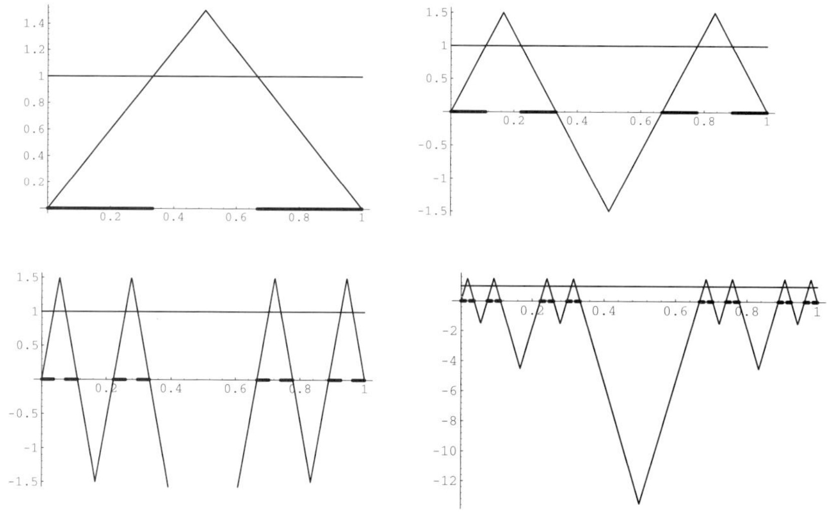

Figure 5-21 Four iterations of the function $f(x) = -3|x - 1/2| + 3/2$. Note the sets Z_1, Z_2, Z_3, and Z_4 (inverse images of [0,1]) are represented as bolder line segments on the x-axis; see Example 5.31

Note that $N(1/2)$ is the unit interval since, if $x = \sum_1^\infty a_n(1/2)^n$, then $x = 0.a_1a_2a_3...$ is the binary decimal expression for x. A little reflection shows that $N(1/3)$ is the half Cantor set, Example 5.05.

We consider other values of b. Certainly, $N(0)$ consists of only the number 0. Also, $N(1)$ is the set of natural numbers. (The expression $\sum_1^\infty 1(1)^n$ is not a number since this series doesn't converge; thus it does not correspond to an element of $N(1)$.) Because of the convergence requirement, it should be clear that the interesting examples of $N(b)$ are when $|b| < 1$.

We leave as an exercise: determine $N(-1/2)$ and $N(-1/3)$ (Problem 5.24).

The definition of $N(b)$ makes sense even if b is a complex number! Again, to avoid convergence problems, we restrict our attention to the case $|b| < 1$ where now $|b|$ denotes the modulus of the complex number b. It is not easy to describe the sets $N(b)$ for complex numbers, but we can get some kind of idea by using computer graphics, Figure 5-22. (For more information; see [47]; an alternate description is found in [5]).

Figure 5-22 shows $N(0.5 + 0.5i)$, , $N((0.5 + 0.5i)^2)$, $N(0.45 + 0.5i)$, and $N((0.17 + 0.68i)^3)$, (top left, top right, lower left and lower right, respectively). The case of $N((0.5 + 0.5i)^2)$ seems of particular interest since it looks like it might be homeomorphic to K^2 of Example 5-7.

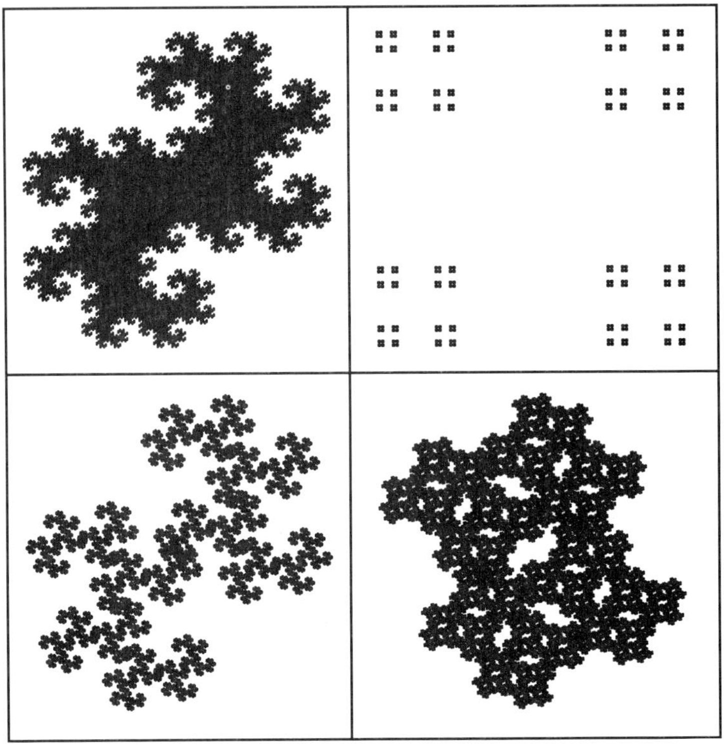

Figure 5-22 Some complex Cantor sets for four choices of base; see the remark on page 157

Indeed, it can be shown, for many values of b, that $N(b)$ is homeomorphic to the standard Cantor set. This gives rise to the following (difficult but interesting) question.

Question 5.33 Consider the set, N, of all subsets of the plane, $N(b)$, where b is a complex number. How many different homeomorphism types of sets are there in N?

*5.6 General topology and Chapter 5

The following proposition shows the inherent power of general topology. It also shows the utility of product topology for infinite products.

Proposition 5.34 *Let X be the two-point space $X = \{0, 1\}$ with the discrete topology. The standard Cantor set is homeomorphic to the product of X with itself, a countable number of times, $\prod_1^\infty X$.*

Proof: A point of $\prod_1^\infty X$ corresponds to a binary string, and we have seen (in the proof of Proposition 5.02) there is a one-to-one correspondence, call it h, between the set of such strings and points of the standard Cantor set. One needs to verify that h is a homeomorphism. Verification of this is an excellent exercise in the use of, and the appreciation for, the product topology. ∎

In a similar way one can see that the middle sevenths Cantor set, Example 5.14, is homeomorphic to a countable product, $\prod_1^\infty Y$, where Y is a four-point set with the discrete topology. One way to prove that the middle sevenths Cantor set is homeomorphic to the standard Cantor set is to prove that $\prod_1^\infty Y$ is homeomorphic to $\prod_1^\infty X$.

The point of view of infinite products makes some properties of the Cantor set, K, easier to understand; in particular, the fact that $K \times K$ is homeomorphic to K, remark, page 143, since the product of two countable products of two-point sets is a countable product of two-point sets.

Viewing the Cantor set as an infinite product can also be helpful in understanding why the Cantor set is homogeneous, Question 5.16, since endpoints of the Cantor set are not particularly distinguished in $\prod_1^\infty X$.

In terms of general topology, the most important result about the Cantor set is found in Proposition 10.54.

5.7 Problems for Chapter 5

5.1 Verify the claim in the proof of remark, page 133, that the standard Cantor set is the same as $\bigcap_{i=1}^{\infty} C_i$.

5.2 Which of the following numbers are in the standard Cantor set: 1/2, 1/3, 1/4, 1/5?

5.3 Using the map described in the proof of Proposition 5.02, together with the map described in proof of Proposition C.13, we can get a one-to-one function from K to I. Show that this is not continuous. Are there any points for which this map is continuous?

5.4 Verify the statement made in remark, page 133, that for the standard Cantor set, K, every point of K is a limit point of K.

5.5 Verify the claim in remark, page 133, that the map f is a homeomorphism.

5.6 Verify that the map, F, of Example 5.05 is a homeomorphism.

5.7 Prove Proposition 5.09.

5.8 Prove Proposition 5.12.

5.9 Verify the details of the proof of Proposition 5.13 that $h(x)$ is a one-to-one correspondence and that $h^{-1}(A) = B$.

5.10 Verify that the map f, defined in Example 5.17, is a homeomorphism of K onto itself.

5.11 Prove the assertion of Example 5.18 that $X = K'$.

5.12 Complete the proof, begun in Example 5.19, that K^2 and K are homeomorphic.

5.13 Prove: If X is homeomorphic to the standard Cantor set K and U is an open non-empty subset of X, then U contains uncountably many points.

5.14 Complete the proof of the assertion of Example 5.20, that K^2 is homeomorphic to S.

5.15 Verify the assertion of Example 5.3 that Q is $K \times K$.

5.16 Verify the claim of Example 5.25 that $K \times I \cup I \times K = \bigcap_{i=1}^{\infty} S_i$.

5.17 Show that Example 5.25, Sierpiński's carpet, S, does not have the fixed-point property.

5.18 Consider the subset of the plane $A = \{(x, y) \in \mathbf{R}^2 : 1 \leq x^2 + y^2 \leq 4\}$.

(a) Prove that A is an annulus.
(b) Prove the general assertion of remark, page 143. (A special case is found in Example 4.52.)

5.19 Verify the assertion of Example 5.23, that A' is homeomorphic to $K \times S^1$.

5.20 Show the subset B of Example 5.26 is homeomorphic to the standard Cantor set.

5.21 Verify the continuity of the Cantor function, F, of Example 5.30.

5.22 Show that the function, F, of Example 5.30 can also be described as follows. Suppose $x \in I$. Write x as a ternary decimal $x = 0.a_1a_2a_3 \ldots$. Let N be the smallest integer such that a_n is 1; if there is no such integer, let $N = \infty$. Let $b_n = a_n/2$ for $n < N$; let $b_N = 1$. Show that $\sum_1^N b_n/2^n$ is well-defined and is equal to $F(x)$.

5.23 Verify the assertion of Example 5.31, that Z_∞ is the ternary Cantor set.

5.24 Referring to Definition 5.32, what is $N(-1/2)$? $N(-1/3)$?

5.25 Let $X \subseteq \mathbf{R}^1$ denote the set of irrational numbers. Show that there is a subset $Y \subseteq X$, with Y homeomorphic to the standard Cantor set. (Hint: Define $Y = \bigcap_{i=1}^{\infty}$ where Y_i is a union of 2^i disjoint closed sub-intervals of I such that Y_i does not contain any of the numbers q/i where $0 \leq q \leq i$.) So Y_1 is a union of two closed intervals in I which do not contain 0, 1/2 or 1.

6. EMBEDDINGS

OVERVIEW: An embedding is a placement of one set into another. We define embedding, and discuss notions of equivalent embeddings and equivalent subsets, including the notion of stable equivalence.

6.1 Basic examples of embeddings

Once again we address the question: When are two subsets of $\mathbf{R^n}$ the same? The fact that Antoine's necklace, Example 5.28, is homeomorphic to the standard Cantor set does not really tell us all we want to know about this subset.

There is a topological distinction that can be made between Antoine's and the standard Cantor set. Roughly, the distinction we will find is not *what the subsets are* (in fact, they are homeomorphic) but *how the are placed* in space.

Definition 6.01 *Suppose* $X \subseteq \mathbf{R^n}$, $Y \subseteq \mathbf{R^m}$, *and* $f: X \to Y$. *We say* f *is an* **embedding of** X **into** Y, *if* f *maps* X *homeomorphically onto its image.*

The definition above is most often used when f is not an onto map since, if f is an embedding of X *onto* Y, then f is, in fact, a homeomorphism.

Example 6.02 For $n \le m$, the standard inclusion map $i: \mathbf{R^n} \to \mathbf{R^m}$, see Definition 3.20, is an embedding, sometimes called the " canonical embedding of $\mathbf{R^n}$ into $\mathbf{R^m}$". ◆

Definition 6.03 *Suppose* $X \subseteq \mathbf{R^n}$, $Y \subseteq \mathbf{R^m}$, *and* $b \in Y$. *Define a function* $i_b: X \to X \times Y$ *by* $i_b(x) = x \times b$. *Similarly, if* $a \in X$, *define a function* $i_a: Y \to X \times Y$ *by* $i_a(y) = a \times y$. *Such maps* i_a *and* i_b *are called* **inclusion maps of the factor into the product.**

Suppose $X \subseteq \mathbf{R^n}$, $Y \subseteq \mathbf{R^m}$, and $b \in Y$, with $b = (b_1, b_2, \ldots, b_m)$. We can write $i_b(x_1, x_2, \ldots, x_n) = (x_1, x_2, \ldots, x_n, b_1, b_2, \ldots, b_m)$. Similarly, if x is a point of X with $a = (a_1, a_2, \ldots, a_n)$, then $i_a(y_1, y_2, \ldots, y_m) = (a_1, a_2, \ldots, a_n, y_1, y_2, \ldots, y_m)$.

Proposition 6.04 *For any fixed $x \in X$, or for any fixed $y \in Y$, the functions i_x or i_y are embeddings.* ▮ *(Problem 6.1)*

Example 6.05 If $X = [-2, 2]$ and $Y = [-1, 1]$ are closed intervals in $\mathbf{R^1}$, then the map $f(x) = x/3$ defines an embedding of X into Y. ◆

Example 6.06 Define $f\colon \mathbf{R^1} \to S^1$ by $f(x) = (\cos(x), \sin(x))$. The restriction $f|_{(0,2\pi)}$ is an embedding; but the restriction to $f|_{[0,2\pi)}$ is not since, otherwise, we would contradict the fact, established in Example 4.28, that these subsets are not homeomorphic. ◆

Definition 6.07 *If $A \subseteq X$, the map $f\colon A \to X$ defined by $f(a) = a$ is called the* **inclusion map**.

The inclusion map is always an embedding (Problem 6.2).

Example 6.08 The function $f(x) = (\arctan(x), 0)$ is an embedding of $\mathbf{R^1}$ into $\mathbf{R^2}$, whose image is the same as the image, under the canonical embedding, of the interval $(-\pi/2, \pi/2)$. ◆

The graph of a continuous function provides us with a source of simple, but useful, embeddings.

Proposition 6.09 *Let $X \subset \mathbf{R^n}$ and let $f\colon X \to \mathbf{R^m}$ be a continuous map. Define a map $G\colon X \to \mathbf{R^{n+m}}$ by $G(x) = x \times f(x)$; then G is an embedding of X.* ▮

To write G in terms of coordinates, write

$$f(x_1, x_2, \ldots,, x_n) =$$

$$(f_1(x_1, x_2, \ldots,, x_n), f_2(x_1, x_2, \ldots,, x_n), \ldots, f_m(x_1, x_2, \ldots, x_n)).$$

Then, $G(x_1, x_2, \ldots,, x_n) = (x_1, x_2, \ldots, x_n,$

$$f_1(x_1, x_2, \ldots,, x_n), f_2(x_1, x_2, \ldots,, x_n), \ldots, f_m(x_1, x_2, \ldots,, x_n)).$$

The image of G is the graph of f (recall Definition 3.27).

The embedding G for a graph of a function defined above is very special in that the *image* of G can be used to determine the function. It is important to remember that, generally, the image of an embedding does *not* determine an embedding. This topic comes up when one studies parametric curves. A given curve has many different parameterizations. In calculus these are often viewed as different motions of a point along this curve. For example $(\cos(t), \sin(t))$ corresponds to motion along unit circle in the counter-clockwise direction with unit velocity; $(\cos(t), -\sin(t))$ corresponds to motion along unit circle in the clockwise direction with velocity 2.

If the curve is the graph of a function $y = f(x)$, then there is a "natural" parameterization using $x = t$ as parameter. For example, if $y = x^2$, we parameterize the graph as (t, t^2). Proposition 6.09 is a generalization of this concept to functions of several variables.

6.2 Equivalent embeddings; equivalent subsets

Example 6.10 The half Cantor set, the thin Cantor set, the fat Cantor set, and the middle sevenths Cantor set (Examples 5.05, 5.06, 5.07, and 5.14) are examples of embeddings of the standard Cantor set into the line. Examples 5.18, 5.19, and 5.20 are examples of embeddings of the Cantor set into the plane. Example 5.26 and Antoine's necklace, Example 5.28, are examples of embeddings of the Cantor set into three-dimensional space. ◆

As defined, the standard Cantor set is a subset of $\mathbf{R}^1$ and Antoine's necklace A is a subset of $\mathbf{R}^3$. We can view the standard Cantor set K as a subset of three-dimensional space, in a natural way, as a subset of the x-axis. In more detail, let f be the restriction to K of the standard inclusion of $\mathbf{R}^1$ into $\mathbf{R}^3$, and let $K' = f(K)$. Since A is homeomorphic to K there is a $g: K \to \mathbf{R}^3$ that g maps K homeomorphically onto A. Both f and g are embeddings of K into $\mathbf{R}^3$.

Here are the sorts of questions we would like to articulate. Are these two *embeddings*, f and g, in some sense, the same? Are these two *subsets* K' and A, in some sense, the same? The next two definitions give us ways of expressing an idea of "sameness."

Definition 6.11 *Let X and Y be subsets of Z, where $Z \subseteq \mathbf{R}^{\mathrm{m}}$. We say X and Y sets are* **equivalent subsets** *of Z, if there is a homeomorphism h of Z to itself such that $h(X) = Y$.*

Often we will be concerned with the case that $Z = \mathbf{R}^{\mathrm{m}}$ for some m.

Equivalent subsets must be homeomorphic. If X and Y are equivalent subsets of Z via a homeomorphism, $h: Z \to Z$, then $h|_X$ is a homeomorphism from X to Y.

Also, it is not difficult to show that "X and Y are equivalent subsets of Z" is an equivalence relation on the set of all subsets of Z.

Definition 6.12 *Suppose $X \subseteq \mathbf{R^n}$ and $Y \subseteq \mathbf{R^m}$. If f and g are two embeddings of X into Y, we say f and g are* **equivalent embeddings** *if there is a homeomorphism $h: Y \to Y$ such that $g = h \circ f$.*

Note carefully the distinction between these two definitions. The first speaks of *subsets* being equivalent, the second speaks of *embeddings* (that is, functions) being equivalent.

The following diagram shows the relations of the spaces and functions of Definition 6.12:

$$\begin{array}{ccc} & X & \\ f_0 \swarrow & & \searrow f_1 \\ Y & \xrightarrow{h} & Y. \end{array}$$

Example 6.13 We give an example of a pair of non-equivalent planar subsets X and Y. Roughly speaking, X is a circle with a line segment attached from the inside, whereas Y is a circle with the line segment attached from the outside.

Let S^1 be the unit circle in the plane and let $A = \{(x, y) \mid 1/2 \leq x \leq 1$ and $y = 0\}$, $B = \{(x, y) \mid 1 \leq x \leq 3/2$ and $y = 0\}$. Consider $X = A \cup S^1$, and $Y = B \cup S^1$; see Figure 6-1.

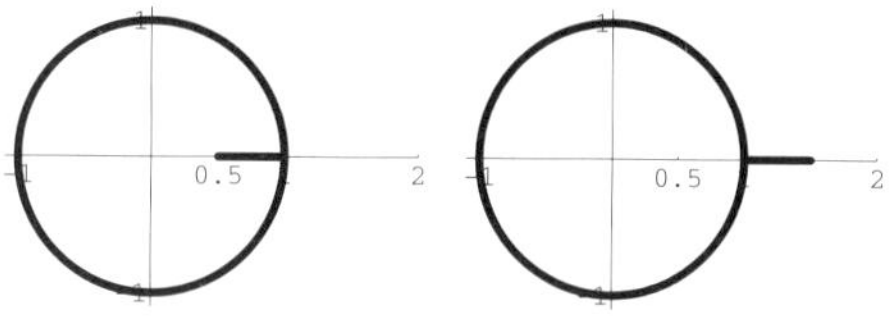

Figure 6-1 Two non-equivalent subsets of the plane. On the left, X, on the right, Y, of Example 6.13.

Using the gluing lemma, Proposition 3.45, one sees that X and Y are homeomorphic. We call this example the "lollypop example" since Y looks something like a flat lollypop. Later, in Example 10.22, we will show these are not equivalent subsets of the plane. ◆

Example 6.14 Consider the lollypop example in the plane, X and Y of Example 6.13. As mentioned, we will show these are not equivalent subsets of the plane. However, if we remove the origin $\vec{0}$ from the plane, then the corresponding subsets *are* equivalent as subsets of $Z = \mathbf{R}^2 - \vec{0}$.

What is notable in our proof, is the strategy used to define the homeomorphism of Z. There are two techniques used to break the problem into smaller problems; these techniques will prove useful in other problems we encounter. The first is to express the desired homeomorphism as a composition of homeomorphisms. The second technique is to write

Z as union of two subsets, define simple maps on each part, and employ the gluing lemma to get a homeomorphism of Z. Consider the homeomorphism g of Z, defined by using polar coordinates $g(r,\theta) = (1/r,\theta)$. This not exactly what we want since $g(X) \neq Y$. The homeomorphism g sends the unit circle to itself, but it sends the subset A to the set $\{(x,y) \in \mathbf{R}^2 \mid 1 \leq x \leq 2 \text{ and } y = 0\}$; see middle figure of Figure 6-2.

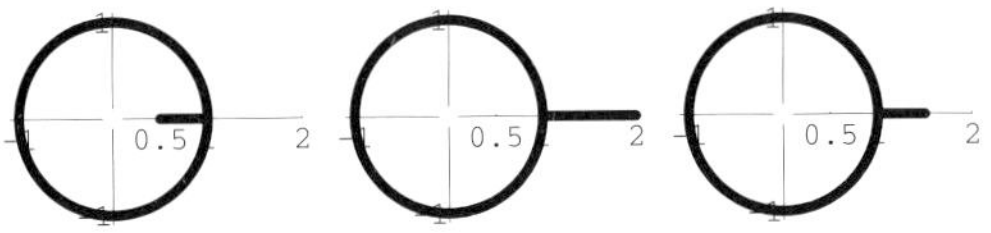

Figure 6-2 Three equivalent subsets of Z, the plane with origin removed; see Example 6.14. On left X, in the middle, $g(X)$, on the right, Y.

To fix this, we define a homeomorphism, s, which is the identity on the unit disk. Outside the unit disk, s shrinks the plane towards the unit disk by a factor of 1/2. Let $Z_1 = \{(r,\theta) \mid 0 < r \leq 1\}$, and let $Z_2 = \{(r,\theta) \mid 1 \leq r\}$. Define a map s of Z

$$s(r,\theta) = \begin{cases} (r,\theta) & (r,\theta) \in Z_1 \\ (\frac{r+1}{2},\theta) & (r,\theta) \in Z_2. \end{cases}$$

It is not difficult to verify that s is a homeomorphism. In fact $s|_{Z_1}$ and $s|_{Z_2}$ are embeddings. Continuity of s can be verified by using the gluing lemma; note that Z_1 and Z_2 are closed subsets of Z. Let $h = s \circ g$; then h is a homeomorphism of Z such that $h(X) = Y$. ◆

The next proposition says that equivalent embeddings give equivalent subsets.

Proposition 6.15 *Suppose $X \subseteq \mathbf{R}^n$ and $Y \subseteq \mathbf{R}^m$. If f and g are equivalent embeddings of X into Y, then $f(X)$ and $g(X)$ are equivalent subsets of Y.*

Proof: If h is the homeomorphism of Y such that $g = h \circ f$, h is a homeomorphism of Y that sends $f(X)$ to $g(X)$. ∎

However, the converse of Proposition 6.15 is not true, as the following example demonstrates.

Example 6.16 Let $X \subseteq \mathbf{R}^1$ be the union of two closed intervals: $X = [0,1] \cup [3,4]$. We give an example of two inequivalent embeddings f and g of X into $\mathbf{R}^1$ where $f(X)$ and $g(X)$ are equivalent subsets of $\mathbf{R}^1$. In fact, we will have $f(X) = g(X)$.

Define $f: X \to \mathbf{R}^1$ to be the inclusion map of X. Roughly, g is the identity on $[3,4]$ but reverses $[0,1]$; see Figure 6-3 for the graph of g. Define $g: X \to \mathbf{R}^1$

$$g(x) = \begin{cases} 1 - x & \text{if } 0 \le x \le 1 \\ x & \text{if } 3 \le x \le 4. \end{cases}$$

One can readily verify that g is an embedding and that $f(X) = g(X)$

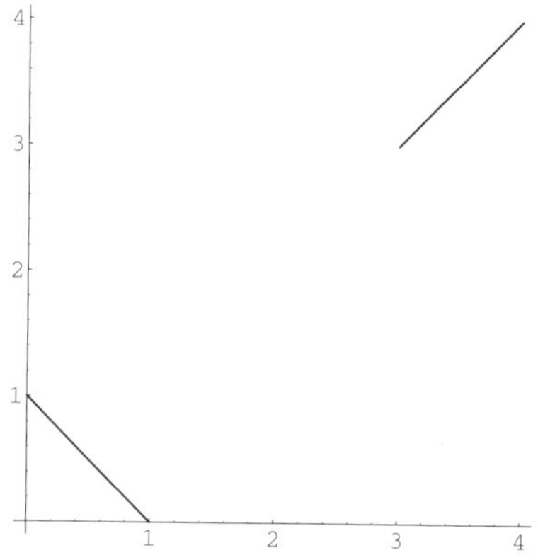

Figure 6-3 A graph of the embedding g of Example 6.16

However, these are not equivalent embeddings. Suppose there were a homeomorphism, $h: \mathbf{R}^1 \to \mathbf{R}^1$, such that $h \circ f = g$. By Proposition 4.23, h is a strictly monotone function. Since f is a restriction of the identity map, then $h \circ f$ would also be strictly monotone. However, g is not monotone since $g(0) = 1$, $g(1) = 0$, and $g(3) = 3$. Thus f and g cannot be equivalent embeddings. ◆

It is worth noting that the argument given in Example 6.16 centered about analysis of a finite set of points—the endpoints of the intervals. Let E be these endpoints: $E = \{0, 1, 3, 4\}$. The crux of the argument is that the embedding of E into $\mathbf{R}^1$, which is the inclusion map, is not equivalent to the embedding g, which is defined by $g(0) = 1$, $g(1) = 0$, $g(3) = 3$ and $g(4) = 4$.

Although there are such inequivalent embeddings of a subset consisting of four points, we next show that all four-point subsets are equivalent subsets. The point here is to emphasize the distinctions between equivalent *subsets* and equivalent *embeddings*.

Suppose that $X = \{a, b, c, d\}$ is a set four distinct points of $\mathbf{R}^1$. We may assume they have been labeled so that $a < b < c < d$. Suppose we have a second set, $X' = \{A, B, C, D\}$, a set four distinct points of $\mathbf{R}^1$, labeled so that $A < B < C < D$. Then X and X' are equivalent subsets of $\mathbf{R}^1$. To prove this, we define a piecewise-linear monotone function $h: \mathbf{R}^1 \to \mathbf{R}^1$ such that $h(X) = X'$. (Problem 6.4)

The following proposition shows one relation between the idea of equivalent subsets and equivalent embeddings. In a sense the distinction between the idea of equivalent subsets and equivalent embeddings of X has to do with self-homeomorphisms of X. This accounts, in part, for our interest in self-homeomorphisms.

Proposition 6.17 *Suppose $X \subseteq \mathbf{R}^n$, $Y \subseteq \mathbf{R}^m$, and f_0 and f_1 are two embeddings of X into Y. Let $X_0 = f_0(X)$ and $X_1 = f_1(X)$, and suppose X_0 and X_1 are equivalent subsets of Y. Then there is a homeomorphism, G of X, such that f_0 and $f_1 \circ G$ are equivalent embeddings of X.*

Proof: Suppose f_0 and f_1 are two embeddings of X into Y, and let $X_0 = f_0(X)$ and $X_1 = f_1(X)$. Suppose X_0 and X_1 are equivalent *subsets* of Y, and let h be a homeomorphism of Y such that $h(X_0) = X_1$. Let $f_1' = h \circ f_0$; then f_0 and f_1' are equivalent embeddings. So equivalent subsets give rise to equivalent embeddings. The distinction here is that the functions f_1 and f_1' may be different even though they have the same image. Consider the function $G = f_1^{-1} \circ f_1' = f_1^{-1} \circ h \circ f_0 in$—this is a homeomorphism of X to itself. We can see that f_0 and $f_1 \circ G$ are equivalent embeddings of X. Diagrammatically, we have changed from the diagram on the left to the diagram on the right:

$$\begin{array}{ccc} & X & \\ f_0 \swarrow & & \searrow f_1 \\ Y & \xrightarrow{h} & Y \end{array} \qquad \begin{array}{ccc} X & \xrightarrow{G} & X \\ \downarrow f_0 & f_1' \searrow & \downarrow f_1 \\ Y & \xrightarrow{h} & Y. \end{array}$$

In these diagrams, the triangular diagram on the left does not commute, but the lower triangle in the diagram on the right does. Now one can verify that f_1 and f_1' are equivalent embeddings if and only if G is the identity map of X. ∎ (Problem 6.5)

Suppose we have two subsets, A and B, of X. How can we show that these are not equivalent subsets? The most common strategy is to use the next proposition.

Proposition 6.18 *If $F: A \to X$ and $G: A \to X$ are equivalent embeddings, then $X - F(A)$ is homeomorphic to $X - G(A)$.* ∎ *(Problem 6.3)*

Example 6.19 In $\mathbf{R}^1$ let $X = (-1, 1)$ and $Y = [-1, 1]$. Let $f: X \to Y$ be the inclusion map, and define $g: X \to Y$ by $g(x) = x/2$. Note that $f(X) = (-1, 1)$ and $g(X) = (-1/2, 1/2)$. We use the Proposition 6.18 to show that f and g are not equivalent embeddings. If f and g were equivalent embeddings, then $Y - f(X)$ would be homeomorphic to $Y - g(X)$. But $Y - f(X) = \{-1, 1\}$ whereas $Y - g(X) = [-1, -1/2] \cup [1/2, 1/2]$. Since there is no one-to-one correspondence of these sets, they are not homeomorphic. ◆

6.3 Some subsets of $\mathbf{R}^3$ which do not embed in $\mathbf{R}^2$

Example 6.20 We describe a simple subset G of $\mathbf{R}^3$ which does not embed in $\mathbf{R}^2$. The set G is sometimes called the "utilities graph." Suppose there are three utilities—gas, water, and electricity—located at points A, B, and C, and three houses located at A', B', and C'. The line segments represent the connections of each utility to each house. These line segments should not intersect, except at the six given points. (Another common name for this subset, which comes from the theory of graphs, is $K(3,3)$.)

To be specific, choose coordinates: $A = (1,0,0)$, $B = (-\frac{1}{2}, \frac{\sqrt{3}}{2}, 0)$, $C = (-\frac{1}{2}, -\frac{\sqrt{3}}{2}, 0)$ and $A' = (0,0,1)$, $B' = (0,0,0)$, $C' = (0,0,-1)$; see Figure 6-4. The points A, B, and C all are equally spaced on the unit circle in the xy-coordinate plane (in fact, they correspond to the three complex cube roots of unity); the points A', B' and C' lie on the z-axis. The set G is the union of the nine line segments $\overline{AA'}$, $\overline{AB'}$, $\overline{AC'}$, $\overline{BA'}$, $\overline{BB'}$, $\overline{BC'}$, $\overline{CA'}$, $\overline{CB'}$, and $\overline{CC'}$; see Figure 6-4.

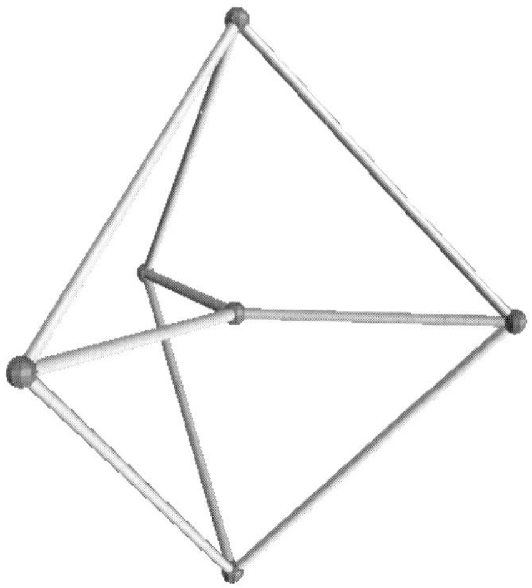

Figure 6-4 The utilities graph in $\mathbf{R}^3$; see Example 6.20. The vertices are shown as small darkened spheres. At the top of the figure is A', at the bottom C', between them, in the middle, is B'.

We outline a proof that there is no embedding of G into the plane. Some details of the proof depend on results which seem intuitively true but which, like the Jordan Curve Theorem, are not proved in this text.

Let θ be the union of the line segments $\overline{AA'}$, $\overline{AB'}$, $\overline{AC'}$, $\overline{BA'}$, $\overline{BB'}$, and $\overline{BC'}$. The subset θ is homeomorphic to the letter θ, in other words, θ is homeomorphic to the union of a circle and a diameter.

Suppose there were an embedding e of G into $\mathbf{R}^2$. What could we say about the image of e? We should first note that e is *not* required to send a line segment of G to a line segment of $\mathbf{R}^2$. If J is one of the line segments of G, then $e(J)$ is a curve in the plane. Since e is an embedding, this curve cannot cross itself. Furthermore, if J and J' are two line different segments of G, the only allowable intersections of $e(J)$ and $e(J')$ are the images of the six points used to define G.

Note that the union of the line segments $\overline{AA'}$, $\overline{A'B}$, $\overline{BC'}$, and $\overline{C'A}$ is homeomorphic to a circle and so, by the Jordan Curve Theorem, the complement of the image, under e, of this set will have two pieces—intuitively these are points inside and an outside the curve.

It may seem clear, but is not easy to prove, that $\mathbf{R}^2 - e(\theta)$ has three components. (We do not prove this here. It is a generalization of the Jordan Curve Theorem, Proposition 4.10; its proof follows from a standard theorem of algebraic topology known as "Alexander Duality", [15, 33].)

In Figure 6-5, we show the image of one such $e(\theta)$ where the regions of $\mathbf{R}^2 - e(\theta)$ are labeled R_1, R_2, and R_3. The point $e(C)$ must be in one of these regions.

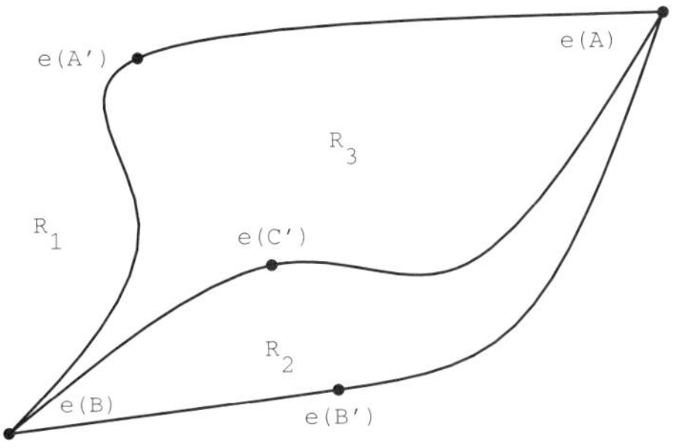

Figure 6-5 A schematic for an attempt to embed the utilities graph in the plane; see Example 6.20. Shown is an image of a possible $e(\theta)$.

Consider possibilities for $e(C)$. If we had an embedding e of G we could connect $e(C)$ to each of the points $e(A')$, $e(B')$, and $e(C')$ by disjoint curves which do not meet $e(\theta)$, except at the points $e(A')$, $e(B')$, and $e(C')$. But this is not possible since each of the three curves that make up $e(\theta)$ lie on the boundary of only two of these components. For example (consult Figure 6-5) could $e(C) \subseteq R_3$? We can join a point of R_3 to either $e(A')$ or $e(C')$ by path in R_3. However any path from a point of R_3 to $e(B')$ would introduce unallowable self-intersections of the image $e(G)$. ◆

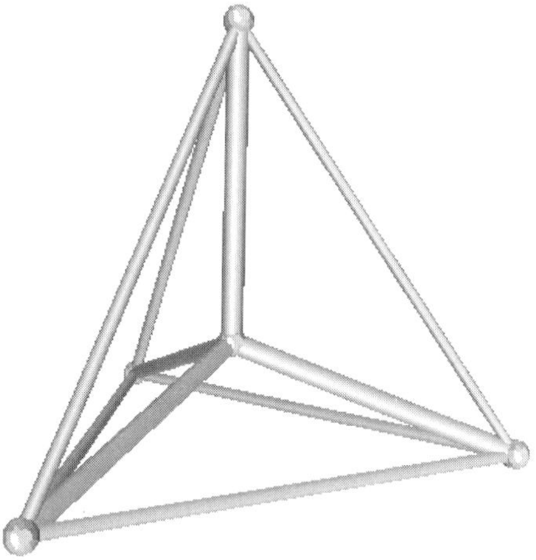

Figure 6-6 This topological graph also does not embed in the plane, it is called the "complete graph on five points", K_5; see the remark on page 171. These points are A, B, C, D, and E. Because of the symmetry of the graph, the exact labeling will not matter for our discussion, but it might be helpful to think of the central vertex as A.

Example 6.21 The topological graph K_5 with five vertices shown in Figure 6-6, called the "complete graph with five points", also does not embed in the plane.

Here is roughly the argument, along the lines of our argument in Example 6.20. Label the center vertex A and the rest, B, C, D, and E. Consider the sub-graph G of K_5, consisting of the vertices B, C, D, and E and all edges between any two of these. If we have an embedding, $f: K_5 \to \mathbf{R}^2$, then $f|_G$ would be an embedding of G into the plane.

It seems clear, but is not easy to prove, that $\mathbf{R}^2 - G$ has four components. This is another generalization of the Jordan Curve Theorem, Proposition 4.10. An example of such an embedding is indicated in Figure 6-7 where the components of $\mathbf{R}^2 - G$ are labeled R_1, R_2, R_3, and R_4.

Now, $f(A)$ must be in one of these regions. Suppose $f(A) \in R_1$. The boundary of R_1 is a simple closed curve γ which is the image under f of the edges of K_5 from C to E, from E to B, and from B to C. By the Jordan Curve Theorem, $\mathbf{R}^2 - \gamma$ has two components, one inside γ, the other outside. We have $f(A)$ on the inside of γ, and $f(D)$ on the

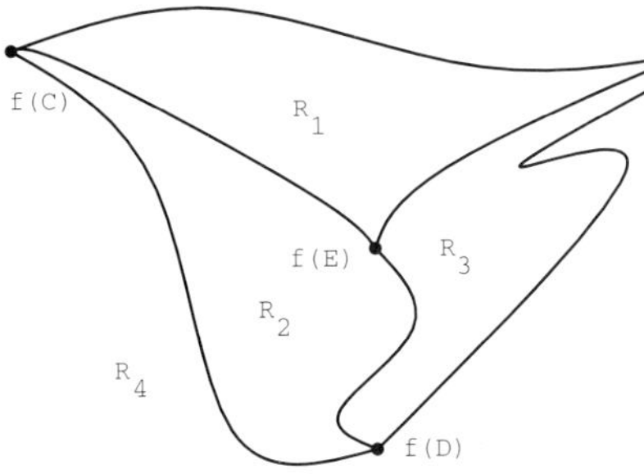

Figure 6-7 A possible embedding into $\mathbf{R}^2$ of the sub-graph G where $G \subseteq K_5$; see sketch of proof in Example 6.21

outside. But then the image the edge of K_5 from A to D will intersect γ, violating the assumption that f is an embedding. ◆

6.4 Stable embeddings

Up to this point, putting a subset of $\mathbf{R}^1$ into $\mathbf{R}^2$ by using the standard inclusion map has little effect on our considerations. However, in the case of equivalence of embeddings, it can have a marked effect as the following example shows.

Example 6.22 We give an example of non-equivalent embeddings f and g into $\mathbf{R}^1$, which become equivalent if we consider them, in a natural way, to be embeddings into $\mathbf{R}^2$.

Consider the embeddings of $X = [0,1] \cup [3,4]$ into $\mathbf{R}^1$, discussed in Example 6.16. Let i denote the standard inclusion map from $\mathbf{R}^1$ to $\mathbf{R}^2$, and define two maps from X to $\mathbf{R}^2$ by $F = i \circ f$ and $G = i \circ g$. Since f, g and i are embeddings, so are F and G. Although f and g are not equivalent embeddings, we show that F and G *are* equivalent embeddings.

As in the proof in Example 6.14, an important aspect is the illustration of general techniques used to construct homeomorphisms. In addition to the two techniques of mentioned in that example, we have a third technique: we find a subset homeomorphic to a Cartesian product, then use this product structure to define a homeomorphism.

Let $A = f([0,1])$, $B = f([3,4])$; then $F(X) = G(X) = A \cup B$. The idea is to define a homeomorphism of the plane which rotates A about its midpoint by an angle of π, yet leaves B fixed. In the plane, rotations are most easily described if the rotation is about the origin. Let A' be the set of points which, in Cartesian coordinates, is described as

$$A' = \{(x, y) \in \mathbf{R}^2 \mid -1/2 \le x \le 1/2 \text{ and } y = 0\}.$$

Define

$$B' = \{(x, y) \in \mathbf{R}^2 \mid 5/2 \le x \le 7/2 \text{ and } y = 0\}.$$

The strategy for defining our homeomorphism is

(a) shift the plane half a unit to the left (then A will be shifted to A' and B shifted to B');
(b) define a map ρ_0 which fixes B' and has the effect of flipping A';
(c) finally, translate back to the original position, sending A' to A and B' to B.

To help keep track of what happens to A we look the endpoints, p and q, of A where $p = (0,0)$ and $q = (1,0)$. To define ρ_0, write the plane as the union of three pieces: a disk of radius 1/2, an annulus between unit circle and a concentric circle of radius 1/2, and all the points on the unit circle and outside of it. Specifically, in polar coordinates, let

$$\begin{aligned} Z_1 &= \{(r,\theta) \mid r \le 1/2\}, \\ Z_2 &= \{(r,\theta) \mid 1/2 \le r \le 1\}, \text{ and} \\ Z_3 &= \{(r,\theta) \mid 1 \le r\}. \end{aligned}$$

Let ρ be the homeomorphism of $\mathbf{R}^2$ which, in polar coordinates, is given by $\rho(r,\theta) = (r, \theta + \pi)$. A homeomorphism of the plane which has the same effect on A' but does not move points outside the standard unit disk is

$$\rho_0(r,\theta) = \begin{cases} (r, \theta + \pi) & (r,\theta) \in Z_1 \\ (r, \theta + 2\pi(1-r)) & (r,\theta) \in Z_2 \\ (r,\theta) & (r,\theta) \in Z_3. \end{cases}$$

To get a feel for the map ρ, note that if C_R is the circle $C_R = \{(r,\theta) \in \mathbf{R}^2 \mid r = R\}$ of radius R, ρ_0 maps each C_R homeomorphically onto itself. The interesting part of ρ_0 happens on points of Z_2; see Figure 6-8.

Let T be the affine translation of the plane which shifts points a distance of 1/2 to the left; $T(x, y) = (x - 1/2, y)$. Let $p' = T(p)$ and $q' = T(q)$. Note that $\rho(p') = q'$ and $\rho(q') = p'$. Define $h = T^{-1} \circ \rho_0 \circ T$. Note that $h(p) = q$ and $h(q) = p$. Then h is a homeomorphism and $h \circ F = G$, showing the two embeddings are equivalent. ◆

Here is a simple observation about embeddings:

Proposition 6.23 *Suppose $X \subseteq \mathbf{R^n}$ and $Y \subseteq \mathbf{R^m}$. Then "there exists an embedding of X into Y" is a topological property of X and a topological property of Y.* ▮

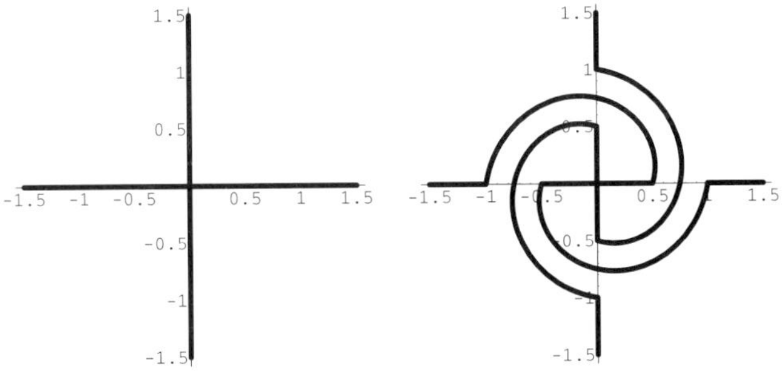

Figure 6-8 To get an understanding of the map ρ_0 of Example 6.22, let X denote the union of the axes, shown as bold lines on the left; the image, $\rho_0(X)$, is shown on the right.

Example 6.24 Admittedly, we have not filled in all details of the proof of Example 6.20. Nevertheless, using Proposition 6.23, we deduce a powerful consequence: $\mathbf{R}^3$ is not homeomorphic to $\mathbf{R}^2$ since the utilities graph embeds in $\mathbf{R}^3$ but not in $\mathbf{R}^2$. ◆

For $n \leq m$, the natural inclusions $\mathbf{R}^\mathbf{n} \subseteq \mathbf{R}^\mathbf{m}$ give rise to an important phenomenon in the topology of $\mathbf{R}^\mathbf{n}$. Recall Examples 6.16 and 6.22. We had two embeddings in the line that were not equivalent, but if we thought of these as embeddings in the plane in a very natural way, they were equivalent. To discuss this more clearly, we need a definition.

Definition 6.25 *Suppose we have a subset X, of some Euclidean space, and two embeddings of X into other Euclidean spaces, $f: X \to \mathbf{R}^\mathbf{n}$ and $g: X \to \mathbf{R}^\mathbf{m}$. The* **embeddings f and g are stably equivalent**, *if, for some dimension k, $i \circ f$ is equivalent to $j \circ g$ where $i: \mathbf{R}^\mathbf{n} \to \mathbf{R}^k$ and $j: \mathbf{R}^\mathbf{m} \to \mathbf{R}^k$ are standard inclusion maps.*

Example 6.26 The embeddings f and g of Example 6.16 are not equivalent, but Example 6.22 shows they are stably equivalent. ◆

We can similarly make a definition of stable equivalence of *subsets.*

Definition 6.27 *Suppose $A \subseteq \mathbf{R}^\mathbf{n}$ and $B \subseteq \mathbf{R}^m$. We say A* **and** *B* **are stably equivalent subsets** *if, for some dimension k, $i(A)$ and $j(B)$ are equivalent subsets of $\mathbf{R}^k$ where $i: \mathbf{R}^\mathbf{n} \to \mathbf{R}^k$ and $j: \mathbf{R}^\mathbf{m} \to \mathbf{R}^k$ are standard inclusion maps.*

Any two embeddings of a finite set of points into $\mathbf{R}^1$ are stably equivalent. We have actually seen a special case of this. Recall Example 6.16 where we had two embeddings, f and g of $X = [0,1] \cup [3,4]$. Consider the set of four points $E = \{0, 1, 3, 4\}$. Let $f' = f|_E$, $g' = f|_E$. A reexamination of Example 6.16 and Example 6.22 will show that f' and g' are inequivalent embeddings of E into $\mathbf{R}^1$, but are stably equivalent (since they become equivalent in $\mathbf{R}^2$) (Problem 6.4).

More generally, since points of $\mathbf{R}^1$ are ordered but, for $1 < n$, points of $\mathbf{R}^n$ are not, one can show:

Proposition 6.28 *Let X be a finite set of three or more points of $\mathbf{R}^n$. Then there are inequivalent embeddings of X into $\mathbf{R}^1$. However, any two embeddings of a finite set X into $\mathbf{R}^2$ are equivalent.* ∎ *(Problem 6.7)*

Proposition 6.29 *Suppose $X \subseteq \mathbf{R}^n$ and $Y \subseteq \mathbf{R}^m$. "Any two embeddings of X into Y are equivalent embeddings" is a topological property of X and of Y.* ∎

6.5 Answering one question and asking many more

Proposition 6.29, together with Proposition 6.28, implies that $\mathbf{R}^1$ and $\mathbf{R}^2$ are not homeomorphic, thus answering Question 4.53.

We have reached an important level in our study of topology. In Chapter 1, we began with some general questions, asking: what is $\mathbf{R}^1$, $\mathbf{R}^2$, $\mathbf{R}^3$? In particular, what are the subsets of these spaces? With the concepts we have developed thus far, we are now able to articulate more interesting, sharper questions. Even though we are not able to answer these questions yet (and many will remain beyond the level of this book), this is significant progress.

Question 6.30 Are any two subsets of $\mathbf{R}^1$, that are homeomorphic to the standard Cantor set, equivalent subsets of $\mathbf{R}^1$? For example, consider Examples 5.05, 5.06, 5.07, and 5.14. Which of these, if any, are equivalent subsets of $\mathbf{R}^1$?

Question 6.31 Are any two subsets of $\mathbf{R}^2$, that are homeomorphic to the standard Cantor set, equivalent subsets of $\mathbf{R}^2$? Consider three subsets of $\mathbf{R}^2$ which are homeomorphic to the Cantor set: Examples 5.18,5.19, and 5.20. Which ones, if any, are equivalent subsets of $\mathbf{R}^2$?

Question 6.32 Are any two subsets of the plane, which are homeomorphic to a circle, equivalent subsets of the plane ?

A "yes" answer to Question 6.32 would imply the Jordan Curve Theorem (Problem 6.21). In fact, it is true that any subsets of the plane homeomorphic to a circle are equivalent. This is a consequence of the Schoenflies theorem.

Question 6.33 Suppose X is a set such that there is at least one embedding of X into $\mathbf{R}^1$. Are any two embeddings of X into $\mathbf{R}^1$ equivalent if we include them into $\mathbf{R}^2$?

Question 6.34 Are there examples of embeddings which are not stably equivalent ?

In fact, of the list above, the only question we will discuss later, in more detail, is the last of these.

Here is an important exercise for testing one's understanding of the material we have considered to this point: add (at least) three more questions to the list above (Problem 6.25).

*6.6 General topology and Chapter 6

General topology, as we have seen, is quite different from the study of subsets of $\mathbf{R}^n$. Some of the examples of topological spaces we have considered cannot be embedded into any $\mathbf{R}^n$. For example, the three-point non-Hausdorff space, Example 1.52. If we had an embedding of a finite topological space into $\mathbf{R}^n$, it would have the discrete topology.

This raises an important question:

Question 6.35 Which topological spaces can embed in $\mathbf{R}^n$ for some n?

Example 6.36 Any subset of $\mathbf{R}^n$ is a metric space, and thus is a Hausdorff space and a first countable space. So the non-Hausdorff Examples *1.8, (see remark page 34), Example 1.82, and the non-first countable $\mathbf{R}^{\mathbf{R}}$, see remark, page 69, are three examples of topological spaces which cannot embed in any $\mathbf{R}^n$. ◆

The Sorgenfry line, Example 1.46, does not embed in $\mathbf{R}^n$. To prove this we consider another topological property of $\mathbf{R}^n$. First, a definition (recall the definition of "basis", Definition 1.66):

Definition 6.37 *Suppose $\{X, \mathcal{T}\}$ is a topological space. We say X* **is second countable** *if there is a countable basis for $\mathcal{T}$.*

For any n, $\mathbf{R}^n$ is second countable. For a countable basis, use the set of all neighborhoods $N_\epsilon(x)$ where ϵ is rational, and each of the n coordinates of x is rational.

A subspace of a second countable space is also second countable.

On the other hand, the Sorgenfry line X is not second countable. Let $\mathcal{B}$ be a basis for X. For any real number a, consider an interval of the form $[a, a+1)$. This is an open subset in this topology. Let $B_a \in \mathcal{B}$ denote some basis element with $a \in B_a \subseteq [a, a+1)$. If $a \neq b$, then $B_a \neq B_b$; for example, if $a < b$ then $a \notin [b, b+1)$ and so $a \notin B_b$. By selecting on subset for each real number, we select an uncountable collection elements of $\mathcal{B}$. Thus any basis for the Sorgenfry line must be uncountable. From this we conclude that there is no embedding of X into any $\mathbf{R}^n$.

A closely associated idea is:

Definition 6.38 *Suppose $\{X, \mathcal{T}\}$ is a topological space. We say X* **is separable** *if X has a countable dense subset.*

A separable metric space, with countable dense subset D, must be second countable—a countable basis consists of neighborhoods of points of D with rational radius. The Sorgenfry line is separable—the rational numbers provide a countable dense subset—but, as remarked above, it is not second countable. It follows that the Sorgenfry line does not embed in *any* metric space.

The significance of second countability is revealed in the following theorem, most often referred to in standard general topology texts, such as [32, 23, 48], as the Urysohn Metrization Theorem:

Proposition 6.39 *If $\{S, \mathcal{T}\}$ is a second countable, regular topological space, then it is possible to define a metric d on S so that the topology obtained from d is $\mathcal{T}$.* ∎

There are metric spaces that are not second countable, for example, $\mathbf{R}^1$ with the discrete metric.

The proof of Proposition 6.39 is beyond the level of our discussion, but we give an outline to illustrate how some of the ideas we have been developing can be used. One shows that one can find an embedding of any second countable, regular topological space into $\mathcal{I}$ where $\mathcal{I}$ is the Hilbert cube, Definition 2.60. By remark page 69, $\mathcal{I}$ is a metric space. A subset of a metric space is a metric space, so we can find a metric for S as desired.

Example 6.40 It is not clear that the projective plane, P^2 (see Definition 1.74) embeds in any $\mathbf{R}^n$. In fact, it does embed in $\mathbf{R}^4$, as seen in Example 18.07. Also, it can be shown, using algebraic topology, that P^2 does not embed in $\mathbf{R}^3$.

More generally, it can be shown that any projective space, P^k, embeds in some $\mathbf{R}^n$ (in fact, in $\mathbf{R}^{2k+1}$. This is a general fact about manifolds. The problem of finding the lowest dimension of Euclidian space for which P^k embeds, is a difficult problem). ◆

The notions of stable equivalence involve, in an essential way, the standard inclusions for $\mathbf{R^n}$ into $\mathbf{R^m}$. So there is no natural abstract setting for these notions.

Finally, we mention one other simple example of an embedding in general topology, which we will use in Section *16.2. Recall the definition of natural inclusions into M_f where M_f is a mapping cylinder.

Proposition 6.41 *If $f: X \to Y$ is a continuous function from one topological space to another with mapping cylinder M_f. Then the natural inclusions of X and of Y into M_f are embeddings.* ∎

6.7 Problems for Chapter 6

6.1 Prove Proposition 6.04.

6.2 Prove the statement of the remark on page 163.

6.3 Prove Proposition 6.18.

6.4 Suppose that $X = \{a, b, c, d\}$ is a set four points of $\mathbf{R}^1$, with $a < b < c < d$. Suppose we also have $X' = \{A, B, C, D\}$ is a set four points of $\mathbf{R}^1$, with $A < B < C < D$. Define a function $h: \mathbf{R}^1 \to \mathbf{R}^1$ such that $h(X) = X'$; see remark page 167.

6.5 Verify the statement of proof of 6.17 that f_1 and f_1' are equivalent embeddings if and only if G is the identity map of X.

6.6 Verify the claim made in Example 6.4 that f' and g' are inequivalent embeddings of E into $\mathbf{R}^1$, but are stably equivalent (since they become equivalent in $\mathbf{R}^2$).

6.7 Prove Proposition 6.28 in the case that X consists of three points.

6.8 In Figure 6-9 one sees a piecewise-linear spiral which limits on a line segment. Also, consider the piecewise-linear topologist's $\sin(1/x)$ curve Example 8.11. Show these subsets are stably equivalent subsets. (By the way this will show these subsets are homeomorphic) (Hint: One idea is to stretch the spiral into $\mathbf{R}^3$ so that its projection into one of the coordinate planes looks like a piecewise-linear version of a $\sin(1/x)$ curve similar to the subset shown in Figure 8-4. You do not need to read the material of Chapter 8, only refer to the figure cited.)

6.9 Let f and g be continuous functions from $\mathbf{R}^1$ to $\mathbf{R}^1$. Let F and G be embeddings of $\mathbf{R}^1$ into $\mathbf{R}^2$, given by $F(x) = (x, f(x))$ and $G(x) = (x, g(x))$. Show that F and G are equivalent embeddings. (Hint: It might help to show this first for the case that G is the function such that $G(x) = 0$ for all $x \in \mathbf{R}^1$.)

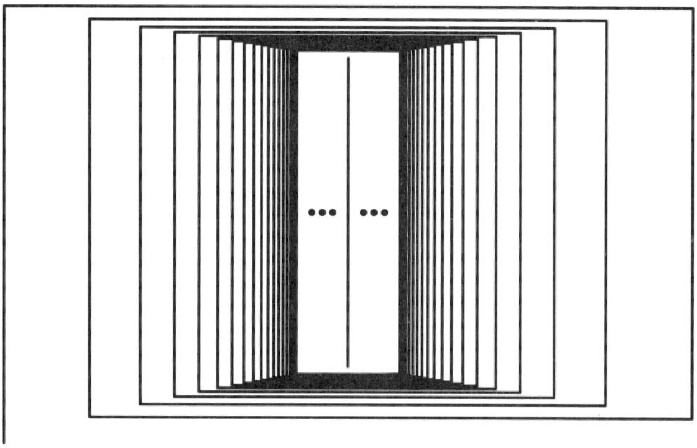

Figure 6-9 Figure for Problem 6.8

6.10 Let $X = \{0\} \cup \{1\} \cup [2, 3]$ and $Y = \{0\} \cup [1, 2] \cup \{3\}$ be subsets of $\mathbf{R}^1$. Show X and Y are not equivalent subsets of $\mathbf{R}^1$.

6.11 Let X and Y be as in Problem 6.10, and let $i: \mathbf{R}^1 \to \mathbf{R}^2$ be the standard inclusion $i(x) = (x, 0)$. Let $X' = i(X)$ and $Y' = i(Y)$. Show that X' and Y' are equivalent subsets of $\mathbf{R}^2$.

6.12 Let $X = \{(x, y) \in \mathbf{R}^2 \mid x = 0 \text{ or } y = 0\}$; X is the union of the coordinate axes of the plane. Let $A = \{(x, y) \in \mathbf{R}^2 \mid -1 \le x \le 1 \text{ and } y = 0\}$. Let $B_0 = \{(x, y) \in \mathbf{R}^2 \mid 0 \le x \le 1 \text{ and } y = 0\}$, $B_1 = \{(x, y) \in \mathbf{R}^2 \mid 0 \le y \le 1 \text{ and } x = 0\}$. Let $B = B_0 \cup B_1$. Are A and B equivalent subsets of X? Supply a proof for your answer.

6.13 Let $i: \mathbf{R}^1 \to \mathbf{R}^2$ be the standard inclusion $i(x) = (x, 0)$, and define $f: \mathbf{R}^1 \to \mathbf{R}^2$ by $f(x) = (\arctan(x), 0)$. Show that these embeddings are not equivalent. (Hint: Suppose there is a homeomorphism h of $\mathbf{R}^2$ such that $i = h \circ f$; consider $h(\pi/2)$.)

6.14 Let $\mathbf{N}$ be the subset of natural numbers in $\mathbf{R}^1$ and $\mathbf{Z}$ the subset of integers. Are these equivalent subsets of $\mathbf{R}^1$?

6.15 Let N be the set of points $(n, 0)$ of the plane where n is a natural number; let Z be the set of points $(z, 0)$ of the plane where z is an integer. Show that N and Z are equivalent subsets of the plane. (Hint: Think about a semicircular motion that rotates (only) the even natural numbers. Somehow define a map of the plane to itself that realizes this as a homeomorphism.)

6.16 Prove that any two embeddings of the closed unit interval into the plane, whose image is a straight line segment, are equivalent.

6.17 Let $\{p_i\}_{i=1}^n$ be a finite collection of n points of $\mathbf{R}^2$. For $i = 1, \ldots n$, let L_i be the line segment with endpoints x_{i-1} and x_i and so that for $L_i \cap L_{i+1} =$

x_{i+1}, for $i = 1, \ldots, n-1$. Then we say L is a "simple polygonal path in the plane."

Show that any two simple polygonal paths in the plane are equivalent subsets. (Hint: Show that any polygonal path in $\mathbf{R}^2$ is equivalent to a closed interval on the x-axis. Try a proof by induction on the number of line segments. A version of a solution of the Problem 6.16 will begin the induction process.)

6.18 Let C be a circle in the plane which passes through $(0,0)$. Let L be a line which does not pass through $(0,0)$. Let $X = \mathbf{R}^2 - \{(0,0)\}$ and $C' = C - \{(0,0)\}$. Show that L and C' are equivalent subsets of X. (Hint: Consider the map $f: X \to X$ defined by $f(\vec{x}) = \frac{1}{|\vec{x}|}\vec{x}$. What would $f(C')$ be if C were a circle of radius $\frac{1}{2}$ with center at $(\frac{1}{2}, 0)$? Alternative hint: Consider using stereographic projection; see Proposition 4.06.)

6.19 Let C be a circle in the plane and let C' be a circle inside C, tangent at one point. Let U be the crescent-shaped region between these two circles. Show that U is homeomorphic to $\mathbf{R}^2$. (Hint: Assume the common point of tangency of C and C' is $(0,0)$, and then consider the hint(s) of Problem 6.18.)

6.20 Prove: If $Z \subseteq \mathbf{R}^k$ and X is homeomorphic to Y, then there is a one-to-one correspondence between the set of equivalence classes of embeddings of X into Z and the set of equivalence classes of embeddings of Y into Z.

6.21 Verify the claim, which follows Question 6.32, that a proof of a "yes" answer to Question 6.32, would give a proof of the Jordan Curve Theorem.

6.22 Define $f(t) = (e^t \cos(t), e^t \cos(t))$. This is a classical curve known as the "logarithmic spiral." A graph of a portion of this curve is shown in Figure 6-10. One can get a better feel for this curve by looking at $f'(t) = (e^{0.1t}\cos(t), e^{0.1t}\cos(t))$; see Figure 6-10. We can view f and f' as two embeddings of $\mathbf{R}^1$ into $\mathbf{R}^2$. Show that these are equivalent embeddings.

6.23 As in Problem 6.22 $f(t) = (e^t \cos(t), e^t \cos(t))$. Let $g(t)$ be defined by $g(t) = (e^t, 0)$; the image of g consists of the positive numbers on the x-axis. Note that f and g are both embeddings of $\mathbf{R}^1$ into $\mathbf{R}^2$. Show that these are equivalent embeddings.

6.24 As in Problem 6.23, define $g(t) = (e^t, 0)$. Also define $g'(t) = (e^{-t}, 0)$. Are these equivalent embeddings?

6.25 Add three questions, in the same style, to the list of Questions 6.30–6.34.

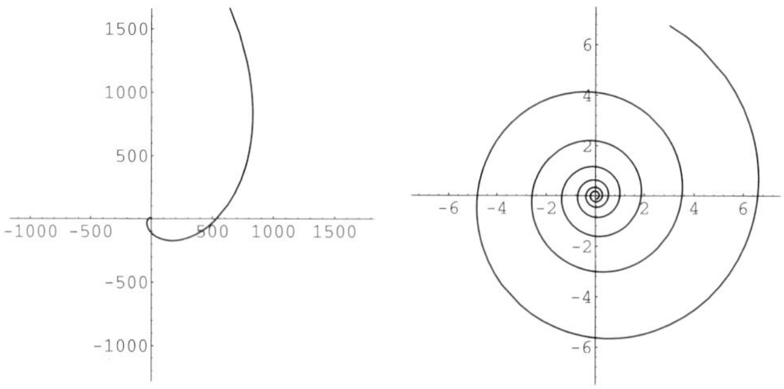

Figure 6-10 On the left, the graph of $f(t) = (e^t \cos(t), e^t \cos(t))$; on the right, the graph of $f'(t) = (e^{0.1t} \cos(t), e^{0.1t} \cos(t))$; see Problems 6.22 and 6.23. In each case, values in the range $-20 \le t \le 20$ are used.

7. CONNECTIVITY

OVERVIEW: We return to the idea of a connected subset, introduced in Chapter 4. We use this concept to capture the idea that a subset has well-defined "pieces" called components. A fundamental result is that any interval is connected, as is $\mathbf{R^n}$, *for any* n.

A cut-point is a point which, if removed, will result in a non-connected subset. We discuss cut-point examples. At the end of this of this chapter we are able to state and prove a result that basically says that there are an uncountable number of topologically distinct subsets of the plane.

7.1 Basic properties of connected subsets

We have briefly encountered the concept of a connected subset, Definition 4.37. We will now study this important concept in more depth.

The following definition is often useful when discussing this concept.

Definition 7.01 *Suppose* $X \subseteq \mathbf{R^n}$. *We say* X **is disconnected** *if* X *can be written as the union of two non-empty disjoint open subsets of* X. *Such a pair of subsets of* X *is called a* **disconnecting partition of** X.

Of course, X is connected if and only if it is not disconnected.

To prove a subset is connected, we need to show that it is impossible for it to be disconnected. As a consequence, if we are doing an elementary proof that a subset is connected, we should expect that the proof is an argument by contradiction. Thus one sees, in many of the proofs which follow, the following pattern: "We now show that X is connected. Assume it is not. Then we can write X as the union of subsets A and B such that ...."

One useful proposition says that a continuous image of a connected subset is a connected subset:

Proposition 7.02 *Suppose $X \subseteq \mathbf{R}^n$, $Y \subseteq \mathbf{R}^m$, and $f: X \to Y$ is a continuous onto map. If X is connected, then Y is connected.*

Proof: We argue by contradiction. Suppose $f: X \to Y$ is a continuous surjection, X is connected, and Y is disconnected. Let U and V be a disconnecting partition of Y. Let $U' = f^{-1}(U)$ and $V' = f^{-1}(V)$. Continuity of f implies that U' and V' are open subsets of X. The fact that f is onto ensures that U' and V' are not empty subsets. By set theory since $U \cap V = \emptyset$, $U' \cap V' = \emptyset$. Thus U' and V' provide a disconnecting partition of X, contradicting the hypothesis that X is connected. ∎

Example 7.03 The standard circle $S^1 \subseteq \mathbf{R}^2$ is a connected subset. Try expressing S^1 as a union of open subsets in various ways. It does not seem that we can have these sets be disjoint, non-empty, with their union all of S^1, but this is not easy to prove directly.

For our proof, we will use Proposition 7.02. Consider the function $f: [0, 2\pi] \to C_1$, defined by $f(x) = (\cos(x), \sin(x))$. By Proposition 4.39, one sees that $[0, 2\pi]$ is connected since any two closed intervals are homeomorphic, Proposition 4.20, and since connectedness is a topological property, Proposition 4.38. Proposition 7.02 implies that the circle is connected. ◆

Example 7.04 Let C_1 denote the unit circle in the plane and C_2 the circle centered at the origin with radius 2. Let $X = C_1 \cup C_2$; see Figure 7-1. We will use connectedness to show that C_1 is not homeomorphic to X.

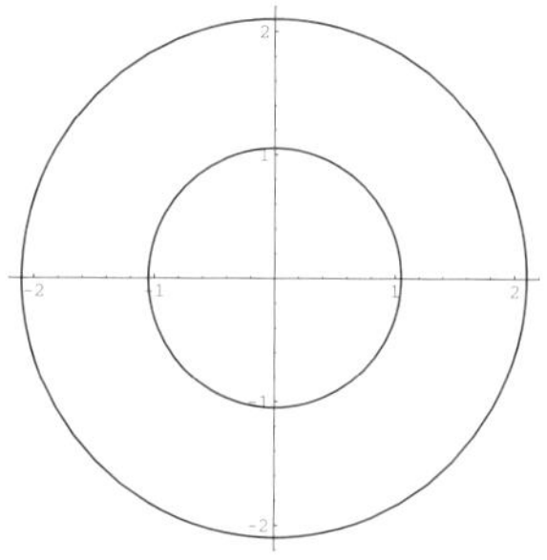

Figure 7-1 A subset X which is a union of two circles, C_1 of radius 1 and C_2 of radius two; see Example 7.04.

In Example 7.03 we showed that C_1 is connected. We will show that X is not connected. Since connectedness is a topological property, X and C_1 are not homeomorphic.

To show X is not connected, note that C_1 and C_2 are both open subsets of X, clearly disjoint, whose union is X. For example, C_1 is an open subset of X since $C_1 = X \cap N_{1.5}((0,0))$. An alternative method is to consider the continuous function $f(x,y) = x^2 + y^2$. Then $C_2 = f^{-1}(\{4\})$, and thus C_2 is a closed subset of X. Since $C_1 = X - C_2$, it follows that C_1 is an open subset of X. Similarly, C_2 is an open subset. ◆

Proposition 7.05 is just a restatement of Proposition 4.39, but we offer a different proof. The proof in Chapter 4 was based on the Intermediate Value Theorem. The purpose of giving a second proof is to expose the "hard work" hidden from view by our use of the Intermediate Value Theorem. Also, once we establish our second proof, then we can legitimately turn things around and show that the Intermediate Value Theorem follows from the connectedness of an interval; see Proposition 7.06. To avoid the appearance of circular reasoning, we formally restate the proposition so that, when we use Proposition 7.05 as a reference, we are not *assuming* the validity of the Intermediate Value Theorem.

Proposition 7.05 *The unit interval $I \subseteq \mathbf{R}^1$ is a connected subset.*

Let us look at an example before attempting a proof. Assume there exists a disconnecting partition $\{A, B\}$ for I. Take a simple choice for A; namely, suppose that A is an open interval $A = (a_0, a_1)$. What then could B be? It couldn't be $[0, a_0] \cup [a_1, 1]$ since this is not an open subset of I. The subset B could not be $[0, a_0) \cup (a_1, 1]$; this is an open subset, however, we do not have $A \cup B = I$ since the points a_1 and a_0 are in neither set. Since $a_1 \notin A$, then $a_1 \in B$. Since B is open, a_1 is contained in a neighborhood $N_\epsilon(a_1)$ which is entirely in B. But if this is the case, since a_1 is a right endpoint of the interval A, we must also have points of A in $N_\epsilon(a_1)$. But then this contradicts our assumption that $A \cap B = \emptyset$. This ends a sketch of the proof for this special choice of A. But what of the general case?

A crucial observation in our example is that the point a_1 is the least upper bound of the subset A. In the general proof, we won't know much specfically about A. However we do have enough information to apply the Least Uppper Bound Axiom.

Proof of Proposition 7.05: We are assuming, by way of contradiction, there are subsets A and B such that $A \neq \emptyset$, $B \neq \emptyset$, A and B are open subsets of I, $A \cup B = I$, and $A \cap B = \emptyset$,

Without loss of generality, we may assume that $1 \in B$. Then 1 is an upper bound for the subset A. Since $A \neq \emptyset$, the Least Upper Bound Property (Definition B.01) asserts there exists a least upper bound for A, call it c.

We first show, by contradiction, that $c \notin B$. If $c \in B$, since B is an open subset of I, we can find an ϵ with $N_\epsilon(c, I) \subseteq B$. As we have found in other examples, it is awkward to have ϵ "too large." In our case, we make sure that $\epsilon < 2c$. Consider $c' = c - \frac{\epsilon}{2}$; this is a point of I since condition for ϵ ensures that $0 < c'$. Now c' is an upper bound of A since $[c', c] \subseteq B$ and c is a upper bound of A. Now we have contradicted the assumption that c is the *least* upper bound of A. We thus conclude that $c \notin B$.

Since $A \cup B = I$, we conclude that $c \in A$. We argue that this, also, is not possible; thus completeing the proof. If $c \in A$ since A is an open subset of I, we can find an ϵ with $N_\epsilon(c, I) \subseteq A$. We may also assume that we have chosen ϵ so small that $\epsilon < \frac{1-c}{2}$. (We know $c \neq 1$ since $1 \in B$.) Consider $c' = c + \frac{\epsilon}{2}$; this is a point of A. However, $c < c'$, contradicting the assumption that c is an upper bound of A. ∎

Here is a generalized version of the Intermediate Value Theorem:

Proposition 7.06 *Suppose $X \subseteq \mathbf{R^n}$ with X connected and $a, b \in X$. Suppose $f\colon X \to \mathbf{R^1}$ is continuous with $f(a) < f(b)$. If c is any number $f(a) < c < f(b)$, then there is an $x_0 \in X$ with $f(x_0) = c$.*

Proof: We will show the contrapositive: if there is no such x_0, then X is not connected. Suppose there is a number c where $f(a) < c < f(b)$ and $f^{-1}(c) = \varnothing$. Let $U = f^{-1}((c, \infty))$, $V = f^{-1}((-\infty, c))$. Now $a \in V$ and $b \in U$, so these are non-empty subsets; each is open since f is continuous. But then U and V give a disconnecting partition of X. ∎

By using our proof of connectedness of an interval, Proposition 7.1 and Proposition 7.06, we obtain a proof of the standard Intermediate Value Theorem as stated in Proposition D.14.

The next proposition is another way of getting new connected subsets.

Proposition 7.07 *Suppose A and B are subsets of $\mathbf{R^n}$ and $A \cap B \neq \varnothing$. If A and B are connected, so is their union $A \cup B$.* ∎ *(Problem 7.4)*

It is not generally true that if $A \cap B = \varnothing$, that $A \cup B$ is disconnected. For example, let A and B be the intervals in the line, $A = [0,\ 1)$, $B = [1, 2]$. By Proposition 7.11, $A \cup B = [0, 2]$ is a connected subset.

Proposition 7.07 can be generalized for unions of more than two subsets.

Proposition 7.08 *Suppose $\{X_\alpha\}_{\alpha \in A}$ is a collection of subsets of X, $X \subseteq \mathbf{R^n}$ such that for all $\alpha \in A$, X_α is a connected subset. Suppose also that $\bigcap_{\alpha \in A} X_\alpha \neq \varnothing$. Then $\bigcup_{\alpha \in A} X_\alpha$ is a connected subset.* ∎

Example 7.09 Sometimes Proposition 7.08 is informally called the "daisy lemma." The idea is that a daisy flower is connected since each petal is connected, and all petals join at the center of the flower. In Figure 7-2, F is the union of 64 circles of radius one each of which contains the origin of the plane. Since a circle is connected, Proposition 7.08 shows that F is a connected subset. ◆

Figure 7-2 A connected subset F which is the union of 64 circles that have a point in common; see Example 7.09 .

Proposition 7.10 *The real line, $\mathbf{R}^1$, is connected.*

Proof: Any closed interval is connected since I is connected and any closed interval is homeomorphic to I, Propositions 7.05 and 4.20. Write $\mathbf{R}^1 = \bigcup_{i=1}^{\infty} [-i, i]$; then apply Proposition 7.08. ∎

We can similarly verify connectedness of other intervals, obtaining: (recall Definition 1.11 for the general notion of interval which includes infinite and half-infinite intervals).

Proposition 7.11 *Let be any interval in $\mathbf{R}^1$; then X is a connected subset.* ∎ *(Problem 7.5)*

Conversely,

Proposition 7.12 *If A is a connected non-empty subset of $\mathbf{R}^1$, then A is an interval (posibly a degenerate closed interval, that is, a single point).* ∎ *(Problem 7.6)*

We next prove that Euclidean spaces are all connected.

Proposition 7.13 *For any n, $\mathbf{R}^n$ is a connected subset.*

Proof: We provide proof for $n = 2$; a proof for general n follows by induction (Problem 7.7). The idea is to use the daisy lemma for subsets consisting of the x-axis and a vertical line.

We use $\mathbf{R}^1$ as an index set. For $\alpha \in \mathbf{R}^1$, define

$$Z_\alpha = \{(x, y) \in \mathbf{R}^2 \mid y = 0 \text{ or } x = \alpha\}.$$

We show that each X_α is connected. Each X_α is a union of two lines. Let $X = \{(x, y) \in \mathbf{R}^2 \mid y = 0\}$; X corresponds to the x-axis and is connected since it is homeomorphic to $\mathbf{R}^1$. Let Y_α be the vertical line: $Y_\alpha = \{(x, y) \in \mathbf{R}^2 \mid x = \alpha\}$. Again, Y_α is connected since it is homeomorphic to $\mathbf{R}^1$. Since $X_\alpha = X \cup Y_\alpha$, and $X \cap Y_\alpha = \{(\alpha, 0)\}$, X_α is connected by Proposition 7.07. Now $\bigcap_{\alpha \in \mathbf{R}^1} X_\alpha \neq \emptyset$ since $\bigcap_{\alpha \in \mathbf{R}^1} X_\alpha = X$. By Proposition 7.08, $\bigcup_{\alpha \in \mathbf{R}^1} X_\alpha$ is connected. Since $\bigcup_{\alpha \in \mathbf{R}^1} X_\alpha = \mathbf{R}^2$, the plane is connected. ∎

One can alternatively prove the result above by considering $\mathbf{R^n}$ to be the union of lines in $\mathbf{R^n}$ which contain the origin. This proof has an advantage in that it does not need an inductive argument. This second proof is so simple; why should we bother with the more complicated proof given in Proposition 7.13 ? One reason is that, by examining the proof of Proposition 7.13, we discover the proof of the next proposition.

Proposition 7.14 *Suppose $A \subseteq \mathbf{R^n}$ and $B \subseteq \mathbf{R^n}$ and that A and B are connected; then $A \times B$ is connected.* ∎ *(Problem 7.8)*

7.2 Components

Example 7.15 Suppose C_3 is the circle in the plane, with center at the origin and with radius 3. Recall the space $X = C_1 \cup C_2$ of Example 7.04, and let $Y = C_1 \cup C_2 \cup C_3$; see Figure 7-3. We ask: is X homeomorphic to Y? Clearly, neither X nor Y is connected, so our argument used in Example 7.04 does not apply.

We need a concept that captures the intuitive idea that X consists of two pieces and Y consists of three. The term "piece" should refer to a connected subset, but not every connected subset is a piece—it may not be large enough. For example, we want to say that the pieces of X are the circles C_1 and C_2. These are connected subsets. There are many other connected subsets of X. For example, we have $N_\epsilon(p, X)$ is a connected subset of X for any $p \in X$ and for any ϵ such that $0 < \epsilon < 1$. (These subsets are connected since they are homeomorphic to an open interval in the line.) ◆

The next definition captures the idea of a piece of a subset by saying that it is a connected subset that is as big as possible, yet still connected.

Definition 7.16 *Let $X \subseteq \mathbf{R^n}$ with $A \subseteq X$. We say A is a* **component of** *X if it is a maximal connected subset of X. By this we mean that, A is connected, and if B is a connected subset with $A \subseteq B \subseteq X$, then $A = B$.*

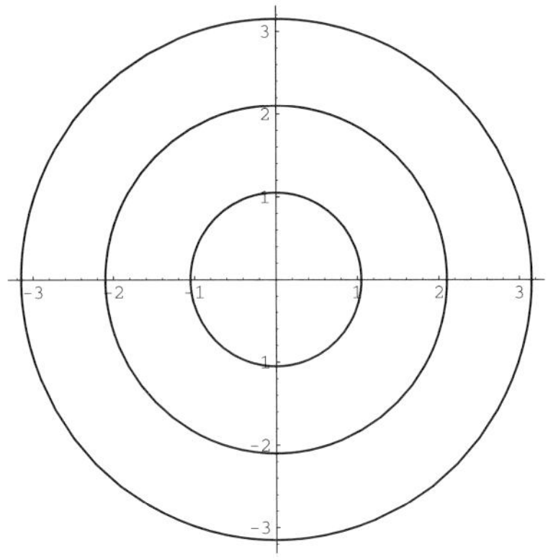

Figure 7-3 A subset Y which is a union of three circles, C_1 of radius 1, C_2 of radius 2, and C_3 or radius 3; see Example 7.15.

We verify that the components of the set X of Example 7.04 are the two circles, C_1 and C_2.

We show that C_1 is a component of X. We have established that C_1 is a connected subset. To show that it is a component, we argue by contradiction. If C_1 is not a maximal connected subset, then there is connected subset of X, call it B, such that $C_1 \subseteq B$ and $C_1 \neq B$. Let $Z = B - C_1$. Then Z is non-empty subset of B, and also $Z \subseteq C_2$. We claim C_1 and Z provide us with a disconnecting partition of B, which contradicts the assumption that B is a connected subset. Certainly $B = C_1 \cup Z$, $C_1 \cap Z = \emptyset$, and C_1 and Z are non-empty sets. Since $N_{3/2}((0, 0), B) = C_1$, C_1 is an open subset of B. Let $U = \mathbf{R}^2 - \overline{N}_{3/2}(0, 0)$; U is an open subset of the plane, and $U \cap B$ is an open subset of B. But $Z = B \cap U$. So Z is an open subset of B. This contradicts the assumption that B is connected. Thus C_1 is a component. One can argue similarly that C_2 is a component of X.

If $X \subseteq \mathbf{R}^\mathbf{n}$, the we can write X as a union of components:

Proposition 7.17 *Consider the relation: $x \sim y$ if and only if x and y are contained in a connected subset of X. This is an equivalence relation, and each equivalence class of this relation is a component of X.*

Proof: That $x \sim x$ follows from the fact that a point is a connected subset. It is clear that if $x \sim y$, then $y \sim x$. To see that $x \sim y$ and $y \sim z$ implies $x \sim z$, let C be a connected subset containing x and y, let K be a connected subset containing y and z; then $C \cup K$ will be a connected subset by the daisy lemma.

Let Z be an equivalence class of X. Let $z_0 \in Z$. Then for any $z \in Z$ there is a connected subset of X containing z and z_0; the daisy lemma implies that X is connected. Suppose there were a connected subset Z'

of X with $Z \subseteq Z'$. If $z' \in Z'$, then $z' \sim z_0$ and so $z' \in Z$. That is, $Z' \subseteq Z$; thus Z is a component. ∎

The next proposition provides tests to see if a subset is a component.

Proposition 7.18 *Let $C \subseteq X \subseteq \mathbf{R^n}$, with $C \neq \emptyset$.*

(a) *If C is connected, an open subset of X, and a closed subset of X, then C is a component of X.*
(b) *If C is a component of X, then C is a closed subset of X.*

Proof: We prove part (b); proof of (a) is left as an exercise, (Problem 7.13). Let C be a component of X. If C is not a closed subset of X, then there is an $x \in X - C$, with x a limit point of C. We assert that $C' = C \cup \{x\}$ is a connected subset. If so, this gives a contradiction since it is a connected subset of X which contains C, and is larger than C.

Suppose C' is not connected; let U and V be open subsets of X which give a disconnecting partition of C'. We may suppose, without loss of generality that $x \in U$. Since x is a limit point of C we can find $p \in U \cap C$. Now let $U' = U \cap C$ and $V' = V \cap C$; U' and V' are open subsets of C since U and V are open subsets of C'. Note that $U = U' - \{x\}$ and $V = V' - \{x\}$. Since $p \in U'$, $U' \neq \emptyset$. Since $x \notin V$ and $V \neq \emptyset$, $V' \neq 0$. But then U' and V' is a disconnecting partition of C, contradicting the assumption that C is connected. ∎

A component does not have to be an open subset. Let $Q \subseteq \mathbf{R}^1$ be the rational numbers. Each one-point subset $\{q\}$, $q \in Q$, is a component of Q. Suppose $q \in C$ with C a component of Q. If $C \neq \{q\}$, then we can find $p \in C$ with $p \neq q$. Assume that $p < q$; it will be clear that our argument can be made as easily in the case $q < p$. Let x be a rational number such that $p < x < q$, and let $U = (-\infty, x) \cap C$ and $V = (x, \infty) \cap C$. Then U and V provide a disconnecting partition of C, contradicting connectedness of C. Thus the components of Q are the one-point subsets. However, a one-point subset, $\{q\}$ cannot be an open subset since, if ϵ is a rational number, then $q + \epsilon/2 \in N_\epsilon(q, Q)$.

Similarly, the components of the standard Cantor set are single-point subsets, (Problem 7.18) and single-point subsets are not open subsets of the Cantor set (Problem 7.19).

Example 7.19 Let $Z = \{(x, y) \in \mathbf{R}^2 \mid x \text{ is an integer, and } y = 0\}$. Each component of Z is a single-point subset. This is easy enough to show directly. It also follows from Proposition 7.18 since points of Z are open and closed subsets. ◆

Using the concept of a component (and recalling the remarks in Example 4.09), we can restate the Jordan Curve Theorem.

Proposition 7.20 Jordan Curve Theorem *If C is a subset of the plane homeomorphic to a circle, then $\mathbf{R}^2 - C$ has two components.* ∎

Example 7.21 As we have mentioned this is a very deep theorem. But we can certainly show some special cases.

Suppose that $C = S^1$ is the unit circle in the plane. We will show that $\mathbf{R}^2 - C$ has two components.

We will use Proposition 7.18. Consider function $g: (\mathbf{R}^2 - C) \to (\mathbf{R}^1 - \{1\})$ defined $g(x, y) = x^2 + y^2$. Let $U = g^{-1}((1, \infty))$ and $V = g^{-1}((-\infty, 1))$. Since $(1, \infty)$ and $(-\infty, 1)$ are open and closed subsets of $\mathbf{R}^1 - \{1\}$, and since g is continuous, we conclude that U and V are open and closed subsets of $\mathbf{R}^2 - C$.

We next prove that each of U and V is connected. It is easy to see V is connected since it is an open two-dimensional ball. By Proposition 4.21 V is homeomorphic to $\mathbf{R}^\mathbf{n}$, and by Proposition 7.13, $\mathbf{R}^\mathbf{n}$ is connected.

We can show that V is connected by showing that V is homeomorphic to $S^1 \times \mathbf{R}^1$. The basic idea is that, using polar coordinates, one sees V is homeomorphic to $S^1 \times (1, \infty)$, which is homeomorphic to $S^1 \times \mathbf{R}^1$, (Problem 7.11). Connectedness would follow since S^1 is connected, Example 7.03; $\mathbf{R}^1$ is connected, Proposition 7.10; and the product of connected subsets is connected, Proposition 7.14.

For another simple illustration of the Jordan Curve Theorem, let P be the perimeter of a rectangle in the plane. Then $\mathbf{R}^2 - P$ has two components (Problem 7.12). ◆

The following proposition is very useful for showing that two subsets are not homeomorphic. It says, roughly, that homeomorphic subsets have the same number of components. Thus, if we find that two subsets have different numbers of components, they cannot be homeomorphic.

Proposition 7.22 *Suppose $X \subseteq \mathbf{R}^\mathbf{n}$, $Y \subseteq \mathbf{R}^\mathbf{m}$, $f: X \to Y$ a homeomorphism. If C is a component of X, then $f(C)$ is a component of Y.*

Moreover, this matching of components by f induces a one-to-one correspondence between the components of X and those of Y.

In particular, "the number of components of X equals N" is a topological property. ∎ *(Problem 7.14)*

Example 7.23 Let $X = \mathbf{R}^1 - \{0\}$. We show that X has two components, intervals $X_- = (-\infty, 0)$ and $X_+ = (0, \infty)$. First, these are open subsets of X, and each of X_- and X_+ are connected, by Proposition 7.07. We see that X_- is a closed subset of X since the complement of X_- in X is X_+, which is an open subset of X. Thus, by Proposition 7.18, X_- is a component of X. Similarly, X_+ is a component of X.

A simple strengthening of the argument of Example 7.23 shows that if $a \in \mathbf{R}^1$, then $\mathbf{R}^1 - \{a\}$ has two components.

But, to illustrate the use of some other propositions, we give an alternative argument. Consider the function $f(x) = x - a$. Then h is a homeomorphism with $h(a) = 0$. Thus $h|_{\mathbf{R}^1-\{a\}}$ is a homeomorphism from $\mathbf{R}^1-\{a\}$ to $\mathbf{R}^1-\{0\}$, and by Proposition 7.22 this function induces a one-to-one correspondence between the components of these sets. By Example 7.23, $\mathbf{R}^1 - \{0\}$ has two components; thus $\mathbf{R}^1 - \{a\}$ has two components. ◆

Example 7.24 We next present yet another proof that the unit circle S^1 and the interval $[0, 2\pi)$ are not homeomorphic. Suppose there were a homeomorphism $h: [0, 2\pi) \to S^1$. Let $p \in (0, 2\pi)$ and let $q = h(p)$. Let $X = [0, 2\pi) - \{p\}$, let $Y = S^1 - \{q\}$. Then $h|_X$ would give a homeomorphism from X onto Y. But Y has one component (it is homeomorphic to an open interval) and X has two components (see Example 7.23), contradicting Proposition 7.22. ◆

As pointed out in the remark on page 52, the set of all non-singular linear maps of $\mathbf{R}^1$ is in one-to-one correspondence with $\mathbf{R}^1 - \{0\}$. This allows us think of a subset of $\mathbf{R}^1 - \{0\}$ as an set of affine maps. In this sense Example 7.23 can be restated: the set of non-singular linear maps of $\mathbf{R}^1$ is not connected; in fact, it has exactly two components.

The set of all non-singular *affine* maps, $y = ax + b$ of $\mathbf{R}^1$, corresponds to all points $(a, b) \in \mathbf{R}^2$ except those on the first axis where $a = 0$. This also has two components: linear functions with positive slope and those with negative slope.

This is our first glimpse of an very important and powerful idea—the idea of a function space. Given two sets X and Y, the corresponding function space is a set whose elements are functions from X to Y. A broad mathematical goal is to define a concept of "open subset" for a function space that will allow us to capture the idea that two functions are close to each other.

For a more substantial example, consider the set of non-singular linear maps of $\mathbf{R}^2$. As noted in the remark on page 55, any linear map can be viewed as a point of $\mathbf{R}^4$. A linear map $F(\vec{x}) = M\vec{x}$ is non-singular if $|M| \neq 0$. Writing $M = \begin{pmatrix} a & b \\ c & d \end{pmatrix}$, $|M| = ad - bd$. Let $X = \{(a, b, c, d) \in \mathbf{R}^4 \mid ad - bd \neq 0\}$; we can now view X as the setŁof non-singular linear maps. Note that X is an open subset of $\mathbf{R}^4$ since the determinant gives a continuous function from $\mathbf{R}^4$ to $\mathbf{R}^1$, and X is the inverse image of the open subset $\mathbf{R}^1 - \{0\}$.

The image of a connected subset is a connected subset, Proposition 7.02, $\mathbf{R}^1 - \{0\}$ is not connected, and the determinant is a continuous function from X onto $\mathbf{R}^1 - \{0\}$. Thus X is not a connected subset. But how many components does X have? We resolve this in the next chapter, in Example 8.19.

7.3 Cut-points

Example 7.25 We present another example of non-homeomorphic subsets, X and Y. Roughly speaking, X looks like the letter "X" and Y looks like the letter "Y"; see Figure 7-4.

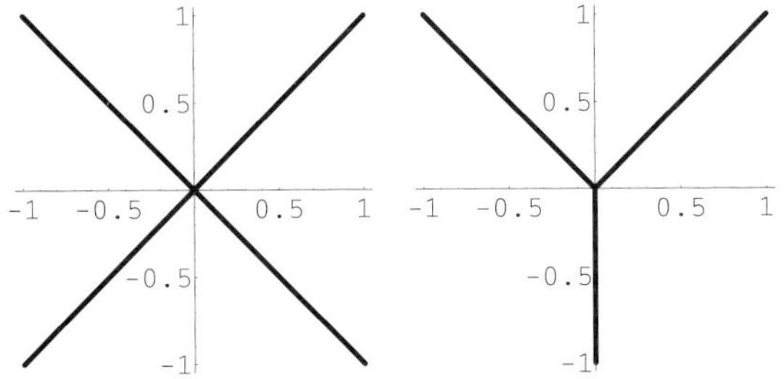

Figure 7-4 Two non-homeomorphic subsets of the plane—X on the left and Y on the right. See Example 7.25

Let $X = \{(x, y) \in \mathbf{R}^2 \mid x = y \text{ or } x = -y\}$; let $Y = \{(x, y) \in \mathbf{R}^2 \mid y = |x|, \text{ or } (x = 0 \text{ and } y \leq 0)\}$. We show that X and Y are not homeomorphic. Suppose there is a homeomorphism $h: X \to Y$. Let $z = h((0,0))$. Then $h|_{X-\{(0,0)\}}$ is a homeomorphism from $X - \{(0,0)\}$ and $Y - \{z\}$. An argument similar to that found in the discussion of Example 7.23 shows that $X - \{(0,0)\}$ has four components. By Proposition 7.22 we see that $Y - \{z\}$ has four components. However, for any $z \in Y$, $Y - \{z\}$ has three components (this only happens if $z = 0$) or two components (otherwise). ◆(Problem 7.16)

The argument of Example 7.25 is so useful that it merits a definition and proposition.

Definition 7.26 *Suppose $X \subseteq \mathbf{R}^n$ and $c \in X$. We say c is a* **cut-point of** X **of order** n, *if $X - \{c\}$ has exactly n components.*

Since this next proposition is used often, we give it the name " the cut-point lemma."

Proposition 7.27 *Suppose $X \subseteq \mathbf{R}^n$, $Y \subseteq \mathbf{R}^m$, $c \in X$, and $f: X \to Y$ is a homeomorphism. Then c is a cut-point of X of order n if and only if $f(c)$ is a cut-point of Y of order n.* ∎ *(Problem 7.17)*

The argument of the Example 7.25 is a milestone of sorts in our exposition to date. Many propositions and definitions are used, directly and indirectly, in the proof. In small steps one journeys up a slope. Each step seems of little significance, yet after many such steps, we may look back and see the valley below in an entirely different way. We have come a long way in our mathematical path. As testament to how far we have come, we are now prepared to answer, in Proposition 7.28, the question of whether there are uncountably many non-homeomorphic subsets of the plane, Question 4.19.

Proposition 7.28 *There is an uncountable collection of subsets of the plane, $\{X_s\}_{s \in S}$, so that, for all s and t in S, X_s is homeomorphic to X_t if and only if $s = t$.*

Proof: The index set S we use is the set of binary strings. By Proposition C.13, S is uncountable.

We begin with a set X which is the union of four lines which all meet at $(0,0)$: $X = \{(x, y) \in \mathbf{R}^2 \mid x = 0, \text{ or } y = 0, \text{ or } x = 2|y|\}$.

Roughly, we describe X_s as follows. Add vertical segments to X, placed along the x-axis according to the string at s. If we find a 1 in the i-th position, we place a vertical line segment A_i crossing at the point $(i, 0)$; if we find 0, we only use the upper half of this line segment B_i; see Figure 7-5.

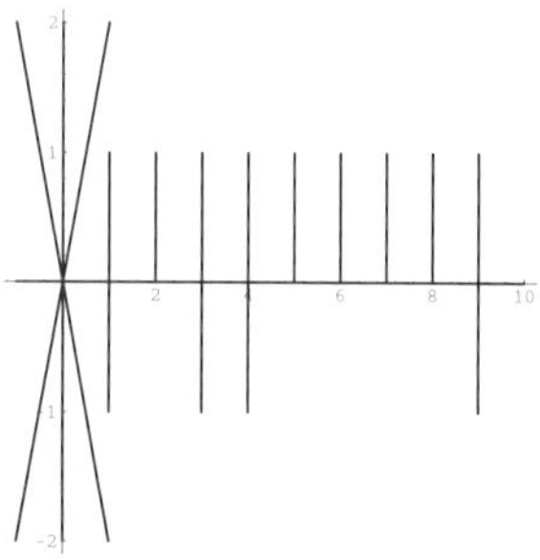

Figure 7-5 A part of a subset of the plane corresponding to X_s where s is the binary string 101100001....

For $i \in \mathbf{N}$, let

$$A_i = \{(x, y) \in \mathbf{R}^2 \mid x = i, \text{ and } 0 \le y \le 1\}; \text{ let}$$

$$B_i = \{(x, y) \in \mathbf{R}^2 \mid x = i, \text{ and } -1 \le y \le 1\}$$

. Now define $X_s = X \cup \bigcup_{i \in N} C_i$ where $C_i = A_i$ if the i-th symbol in the string s is a zero, and $C_i = B_i$ if the i-th symbol of s is a 1. We need to show that if X_s is homeomorphic to X_t, then $s = t$.

Any homeomorphism $f: X_s \to X_t$ must have the property that $f((0,0)) = (0,0)$ since, for both spaces, $(0,0)$ is the only cut-point of order 6.

Let Z be the subset of X_s corresponding to the natural numbers on the x-axis: $Z = \{(n,0) \in \mathbf{R}^2 | n \in \mathbf{N}\}$. A key fact is that, for any string w, a point of X_w is a cut-point of order 3 or of 4 if and only if it is a point of Z. Applying Proposition 7.27, we can conclude that $f(Z) = Z$.

To verify this, we first argue that $f((1,0)) = (1,0)$. We use an indirect proof, and so assume that $f((1,0)) = (n,0)$ with $2 \le n$.

Let $J \subseteq X_s$ be the line segment corresponding to the unit interval on the x-axis: $J = \{(x,y) \in \mathbf{R}^2 \mid y = 0 \text{ and } 0 \le x \le 1\}$. We have a continuous function $f|_J : J \to X_t$. Our plan is to show that there is a point z of the interior of J such that $f(z) = (1,0)$. If so, this gives us a contradiction since z is a cut-point of X_s of order 2 and $(1,0)$ is a cut-point of X_t of order 3 or 4.

It is not hard to verify that $Y - \{(1,0)\}$ consists of three or four components with $(0,0)$ and $(2,0)$ in different components. Since J is an interval, it is connected; since f is continuous, $f(J)$ is a connected subset of Y. We see that $f(J)$ cannot lie in $Y - \{(1,0)\}$, for then $(0,0)$ and $(2,0)$ would lie in a connected subset of $Y - \{(1,0)\}$ and thus lie in the same component. We thus conclude that $f(J) \not\subseteq (Y - \{(1,0)\})$, and so our point z must exist.

Having arrived at a contradiction, we now can conclude that $f((1,0)) = (1,0)$. We have now established that the first symbol of s and t must agree. For example, if the first symbol of s is a 1, then $(1,0)$ is a cut-point of X_s of order 4; $f((1,0))$ is a cut-point of order 4; thus the first symbol of t is a 1. An inductive argument (7.28), completes the proof by showing that for all n, $f((n,0) = (n,0)$. It then follows that two sets of our collection are homeomorphic if and only if they have the same code. ∎

We could construct other proofs of the Proposition 7.28 by using other candidates for our spaces X_s. For example, we could define a subset Y_s as follows. Begin with the x-axis and a binary string. If the i-th entry of the string is zero, we attach $2i + 1$ segments to the point $(i, 0)$; if the $i - th$ entry were 1; then we would attach $2i$ segments to the point $(i,0)$. (See Figure 7-6 for an example.) In this way, points that correspond to a 0 will have an odd number of segments; those that correspond to 1 will have an even number. This subset is a bit harder to define, but the proof is a bit easier. (Problem 7.29)

For the notion of cut-point, we removed a point from a subset. There are occasions when we might want to consider removing more than one point, as the next two examples illustrate.

Example 7.29 We provide a new proof that the circle S^1 is not homeomorphic to the closed interval I. If we remove the endpoints 0 and 1 from I, the result, being an open interval, is connected. However if we remove any two points from S^1, the resulting subset is not connected. In fact, it has two components. ◆

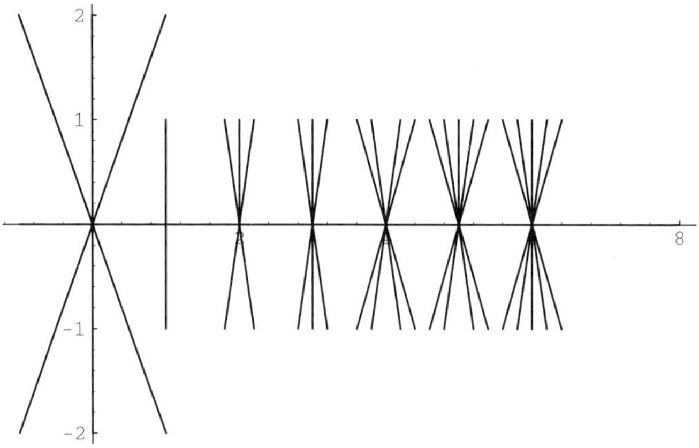

Figure 7-6 A different collection of subsets which code binary strings; see the remark on page 194. In this case the string encoded is 10101....

Example 7.30 We show that Sierpiński's gasket, Example 5.24, is not homeomorphic to Sierpiński's carpet, Example 5.25. In the construction of Sierpiński's gasket, $\mathcal{G}$, is defined to be $\cap_i G_i$ where G_0, is a triangle. Let a and b be the midpoints of two of the sides of G_0. Consider G_1; it is a union of three sub-triangles. It is clear that $G_1 - \{a\} - \{b\}$ is not a connected subset. Since $G_i \subseteq G_j$ for $i > j$, it follows that $\mathcal{G} - \{a\} - \{b\}$ is not connected.

However one can show that if we remove any two points from Sierpiński's carpet that it is connected (Problem 7.31). Thus Sierpiński's gasket is not homeomorphic to Sierpiński's carpet. ◆

*7.4 General topology and Chapter 7

Example 7.31 A discrete topological space with more than two points is never connected and has as many components as points. An indiscrete space is always connected.

Recall the Sorgenfry line, $\mathcal{H}$, obtained from the set $\mathbf{R}^1$, using half-open intervals, Example 1.46. This topological space is not connected since $U = (-\infty, 0)$ and $V = [0, \infty)$ provide a disconnecting partition. ◆

The abstract version of Proposition 7.14 holds. Moreover, it extends to infinite products, but the proof, found in any standard text for general topology, such as [32, 23, 48], is not a simple one:

Proposition 7.32 *Suppose $\{X_\alpha, \mathcal{T}_\alpha\}$ is a collection of connected topological spaces. Then the product space, $\{\prod_\alpha X_\alpha, \prod_\alpha \mathcal{T}_\alpha\}$, is a connected topological space.* ◼

It is important in Proposition 7.32 that we are using the product topology.

One reason the box topology is considered to be bad is that Proposition 7.32 is not true for infinite products if we use the box topology.

Example 7.33 Let X denote the set $\prod_1^\infty \mathbf{R}^1$, a countable product of $\mathbf{R}^1$ with itself, and let $\{X, \mathcal{B}\}$ be the topological space where $\mathcal{B}$ denotes the box topology. Points of X are sequences of real numbers.

Let U be the set of sequences which are bounded by some number (for example, 100, 100/2, 100/3, ...), and let V be the set of unbounded sequences (for example (1, 2, 3, ...)). If $(x_1, x_2, \ldots)$ is a sequence bounded by M, then every sequence in $\prod_i (x_i - \epsilon, x_i + \epsilon)$ is bounded by $M + \epsilon$, and so U is an open set in the box topology. Similarly, V is an open subset. But now U and V provide a disconnecting partition for X. ◆

If $\{X, \mathcal{T}\}$ is a connected space and P a partition of X, then the quotient space X/P will be a connected. This follows from the fact that the quotient map $q: X \to X/P$ is continuous. By the abstract version of Proposition 7.02, the image of a connected space under a continuous map is connected.

For function spaces, one is usually interested in path connectivity; see Sections *8.3 and *15.5.

7.5 Problems for Chapter 7

7.1 Finish the proof of Proposition 7.05 following the suggestion: follow through all the steps of the first part of the proof, adjusting the inequalities and replacing the word "least" by "greatest" and "upper" by lower," etc.

7.2 Finish the proof of Proposition 7.05 following the suggestion. Consider the map $f: I \to I$ defined by $f(x) = 1 - x$. Let $A' = f(A)$, $B' = f(B)$, and $C' = f(C)$. The next step is to apply the argument given in the first part of the proof.

7.3 Let X be the union of the coordinate axes in the plane. That is, $X = \{(x, y) \in \mathbf{R}^2 \mid xy = 0\}$. Verify that X is connected by showing there is a continuous onto map $f: \mathbf{R}^1 \to X$. (We note that an alternative way of showing connectedness of X is by using Proposition 7.07.)

7.4 Prove proposition 7.07.

7.5 Prove Proposition 7.11.

7.6 Prove Proposition 7.12. (Hint: Examine our proof of Theorem 7.10 and focus on the greatest lower bound and least upper bound of the given subset.)

7.7 Finish the proof of Proposition 7.13 by providing the inductive argument.

7.8 Prove Proposition 7.14.

7.9 Show that the empty subset is always a connected subset.

Also show that f $X \subseteq \mathbf{R}^n$ and X is non-empty, then the empty subset of X is never a component of X.

7.10 Let G be the subset of the plane

$$G = \{(0,0), (1/1,0), (1/2,0), (1/3,0), (1/4,0), \ldots\}.$$

. What are the components of G? Give a proof of your claim.

7.11 Verify the claim, in Example 7.21, that V is connected by showing that V is homeomorphic to $S^1 \times \mathbf{R}^1$.

7.12 Prove that the compliment of the perimeter of a rectangle in the plane has two components, as claimed in Example 7.21.

7.13 Prove Proposition 7.18(a).

7.14 Prove Proposition 7.22.

7.15 Let X and Y be the subsets of $\mathbf{R}^1$, defined as follows. $X = [-1,1] - \{0\}$, $Y = [-2,-1) \cup (1,2]$.

(a) Show that X and Y are homeomorphic.
(b) Show that X and Y are not equivalent subsets of $\mathbf{R}^1$. (Hint: Compare the components of their complements.)

7.16 Verify the assertion made in Example 7.25 that for any $z \in Y$, $Y - \{z\}$ has two or three components.

7.17 Prove Proposition 7.27.

7.18 Show that the components of the Cantor set are the one-point subsets; see the remark on page 189.

7.19 Show that the Cantor set contains no open connected subsets; see the remark on page 189.

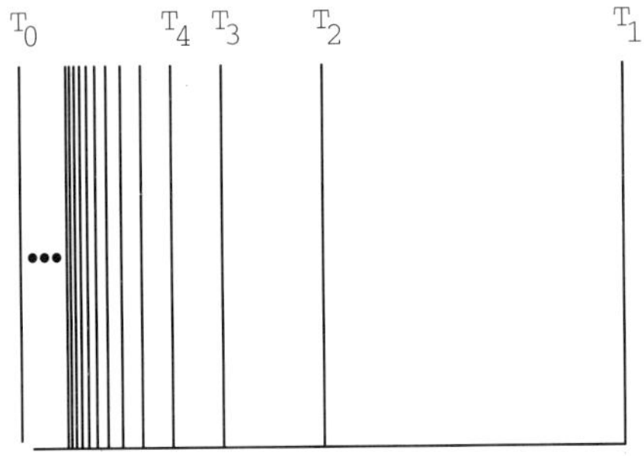

Figure 7-7 The topologist's comb with the point $(0, 0)$ removed; see Problem 7.23.

7.20 Let K be the standard Cantor set. Show that each component of $K \times S^1$ is homeomorphic to S^1.

7.21 Let S^1 be the unit circle in the plane. Show that S^1 does not embed in $\mathbf{R}^1$. (Hint: Use results of Example 7.23.)

7.22 Let $X = \{(x, y) \in \mathbf{R}^2 \mid xy = 0\}$. Prove that X does not embed in $\mathbf{R}^1$.

7.23 Let Z be the subset of the plane obtained by removing $(0, 0)$ from the topologist's comb, Example 2.43; see Figure 7-7. Is Z a connected subset?

7.24 Show that the set A of Example 5.23 is not homeomorphic to the standard Cantor set.

7.25 The order of a cut-point does not have to be finite. Find an example of a subset $X \subseteq \mathbf{R}^2$ which has a point x_0 such that $X - x_0$ has uncountably many components. (Hint: Think of the cone construction, Definition 2.47.)

7.26 Show that if $X \subseteq \mathbf{R}^n$, and X has (at least) two components which are not homeomorphic, then X is not homogeneous.

7.27 Discuss the following proposition. If $X \subseteq \mathbf{R}^n$ and every two components of X are homeomorphic, and if every component of X is homogeneous, then X is homogeneous. By discuss, we mean find a proof, if true, and a counterexample (with proof that it is a counterexample) if the proposition is false.

7.28 Provide details of inductive part of the proof for Proposition 7.28

7.29 Provide the details of the definition and the proof of the remark on page 194.

7.30 If x_0 is an isolated point of A, then $\{x_0\}$ is a component of A (see Definition 1.41).

7.31 Prove the assertion of Example 7.30: If we remove any two points from Sierpiński's carpet, that it is connected. (Hint. Suppose (x, y) and (X, Y) are two points of Sierpiński's carpet; it might be useful to consider cases, the first of which is $x \neq y$ and $X \neq Y$.)

7.32 Show that if X has the fixed-point property, then X is connected.

Note: For the purpose of the next problems, we consider a graph to be a non-empty union of line segments. (That is we specifically rule out a finite set of points as being a graph.) If G is a graph, a subgraph H will be a subset which is the (non-empty) union of some of the line segments that make up G. In the following problems, we ask how many topologically distinct subgraphs are there for a given graph. By this we mean: what is the largest collection of subgraphs of G, so that no two are homeomorphic? For example, if G is a graph consisting of the three edges of the perimeter of a triangle, there are two topologically distinct subgraphs. The subgraph consisting of all three edges (homeomorphic to a circle) and a subgraph consisting of one edge, say E, (homeomorphic to I) are distinct. Any other graph will consist of the union of one edge, or two edges with a point in common, and will be homeomorphic to I.

7.33 Consider the graph k_4 of Figure 7-8. How many topologically distinct subgraphs are there of k_4?

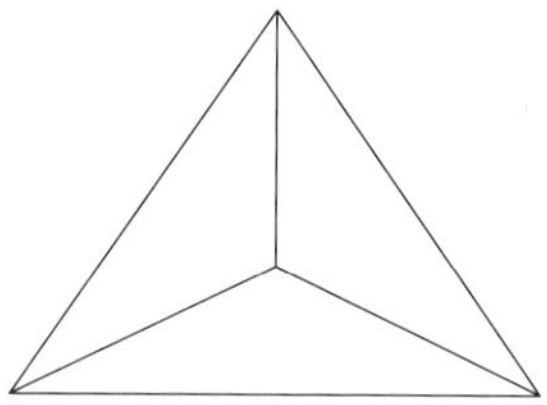

Figure 7-8 The graph k_4, consisting of four points joined by six line segments. See Problem 7.33.

7.34 Consider the utilities graph G of Example 6.20. How many topologically distinct subgraphs are there of G?

7.35 Consider the graph k_5 of Example 6.21. How many topologically distinct subgraphs are there of k_5?

8. PATH CONNECTEDNESS

OVERVIEW: We introduce and study a special kind of connectedness, called path connectedness, and the associated notion of "piece," path component.

8.1 Path connectedness

We introduce a new strategy for analyzing subsets. In space exploration terminology, a probe is an instrument sent into space to send back information. Our strategy is to explore $X \subseteq \mathbf{R}^n$ by using a mathematical probe. Our first probe is a closed interval, I. We send our probe into X, so to speak, by using continuous functions from I into X. By now we know quite a bit about the topology of a closed interval. We seek to apply this knowledge by considering continuous functions from I into our set X. In Chapter 9 we use the integers for a probe; see the remark on page 225. In Chapter 15, we use a circle for our probe.

By way of introduction to this new technique, we offer yet another proof that the plane is a connected set (Proposition 7.13). This proof will also generalize to show the connectivity of $\mathbf{R}^n$ for any n.

Proposition 8.01 $\mathbf{R}^2$ *is a connected set.*

Proof: The proof is indirect. Suppose that $\mathbf{R}^2$ is not connected. Write $\mathbf{R}^2 = A \cup B$, with $A \neq \emptyset$, $B \neq \emptyset$, $A \cap B = \emptyset$, and A and B open subsets of $\mathbf{R}^2$. Since $A \neq \emptyset$, $B \neq \emptyset$, and $A \cap B = \emptyset$ we can find points $a \in A$ and $b \in B$ with $a \neq b$.

Let L be the line segment between a and b. Now L is a connected subset since it is homeomorphic to an interval, and intervals are connected, Proposition 7.11. Let $A' = A \cap L$, and $B' = B \cap L$. Then A' and B' give a disconnecting partition of L, contradicting the connectedness of L. We conclude that the plane is connected. ∎

We next modify this proof and apply to similar examples.

Proposition 8.02

(a) *If* $x \in \mathbf{R^n}$, $n > 1$, *then* $\mathbf{R^n} - \{x\}$ *is a connected subset.*
(b) *If* $n > 0$, *then* S^n *is connected.*

Proof of Proposition 8.02(a): We attempt to prove Proposition 8.02 along the lines of the proof of Proposition 8.01. The proof would begin as before. Given two distinct points, a and b in $\mathbf{R^n} - \{x\}$, let L be the line segment between a and b. The proof would break down in the case that x is on L between a and b since removing x from L would result in a disconnected subset.

If $x \in L$, we find some other connected subset $L' \subseteq \mathbf{R^n} - \{x\}$ which contains a and b. Using this L', we will obtain our contradiction.

Let c be a point of $\mathbf{R^n}$ which does not lie on the line containing L; see Figure 8-1 for a diagram in the plane. Let L_1 be a line segment from a to c; let L_2 be a line segment from c to b. Let $L' = L_1 \cup L_2$. Then L' is a connected set since it is the union of connected sets whose intersection is non-empty.

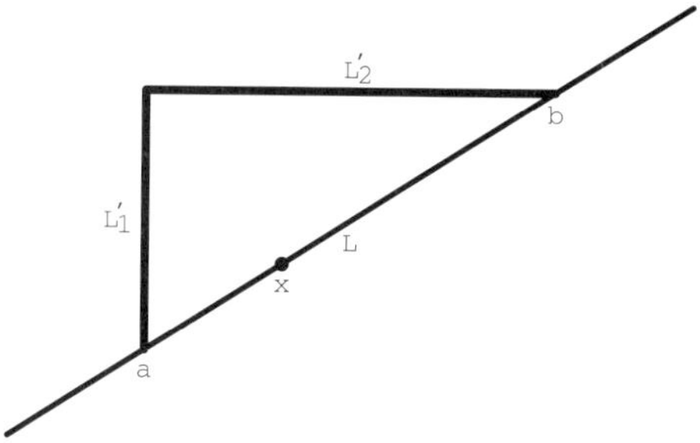

Figure 8-1 Connecting two points, a and b, in the plane while avoiding point x between them; see the proof of Proposition 8.02. The set L' is the union of L_1 and L_2, shown as bold line segments.

Proof of Proposition 8.02(b): Define $r: (\mathbf{R^n} - \{0\}) \to S^n$ by $r(x) = x/|x|$. This is a continuous onto function; thus S^n is the image, under a continuous map of a connected set, and so S^n is connected. ∎

We next provide a new proof of Proposition 6.5, namely:

Proposition 8.03 $\mathbf{R^1}$ *and* $\mathbf{R^2}$ *are not homeomorphic.*

Proof: Suppose there is homeomorphism $f: \mathbf{R^1} \to \mathbf{R^2}$. Then, by Proposition 6.18, $\mathbf{R^1} - \{f^{-1}(0,0)\}$ is homeomorphic to $\mathbf{R^2} - \{(0,0)\}$.

As seen in Example 7.23, $\mathbf{R}^1 - \{f^{-1}(0,0)\}$ is not connected. By Proposition 8.02, $\mathbf{R}^2 - \{(0,0)\}$ is connected. This contradicts the fact that connectedness is a topological property, Proposition 4.38.

Using the terminology of cut points, the proof above can be restated: $\mathbf{R}^2$ has no cut point, but $\mathbf{R}^1$ does (in fact, every point of $\mathbf{R}^1$ is a cut point); thus $\mathbf{R}^1$ cannot be homeomorphic to $\mathbf{R}^2$. ∎

The success of the proofs lead above us to consider a new type of connectedness, called "path connectedness."

Definition 8.04 *Let $X \subseteq \mathbf{R}^n$. A* **path in** *X is a continuous function $f: I \to X$ where I is the unit interval $[0,1] \subseteq \mathbf{R}^1$. If $a, b \in X$, we say f is a* **path in X from a to b** *if f is a path in X with $f(0) = a$ and $f(1) = b$.*

Definition 8.05 *Let $X \subseteq \mathbf{R}^n$, we say X* **is path connected** *if for each pair of points $a, b \in X$, there is a path in X from a to b.*

Regarding the examples we have been looking at:

Proposition 8.06

(a) *For any n, $\mathbf{R}^n$ is path connected.*
(b) *If $x \in \mathbf{R}^n$, $n > 1$ then $\mathbf{R}^n - \{x\}$ is path connected.*
(c) *If $n > 0$, then S^n is path connected.*

Proof: Clearly $\mathbf{R}^n$ is path connected since we may join any two points by a line segment L, or in the case of $\mathbf{R}^n - \{x\}$ by a union of line segments L'. But, be careful here. A line segment is *not* a path—it is a subset and not a function. Similarly, our proof of Proposition 8.02 provides sets that are images of paths. We need to find parameterizations of the subsets L and L'. This is not difficult, but it must be done.

Using vector notation, define $f: I \to \mathbf{R}^2$ by $f(t) = (1-t)\vec{a} + t\vec{b}$. This is a path from a to b.

The subset L' is union of line segments L_1 and L_2. Each of these, separately, can be considered a path. Define f_1 by $f_1(t) = (1-t)\vec{a} + t\vec{c}$, and define f_2 by $f_1(t) = (1-t)\vec{c} + t\vec{b}$. We want to combine these two paths into a single path F whose image is L'. We do this as follows:

$$F(t) = \begin{cases} f_1(2t) & \text{if } 0 \leq t \leq \frac{1}{2} \\ f_2(2t-1) & \text{if } \frac{1}{2} \leq t \leq 1. \end{cases}$$

Here is one way to describe F. Think of t as time, and a path as a journey that begins at time 0 and ends at time 1. Suppose we have two journeys such that the ending point of the first is the beginning point of the second. We combine these into a single journey. We travel the first journey, twice as fast. Then, after taking into account our delayed start, travel the second journey, also at double speed. ∎

This method of "gluing paths together" used in the proof above is a useful one. We formalize this process in the following definition.

Definition 8.07 *Suppose $X \subseteq \mathbf{R^n}$, $f: I \to X$, and $g: I \to X$ are paths in X with $f(1) = g(0)$. Define a path $f * g$ in X, called the* **concatenation of f and g**, *by*

$$(f * g)(t) = \begin{cases} f(2t) & \text{if } 0 \le t \le \frac{1}{2} \\ g(2t-1) & \text{if } \frac{1}{2} \le t \le 1. \end{cases}$$

The proof of Proposition 8.02 can be modified to prove:

Proposition 8.08 *If X is path connected $X \subseteq \mathbf{R^n}$, then it is connected.* ∎ *(Problem 8.1)*

Connectedness does not imply path connectedness, as we see in the following example (a subset of Example 3.30).

Example 8.09 Let

$$\Gamma = \{(x, y) \in \mathbf{R}^2 \mid y = \sin(1/x) \text{ and } 0 < x\}; \text{ let}$$

$$L = \{(x, y) \in \mathbf{R}^2 \mid x = 0 \text{ and } -1 \le y \le 1\}.$$

Note that Γ is the graph of $\sin(1/x)$ for positive x; see Figure 8-2. Let $X = \Gamma \cup L$. This example is frequently referred to as the "topologist's $\sin(1/x)$ curve." ◆

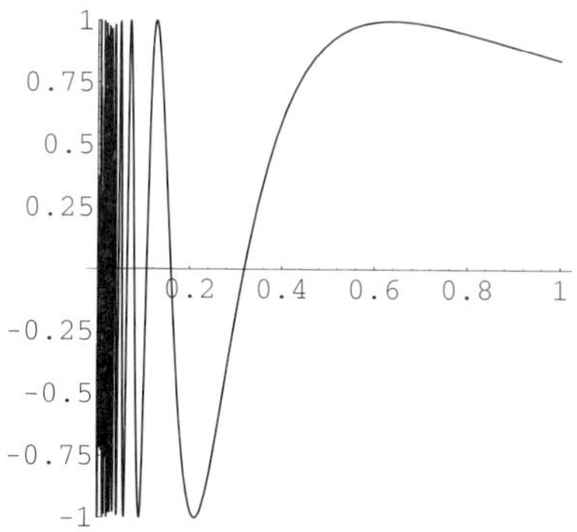

Figure 8-2 The topologist's $\sin(1/x)$ curve, Example 8.09.

Proposition 8.10 *The topologist's* $\sin(1/x)$ *curve X is connected but not path connected.*

Proof: We first show that X is connected. Observe that each of Γ and L are connected subsets. Since Γ is the graph of a function (recall Proposition 3.28), it is homeomorphic to $(0, \infty)$ and thus connected. Also, L is homeomorphic to a closed interval, and so L is connected.

Suppose we could express X as the disjoint union of non-empty open subsets, say U and V. Certainly L cannot meet both of U and V. (If it did, then we could express L as the disjoint union of non-empty open subsets, $U \cap L$ and $V \cap L$, and these two sets would provide a disconnecting partition of L.). We may assume without loss of generality that $L \subseteq U$. We similarly argue that Γ cannot meet both U and V. If $\Gamma \subseteq U$, then $X \subseteq U$, and thus $V = \emptyset$. We thus conclude: L is contained in U, and Γ is contained in V.

If U is any open subset of X which contains L, it must also contain a point (in fact infinitely many points) of Γ since, intuitively, Γ gets closer and closer to L. In fact what we will show is that every point of L is a limit point of Γ. To show this, let $q \in L$, and write $q = (0, y_q)$. Let $N_\epsilon(q, X)$ be a neighborhood of q, with $N_\epsilon(q, X) \subseteq U$.

We find a point, $q_0 \in \Gamma \cap N_\epsilon(q, X)$, as in Figure 8-3.

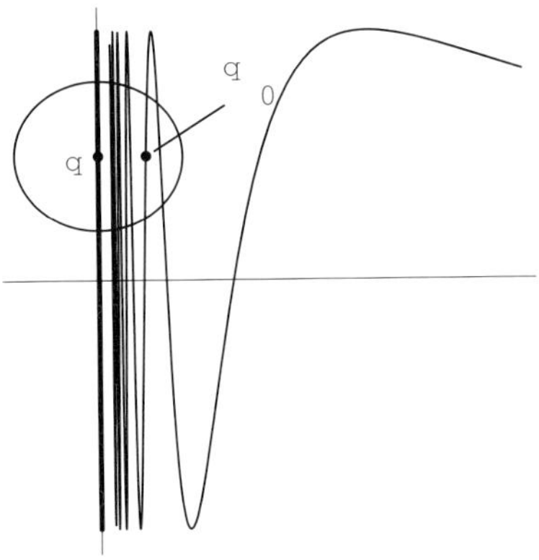

Figure 8-3 A neighborhood of point q, as in the sketch of proof of Proposition 8.10 that the topologist's $\sin(1/x)$ curve is connected.

Studying this figure, there must be a point q_0 with $q_0 = (x_0, y_0)$ where $y_0 = y_q$ and $0 < x_0 < \epsilon$. We see that $\sin(1/x) = y_q$ for infinitely many values of x; in particular, for infinitely many values of $x \in (0, \epsilon)$. We need to specify only one of these. Here are some details for finding q_0. Choose an integer N such that $\frac{1}{2N\pi} < \epsilon$, and consider the interval $[\frac{1}{(2N+2)\pi - \pi/2}, \frac{1}{(2N+2)\pi + \pi/2}]$. On such an interval the function $\sin(1/x)$ takes on all values between -1 and $+1$ at exactly once. We want to locate the value of x in this interval corresponding to y_q. Let $x_0 = 1/(\arcsin(y_q) + (2N+2)\pi)$; then we clearly have let $y_0 = \sin(1/x_0) =$

y_q, $y_0 = q_0$, and $0 < x_0 < \epsilon$.

Now, $q_0 \in \Gamma \subseteq V$ and $q_0 \in N_\epsilon(q, X) \subseteq U$. This contradicts our assumption that $U \cap V = \emptyset$. We have now shown X is connected.

Next, we show X is not path connected.

Each of Γ and L are path connected. However, we show that there are no paths joining points of L to points of Γ. Consider $a \in L$ and $b \in \Gamma$. Intuitively, there cannot be a path in X from b to a since any path we try to draw from b to a intuitively seems as if it must have infinite length. This is because it must travel up to a y-value of $+1$, then down to a y-value of -1 (a distance of at least two), an infinite number of times. This argument depends heavily on the idea of length. In topology, one goal is to avoid such geometric considerations if they are not really essential. We could develop this intuitive sketch into a proof, but we prefer a more topological style.

We show X is not path connected by showing that there is no path, $f: I \to X$ from $f(0) = (0,0) = a$ to $f(1) = (1/2\pi, 0) = b$. Consider the sets $U = f^{-1}(L)$ and $V = f^{-1}(\Gamma)$. We know that V is an open subset of I since Γ is an open subset of X, and f is continuous. Also, $U \neq \emptyset$ since $0 \in U$; similarly, $V \neq \emptyset$ since $1 \in V$. This suggests the following strategy—show that U must be an open subset of I. If so, we will have contradicted the connectedness of I, and our proof will be complete.

Suppose $x \in U$ and let $q = f(x)$; then $q \in L$. We look at the continuity of f at q. Consider $N_{1/2}(q, X)$; there must be a $\delta > 0$ so that $f(N_\delta(x, I)) \subseteq N_{1/2}(q, X)$. We show that, in fact, $f(N_\delta(x, I)) \subseteq L$. This would imply that U is an open subset and our proof would be completed. As we see from Figure 8-3, $N_{1/2}(q, X)$ looks complicated. The projection of $N_{1/2}(q, X)$ onto the x-axis is a lot simpler.

Consider the standard projection $\pi: \mathbf{R}^2 \to \mathbf{R}^1$ given by $\pi(x, y) = x$. Let $J = \pi \circ f(N_\delta(x, I))$. We know that $J \subseteq \pi \circ N_{1/2}(q, X)$ since $f(N_\delta(x, I)) \subseteq N_{1/2}(q, X)$. Let $Z = \pi \circ N_{1/2}(q, X)$; then $J \subseteq Z$. One useful fact is that $N_\delta(x, I)$ is a connected set and, therefore, so is $f(N_\delta(x, I))$. Since J is the image, via a continuous function, of a connected subset, J is connected and thus is an interval. Also, $0 \in J$ since $q \in f(N_\delta(x, I))$. If J is a degenerate interval it must consist of the number 0, in which case we can conclude that $f(N_\delta(x, I)) \subseteq L$. So let us assume that J is a non-degenerate interval, $J = [0, d]$ or $J = [0, d)$, and try to find a contradiction.

Take a close look at Z. Using some examples we can see that the problem: Z is not connected. It has infinitely many pieces which get smaller and smaller as one gets close to 0. Due to this, Z cannot contain a non-degenerate interval which has 0 as a left endpoint. All we need to do now is show there must be a point of $[0, d)$ which is not in Z; that is, show that $J - Z \neq \emptyset$.

Write $q = (0, q_y)$. Our argument has two cases, depending on the value of q_y. In the first case, $q_y < 1/2$. In this case, the local maxima of

Γ are not in $N_{1/2}(q, X)$. In particular, choose an N so that $\frac{1}{\pi/2+2N\pi} < d$. Then $\frac{1}{\pi/2+2N\pi}$ is in $J - Z$. In the second case, $1/2 \leq q_y$. Then the local minima of Γ are not in $N_{1/2}(q, X)$. Choose N so that $\frac{1}{3\pi/2+2N\pi} < d$; then $\frac{1}{3\pi/2+2N\pi}$ is in $J - Z$. ∎

In the proof of Proposition 8.10, we showed there was no path in X between from $a = (0, 0)$ to $b = (1/2\pi, 0)$. An examination of the proof reveals that we did not use any facts about a and b other than $a \in L$ and $b \in \Gamma$. So we actually proved that there is no path in X between a point of L and a point of Γ.

In the proof of Proposition 8.10 we showed that small neighborhoods of points of L in X are not connected subsets (when we project them onto the x-axis we obtain the number 0 together with a collection of open intervals which get smaller and smaller as they get close to 0).

In Example 9.06 we will see another way to prove that the topologist's $\sin(1/x)$ curve is a connected subset.

The topologist's $\sin(1/x)$ curve is a classic example in topology. For describing and analyzing similar examples, it will be convenient to have an alternative model, which in some ways is technically easier to handle.

Example 8.11 The subset of the plane T, shown in Figure 8-4, is a minor variation of Example 8.09. We call it the "piecewise-linear $\sin(1/x)$ curve."

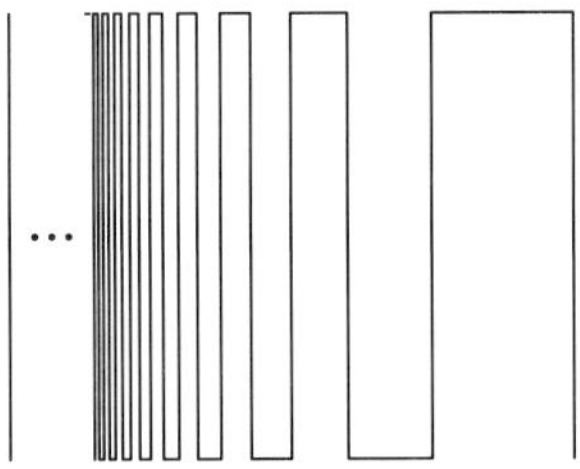

Figure 8-4 A variation of the topologist's $\sin(1/x)$ curve; see Example 8.11.

It is a union of line segments. For any $n \in \mathbf{N}$, let V_n be the vertical line segment with endpoints $(\frac{1}{n}, 1)$ and $(\frac{1}{n}, -1)$. For odd n, let H_n be the horizontal line segment between $(\frac{1}{n}, 1)$ and $(\frac{1}{n+1}, 1)$. For even n, H_n is the horizontal line segment between $(\frac{1}{n}, -1)$ and $(\frac{1}{n+1}, -1)$. Finally, L is the line segment with endpoints $(0, -1)$ and $(0, 1)$, and we define

$$T = L \cup \left(\bigcup_{n=1}^{\infty} V_n \right) \cup \left(\bigcup_{n=1}^{\infty} H_n \right).$$

We leave as an exercise to check that T is connected but not path connected; a proof can be constructed from the proof of Proposition 8.10 with only minor modifications (Problem 8.2). In fact, one can show, (but this is a bit harder, for technical reasons) that T is homeomorphic to X of Example 8.09 (Problem 8.3). Those who enjoy building homeomorphisms might try to show that these two subsets of $\mathbf{R}^2$ are equivalent subsets. ◆

Here is a positive result in which connectedness implies path connectedness. Be very careful when using this proposition. Note that it refers to *open subsets of* $\mathbf{R}^\mathbf{n}$ and is not a general result about open connected subsets of other subsets.

Proposition 8.12 *If U is an open connected subset of $\mathbf{R}^\mathbf{n}$, then it is path connected.*

Proof: As is often the case with connectedness arguments, we argue indirectly. Assume that U is an open connected subset of $\mathbf{R}^\mathbf{n}$, and that U is not path connected. Then there must be two points, $a, b \in U$, such that there does not exist a path from a to b.

Looking ahead to the end of the proof, we need to contradict something. It seems natural to arrive at a contradiction of the assumption that U is connected. Thus we will want, somehow, to find two sets A and B which give a disconnecting partition of U. One of the essential properties of A and B is that they are non-empty sets. Our assumption that U is not path connected gave us the existence to two points, so a reasonable first step is to define our sets so that $a \in A$ and $b \in B$. There is no path from a to b, and we are lead to define:

$$B = \{x \in U \mid \text{there } \textit{does not} \text{ exist a path in } U \text{ from } a \text{ to } x\}.$$

Then $B \neq \emptyset$ since $b \in B$. Now define:

$$A = \{x \in U \mid \text{there } \textit{does} \text{ exist a path in } U \text{ from } a \text{ to } x\}.$$

The constant map $f(t) = a$, for all $t \in I$, is a continuous map which gives a path from a to a. Thus $a \in A$ and, in particular, $A \neq \emptyset$. Clearly, $U = A \cup B$ and $A \cap B = \emptyset$. To arrive at our contradiction, we need to show that A and B are open subsets of U. Since U is open in $\mathbf{R}^\mathbf{n}$, this is equivalent to showing that A and B are open in $\mathbf{R}^\mathbf{n}$, by Proposition 2.19.

We first show that A is an open subset of U. Let $x \in A$. We need to find an ϵ such that $N_\epsilon(x) \subseteq A$. Since U is an open subset of $\mathbf{R}^\mathbf{n}$, there is an ϵ such that $N_\epsilon(x) \subseteq U$. We show that, in fact, $N_\epsilon(x) \subseteq A$. Let $y \in N_\epsilon(x)$. To show that $y \in A$, we find a path in U from a to y. Since $x \in A$, there is a path $f: I \to U$ such that $f(0) = a$ and $f(1) = x$. Join x to y by a straight line segment L. We have $L \subseteq N_\epsilon(x)$. Our idea is to "put f and L together to make a path from a to y." To make this precise, we first express L as a path by a parameterization: $g(t) = (1-t)\vec{x} + t\vec{y}$ for $t \in I$. Since $f(1) = g(0)$, concatenation is defined. Thus $f * g$ is a path in U from a to y. Therefore A is an open subset of U.

The argument that B is an open subset of U follows along similar lines. Since U is an open subset of $\mathbf{R^n}$, there is an ϵ such that $N_\epsilon(x) \subseteq U$. Let $y \in N_\epsilon(x)$. To show that $y \in B$, we show there is no path in U from a to y. We reason indirectly. Suppose there is a path f in U from a to y. Let L be the line segment from y to x. As before, define $g(t) = (1-t)\vec{x} + t\vec{y}$, for $t \in I$. The path $F = f * g$ is in U from a to x. But this contradicts the assumption that $x \in B$. Thus we can conclude that $N_\epsilon(x) \subseteq B$. Therefore B is a open subset.

We have now completed our demonstration that U is not connected; thus obtaining our desired contradiction. ∎

We have the following version of Proposition 6.07.

Proposition 8.13 *Suppose $X \subseteq \mathbf{R^n}$. If X is path connected and f is a continuous function, then $f(X)$ is path connected.* ∎ *(Problem 8.4)*

The proposition above can now be used to show that path connectedness is a topological property.

Proposition 8.14 *Suppose $X \subseteq \mathbf{R^n}$, $Y \subseteq \mathbf{R^m}$, and X and Y are homeomorphic. Then X is path connected if and only if Y is path connected.* ∎ *(Problem 8.5)*

Example 8.15 Proposition 8.14 implies the unit interval I is not homeomorphic to the topologist's $\sin(1/x)$ curve since I is path connected and the topologist's $\sin(1/x)$ curve is not.

We need to show that I *is* path connected. We can do this by parameterizing any closed sub-interval of I as a path. If $a, b \in I$ with $a \le b$, then we can define a path in I from a to b by $f(t) = a + (1-t)b$. ◆

To use the definition of path connectedness, directly, we would have to consider all pairs of points of a space. The following proposition shows that we can focus attention on one point, say $x_0 \in X$, and ask whether we can find paths from x_0 to any other points of X.

Proposition 8.16 *Suppose $X \subseteq \mathbf{R^n}$ and $x_0 \in X$. Then X is path connected if and only if for any $y \in X$ there is a path from x_0 to y.* ∎ *(Problem 8.6)*

8.2 Path components

A component of a subset is a piece of the subset as defined by the concept of connectedness; similarly, we have a definition of a piece of the subset with respect to the concept of path connectedness.

Definition 8.17 *Let $A \subseteq X \subseteq \mathbf{R}^n$; we say A is a* **path component of** *X if A is a maximal path-connected set. This means that A is path connected and if B is path-connected with $A \subseteq B \subseteq X$, then $A = B$.*

Example 8.18 The topologist's $\sin(1/x)$ curve, Example 8.09, has two path components, Γ and L (Problem 8.18). ◆

Example 8.19 In Example 7.23 we noted that $\mathbf{R}^1 - \{0\}$ had two components. In the remark on page 191, we viewed $\mathbf{R}^1 - \{0\}$ as the set of non-singular linear maps of $\mathbf{R}^1$. Clearly, $\mathbf{R}^1 - \{0\}$ has two *path* components: $(-\infty, 0)$ and $(0, \infty)$ (Problem 8.7).

So we can say that the set of non-singular linear maps of $\mathbf{R}^1$ has two path components, one component corresponding to positive slopes, the other to negative slopes. This leads to the next, more interesting topic.

Consider the set of non-singular linear maps of $\mathbf{R}^2$. This corresponds to a subset X of $\mathbf{R}^4$, as in the remark on page 191. Let X_+ be the set of matrices with positive determinant and X_- be the set of matrices with negative determinant. Claim: X has two path components, X_+ and X_-. Since X is an open subset of $\mathbf{R}^4$, this implies that X has two components, by Proposition 8.12.

We will show that X_+ is path connected by showing that we can connect any point in X_+ to the point $(1, 0, 0, 1)$ by a path in X_+, and invoking Proposition 8.16. Note that $(1, 0, 0, 1)$ corresponds to the identity matrix, $\begin{pmatrix} 1 & 0 \\ 0 & 1 \end{pmatrix}$. (Similarly, we show that X_- is path connected by finding a path in X_- connecting any of its points to $(1, 0, 0, -1)$.)

Here is an outline of the idea. Let Y be the subset of X corresponding to *orthogonal* matrices of determinant +1. We begin with $M \in X_+$ and find a path ϕ from M to $M' \in Y$. We then show Y is path connected, using Proposition 8.16, by showing that any point in Y can be connected by a path ψ in Y to the 2×2 identity matrix.

The idea of the definition of ϕ is to define a continuous version of the Gram-Schmidt orthonormalization process. In the plane, this process will take two linearly independent vectors, $\vec{v_1}$ and $\vec{v_2}$, and alter these so as to obtain an orthonormal basis, $\vec{v_1'}$ and $\vec{v_2'}$. Briefly, one first normalizes $\vec{v_1}$, getting unit vector $\vec{v_1'} = (1/|\vec{v_1}|)\vec{v_1}$. One then replaces $\vec{v_2}$ by the orthogonal projection of $\vec{v_2}$ with respect to $\vec{v_1'}$ which is then normalized to obtain unit $\vec{v_2'}$.

To relate this process to matrices, we apply the Gram-Schmidt orthonormalization process to the column vectors of M. The plan is to get a path ϕ_1 in X_+ from M to M_1 where the first column vector of M_1 is a unit vector. Then we plan to find a path ϕ_2 in X_+ from M_1 to M_2 where the first column vector is unchanged but the second column vector is orthogonal to the first. Finally, we find a path ϕ_3 in X_+ from M_2 to an orthogonal matrix, M'.

Write $M \left(\begin{smallmatrix} a & b \\ c & d \end{smallmatrix}\right)$. Our first path will go from M to $M_1 = \left(\begin{smallmatrix} a' & b \\ c' & d \end{smallmatrix}\right)$ where $a' = \frac{a}{\sqrt{a^2+c^2}}$ and $c' = \frac{c}{\sqrt{a^2+c^2}}$. In fact, our path is just a parameterization of the line segment, in $\mathbf{R}^4$, between these two points:

$$\phi_1(t) = \begin{pmatrix} ta & b \\ tc & d \end{pmatrix} + \begin{pmatrix} (1-t)a' & b \\ (1-t)c' & d \end{pmatrix}.$$

To show that $\phi_1(t)$ is a path in X_+, we need to show that, for all $t \in I$, $|\phi_1(t)| > 0$. Note that $|M_1| = (1/\sqrt{a^2+c^2})|M|$, and we have:

$$|\phi_1(t)| = t|M| + (1-t)|M_1| = t|M| + \frac{1-t}{\sqrt{a^2+c^2}}|M| = (t + \frac{1-t}{\sqrt{a^2+c^2}})|M|.$$

Since $M \in X_+$, $|M| > 0$. For $t \in I$, $(t + \frac{1-t}{\sqrt{a^2+c^2}}) > 0$ since this quantity varies linearly between the two positive numbers 1 and $\frac{1}{\sqrt{a^2+c^2}}$. Thus, for all $t \in I$, $|\phi_1(t)| > 0$.

Construction of the paths $\phi_2(t)$ and $\phi_3(t)$ are defined in a similar manner—they are parameterizations of the line segments between the matrix M_1 and M_2 and between M_2 and M'. Details are left as an exercise (Problem 8.8). The map ϕ can be then defined: $\phi = \phi_1 * \phi_2 * \phi_3$.

Let $M \in X_+$, so M is a 2×2 matrix, with $|M| > 0$, which we are viewing as a point of $\mathbf{R}^4$. First (and this will take several sub-steps), we find a path, ϕ in X_+ from M to an orthogonal matrix M'. In general, a 2×2 orthogonal matrix A corresponds to a rotation, or a reflection in a line followed by a rotation. Thus A is has the form of either

$$\begin{pmatrix} \cos(\theta) & -\sin(\theta) \\ \sin(\theta) & \cos(\theta) \end{pmatrix}, \text{ if } |A| > 0, \text{ or } \begin{pmatrix} \cos(\theta) & \sin(\theta) \\ \sin(\theta) & -\cos(\theta) \end{pmatrix}, \text{ if } |A| < 0.$$

So there is a θ_0 with

$$M' = \begin{pmatrix} \cos(\theta_0) & \sin(\theta_0) \\ -\sin(\theta_0) & \cos(\theta_0) \end{pmatrix}.$$

For $0 \le t \le 1$, define

$$M'_t = \begin{pmatrix} \cos((1-t)\theta_0) & \sin((1-t)\theta_0) \\ -\sin((1-t)\theta_0) & \cos((1-t)\theta_0) \end{pmatrix}.$$

If we let $\psi(t) = M'_t$, this will correspond to a path in X_+ from M' to $\left(\begin{smallmatrix} 1 & 0 \\ 0 & 1 \end{smallmatrix}\right)$. Then, using concatenation of paths, Definition 8.07, $\phi * \psi$ will be our path in Y from M to $\left(\begin{smallmatrix} 1 & 0 \\ 0 & 1 \end{smallmatrix}\right)$.

In summary, we can now say that the set of non-singular linear maps of $\mathbf{R}^2$ has two path components, X_+ and X_-, where the sign of subscript is the sign of the determinant of the matrix. ◆

Propositions 7.14 and 7.22 for connected subsets have counterparts for path connectedness.

Proposition 8.20 *Suppose $A \subseteq \mathbf{R^n}$ and $B \subseteq \mathbf{R^n}$, both non-empty. Then A and B are path connected if and only if $A \times B$ is path connected.* ∎ *(Problem 8.20)*

Proposition 8.21 *If f is a homeomorphism from X onto Y, and C is a path component of X, then $f(C)$ is a path component of Y. Moreover, this matching of path components by f induces a one-to-one correspondence between the path components of X and those of Y.*

In particular, "the number of path components of X equals N" is a topological property. ∎ *(Problem 8.14)*

Example 8.22 In the remark on page 211 we showed that the set of non-singular linear maps X of $\mathbf{R}^2$, considered, as a subset of $\mathbf{R}^4$, has two path components, X_+, and X_-.

What about affine maps? The set of non-singular *affine* maps corresponds to a subset, call it Y, of $\mathbf{R}^6$. Is this connected? Rather than modify the proof of the remark on page 211, we suggest a different method.

Suppose $F = M\vec{x} + \vec{B}$ is a non-singular affine map. Note that Y is homeomorphic to $X \times \mathbf{R}^2$ where the second factor corresponds to $\vec{B}$ (Problem 8.9). Since X is not path connected, then neither is $X \times \mathbf{R}^2$, by Proposition 8.20. Also, using Proposition 8.20, we can verify that Y has exactly two path components: $X_+ \times \mathbf{R}^2$, and $X_- \times \mathbf{R}^2$ (Problem 8.10). ◆

Example 8.23 We could alter the piecewise-linear topologist's $\sin(1/x)$ curve, Example 8.11, to get another interesting example, shown in Figure 8-5.

For any $n \in \mathbf{N}$, let V_n be the vertical line segment with endpoints $(\frac{1}{n}, 1)$ and $(\frac{1}{n}, -1)$. For odd n, let H_n be the horizontal line segment between $(\frac{1}{n}, 1)$ and $(\frac{1}{n+1}, 1)$ while, for even n, H_n is the horizontal line segment between $(\frac{1}{n}, -1$ and $(\frac{1}{n+1}, -1)$. We use four more line segments. The first, L, is the line segment, as used before, with endpoints $(0, -1)$ and $(0, 1)$. The other three line segments join together to determine a path from $(0, -1)$ to $(1, -1))$ in the lower half of the plane: M_1 has endpoints $(0, -1)$, and $(0, -4/3)$, M_2 has endpoints $(0, -4/3)$ and $(1, -4/3)$, and M_3 has endpoints $(1, -4/3)$ and $(1, -1))$.

$$T' = L \cup (M_1 \cup M_2 \cup M_3) \cup \left(\bigcup_{n=1}^{\infty} V_n \right) \cup \left(\bigcup_{n=1}^{\infty} H_n \right).$$

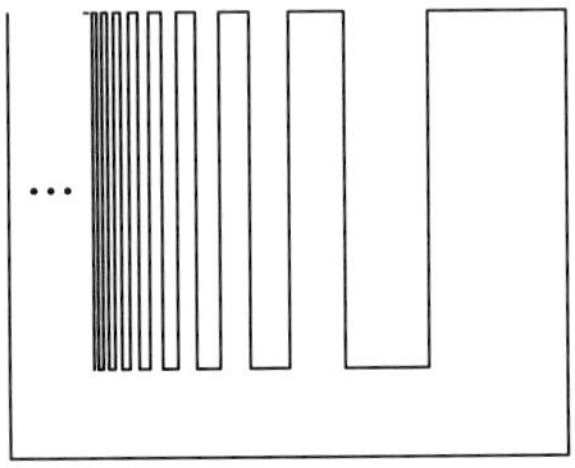

Figure 8-5 A set T', a variation of the topologist's $\sin(1/x)$ curve, called the "closed piecewise-linear, topologist's $\sin(1/x)$ curve"; see Example 8.23.

We call T' the "piecewise-linear closed $\sin(1/x)$ curve."

We note that T' is path connected (Problem 8.11); thus we can conclude that the piecewise-linear $\sin(1/x)$ curve and the piecewise-linear closed $\sin(1/x)$ curve are not homeomorphic.

For the record, the more traditional "closed $\sin(1/x)$ curve" is as follows. (We choose not to use this as it makes details of the exercises more tedious.) We let

$$\Gamma' = \{(x, y) \in \mathbf{R}^2 \mid y = \sin(1/x) \text{ and } 0 < x \leq M \text{ for some number } M\}.$$

Let$L$ be the line segment as above, and let α be the image of a path from $(0, -1)$ to $(M, \sin(1/M))$, which intersects $\Gamma' \cup L$ only in those points. Let $X' = \Gamma' \cup L \cup \alpha$. ◆

Example 8.24 A consequence of The Jordan Curve Theorem, Proposition 4.10, is that if X is a subset of the plane homeomorphic to a circle, then $\mathbf{R}^2 - X$ has two path components. What is the situation for subsets, X, homeomorphic to the line? That is: if $X \subseteq \mathbf{R}^2$ and X is homeomorphic to $\mathbf{R}^1$, how many path components might $\mathbf{R}^2 - X$ have?

Here are two simple examples where the number of path components of $\mathbf{R}^2 - X$ is two and one, respectively.

A line separates the plane into two path components (Problem 8.22).

If $X = \{(x, y) \in \mathbf{R}^2 \mid y = 0 \text{ and } -1 < x < 1\}$, then X is homeomorphic to $\mathbf{R}^1$ and $\mathbf{R}^2 - X$ is path connected (Problem 8.23). ◆

Example 8.25 We describe a subset, $\mathcal{R} \subseteq \mathbf{R}^2$, which is homeomorphic to $\mathbf{R}^1$, whose complement in $\mathbf{R}^2$ has four path components, (Problem 8.4). The use of the symbol $\mathcal{R}$ is notational reference to the mathematician Reeb; the set $\mathcal{R}$ is related to a mathematical structure called a "Reeb foliation."

The set $\mathcal{R}$ (see Figure 8-6) spirals, in the plane towards a pair of concentric circles; denote the smaller of these circles by C_0, the larger by C_1. As we will see, the radius of C_1 is $e^{\pi/2} \approx 4.8$ and the radius of C_0 is $e^{-\pi/2} \approx 0.21$.

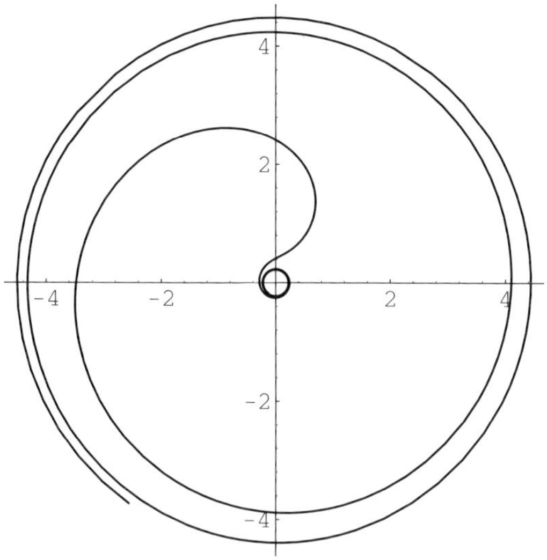

Figure 8-6 The set $\mathcal{R}$ of Example 8.25. For a topologically equivalent subset which is, topologically, more revealing; see Figure 8-7.

Let J be the interval $-\frac{\pi}{2} < x < \frac{\pi}{2}$. We have an embedding of J into the plane as the graph of $y = \sec(x)$: define $\Gamma: J \to \mathbf{R}^2$ by $\Gamma(x) = (x, \sec(x))$. Consider the complex exponential function, $w = \exp(z)$. Write this as a map of $\mathbf{R}^2$, using Cartesian coordinates: $E(x, y) = (e^y \cos(x), e^y \sin(x))$. (This is a version of Euler's formula.) Now define $\mathcal{R} = E \circ \Gamma(J)$.

Let D_0 be the closed disk whose boundary is C_0. Let D_1 denote the points of $\mathbf{R}^2$ on or outside the circle C_1. Then $A = \mathbf{R}^2 - D_0 - D_1$ is an annular region (homeomorphic to the product of a circle and an open interval). We have $\mathcal{R} \subseteq A$. One can show that $A - \mathcal{R}$ has two path components, A_0 and A_1. Then the four path components of $\mathbf{R}^2 - \mathcal{R}$ are A_0, A_1, D_0, and D_1.

Although Figure 8-6 shows $\mathcal{R}$ accurately, it is not very revealing. To get better graphic, we do some rescaling; see Figure 8-7. We have used a non-singular linear map $F(\vec{x}) = M\vec{x}$ where M is a matrix of the form $M = \left(\begin{smallmatrix} a & 0 \\ 0 & b \end{smallmatrix}\right)$. We then define: $\mathcal{R}' = E \circ F \circ \Gamma(J)$. Now we can more clearly see that our curve spirals, in a counter-clockwise direction, to the two limiting circles.

The two subsets $\mathcal{R}$ and $\mathcal{R}'$ are equivalent subsets of $\mathbf{R}^2$ (Problem 8.4). In Figure 8-7, we show $\mathcal{R}'$ for a particular choice of M. ◆

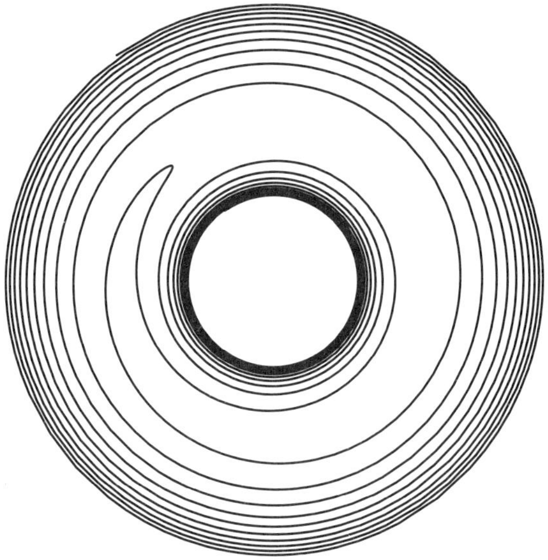

Figure 8-7 The set $\mathcal{R}'$ of Example 8.25.

*8.3 General topology and Chapter 8

As with connectedness, path connectedness in the abstract setting behaves much as expected. Using the product topology, any product of path-connected subsets will be path connected. Also, a quotient space obtained from a path-connected space will be path connected.

The topic of path components in function spaces is an important one that will be discussed later, in Section*15.5, but here is one simple example.

Recall the function space $C(I, \mathbf{R}^1)$, Example 1.70, considered as a metric space.

Proposition 8.26 *Let $C(I, \mathbf{R}^1)$ be the topological space of continuous real-valued functions with domain I; then $C(I, \mathbf{R}^1)$ is path connected.*

Proof: Let $f, g \in C(I, \mathbf{R}^1)$. We can define a path, ϕ from f to g by: $\phi(t) = (1 - t)f(x) + tg(x)$. See Figure 8-8 for an example. ∎

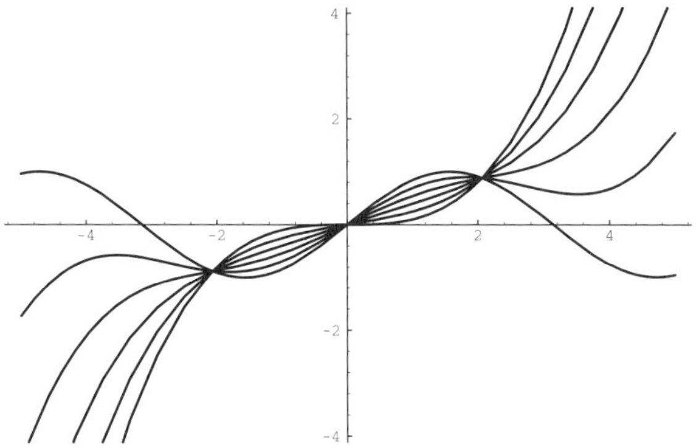

Figure 8-8 Graphs representing a path in $C(I, \mathbf{R}^1)$ from $\sin(x)$ to $x^3/10$ given by $t\sin(x) + (1-t)x^3/10$; see Proof of Proposition 8.26.

8.4 Problems for Chapter 8

8.1 Prove Proposition 8.08.

8.2 Prove the assertion of Example 8.11 that T is connected but not path connected.

8.3 Prove that the set T of Example 8.11 is homeomorphic to X, the topologist's $\sin(1/x)$ curve of Example 8.09.

8.4 Prove Proposition 8.13.

8.5 Prove Proposition 8.14

8.6 Prove Proposition 8.16.

8.7 Verify that $\mathbf{R}^1 - \{0\}$ has two path components, as stated in Example 8.19.

8.8 Fill in details of definitions of $\phi_2(t)$ and $\phi_3(t)$, Example 8.19, following the discussion for $\phi_1(t)$.

8.9 Verify the claim of Example 8.22 that Y is homeomorphic to $X \times \mathbf{R}^2$.

8.10 Verify the claim of Example 8.22 that Y has exactly two path components: $X_+ \times \mathbf{R}^2$ and $X_- \times \mathbf{R}^2$.

8.11 Verify the assertion made in Example 8.23 that the piecewise-linear closed $\sin(1/x)$ curve T' is path connected.

8.12 Show that the topologist's $\sin(1/x)$ curve, Example 8.09, has two path components, Γ and L.

8.13 Show that the topologist's $\sin(1/x)$ curve does not have the fixed-point property.

8.14 Prove Proposition 8.21.

8.15 Show that if $f: \mathbf{R}^1 \to \mathbf{R}^1$ is a continuous function and Γ is the graph of f, then $\mathbf{R}^2 - \Gamma$ has two path components.

8.16 Find an example of a function f, $f: \mathbf{R}^1 \to \mathbf{R}^1$ such that if we let Γ be the graph of f, then $\mathbf{R}^2 - \Gamma$ has more than two path components. (By the previous problem, f will not be continuous.)

Find an example of a function such that $\mathbf{R}^2 - \Gamma$ has infinitely many path components.

Here is a (optional) challenge question: Is there an example of a function such that $\mathbf{R}^2 - \Gamma$ has uncountably many path components?

8.17 Show that the closed $\sin(1/x)$ curve, call it G, is not homeomorphic to the circle, by showing that the closed $\sin(1/x)$ curve has the self-homeomorphism fixed-point property (recall Definition 4.45.) (Hint: Let $z = (0, 1)$ and let $x \in G - z$. How many path components does $G - \{z\}$ have? How many components does $G - \{x\}$ have?) (Remark: One could also use this reasoning for the piecewise-linear closed $\sin(1/x)$ curve.)

8.18 Let $X \subseteq \mathbf{R}^n$. Show that the relation "there is a path in X from a to b" gives an equivalence relation on the set of points of X.

8.19 State and prove a theorem about the union of path-connected subsets.

8.20 Prove Proposition 8.20.

8.21 Let T' be the closed piecewise-linear $\sin(1/x)$ curve, Example 8-5. Show that $\mathbf{R}^2 - T'$ has, at most, two path components. (Note: In Example 9.42, we will show that $\mathbf{R}^2 - T'$ is not path connected. Thus $\mathbf{R}^2 - T'$ must have exactly two components.)

8.22 Verify the assertion of Example 8.24 that a line separates the plane into two path components.

8.23 Verify the assertion of Example 8.24 that if $X = \{(x, y) \in \mathbf{R}^2 \mid y = 0 \text{ and } -1 < x < 1\}$, then X is homeomorphic to $\mathbf{R}^1$ and $\mathbf{R}^2 - X$ is path connected.

8.24 Verify the assertion, made in Example 8.25 that $\mathcal{R}$, is homeomorphic to $\mathbf{R}^1$ and $\mathbf{R}^2 - X$ has four path components.

8.25 Verify the assertion, made in Example 8.25 that $\mathcal{R}$, and $\mathcal{R}'$ are equivalent subsets of $\mathbf{R}^2$.

9. CLOSURE AND LIMIT POINTS

OVERVIEW: We present a collection of basic topics about closed and open subsets, centered about the concept of limit point, defined in Chapter 1. The new concepts of closure, interior, boundary, and convergent sequence do not depend on the concepts of Chapters 2-8; however, the discussion involves examples from some of these chapters.

9.1 The closure operation

We return to the basic topic of closed subsets for an in-depth look. The concept of closure is closely connected to the idea of open subset. In fact, there are some who prefer to use the concept of closure, rather than the concept of open set, as the fundamental primitive notion for developing the notion of topology.

Definition 9.01 *Let $A \subseteq X \subseteq \mathbf{R^n}$. The* **closure of** *$A$* **in** *$X$, denoted $\overline{A}$, is the smallest closed subset of X which contains A. (That is, if B is closed and $A \subseteq B \subseteq \overline{A}$, then $B = \overline{A}$.)*

Sometimes the set X is understood in context. In this case, we denote the closure of A by $\overline{A}$. This notation makes no explicit mention of X. If it becomes important to refer to X, we use the notation $Cl_X(A)$ for the closure of A in X. One of the reasons is that often X is a closed subset of Z. In this case, if $A \subseteq X \subseteq Z$ and X is a closed subset of Z, then $Cl_X(A) = Cl_Z(A)$.

The next two propositions give alternative descriptions of $\overline{A}$.

Proposition 9.02 *Suppose $A \subseteq X \subseteq \mathbf{R^n}$; then $\overline{A}$ is the intersection of all closed subsets of X which contain A.* ▮ *(Problem 9.1)*

Proposition 9.03 *Suppose $A \subseteq X \subseteq \mathbf{R^n}$; then $\overline{A}$ is the union of A and the set of points of X which are limit points of A.* ▮ *(Problem 9.4)*

On of the consequences of either of the propositions above is that, for any subset A, $\overline{A}$ exists and is non-empty if $A \neq \emptyset$.

Example 9.04 In Chapter 1 we used the notation $\overline{N}_\epsilon(p)$ for a closed ball, Definition 1.09. This is consistent with our notation for closure since $\overline{N}_\epsilon(p)$ is the closure in $\mathbf{R^n}$ of $N_\epsilon(p)$. The closure of $N_\epsilon(p)$ would be written as $\overline{N_\epsilon(p)}$ where the line goes over the entire symbol, rather than just the letter N. So we are saying that $\overline{N_\epsilon(p)} = \overline{N}_\epsilon(p)$.

There are several ways one can prove these assertions, using Definition 9.01 or Propositions 9.02, 9.03. ◆(Problems 9.2, 9.3).

A basic fact about closures of connected sets is

Proposition 9.05 *Suppose $A \subseteq X \subseteq \mathbf{R^n}$. If A is connected, then $\overline{A}$, the closure of A in X, is also connected.*

More generally, if A is connected and $A \subseteq B \subseteq \overline{A} \subseteq X$, then B is connected.

Proof: We prove the second statement; it implies the first. Suppose B is not connected. Then there are open subsets U and $V$of B which give a disconnecting partition of B. Now consider the sets $U' = U \cap A$ and $V' = V \cap A$. One of U' or V' must be empty, otherwise U' and V' would give a disconnecting partition of A, contradicting the connectedness of A. Thus A is entirely contained in one of U' or V'. Without loss of generality, we may assume that $A \subseteq V'$. Since $U \neq \emptyset$, then there is a point $x \in B$ with $x \in U$. By Proposition 9.03, we conclude that x is a limit point of A. But this is impossible since the set U is an open subset of B and contains no points of A. ▮

By Proposition 9.03, $\overline{A}$ consists of A together with all its limit points in X. It follows that if $A \subseteq B \subseteq \overline{A}$, then B is a union of A with *some* of its limit points. So our proposition could be restated: if A is a connected subset, and we add to A some of its limit points in X, the result is still a connected subset.

Example 9.06 Consider the topologist's $\sin(1/x)$ curve X, Example 8.09. It is not hard to show that L is the set of limit points in $\mathbf{R}^2$ of Γ, which are not points of Γ. Thus $\overline{\Gamma} = X$. Certainly, Γ is connected; it is the continuous image of a connected set. The continuous map we refer to is $f\colon (0, \infty) \to \mathbf{R}^2$ given by $f(x) = (x, \sin(1/x))$. Applying Proposition 9.05, we have a proof that X is connected. ◆

The next proposition shows a relationship between dense subsets and closure. In fact, it is often used as a definition for dense subset.

Proposition 9.07 *Suppose $A \subseteq X \subseteq \mathbf{R^n}$; then A is dense in X if and only if $Cl_X(A) = X$.* ▮ *(Problem 9.5)*

For example, if Q denotes all rational numbers in $\mathbf{R^1}$, then $\overline{Q} = \mathbf{R^1}$.

Since continuity can be defined in terms of closed subsets, and closure operation determines closed subsets, we can express continuity by using closure, directly, as follows.

Proposition 9.08 *Suppose $X \subseteq \mathbf{R^n}$; $Y \subseteq \mathbf{R^m}$, and $f: X \to Y$. Then f is a continuous function if and only for all subsets $A \subseteq X$, $f(\overline{A}) \subseteq \overline{f(A)}$.* ▮ *(Problem 9.6)*

One can also relate closure and the distance from a point to a set, Definition 3.50:

Proposition 9.09 *Suppose that $X \subseteq \mathbf{R^n}$; then $\overline{X}$ is the set of points $p \in \mathbf{R^n}$ whose distance to X is zero.*

Proof: Let $D_0(X)$ denote the points which are a distance zero from X. We show that, in $\mathbf{R^n} - \overline{X}$, points are a distance greater than zero from X. If $p \in \mathbf{R^n} - \overline{X}$, then since, $\overline{X}$ is closed, there is an $\epsilon > 0$ with $N_\epsilon(p) \cap \overline{X} = \emptyset$, $\delta(p, X) \geq \epsilon$ and, in particular, $\delta(p, X) \neq 0$. Thus $D_0(X) \subseteq \overline{X}$.

On the other hand, by Proposition 3.51, $D_0(X)$ will be a closed subset of $\mathbf{R^n}$ containing X, and so $\overline{X} \subseteq D_0(X)$. Thus we conclude: $\overline{X} = D_0(X)$. ▮

We examine how closure relates of equivalent subsets.

Proposition 9.10 *Suppose $X \subseteq \mathbf{R^n}$, $A \subseteq X$, and $B \subseteq X$. If A and B are equivalent subsets of X, then $\overline{A}$ and $\overline{B}$ are equivalent subsets of X. Also, $\overline{A} - A$ and $\overline{B} - B$ will be equivalent subsets of X.* ▮ *(Problem 9.7)*

Example 9.11 We give an example of embeddings, f and g, which are not stably equivalent embeddings; thus giving an answer to Questions 6.33 and 6.34. Let $X \subseteq \mathbf{R^1}$ be the set $X = [-1, 0) \cup (0, 1]$, and let $f: X \to \mathbf{R^1}$ be the inclusion map. Let $g: X \to \mathbf{R^1}$ be defined by

$$g(x) = \begin{cases} x - 1 & \text{if } x < 0 \\ x + 1 & \text{if } 0 < x. \end{cases}$$

Let $Y = g(X)$; then $Y = [-2, -1) \cup (1, 2]$. Now X and Y are not equivalent subsets of $\mathbf{R^1}$ since one component of $\mathbf{R^1} - X$, namely $\{0\}$, is a single point, but no component of $\mathbf{R^1} - Y$ consists of a single point. It follows that $\mathbf{R^1} - X$ is not homeomorphic to $\mathbf{R^1} - Y$, by Proposition 6.18. Therefore f and g are *not* equivalent embeddings, by Proposition 6.15.

We next show that f and g are not *stably* equivalent. The argument will be by contradiction. If f and g were stably equivalent then, for some n, f' and g' would be equivalent embeddings into $\mathbf{R^n}$ where $f' = i \circ f$ and $g' = i \circ g$ and $i: \mathbf{R^1} \to \mathbf{R^n}$ is the standard inclusion.

Let $X' = f'(X)$ and $Y' = g'(X)$. The argument we have used to show f and g are not equivalent won't work here since, if $2 \leq n$, both $\mathbf{R^n} - X'$ and $\mathbf{R^n} - Y'$ are connected, and, in fact, path connected.

However we can use Proposition 9.10. Since $\overline{X'}$ is homeomorphic to a closed interval, it is connected. But $\overline{Y'}$ is not connected; it is homeomorphic to a disjoint union of two closed intervals. If they are not homeomorphic, then they certainly cannot be equivalent subsets, and so f' and g' cannot be equivalent embeddings. ◆

9.2 Boundary and interior of a subset

If we have $A \subseteq X \subseteq \mathbf{R^n}$, the concept of closure gives us a way of associating to A a (possibly) larger *closed* subset of X. The definition of interior is a way of associating to A a (possibly) smaller *open* subset of X:

Definition 9.12 *Suppose $a \in A \subseteq X \subseteq \mathbf{R^n}$. We say a* **is an interior point of** *A* **in** *X, if there is an open subset U of X with $a \in U \subseteq A$.*

The subset of A consisting of all interior points of A is called the **interior of** *A and is denoted by $int(A, X)$.*

As with closure, if the set X is understood in context, one often uses the notation $int(A)$ for $int(A, X)$.

Proposition 9.13 *Suppose $A \subseteq X \subseteq \mathbf{R^n}$; then*

(a) *$int(A)$ is the largest open subset of X contained in A. By this we mean that $int(A)$ is an open subset of X, and if U is any open subset of X with $int(A) \subseteq U \subseteq A$, then $int(A) = U$.*
(b) *$int(A)$ is the set of all points of A which are not limit points of $X - A$* ▮ *(Problem 9.8)*

Open subsets can be defined in terms of interior points. It follows from Proposition 9.13 that if $U \subseteq X \subseteq \mathbf{R^n}$, then U is an open subset of X if and only if every point of U is an interior point of U.

Definition 9.14 *Suppose $a \in A \subseteq X \subseteq \mathbf{R^n}$; we say a is a* **boundary point** *of A in X if $a \in \overline{A} \cap \overline{X - A}$. (Note: The closures are closures in X.)*

The set of boundary points of A in X is called the **boundary of** *A* **in** *X, denoted $Bd(A, X)$, or, if X is understood, this is abbreviated as $Bd(A)$.*

If $A \subseteq X \subseteq \mathbf{R^n}$, then the boundary of A is a closed subset of X since it is the intersection of closed subsets.

Proposition 9.15 *If $A \subseteq X \subseteq \mathbf{R^n}$, then the interior of A and the boundary of A are disjoint sets, whose union is $\overline{A}$.* ∎ *(Problem 9.10)*

Proposition 9.16 *If $A \subseteq X \subseteq \mathbf{R^n}$, then $\overline{X - A} = X - int(A)$.*
∎ *(Problem 9.11)*

Proposition 9.17 *Suppose $A \subseteq X \subseteq \mathbf{R^n}$; then*

(a) *A is a closed subset of X if and only if $Bd(A, X) \subseteq A$;*
(b) *A is open in X if and only if $Bd(A, X) \cap A = \varnothing$.* ∎ *(Problem 9.12)*

Here are a few basic facts relating closure, boundary, and interior of a subset to Cartesian products.

Proposition 9.18 *Suppose $A \subseteq X \subseteq \mathbf{R^n}$ and $B \subseteq Y \subseteq \mathbf{R^m}$ then*

(a) $Cl_{(X\times Y)}(A \times B) = Cl_X(A) \times Cl_Y(B)$
(b) $int(A \times B) = int(A) \times int(B)$.
(c) $Bd(A \times B) = (Bd(A) \times Cl_Y(B)) \cup (Cl_X(A) \times Bd(B))$ *where* $(Bd(A) \times Cl_Y(B)) \cap (Cl_X(A) \times Bd(B)) = Bd(A) \times Bd(B)$.
∎ *(Problem 9.13)*

If A and B are closed subsets, then $A = Cl_X(A)$ and $B = Cl_Y(B)$, so the first formula of Proposition 9.18(c) simplifies to the more easily remembered

$$Bd(A \times B) = (Bd(A) \times B) \cup (A \times Bd(B)),$$

and

$$(Bd(A) \times B) \cap (A \times Bd(B)) = Bd(A) \times Bd(B). \quad \blacklozenge$$

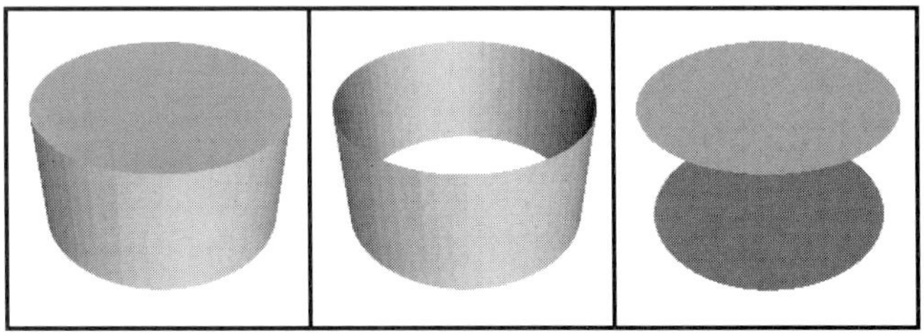

Figure 9-1 An example of the boundary of the product $D^2 \times I$; see Example 9.19. On the left is shown all of $Bd(D^2 \times I)$; the middle shows $Bd(D^2) \times I$; on right is shown $D^2 \times Bd(I)$.

Example 9.19 The formulas of Proposition 9.18(c) would benefit from an example. Suppose D^2 is the closed unit disk in $\mathbf{R}^2$, and I is the unit interval in $\mathbf{R}^1$. Then $Bd(D^2)$ is the unit circle, and $Bd(I)$ is a two-point set consisting of the two numbers 0 and 1. Since D^2 and I are closed subsets, we may use the simplified formula of the remark on page 221.

Now $D^2 \times I$ is a cylindrical solid in $\mathbf{R}^3$ and $Bd(D^2 \times I)$ is the surface of the cylindrical solid; see Figure 9-1, left. Note that $Bd(D^2 \times I)$ is the union of the cylindrical surface, $Bd(D^2)\times I$, which is the side of the solid cylinder, Figure 9-1, middle, and $D^2 \times Bd(I)$, the two disks comprising the top and bottom, Figure 9-1, right. The intersection $Bd(D^2) \times I \cap D^2 \times Bd(I)$ consists of two disjoint circles, one at the top and one at the bottom. ◆

9.3 Dimension

We are now ready to answer another of our basic questions posed at the beginning of Chapter 1, Question 0.6: What does "dimension" mean? The goal is to assign an integer to any subset $X \subseteq \mathbf{R^n}$. Although it is not difficult to define dimension, few of the proofs are simple. So we restrict our investigations to a few topics which involve dimensions 0 and 1.

The problem of formulating a definition is not a simple one. A one-point subset, for example, $P = \{(2,2)\}$, should have dimension 0. A line, such as the x-axis X of $\mathbf{R}^2$, should have dimension 1 and 2-ball, such as D^2, should have dimension 2. But what dimension should $Z = P \cup X \cup D^2$ have? We will wish to say that the dimension of Z is 2, even though this is not apparent from looking at the point $(2,2)$.

It would seem that we want the standard Cantor set to have dimension 0, if for no other reason than dimension 1 seems excessive. Furthermore, we want to have the dimension of a subset to be a topological property. Thus the fat Cantor set and Antoine's necklace should also have dimension 0.

There are several ways to define the dimension of a set; we will present only one, called the "Menger-Urysohn dimension," or sometimes the "small inductive dimension." This definition uses the concept of boundary of a set. We first define dimension as a local property. The definition proceeds inductively with respect to dimension. Because it makes certain proofs easier, it is convenient to begin with the next definition:

Definition 9.20 *We say X has* **dimension** -1 *if and only if* $X = \varnothing$. *We use the notation* $dim(X) = -1$

This definition is traditional, but do not try and find hidden meaning here—in discussing dimension, "X has dimension -1" is just another way of saying that X is empty.

Definition 9.21 *Let $X \subseteq \mathbf{R^n}$, $X \neq \emptyset$, $x \in X$. We say X has* **dimension 0 at** *x if, for any open subset U of X with $x \in U$, there is an open subset V of X such that $x \in V \subseteq U$ and $dim(Bd(V)) = -1$.*

Definition 9.22 *Let $X \subseteq \mathbf{R^n}$. We say X has* **dimension 0** *if X has dimension 0 for every point $x \in X$.*

Example 9.23 If x is an isolated point of X, then X has dimension 0 at x. Any finite subset or $\mathbf{R^n}$ has dimension 0.

But a dimension 0 point does not have to be an isolated point. For example, the standard Cantor set K has dimension 0.

To see this, let $x \in K$ and suppose U is an open subset of K with $x \in U$. Since U is an open subset, we can find an $\epsilon > 0$ with $N_\epsilon(x, K) \subseteq U$. Next, we can find an N such that $\frac{1}{3^N} < \epsilon$. Consider C_N—it is the union of intervals C_{ij}, each of length $\frac{1}{3^N}$. We have $x \in C_{Nj}$ for some j and $C_{Nj} \cap K \subseteq U$. For some k we have $x \in C_{(N+1)k} \subseteq C_{Nj}$. Let a and b be the left and right endpoints of $C_{(N+1)k}$. Let $a' = a - \frac{1}{3^{N+2}}$ and $b' = b + \frac{1}{3^{N+2}}$. We have

$$x - \epsilon < a' < x < b' < x + \epsilon.$$

Thinking of C_{N+1} as obtained from C_N by "erasing middle thirds," we see that $a' \notin C_{N+1}$ and $b' \notin C_{N+1}$, and thus a' and b' are not points of K. Or we can alternatively argue in terms of the ternary decimal expression. For example, suppose $a = 0.0202000\overline{0}$ and $b = 0.0202020\overline{0}$ and $N = 4$. Then $\frac{1}{3^{N+2}} = 0.00001\overline{0}$, so $a' = 0.020122\overline{0}$ and $b' = 0.020210\overline{0}$. Let $J = (a', b')$ and $V = K \cap J$. Then V is an open subset of K, $V \subseteq U$ and since $Bd(J) = \{a', b'\}$ we have $Bd(V, K) = Bd(J, \mathbf{R^1}) \cap K = \emptyset$.

Also, the set of rational numbers in $\mathbf{R^1}$ is a 0 dimensional set. For that matter, the set of irrational numbers of $\mathbf{R^1}$ is a 0 dimensional set, (Problem 9.25). ◆

Definition 9.24 *Let $X \subseteq \mathbf{R^n}$, $x \in X$. We say X has* **dimension 1 at** *x if X does not have dimension 0 at x and, for any open subset U of X which contains x, there is an open subset if $x \in V \subseteq U$, with $Bd(V) \neq \emptyset$ and $Bd(V)$ has dimension 0.*

Definition 9.25 *Let $X \subseteq \mathbf{R^n}$. We say X has* **dimension 1** *if, X has dimension 1 or less, for every point $x \in X$, and at least one point of X has dimension 1.*

Example 9.26 In Definition 9.24 we need to explicitly rule out the possibility that X has dimension 0 at x. Let $X \subseteq \mathbf{R^1}$ be defined to be the number 0 together with a sequence of intervals approaching 0:

$$X = \{0\} \cup \left[\frac{1}{2}, 1\right] \cup \left[\frac{1}{4}, \frac{1}{3}\right] \cup \cdots \cup \left[\frac{1}{2n}, \frac{1}{2n-1}\right] \cup \cdots.$$

Then X has dimension 0 at the number 0 even though, for any open subset V of X containing 0, we can find an open subset U whose boundary consists of a single point. For example, we can take half-open intervals from 0 to a midpoint of one of the intervals

$$U = \left[0, \ \frac{1}{2}\left(\frac{1}{2n} + \frac{1}{2n-1}\right)\right) \cap X$$

where $N_{1/2n}(x, X) \subseteq V$, for some integer n. ◆

Example 9.27 We show $\mathbf{R}^1$ has dimension 1. If U is any open subset of $\mathbf{R}^1$ and any $x \in U$, there is some $N_\epsilon(x) \subseteq U$. Note that $Bd(\overline{N}_\epsilon(x))$ has dimension 0; in fact, $Bd(\overline{N}_\epsilon(x))$ consists of the two numbers $x - \epsilon$, and $x + \epsilon$. This shows that $dim(\mathbf{R}^1) \leq 1$. We next rule out the possibility that $dim(X) \leq 0$. Certainly, $dim(\mathbf{R}^1) \neq -1$ since $\mathbf{R}^1 \neq \emptyset$. So we need to rule out the possibility that $dim(\mathbf{R}^1) = 0$. If $dim(\mathbf{R}^1) = 0$, then taking $x = 0$ and $U = (-1, 1)$ there would be an open subset $V \subseteq U$ with $Bd(V) = \emptyset$. But then $\{V, \mathbf{R}^1 - V\}$ would be a disconnecting partition of $\mathbf{R}^1$, contradicting the connectedness of $\mathbf{R}^1$. ◆

Guided by the argument in the example above, we can show:

Proposition 9.28 *If $X \subseteq \mathbf{R}^n$ where X has dimension 0 and X consists of more than one point, then X is not a connected subset.* ▮ *(Problem 9.28)*

Thus other spaces of dimension 1 are the topologist's $\sin(1/x)$ curve, Sierpiński's carpet, Menger's sponge (Problems 9.26, 9.27).

In general it is easier to show a subset has dimension $\leq n$, so this notion is traditionally used in the general definition:

Definition 9.29 *Let $X \subseteq \mathbf{R}^n$, $x \in X$. We say X has* **dimension less than or equal to** n **at** x, *denoted $dim_x(X) \leq n$, if there is an open subset U of X which contains x and any open subset V of X such that if $x \in V \subseteq U$, then $Bd(V)$ has dimension less than or equal to $n - 1$.*

We say X has **dimension equal to** n **at** x *if $dim_x(X) \leq n$ but $dim_x(X) \nleq n - 1$.*

Definition 9.30 *Let $X \subseteq \mathbf{R}^n$. We say X* **has dimension** n *if X has dimension n or less, for every point $x \in X$, and at least one point of X has dimension n. Dimension of X is denoted $dim(X)$.*

Example 9.31 Suppose $x \in U \subseteq \mathbf{R}^2$ where U is an open subset. We can find an $\epsilon > 0$ so that $N_\epsilon(x) \subseteq U$; $Bd(N_\epsilon(x))$ is a small circle which has dimension ≤ 1. Thus we see that the dimension of $\mathbf{R}^2$ at x is ≤ 2. But is it in fact equal to 2? The answer is yes, but the proof is hard and we omit it. But it is instructive to see what the problem is.

We need to verify that there is no open subset v with $x \in V \subseteq U$ where $dim(Bd(V) \leq 0$. By Proposition 9.28 and since $\mathbf{R^2}$ is connected, we need not worry about the possibility that $dim(Bd(V) = -1$.

But is it possible to find a V with $dim(Bd(V) = 0$? It is not hard to rule out the possibility that $dim(Bd(V)$ consists of a finite set of points. Consider $x = (0,0)$ and $U = N_1(x)$. If $V \subseteq U$ with $Bd(V)$ a finite set F, then $\mathbf{R^2} - F$ would be disconnected since V and $\mathbf{R^2} - \overline{V}$ would give a disconnecting partition. However we have seen that the complement of a finite subset of the plane is a connected subset; in fact, it is path-connected.

But how do we rule out the possibility that $dim(Bd(V)$ might be homeomorphic to a Cantor set, for example? ◆

As one would expect,

Proposition 9.32 $\mathbf{R^n}$ *has dimension* n. ▮

It is not too hard to show that, $dim(\mathbf{R^n}) \leq n$ (Problem 9.29). However, the proof of equality is very hard and we do not attempt it. Note that it would imply that $\mathbf{R^n}$ is not homeomorphic to $\mathbf{R^m}$, if $n \neq m$, settling some of our basic questions posed in Chapter 1.

Less difficult to show is

Proposition 9.33 *Suppose* $X \subseteq \mathbf{R^n}$ *and* $Y \subseteq X$; *then* $dim(Y) \leq dim(X)$. ▮ *(Problem 9.30)*

Assuming that $dim(\mathbf{R^n}) = n$, it follows from Proposition 9.33 that

Proposition 9.34 *Suppose* $X \subseteq \mathbf{R^n}$ *with* $dim(X) < n$. *Then* $int(X) = \emptyset$.

Proof: If $int(X) \neq \emptyset$, there is an $x \in X$ and $\epsilon > 0$ such that $N_\epsilon(x) \subseteq X$. But $N_\epsilon(x)$ is homeomorphic to $\mathbf{R^n}$ which, by Proposition 9.32, has dimension n. Thus we contradict Proposition 9.33. ▮

9.4 Limits of sequences

In $\mathbf{R^n}$ the idea of limit point and limit of sequence are closely related.

Suppose we have a collection of points or $\mathbf{R^n}$, $\{x_n\}_{n \in \mathbf{N}}$. If we view this collection of points as a subset $X \subseteq \mathbf{R^n}$, and nothing else, the role of the index set is just a notational device helping us give names to these points.

However, the natural numbers have an order, and we may transfer this order to $\{x_n\}_{n\in\mathbf{N}}$ and say that x_1 is the first point, x_2 the second, etc. Often this ordering is of significance. If so, we say $\{x_n\}_{n\in\mathbf{N}}$ is a "sequence of points."

This can be formalized by defining a sequence of points of X to be a function $f:\mathbf{N} \to X$ where we denote $f(n)$ by x_n.

Using this point of view, we may look at the use of sequences as probing a space with **N** by using continuous functions. This is in the spirit of our use of the closed interval in the definition of path connectedness. Any function with domain **N** will be continuous—thus any sequence can be thought of as a continuous function. One of the things that makes **N** useful is that it is a countable set. In addition, like the line segment, it has a very useful natural order.

However, in what follows, we will generally use the more informal description: a **sequence of points of** X is a subset $\{x_n\}_{n\in\mathbf{N}}$ where order matters.

Definition 9.35 *Suppose* $X \subseteq \mathbf{R^n}$, $x_0 \in X$, *and* $\{x_n\}_{n\in\mathbf{N}}$ *a sequence of points of* X. *We say the* **sequence** $\{x_n\}$ **converges to** x_0 *if, for any open subset* U *of* X, *with* $x_0 \in U$, *there is an integer* N *such that for all* $n > N$, $x_n \in U$. *In this case we use the notation* $x_n \to x_0$.

Proposition 9.36 *Suppose* $A \subseteq X \subseteq \mathbf{R^n}$, *and* $x_0 \in X$, *with* x_0 *not an isolated point of* A. *Then* x_0 *is a limit point of* A *if and only if there is a sequence of points of* A *which converges to* x_0. ∎ . *(Problem 9.17)*

The following useful proposition says that a limit of a sequence is unique. As a consequence, in the future we will generally speak of *the* limit of a convergent sequence, rather than *a* limit.

Proposition 9.37 *If* $\{x_n\}$ *is a sequence of points of* $X \subseteq \mathbf{R}^n$, *and if* $x_n \to x_0$, *and* $x_n \to x_0'$, *then* $x_0 = x_0'$. ∎ *(Problem 9.33)*

Example 9.38 It is important distinguish between x is a limit point of the *set* $\{x_n\}$, and x is a limit of the convergent *sequence* $\{x_n\}$.

In $\mathbf{R}^1$, let $x_0 = 0$, $x_n = 1/n$ if n is odd, and $x_n = 1$ if n is even. Let $X = \{x_0\} \cup \{x_n\}_{n\in\mathbf{N}}$. In fact, no point of $\mathbf{R}^1$ is a limit of this sequence (Problem 9.31). Here x_0 is a limit point of the *set* $\bigcup_{n\in\mathbf{N}} \{x_n\}$, but it is not a limit of the *sequence* $\{x_n\}_{n\in\mathbf{N}}$.

A second example: Let $x_n = 1/n$ for $n \le 100$; $x_n = 0$ for $100 < n$. Then x_0 is a limit of the *sequence* $\{x_n\}_{n\in\mathbf{N}}$, but not a limit point of the *set* $\bigcup_{n\in\mathbf{N}} \{x_n\}$. (Note $\bigcup_{n\in\mathbf{N}} \{x_n\}$ is a finite set of 101 points, and x_0 is an isolated point of X.) ◆

Proposition 9.36 leads to a useful way of testing continuity of a function. (In general topology, this proposition and, as a consequence,

Proposition 9.39 need to have extra conditions, (see 9.46, 9.47, and 9.48).) The next proposition is in two parts; the first concerns continuity of a function at a point, the second continuity of a function on a subset.

Proposition 9.39 *Suppose* $X \subseteq \mathbf{R^n}$, $Y \subseteq \mathbf{R^m}$, *and* $f: X \to Y$,

(a) *If* $x_0 \in X$, f *is continuous at* x_0 *if and only if the following is true: if* $\{x_n\}$ *is any sequence of points of* X *such that* $x_n \to x_0$, *then* $f(x_n) \to f(x_0)$.
(b) f *is continuous if and only if the following is true: if* $\{x_n\}$ *is any sequence of points of* X *such that* $x_n \to x_0$ *with* $x_0 \in X$, *then* $f(x_n) \to f(x_0)$.

Proof: We prove part (b). Suppose $f: X \to Y$, is continuous; we wish to show that for any sequence of points $\{x_n\}$ of X: if $x_n \to x_0$, then $f(x_n) \to f(x_0)$. We argue by contradiction. Suppose f is continuous and, for some sequence $\{x_n\}$, $x_n \to x_0$, yet $\{f(x_n)\}$ does not converge to $f(x_0)$. If $\{f(x_n)\}$ does not converge to $f(x_0)$, we can find an open subset $U \subseteq Y$ such that $f(x_0) \in U$, and for any $N \in \mathbf{N}$, there is an M, with $M > N$, and $f(x_M) \notin U$. But consider V where $V = f^{-1}(U)$. Since f is continuous, V will be an open subset of X. Certainly, $x_0 \in V$ and, for any $N \in \mathbf{N}$, there will be an M, $M > N$, and $x_M \notin V$. This contradicts the assumption that $x_n \to x_0$.

Next, assume, for any sequence of points $\{x_n\}$ of X: if $x_n \to x_0$, then $f(x_n) \to f(x_0)$. We show that f is continuous by showing that the inverse image of a closed subset is a closed subset (Proposition 3.43.) Let C be a closed subset of Y, and let $K = f^{-1}(C)$. We show K is closed by showing K contains all of its limit points (Proposition 1.39). Let x_0 be a limit point of K. By Proposition 9.36 we can find a sequence of points of K, call them $\{x_n\}$, such that $x_n \to x_0$. The points $\{f(x_n)\}$ will be points of C. Furthermore since $f(x_n) \to f(x_0)$, we conclude that either $f(x_0)$ is an isolated point of C (and in particular a point of C) or a limit point of C. But C is closed and, by Proposition 1.39, C must contain its limit points. Thus, in either case $f(x_0) \in C$; thus $x_0 \in K$, and so f is continuous. ∎

Recall, in Example 4.26, we considered the function $f: [0, 1) \to S^1$, defined by $f(x) = (\cos x, \sin x)$. We showed that f^{-1} is not continuous. Here is yet another proof. Consider $x_0 = (0, 1) \in S^1$, and use polar coordinates for points of S^1. Then $f^{-1}(x_0) = 0$. Let $x_n = (1, 2\pi - 1/n)$; then $f^{-1}(x_n) = 2\pi - 1/n$. We have $x_n \to x_0$, yet it is not true that $f^{-1}(x_n) \to f^{-1}(x_0)$.

A greatest lower bound of a decreasing sequence is a limit of that sequence; we will use this in Example 9.41.

Proposition 9.40 *Suppose $\{a_n\}_{n\in\mathbf{N}}$ is a sequence of numbers with $a_{n+1} < a_n$ for all n, and that G is the greatest lower bound for the set $\bigcup_{n\in\mathbf{N}} \{a_n\}$. Then G is a limit of the sequence $\{a_n\}_{n\in\mathbf{N}}$.* ∎ *(Problem 9.34)*

Example 9.41 Using Proposition 9.39, we can provide a new proof that the topologist's $\sin(1/x)$ curve, X of Example 8.09, is not path connected. Let $a = (0,1)$, $b = (2/\pi, 1)$. We will show there is no path from a to b. Suppose there were a path f from a to b; then we would have $f(0) = a$. If (x, y) is a point of Γ with $0 < x < 2/\pi$, there must be a t with $0 < t < 1$ such that $f(t) = (x, y)$. Furthermore, we can find numbers t_n such that $f(t_n) = (\frac{1}{2\pi n+\pi/2}, 1)$, and, for all n, $t_{n+1} < t_n$. The function $\sin(1/x)$ has a local maximum at $(\frac{1}{2\pi n+\pi/2}, 1)$. Let $x_0 = g.l.b.\ \{t_n\}_{n\in\mathbf{N}}$. Then, by Proposition 9.40. we will have $t_n \to x_0$. Similarly, we can find a sequence of numbers s_n such that $f(s_n) = (\frac{1}{2\pi n+3\pi/2}, -1)$, all n, $s_{n+1} < s_n$, $s_n < t_n$, and $s_n \to x_0$. The function $\sin(1/x)$ has a local minimum at $(\frac{1}{2\pi n+3\pi/2}, -1)$. We leave details of this construction as an exercise (Problem 9.18). We have $f(t_n) \to (0, 1)$ and $f(s_n) \to (0, -1)$. Since f is continuous $f(x_0)$ must be (0, 1), and, yet, must be (0, -1). Thus we have a contradiction since limit points are unique, Proposition 9.37. ◆

Example 9.42 Recall the closed piecewise-linear $\sin(1/x)$ curve T', Example 8-5. We will use techniques of this chapter to help to show that $\mathbf{R}^2 - T'$ is not path connected. We outline the proof leaving details as an exercise (Problem 9.36).

Let $a = (-1/6, -7/6)$, $b = (5/6, -7/6)$. We will show that every path in $\mathbf{R}^2$, from a to b, must contain a point of T'. Suppose there is a path $f: I \to \mathbf{R}^2 - T'$ with $f(0) = a$ and $f(1) = b$. For $n \in \mathbf{N}$, let X_n be the line segment with endpoints $(1/n, -1)$ and $(1/n, -4/3)$; see Figure 9-2.

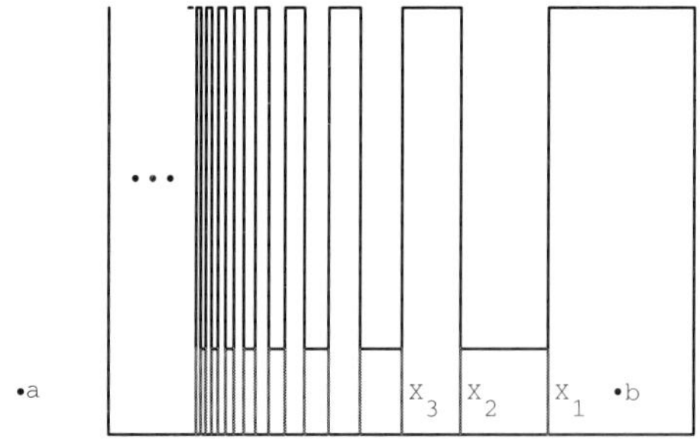

Figure 9-2 Figure for the analysis of the complement of the closed topologist's $\sin(1/x)$ curve; see Example 9.42.

Consider the closed rectangle, $R_1 = [1/2, 1] \times [-4/3, 1]$. The point

b is inside this rectangle. Let B_1 be the boundary of R_1; $B_1 \subseteq T' \cup X_1$. By Example 7.21, $\mathbf{R}^2 - B_1$ has two components. Clearly, a and b are in different components of $\mathbf{R}^2 - B_1$, so there must be a $t_1 \in I$ such that $f(t_1) \in X_1$.

We next argue, inductively, that we can find points $t_n \in I$ with $f(t_n) \in X_n$ and $t_{n+1} \leq t_n$. The set of points, $\{t_n\}_{n \in N}$ is bounded below by 0 and, by the greatest lower bound property, there is a greatest lower bound, t_0, of this set. By Proposition 9.40 we must have $t_n \to t_0$, and by Proposition 9.39, $f(t_n) \to f(t_0)$. Since $f(t_n) \in X_n$, we can argue that $f(t_0) \in L$, contradicting the assumption that $f(I) \cap T' = \emptyset$. ◆

If we take a sequence and consider only some of its members, then, using the inherited ordering, we can obtain a new sequence, called a "sub-sequence." For example, $\{\frac{1}{2n}\}_{n \in \mathbf{N}}$ is a subsequence of $\{\frac{1}{n}\}_{n \in \mathbf{N}}$. Here the n-th term in this subsequence is the $2n$-th term of the sequence.

The following definition captures this notion.

Definition 9.43 *If $f : \mathbf{N} \to \mathbf{N}$ is a strictly monotonically increasing function, the sequence $\{x'_n\}_{n \in \mathbf{N}}$ where $x'_n = x_{f(n)}$ is called a* **sub-sequence of** *$\{x_n\}$. If we let $n_j = f(j)$, we can write our subsequence as $\{x_{n_j}\}$.*

Proposition 9.44 *Suppose $\{x_n\}$ is a sequence of points of $\mathbf{R^n}$, and $\{x_{n_j}\}$ is a subsequence of $\{x_n\}$. If $\{x_n\} \to x_0$, then $\{x_{n_j}\} \to x_0$.* ▮ *(Problem 9.38)*

The following concept provides a test for convergence of a sequence.

Definition 9.45 *Suppose $\{x_i\}$ is a sequence of points of X. We say that $\{x_i\}$ is a* **Cauchy sequence** *if for any $\epsilon > 0$, there is an integer $N \in \mathbf{N}$ such that, if $N \leq m$ and $N \leq n$, then $\delta(x_m, x_n) < \epsilon$.*

If we have a sequence of points of $\mathbf{R^n}$ $\{x_{ni}\}$ with $x_i \to x_0$, then $\{x_{ni}\}$ must be a Cauchy sequence. Given $\epsilon > 0$, we can choose an $N \in \mathbf{N}$ such that for all $N < n$ we have $\delta(x_n, x_0) < \frac{\epsilon}{2}$. For any n and m with $N \leq m$ and $N \leq n$ we have:

$$\delta(x_m, x_n) \leq \delta(x_m, x_0) + \delta(x_0, x_n) < \frac{\epsilon}{2} + \frac{\epsilon}{2} = \epsilon.$$

In the next chapter we will show the converse of the above, Proposition 10.27: If we have a Cauchy sequence of points $\{x_{ni}\}$ of $\mathbf{R^n}$, then there is an $x_0 \in \mathbf{R^n}$ such that with $x_i \to x_0$.

*9.5 General topology and Chapter 9

Example 9.46 The abstract version of Proposition 9.36, that a limit point is a limit of a sequence (if it is not isolated), is not true.

Consider the space of real-valued functions, $\mathbf{R}^{\mathbf{R}}$ with the product topology. Let f_0 be the function $f_0(x) = 0$. Let A be the set of functions f such that $f(x) = 0$ for only a finite number of values for x and, otherwise, $f(x) = 1$. (Recall that $\mathbf{R}^{\mathbf{R}}$ is all functions from $\mathbf{R}^1$ to $\mathbf{R}^1$, not just continuous functions). Then f_0 is a limit point of A. To see this, let U be an open subset with $f_0 \in U$. Then we can find a basic open subset $V \subseteq U$ where, by definition of the product topology, we have $V = \prod_{\alpha \in \mathbf{R}} U_\alpha$ where, for a finite number of points $\alpha_1, \ldots, \alpha_N \in \mathbf{R}^1$, U_{α_i} is an open interval containing 0; otherwise, $U_\alpha = \mathbf{R}^1$. But then

$$g_s(x) = \begin{cases} 0 & \text{if } x = \alpha_i \text{ for some } 1 \le i \le N \\ 1 & \text{if otherwise} \end{cases}$$

is a point of $A \cap U$.

However no sequence of points of A converges to f_0. Suppose there were a sequence, a_i. Let Z_i be the finite set of points for which a_i is zero and let $Z = \bigcup_1^\infty Z_i$. Since Z is a countable set there is a number $p \notin Z$. Let U be the open subset of $\mathbf{R}^{\mathbf{R}}$ consisting of all functions f such that $|f(p)| < 1$. This is an open subset which contains f_0, but does not contain any a_i.

Therefore, even though f_0 is a limit point of A, it is not a limit of a sequence of points of A! ◆

However, with an additional hypothesis, we can get a version of Proposition 9.36:

Proposition 9.47 *Suppose $\{X, \mathcal{T}\}$ is a first countable topological space and $A \subseteq X$ with the induced topology. If x_0 is not an isolated point of A, then x_0 is a limit point of A if and only if there is a sequence of points of A which converges to x_0.* ▮ .

In particular the problems such as in Example 9.46 are of no concern for metric spaces since all metric spaces are first countable.

If we assume first countability, we get a version of Proposition 9.39.

Proposition 9.48 *Suppose $f: X \to Y$ is a function where X and Y are first countable spaces. Then f is continuous if and only if for any sequence of points of X, $\{x_n\}$, if $x_n \to x_0$, then $f(x_n) \to f(x_0)$.* ▮

9.6 Problems for Chapter 9

9.1 Prove Proposition 9.02.

9.2 Prove that closure of the open n-ball in $\mathbf{R^n}$ is the corresponding closed n-disk in $\mathbf{R^n}$; see Example 9.04.

9.3 Prove that closure of the set of all rational numbers in $\mathbf{R^1}$ is all of $\mathbf{R^1}$, as asserted in Example 9.04.

9.4 Prove Proposition 9.03.

9.5 Prove Proposition 9.07.

9.6 Prove Proposition 9.08. (Recall Proposition 3.44.)

9.7 Prove Proposition 9.10.

9.8 Prove Proposition 9.13.

9.9 Suppose $A \subseteq X \subseteq \mathbf{R^n}$. Show that it is not generally the case that $int(\overline{A}) = \overline{int(A)}$.

9.10 Prove Proposition 9.15

9.11 Prove Proposition 9.16.

9.12 Prove Proposition 9.17

9.13 Prove Proposition 9.18.

9.14 Sketch figures which show the relations of Proposition 9.18 for the following:

(a) $[0,1] \times [0,1]$
(b) $[0,1] \times [0,1)$
(c) $[0,1] \times (0,1)$

9.15 Three problems about the boundary of a subset.

(a) In $\mathbf{R^2}$, let X be the union of a circle and a closed disk: $X = S^1 \cup \overline{N_{\frac{1}{2}}((0,0))}$. Show there is a subset $Z \in \mathbf{R^2}$ such that $X = Bd(Z)$.
(b) Show that if X is any closed subset of the plane there is a subset, $Z \in \mathbf{R^2}$, such that $X = Bd(Z)$.

9.16 Suppose that $A \subseteq X \subseteq Y \subseteq \mathbf{R^n}$. Show that

$$Cl_X(A) = Cl_Y(A) \cap X.$$

9.17 Prove Proposition 9.36.

9.18 Verify that the sequences $\{t_n\}$ and $\{s_n\}$, as in Example 9.41, can be found with the stated properties. Also, verify that $t_n \to x_0$.

9.19 Suppose $A \subseteq \mathbf{R^n}$ and $B \subseteq \mathbf{R^n}$:

(a) Show that $\overline{A \cup B} = \overline{A} \cup \overline{B}$.
(b) Show it is not always true that $\overline{A \cap B} = \overline{A} \cap \overline{B}$.

(c) Show that if $A \subseteq B$, then $\overline{A} \subseteq \overline{B}$.
(d) Show that $\overline{(\overline{A})} = \overline{A}$.

9.20 Suppose that $A \subseteq X \subseteq \mathbf{R^n}$. Show that if $x \in X$ is a limit point of A, then any open subset of X which contains x contains infinitely many points of A.

9.21 Let K be the standard Cantor set and let $A = K \cap [0, \frac{1}{3}]$; what is $int(A, K)$?

9.22 Is it true that $\overline{X - A} = X$ if and only if $int(A) \neq \emptyset$?

9.23 Let $Q \subseteq \mathbf{R}^1$ be the set of rational numbers; show that $\overline{Q} = \mathbf{R}^1$.

9.24 Three problems concerning circles in the plane:

(a) Show that if C is a circle in $\mathbf{R}^2$, that $int(C) = \emptyset$.
(b) Find an uncountable collection of circles in $\mathbf{R}^2$ whose union has non-empty interior.
(c) Show that if X is a countable union of circles in $\mathbf{R}^2$, then X has empty interior. (Hint: find a line L that intersects X. Consider how a single circle could intersect L. What can you say about $X \cap L$ given that X has empty interior?)

9.25 Prove, as stated in Example 9.23, that the set of rational numbers in $\mathbf{R}^1$ is a 0-dimensional set and the set of irrational numbers of $\mathbf{R}^1$ is a 0-dimensional set.

9.26 Prove that Sierpiński's carpet has dimension 1.

9.27 Prove that Menger's sponge has dimension 1.

9.28 Prove Proposition 9.28.

9.29 Prove, as claimed in Example 9.3, that $dim(\mathbf{R^n}) \leq n$.

9.30 Prove Proposition 9.33. (Note: One would expect a proof by induction for this.)

9.31 Verify the claim of Example 9.38, that no point of $\mathbf{R}^1$ is a limit of the sequence: $x_0 = 0$; $x_n = 1/n$ if n is odd; $x_n = 1$ if n is even.

9.32 Recall that Antoine's necklace, $A \subseteq \mathbf{R}^3$, is homeomorphic to the standard Cantor set, Example 5-17. The Cantor set has dimension 0, Example 9.23, and dimension 0 is a topological property. Therefore A must have dimension 0. Provide a second proof: Verify directly from the definition of dimension 0 that A has dimension 0.

9.33 Prove Proposition 9.37.

9.34 Prove Proposition 9.40.

9.35 In Example 9.41, verify that $\overline{\Gamma} = X$.

9.36 Provide the details of the proof in Example 9.42. Specifically, show that

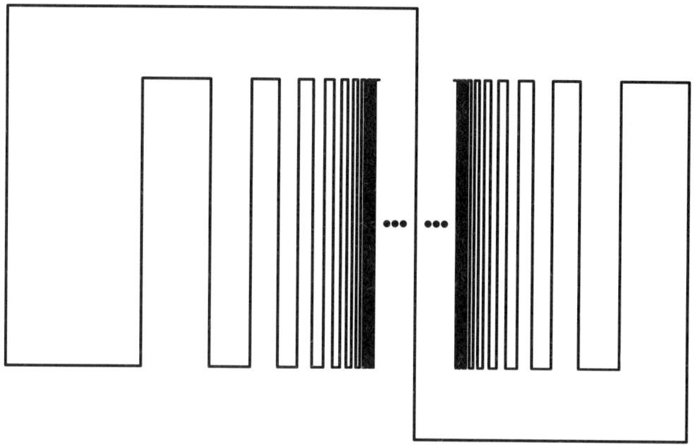

Figure 9-3 Figure for subset X, Problem 9.37.

(a) a and b are in different components of $\mathbf{R}^2 - B_1$.
(b) There must be a $t_1 \in I$ such that $f(t_1) \in X_1$.
(c) We can find points $t_n \in I$ with $f(t_n) \in X_n$ and $t_{n+1} \leq t_n$.
(d) $t_n \to t_0$.
(e) $f(t_0) \in L$.

9.37 Consider the piecewise-linear closed topologist's $\sin(1/x)$ curve, Example 8-5 and the two examples, X and Y, shown in Figures 9-3 and 9-4.

(a) Show that no two of these are homeomorphic
(b) Show $\mathbf{R}^2 - X$ and $\mathbf{R}^2 - Y$ have three path components.
(c) Are $\mathbf{R}^2 - X$ and $\mathbf{R}^2 - Y$ homeomorphic? (Note: this is a difficult problem and goes beyond routine applications of ideas we have considered.)

9.38 Prove Proposition 9.44.

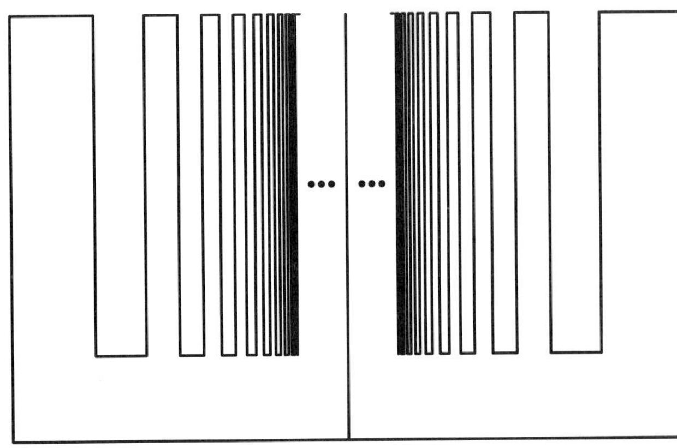

Figure 9-4 Figure for subset Y, Problem 9.37.

10. COMPACTNESS

OVERVIEW: We define and discuss the concept of compactness. A basic important fact is that a closed interval and products of intervals are compact. Many important questions in topology involve proving the existence of a point with certain properties. Often the information given is not enough to allow one to explicitly find the point. Compactness is a crucial property that provides a way of indirectly showing the existence of a point of $\mathbf{R}^n$.

10.1 Closed and bounded subsets

The topological property called "compactness" is meant to convey the concept, in some sense, of "smallness." The requirement that we have a topological property might seem to be a problem. For example, the closed interval $[-1,000,000, 1,000,000]$ might seem very "big," but it is homeomorphic to the closed interval $[\frac{-1}{1,000,000}, \frac{1}{1,000,000}]$. On the other hand, if we say $[\frac{-1}{1,000,000}, \frac{1}{1,000,000}]$ is "small," then it would seem that the corresponding *open* interval $(\frac{-1}{1,000,000}, \frac{1}{1,000,000})$ would also be small. But $(\frac{-1}{1,000,000}, \frac{1}{1,000,000})$ is homeomorphic to $\mathbf{R}^1$, which is certainly should be considered large—it is certainly larger than $[-1,000,000, 1,000,000]$.

We gain insight on how to proceed by trying to solve some problems which, at first; seem to bear little relation to the discussion above.

Example 10.01 Let $D = D^2$ be the closed unit disk in $\mathbf{R}^2$, and let X be the set $X = D - \{(0,0)\}$. The subset X is called a "punctured disk." We wish to show that X and D are not homeomorphic. Our previous techniques fail: these sets have the same number of elements. There is no distinctive point of X, such as a limit point, or a cut point, which has no corresponding point in D. They are both connected and path connected. (To prove path connectedness of X we could adapt the proof used for Proposition 8.02. Alternatively, we could use Proposition 8.02 as follows. Write points of $\mathbf{R}^2 - \vec{0}$ in polar coordinates, and define $f: \mathbf{R}^2 - \vec{0} \to X$ by $f(r,\theta) = (e^{(1-r)^2}, \theta)$; see Figure 10-1. Then X is the image under continuous map f of a path-connected set. For good measure, a third argument is to show that X is homeomorphic to a product of path-connected spaces, $S^1 \times (0,1]$.)

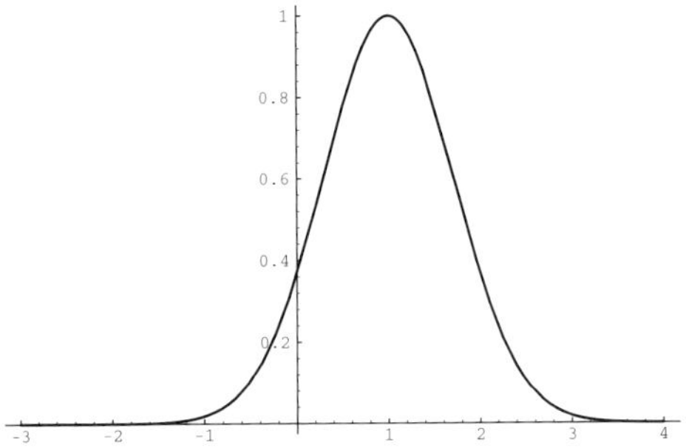

Figure 10-1 The graph of $(e^{(1-r)^2}$; see Example 10.01.

Our argument is by contradiction. Assume there is a homeomorphism $h: D \to X$. Intuitively, we wish to argue somehow that X has a point missing and D does not. This seems paradoxical since, if the point is truly missing, how could we be sure.

The strategy we use is to somehow locate the portion of X where we expect our point to be and show that any search to find it must fail.

Let $b_n = (\frac{1}{n}, 0)$ and $b_0 = (0,0)$. Then $b_n \to b_0$. Let $B = \bigcup_{i=1}^{\infty} \{b_i\}$. Note that $B \subseteq X$. The motivation is that the points of B get close to our "missing" point b_0.

Suppose $x \in X$. If $x \notin B$ there is an open subset of X which contains either no points of B. If $x \in B$, is an open subset of X which contains only one point, x, of B. Thus, no limit point of B is contained in X.

Now, let $A = h^{-1}(B)$; see Figure 10-3. Claim: There is a point a_0 which is a limit point of A in D. Since D is a closed subset of $\mathbf{R}^2$, and since closed subsets must contain all their limit points (Proposition 1.39), it would then follow that $a_0 \in D$. Since h is a homeomorphism, $h(a_0)$ is a limit point of B, Proposition 3.44. This would then give us our contradiction.

We now verify our claim that there is a point a_0 which is a limit point of A. This is a key point; we need to show existence of a point without having much specific information on exactly how to find it. Consider the square,

$$Q = \{(x,y) \in \mathbf{R}^2 : -1 \le x \le 1 \text{ and } 1 \le y \le 1\};$$

then $D \subseteq Q$. Divide Q into four equal sub-squares, Q_1^1, Q_2^1, Q_3^1, and Q_4^1; see Figure 10-2.

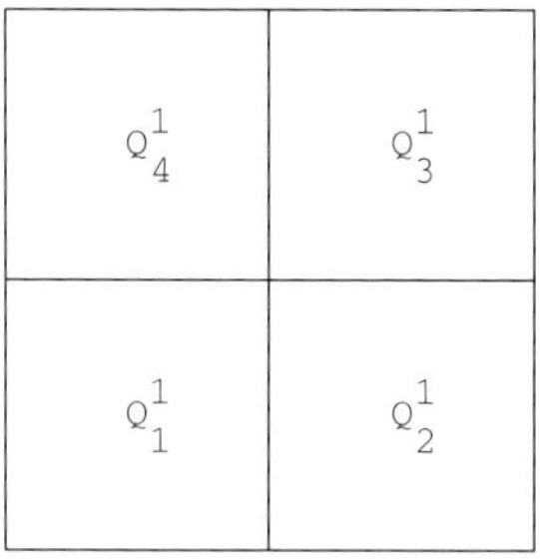

Figure 10-2 Dividing the square Q into four sub-squares; see proof of Example 10.01.

Now (at least) one of these sub-squares, call it Q^1, contains infinitely many points of A. Next, subdivide Q^1 into four equal sub-squares, Q_1^2, Q_2^2, Q_3^2, and Q_4^2. Now (at least) one of these sub-squares, call it Q^2, contains infinitely many points of A. Continuing in this way, we obtain a sequence of squares, Q^i, such that each Q^i contains infinitely many points of A and such that $Q^1 \supseteq Q^2 \supseteq Q^3 \supseteq \ldots$; see Figure 10-3.

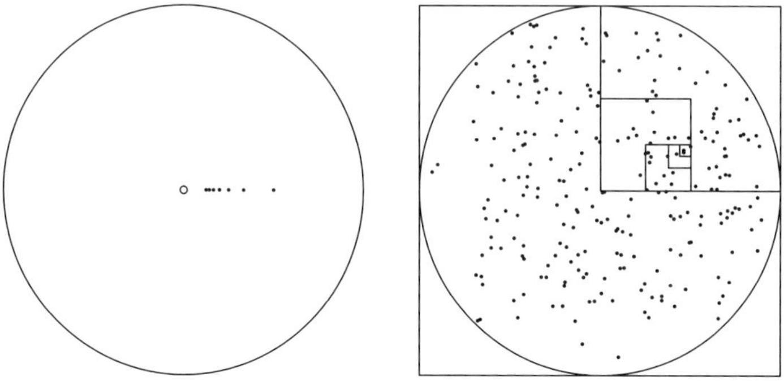

Figure 10-3 On the left, a punctured disk and a subset, B; see Example 10.01. (The small circle near the center indicates that the point $(0,0)$ is *not* a point of X.) On the right, the unit disk with the image $A = h(B)$. Also illustrated is a sequence of squares as used in the proof of Example 10.01.

Consider $\bigcap_1^\infty Q^i$. We will show that $\bigcap_1^\infty Q^i$ is non-empty. By Proposition

5.09, it must consist of only one point, call it q_∞. This must be the a_0 we were searching for: $a_0 = q_\infty$ is a limit point of A.

Write $Q^i = [a_1^{(i)}, b_1^{(i)}] \times [a_2^{(i)}, b_2^{(i)}]$. Let $q_\infty = (\sup_i\{a_1^{(i)}\}, \sup_i\{a_2^{(i)}\})$. We then verify that $\{q_\infty\} = \bigcap_1^\infty Q^i$. Details of this verification are left as an exercise (Problem 10.2). ◆

The argument above is one to remember. We analyze this further and also search for an alternative exposition which is more directly tied to the concept of open subset. The fact that B is an infinite subset of X with no limit point in X, is a key to the argument above. We next discuss the implications of this.

Suppose $B \subseteq X \subseteq \mathbf{R^n}$, B is an infinite subset, with no limit point in X. To begin with, B is a closed subset of X since it contains all its limit points (it has none), and so $X - B$ is an open subset of X.

For each $b \in B$ there exists an $0 < \epsilon_b$ such that $N_{\epsilon_b}(b, X) \cap B = \{b\}$. Thus each point of B is an isolated point of B.

We next define a collection, $\mathcal{G}$, of open subsets of X. We begin by considering $\bigcup_{b \in B} N_{\epsilon_b}(b, X)$. It is possible that $X - B \subseteq \bigcup_{b \in B} N_{\epsilon_b}(b, X)$. If so, we define $\mathcal{G} = \bigcup_{b \in B} \{N_{\epsilon_b}(b, X)\}$; otherwise we enlarge this collection, by adding $X - B$, and define $\mathcal{G} = \bigcup_{b \in B} \{N_{\epsilon_b}(b, X)\} \cup \{(X - B)\}$. We note that $X = \bigcup_{U \in \mathcal{G}} U$. But if we remove any of these sets, we do not get all of X. This is a critical observation. This analysis provides motivation for the next two definitions.

Definition 10.02 *Let $X \subseteq \mathbf{R^n}$. A collection of open subsets of X, $\mathcal{G} = \{U_\alpha\}_{\alpha \in A}$, is called an* **open cover of** *X if $X = \bigcup_{\alpha \in A} U_\alpha$.*

A cover is called a **finite cover** *if the index set A is a finite set.*

Suppose we have a sub-collection of $\{U_\alpha\}_{\alpha \in A}$, say $\mathcal{G}' = \{U_\beta\}_{\beta \in B}$, where $B \subseteq A$. If $\mathcal{G}'$ is also an open cover of X, we say $\mathcal{G}'$ is a **subcover of** *U.*

Definition 10.03 *Suppose $X \subseteq \mathbf{R^n}$. We say X is* **compact** *if every open cover of X has a finite subcover.*

Proposition 10.04 *Suppose $X \subseteq \mathbf{R^n}$. The statement "X is compact" is a topological property.* ▮ *(Problem 10.1)*

Proposition 10.05 *If $X \subseteq \mathbf{R^n}$ and X is a finite set, then X is compact.*

Proof: Write $X = \{x_1, x_2, \ldots, x_N\}$. Suppose that $\{U_\alpha\}_{\alpha \in A}$ is an open cover of X. For each i, $1 \le i \le N$, we can find an $\alpha_i \in A$ such that $x_i \in U_{\alpha_i}$. Thus $X = \bigcup_1^N U_{\alpha_i}$, and the collection of subsets $\{U_{\alpha_i}\}_{i=1}^{i=N}$ is a finite subcover of $\{U_\alpha\}_{\alpha \in A}$. Therefore X is compact. ▮

Here is another argument for the proposition above. If X is finite, then any subset is open. Thus there are 2^N distinct open subsets. So if $\{U_\alpha\}$ is any open subcover of X we can obtain a finite subcover by eliminating all repeated entries in the collection $\{U_\alpha\}$.

Example 10.06 The line $\mathbf{R}^1$ is not compact. Let U_n be the open interval $(n-1, n+1)$ where $n \in \mathbf{Z}$. Then $\mathcal{G} = \{U_n\}_{n \in \mathbf{Z}}$ is an open cover of $\mathbf{R}^1$. If we remove any of these intervals, say U_k, from $\mathcal{G}$, we do not have a cover since the midpoint, k, of each interval $(k-1, k+1)$ is contained only in U_k. Thus this open cover has no finite subcover. ◆

Example 10.07 Let S be the open square defined

$$S = \{(x, y) \in \mathbf{R}^2 \mid 0 < x < 1 \text{ and } 0 < y < 1\}.$$

Then S is not compact. For $n = 3, 4, 5, \ldots$ let

$$U_n = \{(x, y) \mid 1/n < x < 1 - 1/n \text{ and } 1/n < y < 1 - 1/n\}.$$

Then $\{U_n\}$ is an open cover of S with no finite subcover.

The argument is by contradiction. Suppose there is a finite subcover consisting of k subsets, $\{U_{n_j}\}_{j=1}^{j=k}$. Let N be the largest of the subscripts, n_j. If $n \leq m$, then $U_n \subseteq U_m$. Thus $U_N = \bigcup_{j=1}^{j=k} U_{n_j}$. But $\{U_{n_j}\}_{j=1}^{j=k}$ is not a cover of S since $U_N \neq S$. ◆

Example 10.08 Let X the punctured disk of Example 10.01; then X is not compact. Let $U_n = \{(x, y) \in X \mid x^2 + y^2 > \frac{1}{n}\}$, for $n > 2$. Then $\{U_n\}$ is an open cover of X that has no finite subcover. The argument is similar to that given in Example 10.07. ◆

The following subset is useful in discussing compactness.

Definition 10.09 *A* **standard n-dimensional cube of size** M *is* $Q_M^n = \{(x_1, x_2, \ldots, x_n) \in \mathbf{R^n} \mid$ *for all* i, $1 \leq i \leq n$, $-M \leq x_i \leq M\}$. *If the size is not important, we simply refer to this subset as a standard n-dimensional cube.*

Since Q_M^n is the Cartesian product of n copies of the interval $[-M, M]$, Proposition 4.51 implies that, for any positive numbers, M and M', Q_M^n is homeomorphic to $Q_{M'}^n$.

The word "cube" in the definition above is most appropriate for the case $n = 3$. For $n = 2$, Q_M^2 is, in fact, a square, and for $n = 1$, Q_M^1 is the closed interval $[-M, M]$. In dimension 4 (or greater) the term "hypercube" is sometimes used.

Proposition 10.10 *Let Q be a standard n-dimensional cube; then Q is a compact subset of* $\mathbf{R^n}$.

Proof: The argument, by contradiction, is inspired by our proof for Example 10.01 where we were considering squares in $\mathbf{R}^2$.

Suppose Q is not compact. Then there is an open cover of Q, $\mathcal{G} = \{U_\alpha\}_{\alpha \in A}$, such that $\mathcal{G}$ has no finite subcover.

Suppose Q is a cube of size M; Q has edge length $2M$. Divide Q into 2^n equal sub-cubes with edges of length M. Suppose no finite subcover of $\{U_\alpha\}_{\alpha \in A}$ exists. Next, for one of these sub-cubes, call it Q^1, the collection $\mathcal{G}^1 = \{Q^1 \cap U_\alpha\}$ is an open cover of Q^1, with no finite subcover. Then subdivide Q^1 into 2^n equal sub-cubes and find one of these, call it Q^2, such that $\mathcal{G}^2 = \{Q^2 \cap U_\alpha\}$ is an open cover of Q^2, with no finite subcover. Continuing in this way we can define a sequence of sub-cubes Q^i such that $Q^1 \supseteq Q^2 \supseteq Q^3 \supseteq \dots$.

Next, we assert that $\bigcap_1^\infty Q^i$ consists of a single point, q_∞. The argument is similar to that of Example 10.01. (Remark: All we really need is the assertion that $\bigcap_1^\infty Q^i$ is non-empty.) Write

$$Q^i = [a_1^{(i)}, b_1^{(i)}] \times [a_2^{(i)}, b_2^{(i)}] \times \cdots \times [a_n^{(i)}, b_n^{(i)}].$$

Think of the point $(a_1^{(i)}, a_2^{(i)}, \dots, a_2^{(i)})$ as a lower corner of Q^i; q_∞ is to be the limit of this sequence. Let $q_\infty = (\sup_i\{a_1^{(i)}\}, \sup_i\{a_2^{(i)}\}, \dots, \sup_i\{a_n^{(i)}\})$. As in the proof of Example 10.01, one can then see that $\{q_\infty\} = \bigcap_1^\infty Q^i$.

Since $\{U_\alpha\}_{\alpha \in A}$ is a cover, every point of Q is in some element of $\{U_\alpha\}_{\alpha \in A}$. In particular, there is an $\alpha_\infty \in A$ such that $q_\infty \in U_{\alpha_\infty}$. Since U_{α_∞} is an open subset of Q, it follows that we can find an $\epsilon > 0$ so that $N_\epsilon(q_\infty, Q) \subseteq U_{\alpha_\infty}$. In general, for any n, Q^n has edge length $2M/2^n = M/2^{n-1}$, and diameter of Q^n is $D_n = \sqrt{n(M/2^{n-1})^2} = \sqrt{n}M/2^{n-1}$. We can find some N such that $D_N < \epsilon$. We then have $Q^N \subseteq N_\epsilon(q_\infty, Q) \subseteq U_{\alpha_\infty}$. But then the collection consisting of the single set $Q^N \cap W_{\alpha_\infty}$ is a finite open subcover of Q^N, contradicting the definition of $\mathcal{G}^N$. Thus Q is compact. ∎

We can reformulate our proof that the spaces of Example 10.01 are not homeomorphic. The disk D is homeomorphic to the square Q, so D must be compact. However, X is not compact, as noted in Example 10.08. Thus since compactness is a topological property, D and X are not homeomorphic.

The following two useful propositions will allow us to obtain more examples of compact subsets.

Proposition 10.11 *Suppose $X \subseteq \mathbf{R}^n$, X compact, and $f: X \to Y$ is continuous. Then the image of f, $f(X)$, is compact.* ∎ *(Problem 10.3)*

Proposition 10.12 *If X is compact and C is a closed subset of X, then C is compact.*

Proof: Suppose $\mathcal{G} = \{U_\alpha\}_{\alpha \in A}$ is an open covering of C. For each U_α, there an open subset W_α of X such that $U_\alpha = X \cap W_\alpha$. Then $\mathcal{H} = \{W_\alpha\}_{\alpha \in A} \cup (X - C)$ is an open cover of X. Since X is compact, we can find a finite subcover, $\mathcal{H}'$ of $\mathcal{H}$. Intersecting the sets of $\mathcal{H}'$ with C, we obtain a finite sub-collection of our original $\mathcal{G}$; thus giving a finite subcover of $\mathcal{G}$. ∎

Example 10.13 Here is yet another proof that the circle is not homeomorphic to a half-open interval, Example 4.28.

The half-open interval is not compact. We can see this directly. Consider the interval $[0, 2\pi)$ and the covering $\{U_n\}_{n \in A}$ where for $n \in \mathbf{N}$, $U_n = [0, 2\pi - \frac{1}{n})$; this has no finite subcover.

On the other hand, the unit circle S^1 is compact. The square Q_1^2 is compact, by Proposition 10.10. Since S^1 is a closed subset of Q_1^2, S^1 must be compact. ◆

Example 10.14 We show that the circle, S^1, does not embed into $\mathbf{R}^1$. As in the example above, we see that S^1 is compact; it is also connected, Example 7.03. If there were an embedding, $f: S^1 \to \mathbf{R}^1$, then the image, $f(S^1)$ would be compact and connected and, of course, homeomorphic to S^1. By Proposition 7.12, the only connected subsets of $\mathbf{R}^1$ are intervals and points. The only compact interval is a closed interval. We now have a contradiction since S^1 is not homeomorphic to I, Example 4.44. ◆

The following proposition is useful for considerations of compactness with Cartesian products.

Proposition 10.15 *Suppose $X \subseteq \mathbf{R^n}$, $Y \subseteq \mathbf{R^m}$, with X compact. If $y_0 \in Y$ and $U \subseteq X \times Y$ is an open subset with $X \times y_0 \subseteq U$, then there is an open subset V of Y such that $X \times V \subseteq U$.*

Proof: Let $X_0 = X \times y_0$; X_0 is homeomorphic to X and thus is compact. For each $p \in X_0$ we can find an open subset $U_p \subseteq X$ and open subset $V_p \subseteq Y$ such that $U_p \times V_p \subseteq U$, by the remark on page 63. The collection $\{U_p\}_{p \in X}$ is an open cover of X and, by compactness,Ł has a finite subcover $\{U_{p_i}\}_{i=1}^{i=n}$. Let $V = \bigcap_{i=1}^{i=n} V_{p_i}$. Then we have

$$X \times V \subseteq \bigcup_{i=1}^{i=n} (U_{p_i} \times V_{p_i}) \subseteq U. \quad ∎$$

Example 10.16 We are now prepared to finish the proof of Proposition 3.49 by showing continuity of F' at the origin.

Let $N_\epsilon(\vec{0})$ be a neighborhood of the origin. Since $F(\vec{0}) = \vec{0}$, we wish to find a δ such that $F^{-1}(N_\epsilon(\vec{0})) \subseteq N_\delta(\vec{0})$.

Recall $\gamma\colon [0,\infty) \times \mathbf{R^1} \to \mathbf{R^2}$ is defined by $\gamma(r,\theta) = (r,\theta)$ where we are using polar coordinates for $\mathbf{R^2}$. Let $\gamma' = \gamma|_{[0,\infty)\times[0,2\pi]}$; then we have

$$\begin{array}{ccc} [0,\infty)\times[0,2\pi] & \xrightarrow{F} & (\mathbf{R^1})\times[0,2\pi] \\ \big\downarrow \gamma' & & \big\downarrow \gamma \\ \mathbf{R^2} & \xrightarrow{F'} & \mathbf{R^2}. \end{array}$$

Let $U = (\gamma' \circ F)^{-1} N_\epsilon(\vec{0})$. Since $\gamma' \circ F$ is continuous, U is an open subset of $[0,2\pi] \times [0,\infty)$ containing $[0,2\pi] \times 0$. By Proposition 10.15 we can find a δ such that $[0,2\pi]\times[0,\delta) \subseteq U$. Now $\gamma'([0,2\pi]\times[0,\delta)) = N_\delta(\vec{0})$, and $F^{-1}(N_\epsilon(\vec{0})) \subseteq N_\delta(\vec{0})$. ◆

Proposition 10.18 is another common test we use to see if a set is compact.

Definition 10.17 *Let $A \subseteq \mathbf{R^n}$. We say A is* **bounded** *if there is a number M such that $A \subseteq Q_M$ where Q_M is a standard n-dimensional cube.*

Proposition 10.18 *If $A \subseteq \mathbf{R^n}$, then A is compact if and only if A is a closed and bounded subset of $\mathbf{R^n}$.*

Proof: Suppose A is a closed and bounded subset. Then $A \subseteq Q_M^n$ for some M. By Proposition 10.10, Q_M^n is compact. We conclude that A, a closed subset of a compact subset, is compact, Proposition 10.12.

Next, we prove that if A is compact, then it is a closed and bounded subset. First, we show that A must be bounded. For $k > 0$, let U^k be the open n-cube centered at the origin defined:

$$U_k = \{(x_1, x_2, \ldots, x_n) \in \mathbf{R^n} \mid -k < x_i < k, \text{ for } 1 \le i \le n\}.$$

Let $U_k' = U_k \cap A$. If A is not bounded, then it is not contained in any U_k. But then $\{U_k'\}$ is an open cover of A, with no finite subcover.

Next, we show that A must be closed. If not, then, by Proposition 1.39, there is a limit point of A, call it x_0, with $x_0 \notin A$. For $n \in \mathbf{N}$, let

$$U_n = \{x \in A\colon \delta(x, x_0) > \frac{1}{n}\}.$$

Then, as in Example 10.08, we can see that $\{U_n\}_{n\in\mathbf{N}}$ is a covering of A with no finite subcover. Suppose there were a finite subcover $\{U_{n_i}\}$, and let N be the maximum of all the natural numbers $\{n_i\}$. Then for all i we would have $U_{n_i} \subseteq U_N$. Since $\{U_{n_i}\}$ is a cover of A, it follows that $A \subseteq U_N$. Let $U = N_{\frac{1}{N}}(x_0)$. Then U is an open set containing x_0; yet $U - \{x_0\}$ contains no points of U_N. Thus $U - \{x_0\}$ contains no points of A. This contradicts our assumption that x_0 is a limit point of A. We conclude that if x_0 is a limit point of A, then it must be a point of A. Thus A must be closed. ∎

We can use Proposition 10.18 to show $[0, 2\pi)$ is not compact. Although $[0, 2\pi)$ is a bounded subset of the plane, it is not a closed subset since there is a limit point, namely, $2\pi \in [0, 2\pi)$, which is not a point of $[0, 2\pi)$.

As a practical matter, given a specified subset of $\mathbf{R^n}$, it is usually easiest to verify compactness by using Proposition 10.18 since it is usually evident that a given subset is bounded. Also, we have developed many techniques for deciding if a subset is closed.

However, if the subset is described only in general terms, such as "the image of some continuous, but unknown, function," then our definition of compactness might well be the easiest formulation of compactness to use.

10.2 Properties and examples of compactness

For subsets of $\mathbf{R^n}$, products of compact subsets are not difficult to analyze, as in the next proposition. A corresponding statement in general topology is an important result with a difficult proof, (10.49)

Proposition 10.19 *Suppose $X \subseteq \mathbf{R^n}$, $Y \subseteq \mathbf{R^m}$, and X and Y are both compact. Then $X \times Y$ is compact.*

Proof: By proposition 10.18, each of X and Y are closed and bounded subsets. It follows that $X \times Y$ is a closed subset, by Proposition 2.46 and bounded subset (Problem 10.4), and thus compact. ∎

Example 10.20 The following example illustrates some aspects of closed subsets and bounded subsets.

Let X be the punctured disk of Example 10.01; using polar coordinates, $X = \{(\rho, \theta) \mid 0 < \rho \leq 1\}$. Let Y be the complement of an open unit disk, $Y = \{(\rho, \theta) \mid 1 \leq \rho\}$.

Then X is homeomorphic to Y. In fact, a homeomorphism is given by: $h((\rho, \theta)) = (1/\rho, \theta)$. Note that X is not closed, but it is bounded. Y is closed, but it is not bounded. ◆

If $A \subseteq \mathbf{R^n}$ is a bounded subset, A might not be compact. However, the closure of A, $\overline{A}$, *is* compact. Since A is bounded, for some standard cube Q_M, $A \subseteq Q_M$. We know Q_M is a closed subset and thus, by Proposition 9.02, $\overline{A} \subseteq Q_M$. Thus $\overline{A}$ is bounded as well as closed, and so $\overline{A}$ is compact.

Example 10.21 Let C be a circle in $\mathbf{R}^2$, with center p and radius r. We can now express what we mean by a point being "inside" (or "outside") of C, using only topological properties.

As noted in Example 7.21, $\mathbf{R}^2 - C$ has two components. Let B be the component of $\mathbf{R}^2 - C$ which contains the center p. Let U be the other component. Since B is bounded, the closure of B in $\mathbf{R}^2$ is compact. The closure of U in $\mathbf{R}^2$ is not compact since it is not bounded.

We say a point "q lies outside the circle C" if it lies in the component of the complement of C whose closure is not compact, and "q lies inside the circle C" if it lies in the component of the complement of C whose closure is compact.

Another way of distinguishing these inside and outside sets will be considered later (Problem 15.13) ◆

Example 10.22 Using the notions of inside and outside of a circle as in Example 10.21, we can now prove the assertion of 6.13, the lollypop example, that X and Y are not equivalent subsets of the plane.

Suppose there were a homeomorphism h of the plane with $h(X) = Y$. Let $x = (1/2, 0)$ and let $y = (2, 0)$. We first show that $h(x) = y$. Classify points of X and Y according to cut-point type. The set of cut points of order 2 of X is $A - \{x\}$. The set of cut points of order 2 of Y is $B - \{y\}$. So we have $h(A - \{x\}) = B - \{y\}$. By Proposition 9.08, we have $h(\overline{A - \{x\}}) \subseteq \overline{B - \{y\}}$, or $h(A) = B$; thus $h(x) = y$.

For any homeomorphism h of the plane with $h(S^1) = S^1$, $h(x)$ is inside S^1 if and only if x is inside C (Problem 10.5). But x is inside of S^1 and y is outside of S^1; this is our contradiction. We conclude that these subsets are not equivalent. ◆

Another useful fact about compact subsets is the following.

Proposition 10.23 *If $\{X_i\}$ is a finite collection of compact subsets of $\mathbf{R}^n$, then $\bigcup_i X_i$ is a compact subset of $\mathbf{R}^n$.*

Proof: This could be proven directly from the definition of compactness. Another way to prove it is by using Proposition 10.18. If each of X_i is compact, then each is closed and bounded. Clearly, the union is bounded; take the maximum of the bound for each. Also, the union of a finite number of closed subsets is a closed subset. Thus $\bigcup_i X_i$ is closed and bounded and thus compact. ▮

Subsets of the line are especially important, so it is worthwhile noting the following proposition. The proof is a consequence of Proposition 10.18, together with the least upper bound axiom, the greatest lower bound axiom, and the fact that closed sets must contain all their limit points.

Proposition 10.24 *If X is a compact non-empty subset of $\mathbf{R^1}$, then* $\sup(\{x \in X\})$ *and* $\inf(\{x \in X)\}$ *both exist and are points of X.* ▮ *(Problem 10.6)*

Here is another way to express compactness. Recall that, in our proof of Example 10.01, we used a subset B of the punctured disk X which was infinite with no limit point in X.

Proposition 10.25 *Let $X \subseteq \mathbf{R^n}$; then X is compact if and only if every infinite subset of X has a limit point, which is in X.*

Proof: We first show, indirectly, that if X is compact, then every infinite subset of X has a limit point which is in X.

Suppose X is compact, yet $A \subseteq X$ is an infinite set with no limit point in X. Suppose $x \in X - A$. Since x is not a limit point of A, there is an open subset U_x of X which contains no points of A. If $x \in A$, then there is an open subset, $U_x \subseteq X$, which contains exactly one point, x, of A. Let $\mathcal{G} = \{U_x\}_{x \in X}$. Then $\mathcal{G}$ is certainly a cover of X. It cannot have a finite subcover since each element of $\mathcal{G}$ contains at most one point of A, contradicting the assumption that A is infinite.

Next, suppose every infinite subset of X has a limit point, which is in X. We will show X is a closed subset of $\mathbf{R^n}$ by showing that if x_0 is a limit point of X, then $x_0 \in X$. For any natural number, $n \in \mathbf{N}$, there must be a point, $x_n \in N_{1/n}(x_0, X)$ with $x_n \neq x_0$. The set $S = \{x_n\}_{n \in \mathbf{N}}$, although it may have some repeated elements, is clearly an infinite subset of X. Then x_0 is a limit point of S and, in fact, $x_n \to x_0$. By Proposition 9.37 it is the only limit point of S. Our hypothesis now implies that $x_0 \in X$.

Thus X is a closed subset of $\mathbf{R^n}$; we next show that X must be a bounded subset. If it were not bounded, then, for $n \in \mathbf{N}$, we could find points, $x_n \in X$ with $n < |x_n|$; here $|x_n|$ denotes length of vector. Then $\{x_n\}_{n \in \mathbf{N}}$ would be an infinite subset of X with no limit point (Problem 10.12), giving us our contradiction. ▮

We outline another proof that, if X is compact, every infinite subset of X has a limit point which is in X, modeled on the argument in Example 10.01. Suppose $A \subseteq X \subseteq \mathbf{R^n}$ where A is an infinite set. Since X is compact, it is bounded and so contained in some sub-cube of size M. Let $C_0 = Q_M$. Divide Q_M into 2^n sub-cubes of size $\frac{M}{2}$. One of these, call it C_1, must contain infinitely many points of A. Next subdivide C_1 into 2^n sub-cubes of size $\frac{M}{4}$; one of these, call it C_2, must contain infinitely many points of A. Continuing in this way we obtain a sequence of cubes, C_i of size $\frac{M}{2^i}$. The proof concludes with a demonstration that $\bigcap_i C_i$ is a single point which is a limit point of A (Problem 10.13).

With a variation of this proof we can show that

Proposition 10.26 *If A is an uncountable subset of $\mathbf{R^n}$, then it has a limit point.* ▮ *(Problem 10.14)*

We leave details of this argument as exercise; here is the how a proof begins. Write $\mathbf{R^n}$ as a countable union of congruent n-cubes with edge length 1, using points with integer coefficients for vertices. Since A is uncountable, one of these, call it C_1, must contain uncountably many points of A. Next, divide C_1 into 2^n equal sub-cubes of edge length 1/2, etc.

Note that in the proof of Proposition 10.25 we have shown that if X is an infinite bounded set, then X has a limit point.

We can use this to show:

Proposition 10.27 *If $\{x_i\}_{i \in \mathbf{N}}$ is a Cauchy sequence of points of $\mathbf{R^n}$, then there is an x_0 such that with $x_i \to x_0$.*

Proof: If $\{x_i\}_{i \in \mathbf{N}}$ is a Cauchy sequence of points, then $\{x_i\}_{i \in \mathbf{N}}$ is an infinite set. So all we need show is that $\{x_i\}_{i \in \mathbf{N}}$ is a bounded subset of $\mathbf{R^n}$.

If we chose $\epsilon = 1$, then since $\{x_i\}_{i \in \mathbf{N}}$ is a Cauchy sequence, we can find an M_1 so that if $M_1 \le n, m$. then $\delta(x_n, x_m) < 1$. Letting $|x_n|$ denote the length of x_n viewed as a vector, let $M = \max_{i=1,\ldots,M_1} |x_i|$. For any $M_1 < n$, we have $|x_n| \le |x_{M_1}| + |x_n - x_{M_1}| \le M + 1$. Thus $\{x_i\}_{i \in \mathbf{N}} \subseteq Q_{M+1}$, and so $\{x_i\}_{i \in \mathbf{N}}$ is bounded. ∎

A basic theorem of calculus of real-valued functions is that any continuous function defined on a closed interval has a maximum and a minimum. We now generalize this theorem.

Proposition 10.28 *If $X \subseteq \mathbf{R^n}$, X compact, and $f: X \to \mathbf{R^1}$ is a continuous function, then there is an $x_{max} \in X$ such that, for all $x \in X$, $f(x) \le f(x_{max})$. Also, there is a $x_{min} \in X$ such that, for all $x \in X$, $f(x_{min}) \le f(x)$.*

Proof: Let $x_1 \in X$. If there were no maximum, then we can find points $x_n \in X$ where $f(x_n) + 1 < f(x_{n+1})$ for all $n \in \mathbf{N}$. In particular, note that the set of points $\{f(x_n)\}_{n \in \mathbf{N}}$ does not have a limit point,(Problem 10.7). The set $A = \{x_n\}_{n \in \mathbf{N}}$ is an infinite subset of X and, by Proposition 10.12, there is a limit point of A, call it x_0 where $x_0 \in X$. By Proposition 9.36, there must be a sequence of points of A, $\{a_j\}$, with $a_j \to x_0$. Since f is continuous, $f(a_j) \to f(x_0)$. This contradicts our assertion that $\{f(x_n)\}_{n \in \mathbf{N}}$ has no limit point.

This proves the first statement. The second statement follows from the first by considering the function $-f$. ∎

Review our proof of Example 4.28, that the circle and the half-open interval are not homeomorphic. Using the definitions and propositions from this chapter, we can restate this argument. Suppose there is a continuous map from the circle to the half-open interval, say $F : S^1 \to [0, 2\pi)$. The circle is compact since it is closed and bounded, so F must have a maximum on S^1. But then F could not be onto; thus F could not be a homeomorphism.

Another consequence of compactness encountered in calculus is the uniform continuity theorem. One standard way to prove this is using the Lebesgue number:

Proposition 10.29 *Suppose $X \subseteq \mathbf{R^n}$ with X compact and suppose that $\mathcal{U} = \{U_\alpha\}_{\alpha \in A}$ is an open cover of X. Then there is a $\delta > 0$ such that for any $x \in X$, $N_\delta(x, X) \subseteq U_\alpha$ for some $\alpha \in A$.*

Proof: We will use an indirect proof. If the proposition is not true, then for any natural number $n \in \mathbf{N}$, there is a point $x_n \in X$ such that $N_{1/n}(x_n, X) \nsubseteq U_\alpha$ for all $\alpha \in A$.

It may be that the collection $\{x_n\}_{n \in N}$ is finite, in which case let x_0 denote one of the points such that $x_0 = x_k$ for infinitely many values of k. If $\{x_n\}_{n \in N}$ is infinite, let $x_0 \in X$ be a limit point of $\{x_n\}_{n \in N}$, which exists by Proposition 10.25. In either case, we note that x_0 is a limit point of the sequence $\{x_n\}_{n \in N}$.

Since $\mathcal{U}$ is a cover, there is some α_0 such that $x_0 \in U_{\alpha_0}$. Since U_{α_0} is open there is some δ_0 such that $N_{\delta_0}(x, X) \subseteq U_{\alpha_0}$. We may find an $M \in \mathbf{N}$ such that $1/M < \delta_0/2$ and $\delta(x_0, x_M) < \delta_0/2$. By the triangle inequality, we see that $N_{1/M}(x_M, X) \subseteq U_{\alpha_0}$, contradicting our assumption that $N_{1/M}(x_M, X) \nsubseteq U_\alpha$ for all $\alpha \in A$. ∎

Proposition 10.30 *(Uniform continuity) Suppose $X \subseteq \mathbf{R^n}$, $Y \subseteq \mathbf{R^m}$, X compact and $f: X \to Y$ continuous. Then given any $\epsilon > 0$ there exists a $\delta > 0$ such that for $x, y \in X$, if $\delta(x, y) < \delta$ then $\delta(f(x), f(y)) < \epsilon$.*

Proof: For $\epsilon > 0$, consider the covering $\mathcal{U} = \{N_{\epsilon/2}(y, Y)\}_{y \in Y}$. Consider the open cover of X given by $f^{-1}(\mathcal{U}) = \{f^{-1}(U)\}_{U \in \mathcal{U}}$. Let δ_L be the Lebesgue number of $f^{-1}(\mathcal{U})$. For any $x, y \in X$, if $\delta(x, y) < \delta_L$ then $x, y \in f^{-1}(N_{\epsilon/2}(y_0, Y))$ for some $y_0 \in Y$. Then $f(x), f(y) \in N_{\epsilon/2}(y_0, Y))$, and so $\delta(f(x), f(y)) < \epsilon$. ∎

10.3 Distance between subsets

Given two subsets A and B of $\mathbf{R^n}$ we would like to have some sort of topological measure of how different A and B are.

Using Proposition 10.28, we can define a useful notion of the distance between compact subsets. This notion begins with the notion of distance $\delta(p, X)$ from a point p to a subset X in $\mathbf{R^n}$, Definition 3.50, and uses the idea of the d-expansion of a subset:

Definition 10.31 *Suppose $X \subseteq \mathbf{R^n}$ and $0 < d$. The d-**expansion** of X is the set $X^d = \{p \in \mathbf{R^n} \mid \delta(\mathbf{p}, \mathbf{X}) \leq \mathbf{d}\}$.*

Recall that $\delta(p, X)$ is a continuous function from $\mathbf{R^n}$ to $\mathbf{R^1}$, Proposition 3.51. This implies that X^d is a closed subset since it is the inverse image of $[0, d]$ under this continuous map.

Definition 10.32 *Suppose $x \subseteq \mathbf{R^n}$ and A and B are compact subsets of X. Define the **Hausdorff distance** $\delta(A, B)$ **between** A **and** B to be the smallest number d such that $B \subseteq A^d$ and $A \subseteq B^d$.*

Example 10.33 In $\mathbf{R^2}$ let $A = \overline{N}_1((0,0))$ and $B = \overline{N}_2((4,0))$; see Figure 10-4. Then $A^5 = \overline{N}_1((6,0))$ is the smallest expansion of A that contains B. Also, $B^3 = N_5((4,0))$ is the smallest expansion of B that contains A. Thus $\delta(A, B) = 3$.

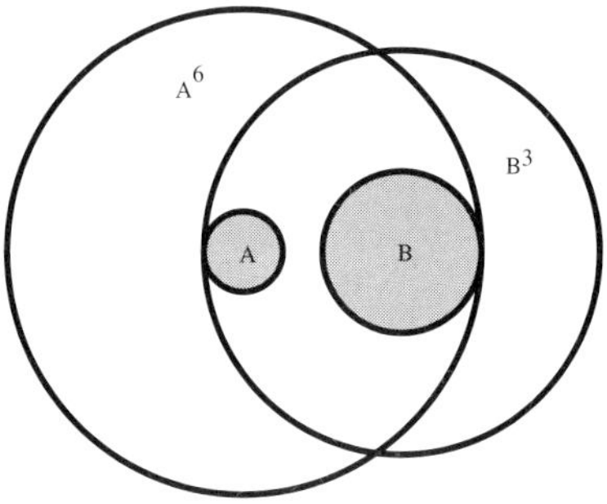

Figure 10-4 Two sets, A and B, of Hausdorff distance 3; see Example 10.33.

Note that $\delta(A, B)$ is not always defined. For example, if A is a line in $\mathbf{R^2}$ and $B = \{p\}$ where $p \notin L$, then $\delta(A, B)$ is not defined. However, if A and B are bounded subsets of $\mathbf{R^n}$, then $\delta(A, B)$ will be defined (Problem 10.15).

Note also that $\delta(N_1(\vec{0}), \overline{N}_1(\vec{0})) = 0$. In fact, more generally, for any bounded subset A, $\delta(A, \overline{A}) = 0$. Thus, in general, we can have $A \neq B$ yet have $\delta(A, B) = 0$. So this concept of distance does not have a basic property of distance, the positive definite property mentioned in Definition 1.03. ◆

However, if we consider *compact* subsets, then $\delta(A, B)$ has very nice properties:

The interest in the Hausdorff distance is that it satisfies the properties of a metric, for any compact subsets, A, B, and C of X (Problem 10.8):

1. $\delta(A, B) \geq 0$. Also, $\delta(A, B) = 0$ only if $A = B$.

2. $\delta(A, B) = \delta(B, A)$.

3. $\delta(A, C) \leq \delta(A, B) + \delta(B, C)$ "triangle inequality."

To show the triangle inequality, first note that, for any $\epsilon > 0$, if $a \in A$, there is a $b \in B$ with $\delta(a, b) \leq \delta(A, B) + \epsilon$. Similarly, there is a $c \in C$ with $\delta(b, c) \leq \delta(B, C) + \epsilon$. Thus $A \subseteq C^d$ where $d = \delta(A, B) + \delta(B, C) + 2\epsilon$. Similarly, $C \subseteq A^d$. So we can see that for *any* $\epsilon > 0$:

$$\delta(A, C) \leq \delta(A, B) + \delta(B, C) + 2\epsilon.$$

From this, the triangle inequality follows.

10.4 Continua

Although compactness and connectedness are unrelated properties, they are both powerful properties which many simple and useful spaces possess.

Definition 10.34 *Let $X \subseteq \mathbf{R^n}$, X is called a* **continuum** *(or* **compactum***) if X is a compact and connected. If $X \subseteq Y$, X is a continuum, then X is called a* **subcontinuum** *of Y. A continuum is called a* **non-degenerate continuum** *if it contains more than one point.*

By the way, the plural of "compactum" is "compacta."
Proposition 10.11 and Proposition 7.02 imply:

Proposition 10.35 *If $X \subseteq \mathbf{R^n}$, $Y \subseteq \mathbf{R^m}$ and $f: X \to Y$ is a continuous function, then if X is a continuum, then so is $f(X)$.* ▮

A restatement of Proposition 7.12 is

Proposition 10.36 *If $X \subseteq \mathbf{R^1}$ is a non-degenerate continuum, then X is a closed interval.* ▮ *(Problem 10.9)*

In Proposition 7.28, we showed there are uncountably many different subsets of the plane. All those subsets given are connected subsets. However, none are compact since they are unbounded sets; thus they are not subcontinua. You are invited to find compact variations and thus show that there are uncountably many distinct *subcontinua* of the plane:

Proposition 10.37 *There is an uncountable set S and a collection of subcontinua of the plane $\{X_s\}_{s \in S}$, so that for all s and t in S, X_s is homeomorphic to X_t if and only if $s = t$.* ▮ *(Problem 10.10)*

So even if we focus our attention to continuua of the plane, Proposition 10.37 says there is a bountiful supply of examples to study.

If one ever to needs prove that a given function is a homeomorphism of a subset X which is compact, the following theorem is most helpful.

Proposition 10.38 *If $X \subseteq \mathbf{R}^n$ is compact and $f: X \to Y$ is a continuous function which is a one-to-one correspondence, then f is a homeomorphism.* ∎ *(Problem 10.11)*

For example, consider the various arguments needed in Chapter 5, where we showed that many seemingly different examples were homeomorphic to the standard Cantor set. We could use Proposition 10.38 to shorten many of these proofs. The standard Cantor set is compact; clearly, it is bounded and, as we have shown in Proposition 5.03, it is a closed subset. Once we have a function $f: K \to X$ which is continuous and a one-to-one correspondence, then it is a homeomorphism by the above proposition.

In Chapter 5 we encountered some rather strange subsets: Sierpiński's carpet and Menger's sponge. The significance of these examples is explained in the next two propositions.

Proposition 10.39 *If $X \subseteq \mathbf{R}^2$, and X is compact and of dimension 1, there is embedding of X into Sierpiński's carpet S.*

Proof: Recall the construction of Sierpiński's carpet S where $Q = [0,1] \times [0,1]$, and we defined subsets $S_i \subseteq Q$ so that $S = \bigcap_i S_i$. Our strategy will be to define a sequence of embeddings, $h_i: X \to S_i$, then define our embedding as the limit of this sequence.

Since X is compact, it is bounded and thus contained in some square $Q_M = [-M, M] \times [-M, M]$. Define a homeomorphism $\sigma_0: Q_M \to Q$ that shrinks the large cube to 1 with sides of length one and shifts so the lower left-hand corner is sent to the origin: by $H(x, y) = (\frac{x}{2M} + \frac{1}{2}, \frac{y}{2M} + \frac{1}{2})$. Let $h_0 = \sigma_0|_X$, and define $X_0 = h_0(X)$.

We can find a closed rectangle R in $int(Q) - X_0$. Since X_0 is a closed subset of Q, $Q - X_0$ is an open subset of Q. Since X has dimension 1, Proposition 9.34 implies that $Q - X_0$ is non-empty. It follows (see Proposition 2.45) that we can find a closed rectangle R with $R \subseteq int(Q)$ and $R \cap X_0 = \emptyset$; see Figure 10-5.

Let $A_1 = Q - int(R)$, then $X_0 \subseteq A_1$. By Example 4.52, we can find a homeomorphism $\sigma_1: A_1 \to S_1$. Define $h_1 = \sigma_1 \circ h_0$. We now have completed our first step—we have defined an embedding $h_1: X \to S_1$. Let $X_1 = h_1(X)$; see Figure 10-6.

Note that σ_1, defined in the proof of Example 4.52, is the identity on the edges of Q. This property of our construction will be useful in further steps.

The next step is to define an embedding of $h_2: X \to S_2$. We will need some additional subsets of S_1 and S_2.

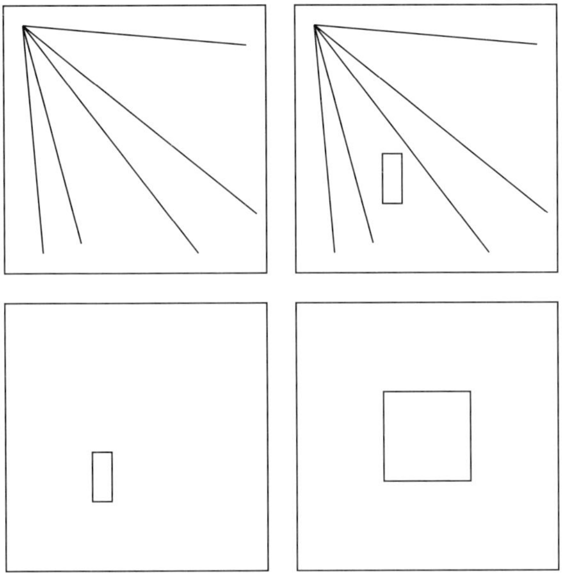

Figure 10-5 For this figure and the related Figures 10-6, 10-8, and 10-8, X is the union of five line segments intersecting at a common endpoint; see Example 10.39. Upper left shows $X_0 \subseteq Q$. Upper right shows rectangle $R \subseteq (int(Q) - X_0)$. Lower right shows $A_1 = Q - int(R)$, which is homeomorphic via σ_0 to S_1, lower right.

We will write S_1 as the union of eight sub-squares. Write I as the union of three intervals, $I_0 = [0, \frac{1}{3}]$, $I_1 = [\frac{1}{3}, \frac{2}{3}]$, and $I_2 = [\frac{2}{3}, 1]$. Note that $I_i = [\frac{i}{3}, \frac{i+1}{3}]$, for $i = 0, 1, 2$. Let $I_{ij} = I_i \times I_j$. Now S_1 is the union of the eight squares: I_{00}, I_{01}, I_{02}, I_{10}, I_{12}, I_{20}, I_{21}, I_{22}. (That is, all squares except the center square I_{11}.)

Let J_i be the middle third interval in each I_i: $J_i = [\frac{i}{3} + \frac{1}{9}, \frac{i+1}{3} - \frac{1}{9}]$. Define $J_{ij} = J_i \times J_j$. Then for all i, j with $i, j = 0, 1, 2$ we will have $J_{ij} \subseteq I_{ij}$. Then $S_2 = \bigcup_{i,j}(I_{ij} - int(J_{ij}))$.

We now apply our argument above to the sub-squares I_{ij}, $i, j = 0, 1, 2$. In each I_{ij}, we can find a closed rectangle, $R_{ij} \subset int(I_{ij})$ such that $R_{ij} \cap X_1 = \varnothing$; thus $(X_1 \cap I_{ij}) \subseteq (I_{ij} - int(R_{ij}))$. Let $A_2 = S_1 - \bigcup_{i,j} int(R_{ij})$; see Figure 10-7.

For each $i, j = 0, 1, 2$ we can find a homeomorphism of $I_{ij} - int(R_{ij})$ onto $(I_{ij} - int(J_{ij}))$ which is fixed on the edges of I_{ij}. Using the gluing lemma, we may obtain a homeomorphism of $\sigma_2 : A_2 \to S_2$. We now define $h_2 = \sigma_2 \circ h_1$, and let $X_2 = h_2(X)$; see Figure 10-8.

We continue, inductively, to define an embedding $h_i : X \to S_i$ by dividing S_i into 8^i sub-squares. We now define an embedding of $h : X \to S$

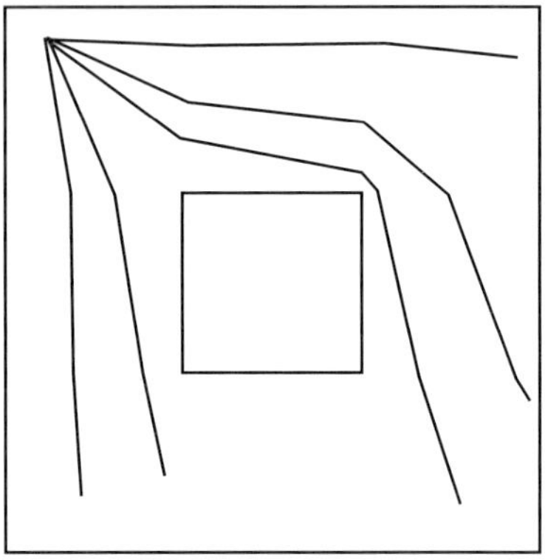

Figure 10-6 The second step for embedding X into Sierpiński's carpet S; see Example 10.39. Shown is $X_1 \subseteq S_1$.

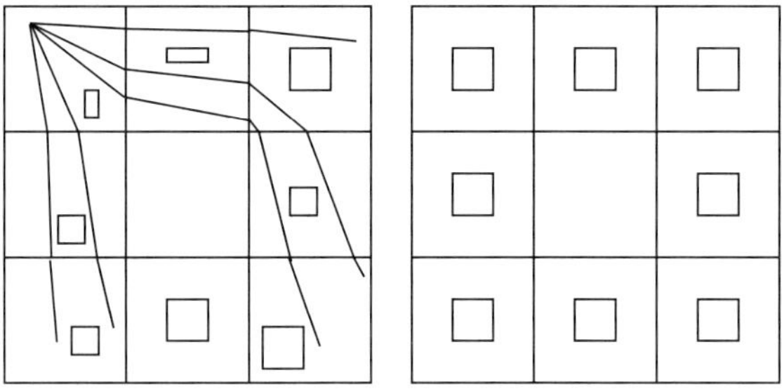

Figure 10-7 On the left, A_2, which is S_1 with eight rectangles, which miss X_0, removed; see Example 10.39. This is homeomorphic, via σ_2 to S_2, shown on the right.

by $h(x) = \lim_{i\to\infty} h_i(x)$. We leave some of the details of verification as an exercise (Problem 10.17), but here is an outline. We note that $h(x)$ is defined since, for fixed x, we can show that $\{h_i(x)\}$ is a Cauchy sequence. If $x_1, x_2 \in X$, then for large enough i, x_1 and x_2 will lie in disjoint sub-squares of S_i. Since each H_i maps these sub-squares to themselves, one can see that h will be one-to-one. Once we verify that h is continuous, we will know that h is an embedding, by Proposition 10.38. ∎

In a similar way one can show:

Proposition 10.40 *If $X \subseteq \mathbf{R}^3$, and X is compact and of dimension 1, there is embedding of X into Menger's sponge, $\mathcal{M}$.* ∎ *(Problem 10.18)*

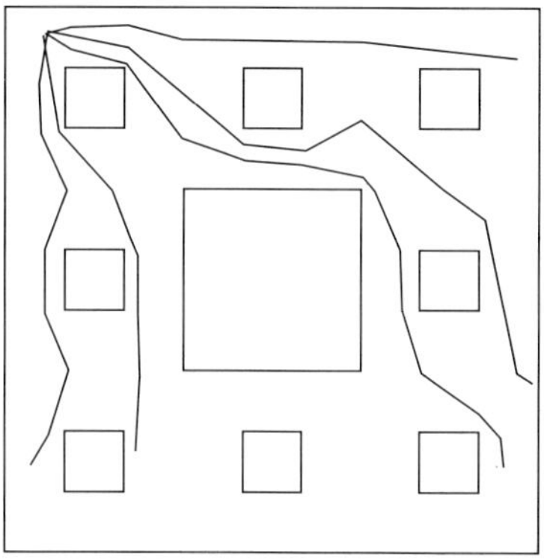

Figure 10-8 The third step for embedding X into Sierpiński's carpet S; see Example 10.39. Shown is $X_2 \subseteq S_2$.

Using these two propositions, we can show that Menger's sponge, $\mathcal{M}$, is not homeomorphic to Sierpiński's carpet, S. Recall the utilities graph G, Example 6.20; this is a compact one-dimensional subset of $\mathbf{R}^3$, which does not embed in $\mathbf{R}^2$. By Proposition 10.40, it does embed in $\mathcal{M}$. (If you have the time and patience, it is not hard to find such an embedding of G into $\mathcal{M}$, without using this proposition.) If $\mathcal{M}$ were homeomorphic to S, then we would get an embedding of G into the plane.

Compact subsets are closed, so it is natural to try and express compactness in terms of closed subsets. The definition of compactness was given in terms of open sets and unions of open sets. Using basic set theory, in particular DeMorgan's rules for subsets, we can reformulate this in terms of closed sets and intersections of closed sets. This is done in the next proposition and its accompanying definitions.

Definition 10.41 *Let $\{C_\alpha\}_{\alpha \in A}$ be a collection of non-empty subsets of X. We say $\{C_\alpha\}_{\alpha \in A}$ has the* **finite intersection property** *if any finite sub-collection of $\{C_\alpha\}_{\alpha \in A}$ has the property: the intersection of all the sets of this sub-collection is non-empty. That is, if $B \subseteq A$ and B is finite, then $\bigcap_{\alpha \in B} C_\alpha \neq \phi$.*

Proposition 10.42

Let $X \subseteq \mathbf{R^n}$, then the following statements are equivalent:

(a) *X is compact.*
(b) *If a non-empty collection of closed sets of X has the finite intersection property, then the intersection of all the sets is non-empty.* ∎
(Problem 10.19)

Example 10.43 In the line consider the half-infinite intervals $C_n = [n, \infty)$ for $n \in \mathbf{N}$. The collection $\{C_n\}_{n \in \mathbf{N}}$ is a collection of closed sets which satisfies the finite intersection property, yet $\bigcap_{n \in \mathbf{N}} C_n = \emptyset$. This implies that $\mathbf{R}^1$ is not compact. ◆

A simple collections of sets which satisfies the finite intersection property is a nested collection of sets.

Definition 10.44 *Let $\{A_n\}_{n \in \mathbf{N}}$ be a collection of sets; we say that $\{A_n\}$ is a* **nested collection,** *if $A_1 \supseteq A_2 \supseteq A_3 \supseteq \dots$.*

Proposition 10.45 is a special consequence of Proposition 10.42, which is easier to remember.

Proposition 10.45 *If X is compact and $\{A_n\}_{n \in \mathbf{N}}$ is a collection of nested, closed, non-empty sets, then $\bigcap_{n \in \mathbf{N}} A_n \neq \emptyset$.* ▮ *(Problem 10.20)*

Do not get the impression that basic topology problems in the plane are largely solved. This is certainly not the case.

We end this section with a statement of probably the most important currently unsolved problem of plane topology.

Question 10.46 Suppose X is a continuum in the plane. If the complement of X in the plane is connected, does X have the fixed-point property?

*10.5 General Topology and Chapter 10

The idea of a set being bounded makes sense in any metric space.

Definition 10.47 *Let $\{S, d\}$ be a metric space, $A \subseteq S$. We say A is bounded if there is a number M such that for all $x, y \in A$, $d(x, y) < M$.*

In general topology we have seen examples of topological spaces for which no metric can be defined. So there is no abstract version of the proposition that compact subsets are those that are closed and bounded, Proposition 10.18. A proof of the next proposition can be made from modifications of the arguments of Section 10.1.

Proposition 10.48 *Let $\{S, d\}$ be a complete metric space. Then $A \subseteq S$ is compact if and only if and only if it is closed and bounded.* ▮

For general products of topological spaces, we have the following version of Proposition 10.19. This is not an easy theorem to prove. It is usually called the Tychonoff Theorem; a proof can be found in any general topology text such as [32, 23, 48]. There are a number of proofs and all of them use, in an essential way, the Axiom of Choice; see Definition A.43, or something equivalent to it.

Proposition 10.49 *Using the product topology, the product of any number of compact spaces is a compact space.* ∎

Compactness is an important consideration in the study of function spaces. We have looked at some ways to obtain a topology for a function space. Affine maps from $\mathbf{R^n}$ to $\mathbf{R^m}$ correspond to $m \times n$ matrices and thus to points in $\mathbf{R^{nm}}$, and we can, and have, used the topology of $\mathbf{R^{nm}}$ to get a topology for this space, and its subsets such as in Example 8.19. For maps of one compact metric space into another metric space, one can define a metric for the function space, as we did in Example 1.70, and obtain a topology that way.

Using the product topology for the space of functions from X to Y, Y^X, Example *2.7, gives us an interesting space. But in many ways it is unsatisfying since Y^X consists of all functions from X to Y, not just continuous functions. The product topology for Y^X depends on the topology of Y but not the topology of X.

If we are interested in a most general, useful topology for the set of continuous functions, from one topological space to another, the next definition is the most successful. It not immediately obvious, but it does generalize the examples mentioned above. We need compactness to define it, but we need one more condition, to be discussed in the next section, the remark on page 263, to get a really useful topology. (Recall the definition of basis, Definition 1.66.)

Definition 10.50 *Suppose $\{X, \mathcal{T}\}$ and $\{Y, \mathcal{T}'\}$ are topological spaces. Let $C(X,Y)$ denote the subset of Y^X of continuous functions from X to Y. If C is a compact subset of X, and V is an open subset of Y, write $B_{C\to V} = \{f \in C(X,Y) \mid f(C) \subseteq V\}$.*

We define a topology $\mathcal{O}$ for $C(X,Y)$ by using the subsets $B_{C\to V}$ as a basis; this is called the **compact-open topology** *of $C(X,Y)$.*

On a different topic concerning compactness and general topology, the next proposition gives a list of simple conditions that assure us that a metric space is homeomorphic to the standard Cantor set. This accounts for the mathematical ubiquity of the Cantor set. Many different topological constructions involve infinite processes which generate sets with these properties.

Definition 10.51 *Let $\{X, \mathcal{T}\}$ be a topological space. We say that X is totally disconnected if each component of X is a single point.*

Definition 10.52 *Let $\{X, \mathcal{T}\}$ be a topological space. We say that X is a perfect set if every point of X is a limit point of X.*

Example 10.53 Any topological space with the discrete topology will be totally disconnected and not perfect.

Any topological space with more than one point with the indiscrete topology will not be totally disconnected, but will be perfect.

The Sorgenfry line, Example 1.46, is totally disconnected and perfect, but is not a metric space; see the remark on page 177.

The standard Cantor set is perfect and totally disconnected. ◆

Proposition 10.54 *If $\{X, d\}$ is a non-empty metric space that is compact, perfect, and totally disconnected, then it is homeomorphic to the standard Cantor set.* ■

A proof of the proposition above can be found in several texts for general topology such as [11, 42]. One could use Proposition 10.54 to show that subsets defined in Chapter 5, such as $K \times K$ and Antoine's necklace, to name a few, are homeomorphic to the standard Cantor set K.

10.6 Problems for Chapter 10

10.1 Prove Proposition 10.04.

10.2 Verify the assertions of Proposition 10.10 that $q_\infty = \bigcap_1^\infty Q^i$, and that q_∞ is a limit point of A.

10.3 Prove Proposition 10.11.

10.4 Prove the assertion in the proof of Proposition 10.19 that the product of bounded subsets is a bounded subset.

10.5 Prove the assertion of Example 10.22: Let h be a homeomorphism of the plane with $h(C) = C$; then $h(x)$ is inside C if and only if x is inside C.

10.6 Prove Proposition 10.24.

10.7 Verify the assertion in the proof of Proposition 10.28 that we can find points $x_n \in X$ where $f(x_n) + 1 < f(x_{n+1})$ for all $n \in \mathbf{N}$. Also show that the set of points $\{f(x_n)\}_{n \in \mathbf{N}}$ does not have a limit point.

10.8 Verify the properties of the Hausdorff distance listed in the remark on page 248.

10.9 Prove Proposition 10.36.

10.10 Prove Proposition 10.37.

10.11 Prove Proposition 10.38. (Hint: Think of continuity in terms of closed subsets.)

10.12 Verify the claim of Proposition 10.25 that $\{\vec{x}_n\}_{n\in N}$ is an infinite subset of X with no limit point.

10.13 Complete the details of the proof, in the remark on page 245: if X is compact, every infinite subset of X has a limit point which is in X.

10.14 Complete the details of the proof, in the remark on page 245: if A is an uncountable subset of $\mathbf{R^n}$, then it has a limit point.

10.15 Show, as stated in Example 10.33, that if A and B are bounded subsets of $\mathbf{R^n}$, then $\delta(A, B)$ will be defined.

10.16 As mentioned in the remark on page 55, we can view a 2×2 matrix as a point in $\mathbf{R}^4$.

(a) Let X correspond to all 2×2 matrices with determinant 1. Prove that X is not compact.
(b) Let Y correspond to all 2×2 orthogonal matrices. Prove that Y is compact.

10.17 Provide the details needed for Proposition 10.39 for the proofs that

(a) $\{h_i(x)\}$ is a Cauchy sequence.
(b) h is one-to-one.

10.18 Prove Problem 10.40, modeled on the proof of Proposition 10.39.

10.19 Prove Proposition 10.42.

10.20 Prove Proposition 10.45.

10.21 We define a compact subset of topologist's $\sin(1/x)$ curve, Example 8.09, by considering the portion of the graph of $\sin(1/x)$ with domain (0,1]: Let $\Gamma' = \{(x, y) \in \mathbf{R}^2 : y = \sin(1/x) \text{ and } 0 < x \leq 1\}$; let $L = \{(x, y) \in \mathbf{R}^2 : x = 0 \text{ and } -1 \leq y \leq 1\}$. Let $X' = \Gamma' \cup L$.

Show that X' has the fixed point property. (Hint: What can $f(L)$ be?)

11. LOCAL CONNECTIVITY

OVERVIEW: A local property is one which depends only on topological properties of a neighborhood of a point. We discuss one such property, local connectedness.

11.1 Local connectedness

A property is termed "local," if it is defined in a neighborhood of a point without regard to points not in that neighborhood. One example we have previously seen of a local property is the dimension of a subset at a point, Definition 9.29. The focus of this chapter is a local property, local connectedness.

Recall the topologist's comb X, Example 2.43, and the shrinking comb Y, Example 2.44. We would like to show that they are not homeomorphic. They are compacta, so connectedness and compactness does not distinguish them. Nor does classification of the cut-points distinguish them. Both X and Y have infinitly many cut-points of orders 1, 2, and 3.

What is unusual about the topologist's comb is that, for points $x \in L - (0,0)$, and for small enough ϵ, $N_\epsilon(x, X)$ is a disconnected set. However, this distinction, as stated in terms of $N_\epsilon(x, X)$, is not a topological property.

Example 11.01 Let P_n denote the perimeter of a square in $\mathbf{R}^2$ with corners at

$$(\frac{1}{n}, \frac{1}{n}), (-\frac{1}{n}, \frac{1}{n}), (-\frac{1}{n}, -\frac{1}{n}), \text{ and } (\frac{1}{n}, -\frac{1}{n}).$$

Let C_n be the circle with center $(0,0)$ and radius $1/n$, and let L be the line segment between $(-1,0)$ and $(1,0)$. Consider

$$P = (\bigcup_{n\in\mathbf{N}} P_n) \cup L \text{ and } Q = (\bigcup_{n\in\mathbf{N}} C_n) \cup L;$$

see Figure 11-1.

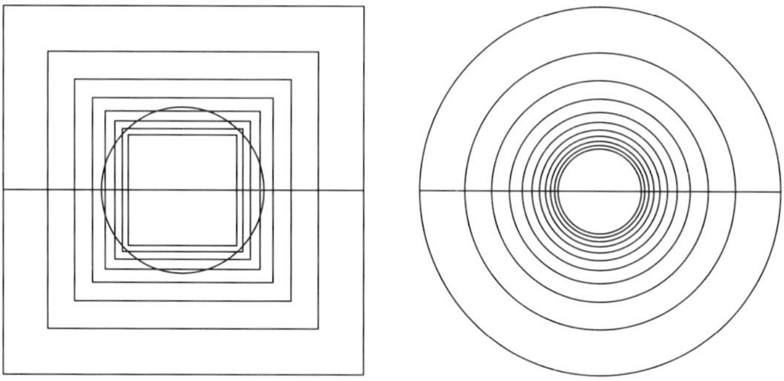

Figure 11-1 On the left, a portion of P, with a small circle shown, indicating the boundary of $N_\epsilon((0,0))$, for small ϵ. Near the top and the bottom of this neighborhood one sees components of $N_\epsilon((0,0),P)$, open, horizontal line segments, showing that $N_\epsilon((0,0))$ is not connected. On the right, a portion of Q. For any ϵ, $N_\epsilon((0,0),Q)$ is connected. See Example 11.01.

We can see that $N_\epsilon((0,0),P)$ will always be disconnected for small enough ϵ. Given a small ϵ, we can find some n such that squares P_n have corners outside the circle C_n, yet with upper and lower edges meeting $N_\epsilon((0,0),P)$. These portions of the upper and lower edges within $N_\epsilon((0,0),P)$ are distinct components of $N_\epsilon((0,0),P)$. On the other hand, for any ϵ, $N_\epsilon((0,0),Q)$ is connected. Yet P is homeomorphic to Q by a homeomorphism that takes $(0,0)$ to itself. So perhaps the property we really want is something like: there are *some* small connected open subsets about $(0,0)$. This is the idea, but "small" is relative. So we must somehow express this indirectly. A resolution of our problem is found in the next definition. ◆

Definition 11.02 *Let $X \subseteq \mathbf{R^n}$, $x \in X$. We say X* **is locally connected at** x *if, for every open subset U of X with $x \in U$, there exists an open subset V of X such that $x \in V$, $V \subseteq U$ and V is connected. We say that X is* **locally connected** *if X is locally connected at every point of X.*

Using this concept we will can distinguish X from Y (Problem 11.4).

Example 11.03 The subsets P and Q are both locally connected at $(0,0)$ (Problem 11.1). ◆

Proposition 11.04 *The property "X is locally connected" is a topological property.* ▮

Example 11.05 Every open subset of $\mathbf{R^n}$ is locally connected. Suppose U is an open subset of $\mathbf{R^n}$. For each $x \in U$, there is an ϵ such that $N_\epsilon(x) \subseteq U$. Now, $N_\epsilon(x)$ is connected since it is homeomorphic to $\mathbf{R^n}$, which is connected, by Proposition 7.13. ◆

Example 11.06 Consider the topologist's $\sin(1/x)$ curve, Example 3-2. Note that X is locally connected at points of Γ. The map $G: J \to X$ defined by $G(x) = (x, \sin(1/x))$ gives an embedding of $(0, \infty)$ into the open half-plane $\overset{\circ}{R}{}^2_+$, Proposition 6.09. In particular, Γ is homeomorphic to $\mathbf{R^1}$, which is locally connected, by Example 11.05. Thus Γ is locally connected since locally connectedness is a topological property.

However, X is not a locally connected subset. Using the remark on page 206, one can show that local connectedness does not hold at points of L (Problem 11.3).

Similarly, closed topologist's $\sin(1/x)$ curve, Example 8.09, is not locally connected. In particular, it is not homeomorphic to the circle S^1. ◆

Regarding spaces that are not locally connected, Proposition 11.08 is sometimes useful.

Proposition 11.07 *Suppose $X \subseteq \mathbf{R^n}$, $Y \subseteq \mathbf{R^m}$, $x \in X$, and $f: X \to Y$ is a homeomorphism. Then, X is locally connected at x if and only if Y is locally connected at $f(x)$.* ▮

A corollary of the proposition above is the following:

Proposition 11.08 *Suppose $X \subseteq \mathbf{R^n}$, $Y \subseteq \mathbf{R^k}$ and $f: X \to Y$ is a homeomorphism. Let $A = \{x \in X \mid X$ is locally connected at $x\}$, and let $B = \{y \in Y \mid Y$ is locally connected at $y\}$. Then A and B are homeomorphic. In fact, $f|_A$ is a homeomorphism from A to B.* ▮

Example 11.09 Recall the definition of the topologist's comb X, Example 2.43. We consider a variation of this, the double comb; see Figure 11-2.

Recall we wrote $X = I \cup T_0 \cup (\bigcup_1^\infty T_n)$ where T_0 is the tooth at 0, and T_n is the tooth at $1/n$:

$$I = \{(x, y) \in \mathbf{R}^2 \mid 0 \le x \le 1 \text{ and } y = 0\},$$
$$T_0 = \{(x, y) \in \mathbf{R}^2 \mid x = 0 \text{ and } 0 \le y \le 1\},$$
$$T_n = \{(x, y) \in \mathbf{R}^2 \mid x = \frac{1}{n} \text{ and } 0 \le y \le 1\}.$$

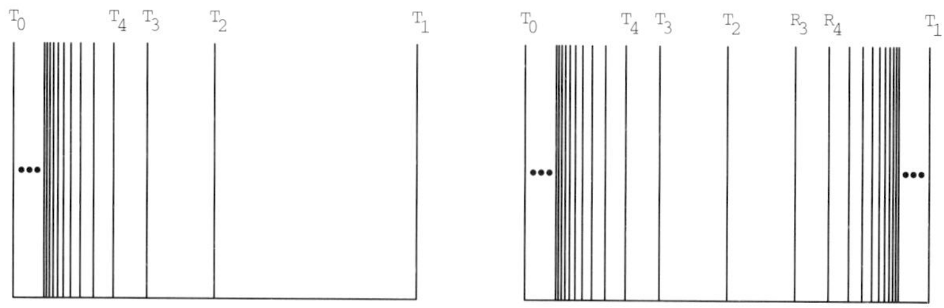

Figure 11-2 On the left, the topologist's comb X. On the right, the double comb Z; see Example 11.09.

For n with $3 \leq n$ we define

$$R_n = \{(x, y) \in \mathbf{R}^2 \mid x = 1 - \frac{1}{n} \text{ and } 0 \leq y \leq 1\}.$$

That is, the R_n's are teeth of a comb which get closer and closer to T_1 from the right; now let

$$Z = I \cup T_0 \cup (\bigcup_1^{\infty} T_n) \cup (\bigcup_3^{\infty} R_n);$$

see Figure 11-2. We show that X and Z are not homeomorphic.

Neither X nor Z is locally connected, but we can use Proposition 11.08 to distinguish X and Z. We verify that the set of points of X where X is not locally connected is the subset T_0; the set of points of Z where Z is not locally connected is the subset $T_0 \cup T_1$. But T_0 and $T_0 \cup T_1$ are not homeomorphic since T_0 is connected and $T_0 \cup T_1$ is not connected. ◆

Example 11.10 The standard Cantor set K is not locally connected. In fact, it is not locally connected at any of its points. Each point of K is a component of K, the remark on page 189. So for any open subset U of K, each component of U must also be a single point. Every open subset of K contains at least two points—in fact, every point of K is a limit point of points of K, the remark on page 189. Thus every open subset of K is disconnected. Therefore it is not locally connected at any of its points. ◆

Example 11.11 Recall Sierpiński's carpet S, Example 5.25. This is not a locally connected subset.

For example, consider the point $(\frac{1}{2}, 0)$. Let $R = (\frac{1}{3}, \frac{2}{3}) \times (-2, 2)$, and let $U = R \cap S$; U is an open subset of S, and we can check that $U = (\frac{1}{3}, \frac{2}{3}) \times K$. We prove S is not locally connected by showing that, in fact, U contains no connected open subsets. Suppose there were an open connected subset, V with $V \subseteq U$. Consider $p: U \to K$ obtained as the restriction to U of the projection $(x, y) \to y$. By Proposition 3.36, $p(U)$ is an open subset of K; by Proposition 7.02, $p(U)$ is also connected. We have a contradiction, since K contains no connected open subsets (Problem 7.19).

On the other hand, one can see that S is locally connected at some points, for example, at $(0,0)$ (Problem 11.9).

Since S is locally connected at some points and not at others, it follows that Sierpiński's carpet is not homogeneous. ◆

Example 11.12 The Cantor swirl is an interesting subset of $\mathbf{R}^2$ and is the subject of several exercises; see Figure 11-3. It is compact, connected, and is not locally connected (Problem 11.13).

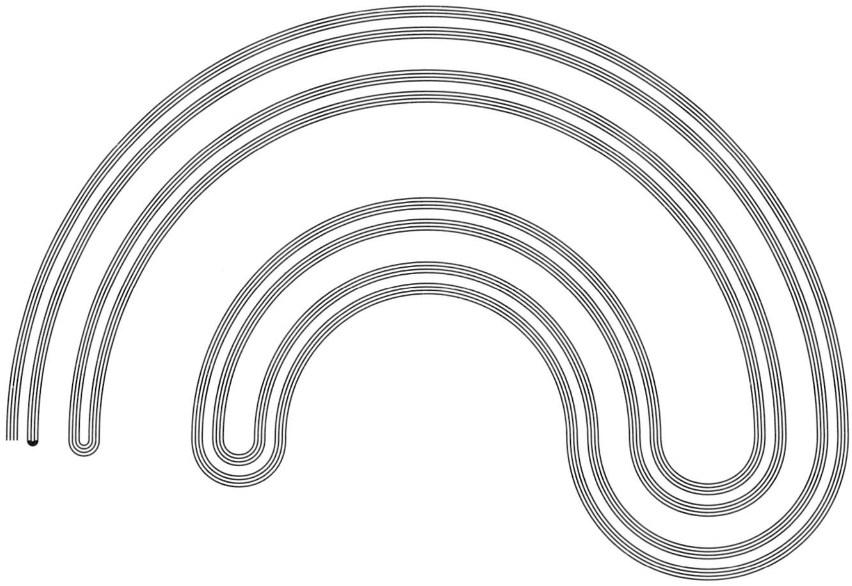

Figure 11-3 The Cantor swirl; see Example 11.12.

Let X denote the x-axis of the plane: $X = \{(x, y) : y = 0\}$. For $A \subseteq X$, let $S^+(c, A)$ be the union of semicircles in the upper half-plane ($\{(x, y) \mid 0 \le y\}$ with center $(c, 0)$ and endpoints in A. Let $S^-(c, A)$ correspond to the union of semicircles in the lower half-plane with center $(c, 0)$ and endpoints in A.

Let $S_0 = S^+(\frac{1}{2}, K')$, where $K' = i(K)$ where K is the Cantor set, and i the standard inclusion of the line into the plane. Let $S_1 = S^-(\frac{3}{4}, K'')$ where $K'' = i(K \cap (\frac{2}{3}, 1))$. Let f be the similarity of $\mathbf{R}^2$: $f(x, y) = (\frac{x}{3}, \frac{y}{3})$.

For $i > 1$ define $S_i = f^{(n)}(S_{i-1})$ where $f^{(n)}$, denotes the n-th iteration of f, Definition 3.24. Note that $S_2 \cap X$ is the image of the "second quarter of the Cantor set," which is the same as the last half of the Cantor set, multiplied by $\frac{1}{3}$ and, more generally, $S_i \cap X = f^{(n)}(K')$.

We call set $Z = \bigcup_i S_i$ the "Cantor swirl." ◆

*11.2 General Topology and Chapter 11

In general topology, an important local property is "local compactness."

Definition 11.13 *Suppose $\{X, \mathcal{T}\}$ is a topological space. We say X is* **locally compact** *if for every $x \in X$ and open set, $U \in \mathcal{T}$, with $x \in U$, there is a $V \in U$ with $\overline{V}$ compact where $\overline{V}$ denotes closure in X.*

Certainly $\mathbf{R^n}$ is locally compact since for any open subset U of $\mathbf{R^n}$ and any $x \in U$ we can find ϵ with $N_\epsilon(x) \subseteq U$, and $\overline{N_\epsilon(x)}$, is closed and bounded; thus compact. Similarly, any subset of $\mathbf{R^n}$ is locally compact. Furthermore, any metric space is locally compact.

Example 11.14 On the other hand, the Sorgenfry line, $\{\mathbf{R^1}, \mathcal{H}\}$, Example 1.46, is not locally compact. Let $x = 0$, $U = [0, 1)$, and let V be an open subset of $\mathcal{H}$ with $x \in V$ and $\overline{V}$ is compact. Since V is open we can find a number c with $[0, c) \subseteq V$. Let $U_0 = (\mathbf{R^1} - [0, c)) \cap \overline{V}$; U_0 is an open subset of $\overline{V}$. For $i \in N$, let $U_i = [0, \frac{i}{i+1}c)$. Then $\{U_i\}_{i=0}^{\infty}$ is an open cover of $\overline{V}$ with no finite subcover. So $\overline{V}$ is not compact, and we conclude that the Sorgenfry line is not locally compact.

This gives us yet another way, see the remark on page 176, of showing that $\mathbf{R^1}$ with the topology $\mathcal{H}$, cannot be a metric space. ◆

Local compactness is important in general topology of functions spaces. Recall the compact-open topology $\mathcal{O}$ defined for $C(X, Y)$, Definition 10.50. Suppose V is an open subset of Y and $x \in X$ if f is continuous at x, there is an open subset, U of X with $f(U) \subseteq V$. To get an open subset of $\mathcal{O}$ we need to find a compact subset C with $f(C) \subseteq V$. It seems fairly clear $\mathcal{O}$ won't have many open sets unless we have a reasonably large collection of compact subsets. Thus one usually finds that, in the study of function spaces, one assumes X is locally compact.

Recall that in topology we are trying to abstract a notion of nearness. If x and x' are in an open set, then in some sense they are close. Considering the space of continuous functions from X to Y, we would hope that we could say: if x is close to x' and f is close to f', then

$f(x)$ is close to $f(x)'$ in $C(X, Y)$. The next Proposition says that this is true, if X is locally compact and Hausdorff; see [14, 32] for a proof and more information.

Proposition 11.15 *Suppose $\{X, \mathcal{T}\}$ and $\{Y, \mathcal{T}'\}$ are topological spaces with X locally compact and Hausdorff. Let $C(X, Y)$ have the compact-open topology, then the map*

$$e : X \times C(X, Y) \to Y \text{ defined by } e(x, f) = f(x)$$

is continuous. ∎

The map e is called the "evaluation map."

11.3 Problems for Chapter 11

11.1 Show that the subsets P and Q of Example 11.03 are both locally connected at $(0, 0)$.

11.2 Show that the subset Q of Example 11.03 is locally connected.

11.3 Verify the details of the claim in Example 11.06 that topologist's $\sin(1/x)$ curve is not locally connected at points of L.

11.4 Show that the topologist's comb and the shrinking comb, X and Y of Example 11.01, are not homeomorphic, by showing that X is not locally connected but Y is.

11.5 Recall the topologist's comb X, Example 2.43, and the shrinking comb Y, Example 2.44. Is $X - T_0$ homeomorphic to $Y - (0, 0)$?

11.6 Let Q denote the set of points in the plane both of whose coordinates is a rational number. Is $X - Q$ homeomorphic to $Y - Q$ Is $X \cap Q$ homeomorphic to $Y \cap Q$?

11.7 Suppose that X is a locally connected space and that $f: X \to Y$ is a continuous function. Is it necessarily true that $f(X)$ is locally connected?

11.8 Suppose that $X \subseteq \mathbf{R}^n$ with X compact and locally connected. Show X has finitely many components

11.9 Show that Sierpiński's carpet S is locally connected at $(0, 0)$; see Example 11.11.

11.10 What is the subset of points of Sierpiński's carpet at which it is locally connected? (See Example 11.11.)

11.11 Show $S_0 = S^+(\frac{1}{2}, K') = K'$ where these sets are defined in Example 11.12.

11.12 Show the Cantor swirl, Example 11.12, is compact and connected.

11.13 Show the Cantor swirl, Example 11.12, is not locally connected.

11.14 How many path components does the Cantor swirl, Example 11.12, have?

11.15 Show the Cantor swirl, Example 11.12, is the closure of the path component which contains the point $(0, 0)$.

Part II

ADVANCED TOPICS

These last chapters cover some more advanced topics. The purpose is to give some idea of what questions motivate further investigation. In each chapter we look at a few topics in detail to show the reader how the background obtained in the study of the basic material is applied in the study of these topics. As with the Basic Topics, each chapter ends with an optional brief section relating the material in that chapter to general topology.

There is far less detail in these sections and some problems are a higher degree of difficulty compared with material in the Basic Topics. These advanced chapters generally depend on all the basic material. Here is a guide to the dependencies of these later chapters with each other.

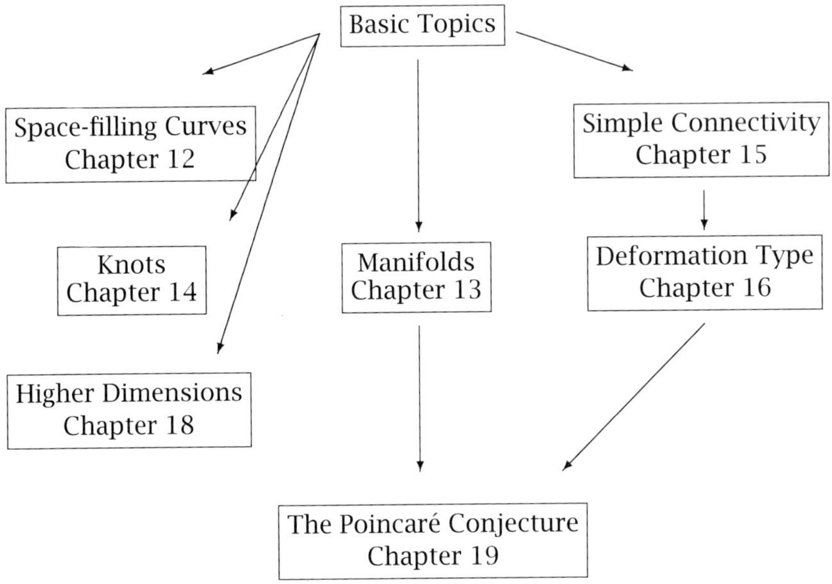

For the general topology sections here are some brief descriptions:

***12.2** We state a very general theorem about topological spaces that are images of continuous maps of I

***13.6** We discuss a definition of manifold from the point of view of general topology, and relate these two viewpoints. We also discuss an abstract notion of smooth manifold

***14.3** The advanced topic in knot theory are covered in Chapter 18.

***15.5** We relate the idea of homotopy to a path in a function space. We define the fundamental group and relate this to simple connectivity.

***16.2** We define the notion of homotopy type and relate it to the concept of deformation type of Chapter 16.

***17.3** We discuss cell complex, a generalization of simplicial complex.

***18.5** The natural topic to discuss here, infinite dimensional space, has already been discussed as the topic of infinite products, especially the Hilbert cube.

***19.2** A brief mention of the smooth Poincaré Conjecture.

12. SPACE-FILLING CURVES

OVERVIEW: We show that there is a continuous map of the line onto *the plane, called a space-filling curve.*

12.1 An example of a space-filling curve

There is a one-to-one correspondence between the line and the plane, Proposition C.12. On the other hand, the line and the plane are not homeomorphic, Propositions 6.5 and 8.03.

However, there is a continuous function from the unit interval *onto* the square! There are many interesting methods of constructing such a function as a limit of continuous functions from the interval to the square, [40]. The method we present here is a bit different from the usual presentations.

Example 12.01 We wish to define a continuous function from I to the closed square, $Q = [0, 1] \times [0, 1]$.

We divide the interval I into four consecutive, congruent subintervals,

$$I_1^1 = [0, 1/4],\ I_2^1 = [1/4, 1/2],\ I_3^1 = [1/2, 3/4],\ \text{and}\ I_4^1 = [3/4, 1].$$

Note that $I_i^1 \cap I_{i+1}^1$ is a single point, for $1 \leq i < 4$. Also $0 \in I_1^1$. Continue this process of subdivision. For $n \in \mathbf{N}$, divide I into 4^n equal subintervals such that, for $1 \leq i < 4^n$, we have $I_i^n \cap I_{i+1}^n$ is a single point. Also, $0 \in I_1^n$. Furthermore,, the order of these intervals is consistent with the order of $\mathbf{R}^1$: if $i < j$ and $x \in I_i^n$, and $y \in I_j^n$ then $x \leq y$. Also, $x = y$ only if $j = i + 1$ and $\{x\} = \{y\} = I_i^n \cap I_{i+1}^n$.

Divide Q into four congruent squares which we denote by Q_1^1, Q_2^1 Q_3^1, and Q_4^1. We want to number these so that they fit together consistently with the intervals. Label so that Q_i^1 and Q_{i+1}^1 share an edge, if $1 \le i < 4$; as shown in Figure 12-1.

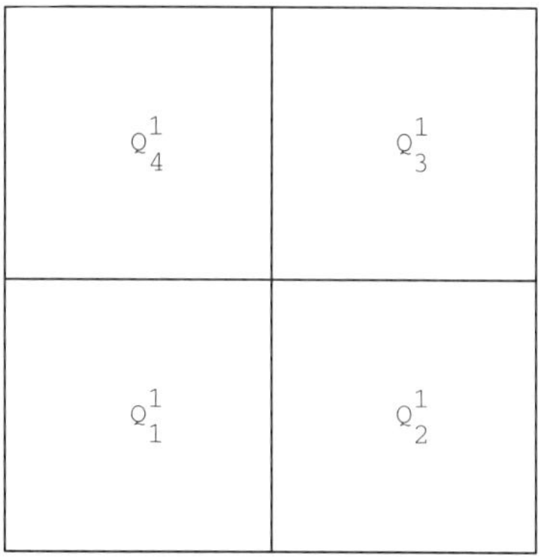

Figure 12-1 Dividing the square into four sub-squares; see Example 12.01.

We continue this process, so that at the n-th stage we divide Q into 4^n equal squares, Q_i^n where $1 \le i \le 4^n$. Label so that, for $1 \le i < 4^n$, Q_i^n and Q_{i+1}^n share an edge. Furthermore, we require the first four sub-squares of Q_i^n to lie in the first square Q_1^{n-1}, the second four sub-squares of Q_i^n to lie in the square Q_2^{n-1}, etc; see Figure 12-2. We also insist, for any n, that the sub-square Q_1^n contains the point $(0, 0)$.

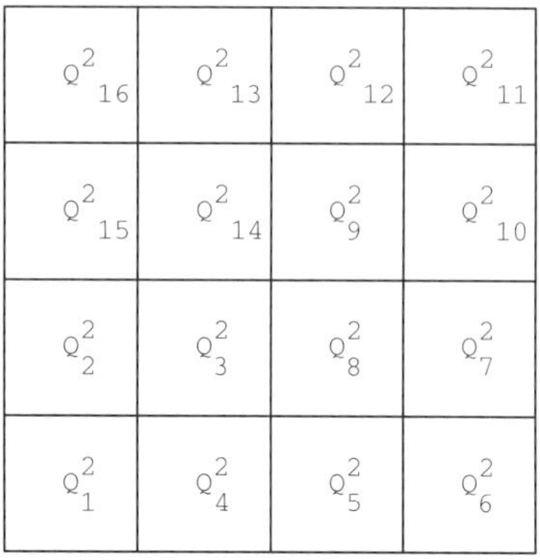

Figure 12-2 Dividing the square into 16 sub-squares; see Example 12.01.

One can argue inductively that this ordering of sub-squares is uniquely determined. For example, we show how the ordering of

Figure 12-1 and the other conditions determines the ordering shown in Figure 12-2. We are given that Q_1^2 must contain the lower left corner. The sub-squares Q_1^2, Q_2^2, Q_3^2, and Q_4^2 must fill up Q_1^1, and Q_4^2 must meet $Q_5^2 \subseteq Q_2^1$. Thus Q_4^2 must either be the square to the right of Q_1^2 or the diagonally opposite sub-square which lies above and to the right. But it could not be this diagonally opposite sub-square since then that would force Q_2^2 and Q_3^2 to be diagonally opposite sub-squares, violating our condition that consecutively labeled sub-squares must share an edge. The only choice for Q_4^2 is the one shown in Figure 12-2. This, in turn, forces us to label Q_5^2 as shown. A similar argument next shows that the only possible choice for the sub-square Q_8^2 is the one shown, etc. Note that if $I_j^{n+1} \subseteq I_i^n$, then $Q_j^{n+1} \subseteq Q_i^n$.

Suppose that $x_0 \in I$; then we can write $x_0 = \bigcap_{k=1}^{\infty} I_{n_k}^k$ for certain choices of subscripts n_k. Define $f(x_0)$ by $\{f(x_0)\} = \bigcap_{k=1}^{\infty} Q_{n_k}^k$. We leave as an exercise that this gives a well-defined function (Problem 12.3). Assuming it is well-defined, we show that f is continuous.

Let U be open in Q, with $x_0 \in f^{-1}(U)$. Let $y_0 = f(x_0)$. Since $\{x_0\} = \bigcap_{k=1}^{\infty} I_{n_k}^k$, we have $\{y_0\} = \bigcap_{k=1}^{\infty} Q_{n_k}^k$. Since the squares become smaller with increasing k, we can find an integer, N, such that

$$Q_{n_{N-1}}^N \cup Q_{n_N}^N \cup Q_{n_{N+1}}^N \subseteq U.$$

It would then follow that

$$f(I_{n_{N-1}}^N \cup I_{n_N}^N \cup I_{n_{N+1}}^N) \subseteq U,$$

and thus the interior of $I_{n_{N-1}}^N \cup I_{n_N}^N \cup I_{n_{N+1}}^N$ is an open set containing x_0 which is contained in $f^{-1}(U)$. Thus $f^{-1}(U)$ is an open subset of I.

The function above is not one-to-one. If we had a continuous one-to-one map of I onto Q, then since I is compact, f would be a homeomorphism, contradicting the fact that Q and I are not homeomorphic.
◆

One can modify the proof of Example 12.01 and obtain the following two variations.

Proposition 12.02 *There is a continuous map of* $\mathbf{R}^1$ *onto* $\mathbf{R}^n$ *for any* $n > 1$. ▮ *(Problem 12.4)*

On the other hand, there is no continuous map of I onto $\mathbf{R}^2$ since the image of such a map would have to be compact, and $\mathbf{R}^2$ is not compact.

Proposition 12.03 *There is a continuous map of* S^1 *onto* S^2. ▮ *(Problem 12.5)*

The reason that all these examples might seem surprising is that familiar maps such as piecewise-linear maps (see Definition 3.47) from the interval to the square are not onto functions.

That is, if $f: \mathbf{R}^1 \to Q$ is piecewise linear, then $f(\mathbf{R}^1) \neq Q$. Each of the sets $S_i = f([t_i, t_{i+1}])$ is a line segment in the plane. There may be some segments which are vertical, but the rest have a slope m_i. Let L be the line in $\mathbf{R}^2$ which passes through the center of Q, $(1/2, 1/2)$, and has slope M which is *not* equal to any m_i and let L_0 be the line segment $L_0 = L \cap Q$.

Since L_0 meets each line segment S_i in at most one point, it follows that $L_0 \cap f(\mathbf{R}^1)$ consists of a countable number of points. But since an interval must have *uncountably many points*, we conclude that there is a point of $L_0 \notin f(\mathbf{R}^1)$, and so $f(\mathbf{R}^1) \neq Q$.

*12.2 General Topology and Chapter 12

The discussion of Chapter 12 leads one to ask the question:

Question 12.04 Suppose $X \subseteq \mathbf{R}^n$. When is there some continuous *onto* function $f: I \to X$?

What properties must X have? Certainly, X must be compact, connected, in fact, path connected.

The most basic result along these lines is generally known as the "Hahn-Mazurkiewicz Theorem," [32, 41, 48]:

Proposition 12.05 *Suppose X is a Hausdorff space. Then there is a continuous onto function $f: I \to X$ if and only if X is a compact, connected, locally connected metric space.* ∎

In the literature, this set of conditions has a name:

Definition 12.06 *A metric space, X, is called a* **Peano space** *if it is compact, connected, and locally connected.*

12.3 Problems for Chapter 12

12.1 Find some points at which the function of Example 12.01 is not one-to-one. (Hint: There is a countably infinite set for which the function is four-to-one.)

12.2 Is there any point, $q \in Q$, such that $f^{-1}(q)$ has an infinite number of points where f is as in Example 12.01?

12.3 Show that the function f defined in Example 12.01 is well-defined. (Hint: Possible problems occur when intervals overlap.)

12.4 Prove Proposition 12.02.

12.5 Prove Proposition 12.03.

12.6 Suppose $f: I \to Q$ is a continuous map of the unit interval onto the square. Let D be the set of points of Q which are the image of more than one point. Show D is a dense subset of Q.

12.7 Prove that $\mathbf{R}^2$ is not the union of images of any countable collection of piecewise-linear maps of $\mathbf{R}^1$ to $\mathbf{R}^2$. (Hint: Use proof in the remark on page 273 as guide.)

13. MANIFOLDS

OVERVIEW: A manifold is a subset $X \subseteq \mathbf{R^n}$ locally homeomorphic to $\mathbf{R^k}$, for some k. Manifolds form a collection of familiar of geometric objects such as curves, surfaces, three-dimensional regions, encountered in the study of calculus, and constitute an important class of objects to study in many branches of mathematics. We also discuss, briefly, the notion of a smooth manifold.

13.1 Some basic properties of manifolds

Manifolds provide a simple, but rich, collection of objects, which relate to important topics in a number of areas of mathematics.

We begin with the definition of a manifold. Recall that R^k_+ consists of points of $\mathbf{R^k}$ whose first coordinate is greater than or equal to zero. Then $Bd(R^k_+, \mathbf{R^k}) = \{(x_1, \ldots, x_k) \in \mathbf{R^k} \mid x_1 = 0\}$. Certainly, $Bd(R^k_+, \mathbf{R^k})$ is homeomorphic to $\mathbf{R^{k-1}}$, but recall that the standard inclusion $\mathbf{R^{k-1}} \subseteq \mathbf{R^k}$ is as the subset with the k-th coordinate zero, rather than the first coordinate.

Definition 13.01 *Let $X \subseteq \mathbf{R^n}$; X is a* **manifold with boundary, of dimension** *k, if for every point $x \in X$, there is an open subset $U \subseteq X$ such that either*

(a) *U is homeomorphic to $\mathbf{R^k}$, or*
(b) *U is homeomorphic to R^k_+, and x corresponds to a point of $Bd(R^k_+, \mathbf{R^k})$.*

If X is an empty set, then, in a vacuous way, it is an n-dimensional manifold for any n. As peculiar at this may seem, it is traditional to allow this usage.

Propositions 13.06 and 13.07 are important basic results. Complete proofs are beyond the scope of this text; a proof is found in most texts on algebraic topology such as [15, 33, 44]. We sketch the proofs based on Proposition 13.02, which we assume to be true without proof.

Proposition 13.02 *Suppose $X \subseteq \mathbf{R^n}$ with X an open subset, $x \in X$ and $f: X \to \mathbf{R^n}$ an embedding. Then $f(x)$ is an interior point of $f(X)$. More precisely, $f(x) \in int(f(X), \mathbf{R^n})$.* ∎

Since $f(X)$ will be an open subset of $\mathbf{R^n}$ if every point is an interior point, we conclude that:

Proposition 13.03 *Suppose $X \subseteq \mathbf{R^n}$ with X a non-empty open subset. Suppose $Y \subseteq \mathbf{R^n}$, and $f: X \to Y$ is a homeomorphism. Then Y is an open subset of $\mathbf{R^n}$.* ∎

Proposition 13.04 *Suppose U is a non-empty open subset of $\mathbf{R^m}$. There is no embedding of U into $\mathbf{R^n}$ if $m > n$.*

Proof: Suppose there were an embedding $f: U \to \mathbf{R^n}$. Let $i: \mathbf{R^n} \to \mathbf{R^m}$ be the standard inclusion. As in Problem 2.37, $i(U)$ is not an open subset of $\mathbf{R^m}$. But by Proposition $i(U)$ must be an open subset, giving us a contradiction. ∎

A homeomorphism is an embedding. So, Proposition 13.03, known as "invariance of domain", implies answers to some of our most basic questions:

Proposition 13.05 *$\mathbf{R^n}$ and $\mathbf{R^m}$ are not homeomorphic if $n \neq m$.* ∎

With use of Proposition 13.02, we have shown the invariance of domain in the case that m and n are less than or equal to two, Proposition 6.5 or 8.03. In Proposition 15.31, we extend this to include dimension 3. Proposition 6.5 implies that a 1-dimensional manifold cannot be homeomorphic to a two-dimensional manifold (Problem 13.1).

The dimension of a manifold is a topological invariant:

Proposition 13.06 *If $X \subseteq \mathbf{R^k}$ is a n-dimensional manifold and $Y \subseteq \mathbf{R^{k'}}$ is an m-dimensional manifold with $n \neq m$, then X and Y are not homeomorphic.*

Proof: We can assume without loss of generality that $m > n$. Suppose there is a homeomorphism $f : X \to Y$. Choose an $x \in X$; let $y = f(x)$. There is an open subset $V \subseteq Y$ with $y \in Y$, and a homeomorphism $g : V \to \mathbf{R^n}$. Also, there is an open subset $U \subseteq X$ with $x \in X$, and a homeomorphism $h : U \to \mathbf{R^m}$. Now, $U' = U \cap f^{-1}(V)$ will be an open subset of X, non-empty since $x \in U \cap f^{-1}(V)$. Let

$Z = h(U \cap f^{-1}(V))$; this is an open subset of $\mathbf{R^m}$. Now we have an embedding, $f' = g \circ f \circ h|_{U'}^{-1}$ of an open subset of $\mathbf{R^m}$ into $\mathbf{R^n}$, contradicting Proposition 13.04; see the diagram below.

$$\begin{array}{ccc} X \supseteq U \supseteq U' & \xrightarrow{f} & V \subseteq Y \\ \big\downarrow h|_{U'} & & \big\downarrow g \\ \mathbf{R^m} \supseteq Z & \xrightarrow{f'} & \mathbf{R^n} \end{array} \blacksquare$$

Proposition 13.07 *Conditions (a) and (b) in the Definition 13.01 are mutually exclusive.*

Proof: Suppose a point $X \subseteq \mathbf{R^n}$ is a k-manifold and that $x \in X$ satisfies both (a) and (b) of Definition 13.01. Then we have two open subsets, U and V, containing x, a homeomorphism $f: U \to \mathbf{R^k}$, and a homeomorphism $g: V \to R^k_+$ with $g(x) \subseteq Bd(R^k_+)$. Let $Z = f(U \cap V)$. Since $U \cap V$ is an open subset of X and f is a homeomorphism, Z is an open subset of $\mathbf{R^k}$. Consider $g \circ f^{-1}: Z \to \mathbf{R^k}$; this map is an embedding of Z, and so $Z' = g \circ f^{-1}(Z)$ is an open subset of $\mathbf{R^k}$. Also, $g(x) \in Z'$. For some $\epsilon > 0$, $N_\epsilon(g(x)) \subseteq Z'$. By condition (b) we may write $g(x) = (0, x_2, \ldots, x_k)$. Then we will have $(-\epsilon/1, x_2, \ldots, x_k) \in Z' \subseteq g(V)$, contradicting the assumption that $g(V) = R^k_+$. ∎

We can use the result above to make the following two definitions.

Definition 13.08 *If $X \subseteq \mathbf{R^k}$ is a manifold with boundary, then the* **interior** *of X, $Int(X)$, is defined by*

$$Int(X) = \{x \in X\text{: there is an open subset } U \subseteq X \text{ with } U \text{ homeomorphic to } \mathbf{R^k}\}.$$

A point of $Int(X)$ is called an **interior point** *of the manifold.*

Definition 13.09 *If $X \subseteq \mathbf{R^k}$ is a manifold with boundary, then the* **boundary of** *X is defined:*

$$\partial X = \{x \in X\text{: there is an open subset } U \subseteq X, \text{ with } U \text{ homeomorphic to } R^k_+ \text{ by a homeomorphism that sends } x \text{ to a point of } Bd(R^k_+, \mathbf{R^k})\}.$$

A point of ∂X is called a **boundary point** *of the manifold.*

Do not confuse boundary of a *set*, Definition 9.14, and boundary of a *manifold*, Definition 13.09. Similarly, we have interior of a *set*, Definition 9.12, and interior of a *manifold*, Definition 13.08. We use a capital I in "Int" when referring to interior of a manifold and the lowercase i in int for interior of a subset. This is a potential source of confusion but unavoidable since these are all standard definitions in wide-spread use. Some topologists use the term "frontier" of a subset to denote the boundary of the subset.

Example 13.10 Consider the standard unit ball in the plane: $D^2 = \{(x,y) \in \mathbf{R}^2 \mid x^2 + y^2 \leq 1\}$. The boundary of the *set* D^2 and the boundary of the *manifold* are the same, namely, the unit circle in the plane. Also, the interior of the *set* D^2 and the interior of the *manifold* D^2 are the same, namely, $\{(x,y) \in \mathbf{R}^2 : x^2 + y^2 < 1\}$. So, $Bd(D) = \partial D$ and $int(D) = Int(D)$ This example explains why mathematicians would use the same word for two distinct concepts.

On the other hand, there are times when the two uses of boundary and interior do not agree. The standard inclusion of $\mathbf{R}^3$ into $\mathbf{R}^3$ sends D^2 to $D = \{(x,y,z) \in \mathbf{R}^3 : x^2+y^2 \leq 1 \text{ and } z = 0\}$. The boundary of the manifold D is the circle $\{(x,y,z) \in \mathbf{R}^3 \mid x^2+y^2 = 1 \text{ and } z = 0\}$ and the interior of the manifold D is $\{(x,y,z) \in \mathbf{R}^3 \mid x^2 + y^2 < 1 \text{ and } z = 0\}$. The interior of the set D, $int(D, \mathbf{R}^3)$, is the empty set, and the boundary of the set D $Bd(D, \mathbf{R}^3)$, is all of D, so in this case $Bd(D) \neq \partial D$ and $int(D) \neq Int(D)$. ◆

Example 13.11 Any interval in $\mathbf{R}^1$ (open, closed, half-open) is a 1-dimensional manifold. ◆

Example 13.12 For each k, S^k is a k-dimensional manifold. Also, $\mathbf{R}^k$ is a k-dimensional manifold, as well as any open subset of $\mathbf{R}^k$ All these examples have an empty boundary. ◆

Example 13.13 For each k, D^k is a k-dimensional manifold and $\partial D^k = S^{k-1}$ (Problem 13.2). ◆

Example 13.14 Use polar coordinates, and let $A = \{(\rho, \theta) \in \mathbf{R}^2 \mid 1 \leq \rho \leq 2\}$. This 2-dimensional manifold is an annulus; see Definition 5.21. Here ∂A consists of two disjoint circles: $\rho = 1$ and $\rho = 2$.

Using spherical coordinates, we could let $M = \{(\rho, \theta, \phi\} \in \mathbf{R}^3 \mid 1 \leq \rho \leq 2$. This is a 3-dimensional manifold, and ∂M is the disjoint union of two concentric spheres: $\rho = 1$ and $\rho = 2$. ◆

Using Proposition 13.07 one can show that:

Proposition 13.15 *If X is an n-dimensional manifold, then $int(X)$ is an n-dimensional manifold, and ∂X is an $(n-1)$-dimensional manifold (possibly empty).* ▮ *(Problem 13.3)*

Consider that ∂D^2 is a circle, and the circle has empty boundary. Similarly, ∂D^3 is a sphere, and the boundary of this sphere is the empty set. In general, we have the following.

Proposition 13.16 *If X is a manifold, then $\partial(\partial X) = \emptyset$.* ▮

Since a k-manifold is locally homeomorphic to $\mathbf{R}^k$ or R^k_+, and since both of these are locally connected:

Proposition 13.17 *If $X \subseteq \mathbf{R^k}$ is an n-manifold, it is locally connected (Problem 13.5).* ∎

Proposition 13.17 implies that the topologist's $\sin(1/x)$ curve is not a manifold.

Here is a basic question about manifolds:

Question 13.18 For a given natural number, n, how many n-dimensional manifolds are there? By this we mean—how many homeomorphism classes of such manifolds are there? Are there uncountably many homeomorphism classes of manifolds ?

The 1-dimensional manifolds are familiar subsets. An example of a compact, disconnected, 1-dimensional manifold is a union of two disjoint circles. By considering unions of more and more circles, one sees that there are infinitely many 1-dimensional manifolds, no two of which are homeomorphic. An example of a connected, 1-dimensional manifold which is not compact is $\mathbf{R^1}$. An example of a compact, connected, 1-dimensional manifold with non-empty boundary is I.

The next proposition assures us that closed, connected, 1-dimensional manifolds are circles. The proof, although not difficult, is too long to present here.

Proposition 13.19 *If X is a connected, compact, 1-dimensional manifold with empty boundary, then X is homeomorphic to a circle.* ∎

One of the goals of topology is to capture the nature of certain subsets terms of topological properties. Proposition 13.19 is a statement of this sort. We cannot extend this result to the two-dimensional sphere easily since there are compact, connected, 2-dimensional manifolds which are not homeomorphic to S^2. One example is the "torus," discussed in the next section; see Definition 13.24.

Regarding Question 13.18, for the case $n = 1$, there is an infinite collection of distinct 1-dimensional manifolds—one could just consider the collection, $\{M_j\}$ where M_j consists of j disjoint circles. But this is a countable collection. Is there an uncountable number of distinct 1-dimensional manifolds? We will sketch the argument that there are not.

The 1-dimensional manifolds are not difficult to understand. One can improve Proposition 13.19. A connected 1-dimensional manifold is homeomorphic to the circle or an interval. A component of an n-dimensional manifold is an n-dimensional manifold. So any 1-dimensional manifold is a union of components, each of which is homeomorphic to a circle or an interval.

So the question of whether or not there are uncountably many distinct 1-dimensional manifolds hinges on whether a 1-dimensional manifold can have an uncountable number of components. If $M \subseteq \mathbf{R^n}$ is a

1-dimensional manifold $x \in M$, there is an open subset $U_x \subseteq \mathbf{R^n}$ such that $U_x \cap M$ is connected. In particular, U_x is a subset of only one component of M. Although the collection $\{U_x\}_{x \in M}$ is uncountable, it is an open cover of M and so has a countable subcover. (This follows from the fact that if U is an open subset of $\mathbf{R^n}$ we can find $N_\epsilon(p) \subseteq U$ such that ϵ is rational, and all coordinates of p are rational). The union of elements of this subcover contains M and must be contained in the union of a countable number of components of M. We now have a contradiction of the assumption that M has an uncountable number of components.

13.2 The torus

In discussing the torus, and often other manifolds, it is useful to define a special kind of continuous function, an "immersion." Although it makes sense to consider immersions for arbitrary subsets of $\mathbf{R^n}$, immersions are most often used to discuss continuous functions from one manifold to another. An immersion is a function which is locally an embedding:

Definition 13.20 *Suppose $X \subseteq \mathbf{R^n}$, $Y \subseteq \mathbf{R^m}$ and $f: X \to Y$ is a continuous function. We say that f is an* **immersion** *of X into Y if, for every $x \in X$ there is an open subset of X with $x \in X$ such that $f|_U$ is an embedding of U into Y.*

Example 13.21 Let $X = \mathbf{R^1}$, and $Y = S^1$. The map $f: X \to Y$ defined by $f(x) = (\cos(x), \sin(x))$ is an immersion. For $x \in \mathbf{R^1}$, if we consider $J = (x - 2\pi, x + 2\pi)$, $f|_J$ is an embedding of J into Y. ◆

Example 13.22 Many smooth parametric curves in the plane are images of immersions. Some typical examples are seen in Figure 13-1. (On the left in this figure we see the parametric plot of $f(x) = (\sin 3x \cos x, \cos 5x \sin x)$; on the right, is a parametric plot for $g(x) = (\sin \sqrt{2}x \cos x, \cos \pi x \sin x)$ for $0 \le x \le 30$.) Since these curves have self-crossings, f and g are not embeddings. However, they are immersions. This follows from a theorem of calculus known as the "implicit function theorem"; see Propositions D.23 and D.24. ◆

Example 13.23 The torus in $\mathbf{R^3}$ is an important 2-dimensional manifold; see Figure 13-2. Informally, we an speak of the torus as the "surface of a donut" $D_{(a,b)}$ where the thickness of the donut is 2b and size of the hole is $2(a - b)$. Of course, we need $b < a$. We may describe a torus as a surface obtained by spinning a circle about the z-axis. This type

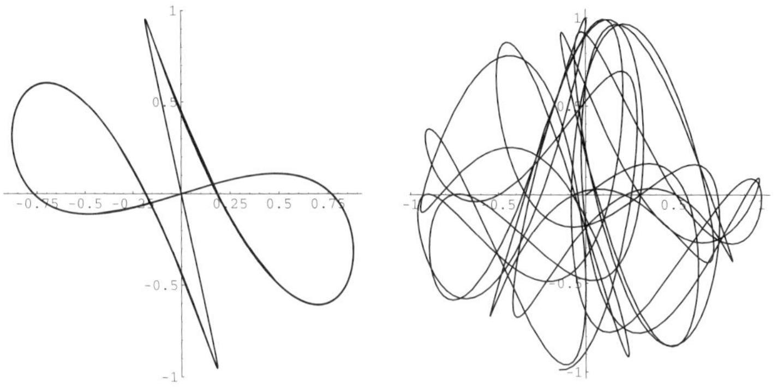

Figure 13-1 Two examples showing images of immersions of $\mathbf{R}^1$ into $\mathbf{R}^2$. The curve on the left is traced infinitely many times, the curve on the right has no such periodicity. Because these curves have self-intersections, they are not images of an embedding of $\mathbf{R}^1$.

of construction will be used for other examples such as the Möbius band,Example 13.28, the Klein bottle, Example 18.06, the projective plane Example 18.07 and twist-spun knots, Example 18.4. Let C be the circle in the xy-plane, with center at the origin and radius a. The circle C will correspond to the "core circle or the donut." This circle is used for the construction but is not a subset of the torus. Using cylindrical coordinates, we describe C as all points (r, θ, z) with $r = a$, $z = 0$.

For any angle α, consider the page H^2_α (recall Definition 2.10). In cylindrical coordinates, let $m(\alpha)$ be the circle in H^2_α with center $(a, \alpha, 0)$ and radius b, and let $D(\alpha)$ be the closed disk in H^2_α with the same center and radius. Let $T_{(a,b)} = \bigcup_{0\leq\alpha<2\pi} m(\alpha)$, and $D_{(a,b)} = \bigcup_{0\leq\alpha<2\pi} D(\alpha)$.

You may think of the letter D of $D_{(a,b)}$, here as referring to donut. The circles $m(\alpha)$, will later be referred to as meridians.

The torus is the image of a continuous map that will play an important role in later chapters. Let a and b be numbers with $0 < b < a$. Define a function, $P: \mathbf{R}^2 \to \mathbf{R}^3$ (using Cartesian coordinates, this time) by

$$P(x, y) = \Big((a + b\cos(x))\cos(y), (a + b\cos(x))\sin(y), b\sin(x)\Big).$$

The map P is an immersion, and the image of this map is $T_{(a,b)}$.

No particular choice of a and b is mathematically special, but for the sake of being specific we choose to focus on $T_{(2,1)}$. ◆

Definition 13.24 *The* **standard torus** T^2 *in* $\mathbf{R}^3$ *is defined to be* $T_{(2,1)}$. *The* **standard covering map** $P: \mathbf{R}^2 \to T^2$ *is the map described above, with* $a = 2$ *and* $b = 1$.

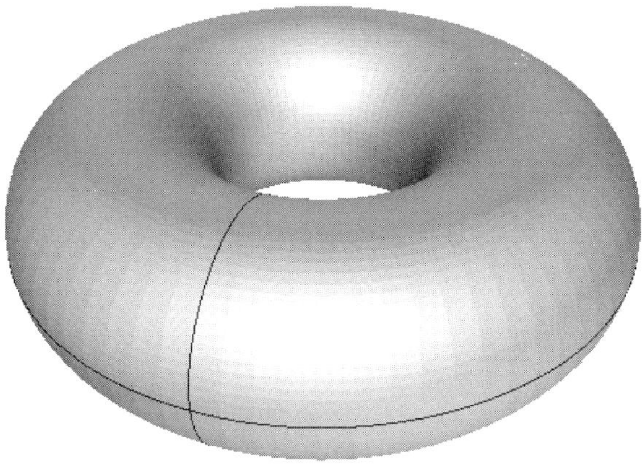

Figure 13-2 The standard torus in $\mathbf{R}^3$; see Definition 13.24. Also shown are a meridian and longitude circles; see Example 13.26. The meridian goes through the hole and the longitude goes around the hole.

The subset $D_{(2,1)}$ is called the **standard solid torus** *in* $\mathbf{R}^3$. *The circle $C \subseteq D_{(2,1)}$ is called the* **core** *of the standard solid torus.*

The standard solid torus $D_{(2,1)}$ in $\mathbf{R}^3$ is a 3-dimensional manifold, and $\partial D_{(2,1)} = T^2$. Later, we will examine the torus as a subset of $\mathbf{R}^4$, Definition 18.01.

Example 13.25 We examine the map P in detail. Certainly, P is not a one-to-one function. In fact, if p and q are integers, then $P(p(2\pi), q(2\pi)) = (3, 0, 0)$. More generally, for any real numbers x and y and integers p and q, we have $P(x+p(2\pi), y+q(2\pi)) = P(x, y)$. Thus every point of $T_{(a,b)}$ is the image of infinitely many points; see Figure 13-3. Note however that P *is* one-to-one on the subset $[0, 2\pi) \times [0, 2\pi)$. ◆

Example 13.26 We can use the map P to describe interesting subsets of the torus. Suppose that $l_{(A,B,C)}$ is the line in the plane with equation $Ax + By = C$ (with A and B not both zero). Then we can define $L_{(A,B,C)} = P(l_{(A,B,C)})$. The sets $L_{(1,0,0)}$ and $L_{(0,1,0)}$ are circles on the torus. Generally, any of the circles $L_{(1,0,C)}$ are called a "longitude" of the torus, a circle $L_{(0,1,C)}$ is call a "meridian" of the torus; see Figure 13-2

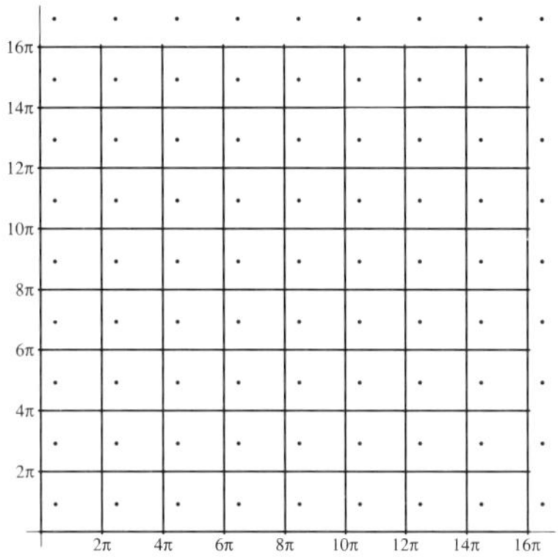

Figure 13-3 The inverse image via P of a point on the torus; see Example 13.25.

In our figures, we can verify visually the type of curve using a grid for the plane. The image under P of such a rectilinear grid in the plane is a curvilinear grid on the torus. For example, the line and grid shown in Figure 13-4 transfer to the circle and circular grid, respectively, as shown in the bottom figure of Figure 13-5.

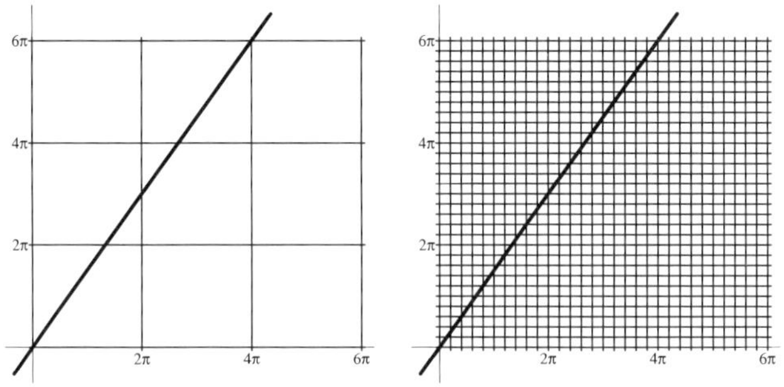

Figure 13-4 The line $l_{(3,-2,0)}$; see Example 13.26. On the left, a coarse grid where every square is mapped onto T^2. The finer grid on the right is used for Figure 13-5.

For any two *rational* numbers, r and q, $L_{(r,q,0)}$ is homeomorphic to a circle (Problem 13.12). Such a set can be described as the image, under P, of a line in the plane through the origin with rational slope. In Figure 13-5 we see the curve $L_{(3,-2,0)}$. Here $l_{(3,-2,0)}$ is a line with slope $3/2$ and thus traverses three squares in the y-direction for every two squares in the x-direction; see Figure 13-4. On the torus we can verify visually

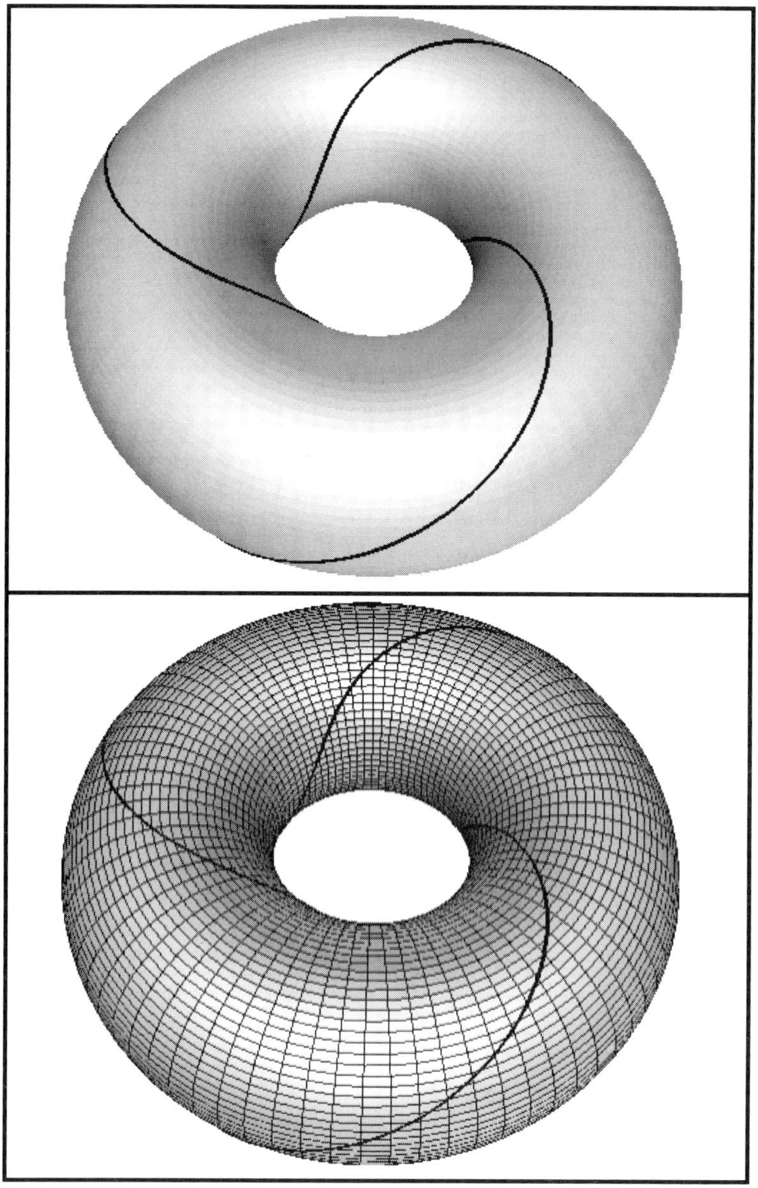

Figure 13-5 The circle $L_{(3,-2,0)}$; see Example 13.26. The lower image with a curvilinear grid shows the slope graphically.

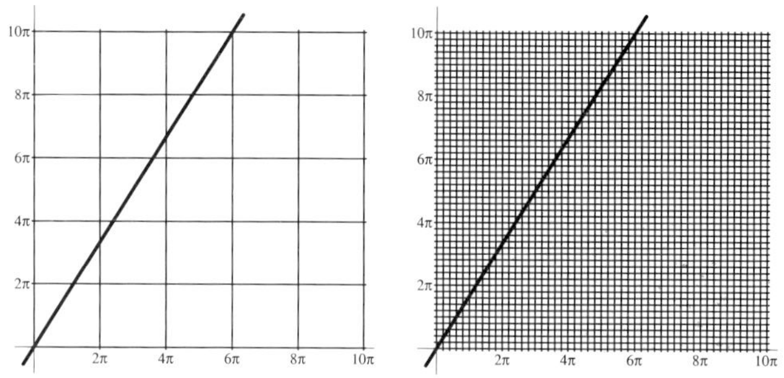

Figure 13-6 The line $l_{(5,-3,0)}$; see Example 13.26. On the left, a coarse grid where every square is mapped onto T^2. The finer grid on the right is used for Figure 13-7.

the type of curve by noting that it traverses the curvilinear grid three units in a meridian direction for every two in the longitude direction, Figure 13-5. Similarly, one has the curve $L_{(5,-3,0)}$; see Figures 13-6 and 13-7. ◆

Example 13.27 One gets interesting subsets of the torus by taking lines with irrational slopes. Consider a line $l = mx$, with irrational slope; let $L = P(l)$. Claim: The curve L is dense in T^2; see Definition 1.35 for definition of "dense."

As an example, we focus at one point, $Y = (3, 0, 0)$ of T^2, and leave as an exercise proof for an arbitrary point of T^2 (Problem 13.13). Let U be an open subset of T^2 containing Y; we show that there is a point of L in U.

Using P^{-1}, we transfer this problem to a problem in the plane. Let $\tilde{Y} = P^{-1}(Y)$; $\tilde{Y} = \{(p2\pi, q2\pi\}_{p,q \in \mathbf{Z}}$. In Figure 13-8, we show the line l; $\tilde{Y}$ corresponds to the grid points. We need to show that we can find points of l arbitrarily close to a grid point.

For any integer N, the line l intersects the vertical line $x = N2\pi$ at $(N2\pi, mN2\pi)$. Since m is irrational, $mN2\pi$ is an irrational multiple of 2π. So we need to show that, given $\epsilon > 0$, we can find an N so that, for some Q, the distance from $P(N2\pi, mN2\pi)$ and $P(N2\pi, Q2\pi)$ is less than ϵ. This follows from the proof of Example 1.43. There we showed that we can find multiples of angle $mN2\pi$ arbitrarily close to a multiple of 2π.

Not only is L a dense subset of T^2; it has some other interesting properties. Further analysis shows (Problem 13.14), that L fails to be locally connected at every point! In particular, this shows that L is not a manifold, even though it is the image of a manifold and $P|_l$ is an immersion. ◆

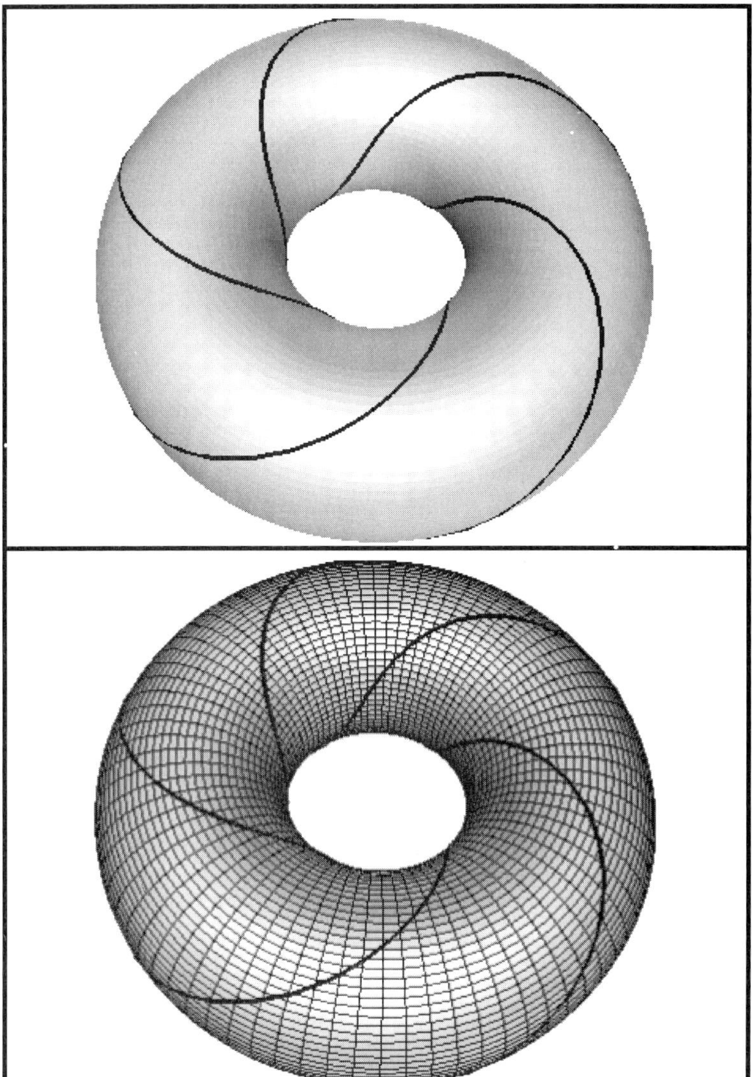

Figure 13-7 The circle $L_{(5,-3,0)}$; see Example 13.26. The lower image with a curvilinear grid shows the slope graphically.

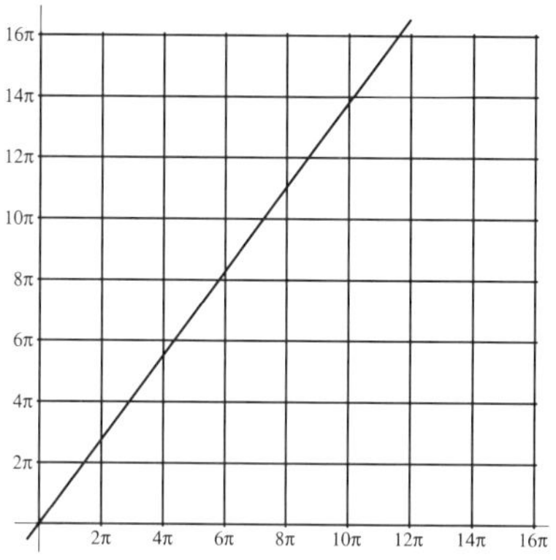

Figure 13-8 A line with an irrational slope; see Example 13.26.

13.3 Some other manifolds

Example 13.28 We describe an interesting subset of $\mathbf{R}^3$, the Möbius band. A rough description of this is that the Möbius band is a band with a half-twist. The Möbius band in $\mathbf{R}^3$ will be described as a union of line segments whose endpoints lie on a circle.

Consider the circle $L(2,-1,0)$ in $\mathbf{R}^3$, defined in Example 13.26. Here we are simply thinking of this circle as a subset of $\mathbf{R}^3$, rather than a subset of the standard torus. Clearly, each page H_α of $\mathbf{R}^3$ intersects $L(2,-1,0)$ at exactly two points. For any $0 \le \alpha \le 2\pi$, let I_α be the line segment in H_α with those two points as endpoints. (We note that $I_0 = I_{2\pi}$.). Define: $M = \bigcup_{0\le\alpha\le 2\pi} I_\alpha$; see Figure 13-9.

We will call M the "standard Möbius band in $\mathbf{R}^3$." Generally, any subset of some $\mathbf{R}^n$ homeomorphic to the standard Möbius band will be called a "Möbius band." It is clear from the Figure 13-9 that M is a 2-dimensional manifold, and $\partial M = L(2,-1,0)$.

We could similarly consider the circle $L(2,1,0)$. This also meets each H_α at two points. For any $0 \le \alpha \le 2\pi$, let J_α be the line segment in H_α between these points. We let M' be the union of these line segments: $M' = \bigcup_{0\le\alpha\le 2\pi} J_\alpha$; see Figure 13-10.

We can describe M' also as a band with a half-twist, but this half-twist is in the opposite direction, as if seen in a mirror. Clearly, M and M' are homeomorphic; in fact, by a homeomorphism that takes I_α to J_α. Consider $m\colon \mathbf{R}^3 \to \mathbf{R}^3$ defined by $m(x,y,z) = (x, y-z)$. We can see that m is a homeomorphism; in fact, a rigid motion, of $\mathbf{R}^3$ which takes M to M'. Sometimes m is described as a mirror reflection in the

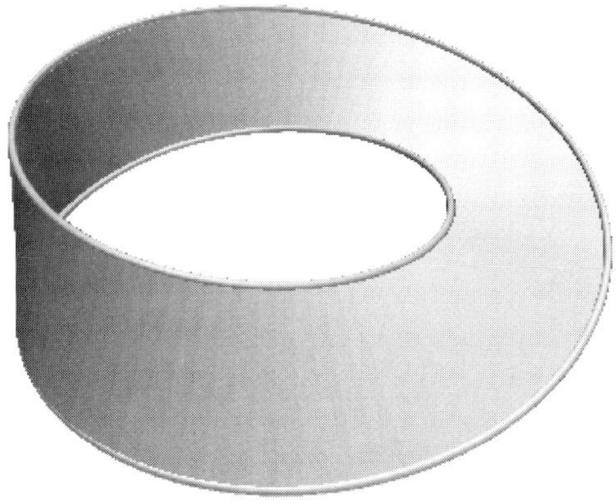

Figure 13-9 The standard Möbius band in $\mathbf{R}^3$; see Example 13.28.

xy-plane. We call M' the "mirror image" of M.

In Figure 13-11 we show Möbius bands constructed by using the circles $L(3,-1,0)$ and $L(3,1,0)$. The fact that the Möbius bands of Figure 13-10 and Figure 13-11 look so different from the standard Möbius band, or its mirror image, is a reflection of the fact that these are not equivalent subsets of $\mathbf{R}^3$. For example, a strategy to show that the Möbius bands of Figure 13-11 are not equivalent subsets is to show that knotted circles which correspond to the boundaries are not equivalent knots. It is not easy to prove this, and we offer no proof here. In Chapter 14 we discuss the idea of equivalence of knotted circles. ◆

Example 13.29 Another description of the torus uses the idea of distance from a point to a set, Definition 1.4. Let C be the circle, as in Example 13.23—centered at the origin, in the xy-coordinate plane, of radius 2. Then T^2 consists of those points in $\mathbf{R}^3$ whose distance to C is 1; $D_{(2,1)}$ consists of those points whose distance to C is less than or equal to 1. More formally, consider the map $f:\mathbf{R}^3 \to R^1_+$ defined by $f(\vec{x}) = \delta(\vec{x}, C)$. Then $f^{-1}\big((-\infty,1]\big) = D_{(2,1)}$, and $f^{-1}(1) = T^2$.

Let $U = f^{-1}\big((1,\infty)\big)$, $V = f^{-1}\big([0,1)\big)$. It is not hard to show that both U and V are connected. In fact, they are path connected (Problem 13.15). It follows then that $\mathbf{R}^3 - T^2$ has two components: V, which is bounded, and U, which is unbounded.

Also, $f^{-1}\big([1,\infty)\big) = \mathbf{R}^3 - Int(D_{(2,1)}) = U \cup T^2$ is an interesting noncompact 3-dimensional manifold whose boundary is the torus T^2.

Let W be a ball of radius 4 with the interior of the standard solid torus removed: $W = \overline{N}_4(\vec{0}, \mathbf{R}^3) - V$. Then W is a compact 3-dimensional manifold whose boundary has two components—one of which is home-

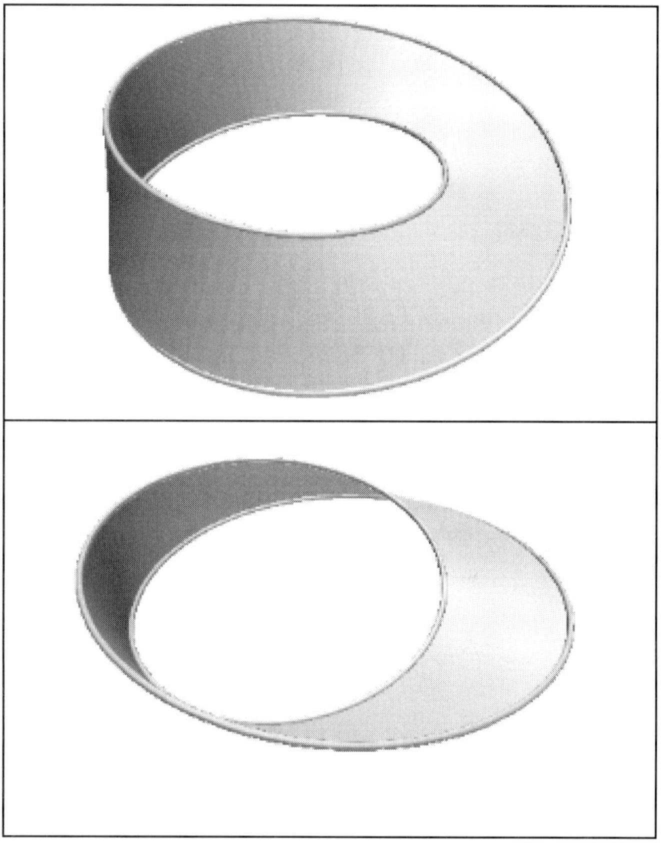

Figure 13-10 At the top is the standard Möbius band M; below is a mirror image, M'; see Example 13.28.

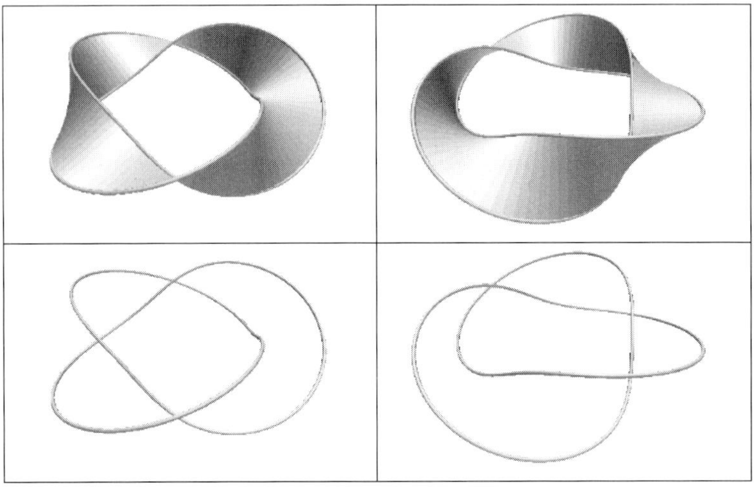

Figure 13-11 Two other embeddings of the Möbius band into $\mathbf{R}^3$. These are mirror images of each other, but are not equivalent subsets to those shown in Figure 13-10; see Example 13.28.

omorphic to a sphere and the other to a torus. ◆

Example 13.30 We say that the torus has a "hole" (thinking of the hole in the donut). Some other examples of manifolds of dimensions 2 and 3 are variations of where we allow several holes.

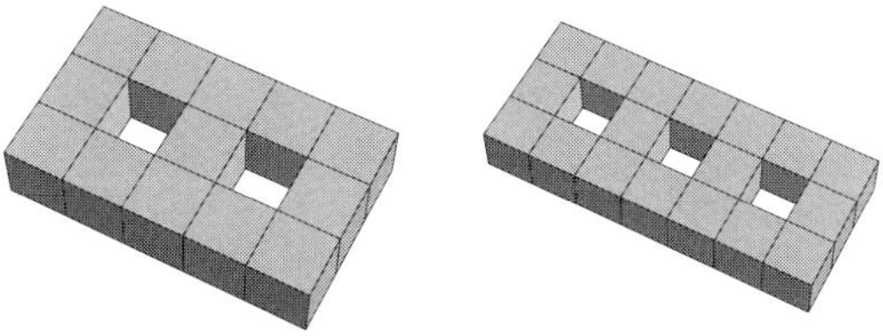

Figure 13-12 On the left, a solid torus with two holes; on the right, a solid torus with three holes; see Example 13.30.

Picture "bricks," as in Figure 13-12, with two and three holes. Considered topologically, these are called the "solid torus with two holes" and the "solid torus with three holes," respectively. The surfaces of these bricks are called the "torus with two holes" and the "torus with three holes," respectively. A more traditional, smoother version of a torus with three holes is shown in Figure 13-13.

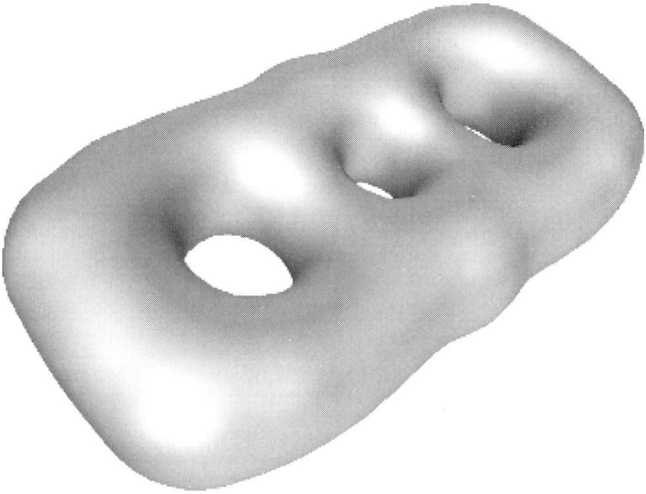

Figure 13-13 A smoother version of a torus with three holes; see Example 13.30.

More generally, we can have an "torus with n holes" and the "solid torus with n holes." The solid torus with n holes is a 3-dimensional manifold whose boundary is the torus with n holes. In contrast to the torus, it is not easy (nor profitable for us) to describe these examples with equations. ◆

Question 13.31 What is a mathematical definition of a "hole," as used above? Is there a topological property corresponding to the intuitive notion of the number of holes of X so that the torus has one hole, etc?

The answer to the questions above is the entry point of the subject of algebraic topology, known as "homology theory" and is beyond the scope of this book. However, we can give the flavor of the approach. Consider a circle K which is a subset of the surface being examined. We say that the circle is "trivial" if it is the boundary of an embedded disk in the surface. The idea is that a hole can be detected by a non-trivial circle that goes around the hole. A pair of circles are called "algebraically trivial" if they are disjoint and their union is the boundary of an embedded 2-manifold in the surface. A collection of circles in the surface is called "algebraically trivial" if the circles are disjoint and their union is the boundary of an embedded 2-manifold in the surface. For example, a pair of circles that is the boundary of an embedded annulus in the surface is an algebraically trivial pair.

Any two disjoint meridians of a torus are algebraically equivalent; any two disjoint longitudes of the torus are algebraically equivalent. However, a longitude and a meridian are not algebraically equivalent since they cannot be disjoint.

A torus with two holes has a pair of disjoint non-trivial curves which is not algebraically trivial, but does not have a set of three disjoint non-trivial which is algebraically trivial, with no two being algebraically trivial. On the other hand, a torus with three holes has three disjoint curves so that no sub-collection is algebraically trivial (find such a set), but does not have four disjoint non-trivial curves, so that no sub-collection is algebraically trivial.

For further discussion of manifolds, we introduce a definition:

Definition 13.32 *If $M \subseteq \mathbf{R^n}$ is a manifold which is compact with empty boundary, we say the manifold M is* **closed**.

Once again, we see that words are used for manifolds which do not have the same meaning as for subsets. If M is a closed manifold and $M \subseteq X \subseteq \mathbf{R^n}$, then M is also a closed subset of X since all compact subsets are closed subsets, Proposition 10.12. So, "closed" as a manifold implies "closed" as a subset. However a closed *subset* which is a manifold need not be a closed *manifold*. The closed n-ball D^n, $0 < n$, is a compact subset of $\mathbf{R^n}$, which is not a closed manifold since $\partial D^n = S^{n-1} \neq \emptyset$.

There are other similarities between the notion of the boundary of a set and the boundary of a manifold. Compare the result below to similar statements about the boundary of a set (Proposition 9.18(c)); also consult the remark on page 222 and Figure 9-1.

Proposition 13.33 *Suppose that M is an m-manifold, and N is an n-manifold. Then $Int(M \times N) = Int(M) \times Int(N)$.*

Let $A = M \times \partial N$, and $B = \partial M \times N$. Then $M \times N$ is an $(n+m)$-manifold, and $\partial(M \times N)$ is the union of A and B. Furthermore, $A \cap B = \partial M \times \partial N$. ▮ *(Problem 13.4)*

The proof of Proposition 13.33 is not too difficult, if either $\partial M = \emptyset$ or $\partial N = \emptyset$.

We can obtain a bountiful supply of manifolds (generally non-compact) by using the next proposition.

Proposition 13.34 *If X is an n-dimensional manifold, and C is a closed subset of $int(X)$, then $X - C$ is an n-dimensional manifold.* ▮

In particular, any open subset of $\mathbf{R^n}$ is an n-dimensional manifold. Also, if we remove a finite number of points from a torus with n holes, we get a two-dimensional manifold. Suppose K is a subset of $\mathbf{R}^3$ such that K is homeomorphic to S^1. Examples where K is "knotted" give rise to interesting three-dimensional manifolds by considering $\mathbf{R}^3 - K$. This will be discussed in more detail in Chapter 14.

13.4 Further properties of manifolds

Proposition 13.35 *If $M \subseteq \mathbf{R}^n$ is a connected k-dimensional manifold, then M is path connected.*

Proof: Let $p \in M$. Let

$$\begin{aligned} U &= \{q \in M \mid \text{there is a path in } M \text{ from } p \text{ to } q\}, \text{ and} \\ V &= \{q \in M \mid \text{there is no path in } M \text{ from } p \text{ to } q\}. \end{aligned}$$

Our goal is to show that $V = \emptyset$.

Certainly, $U \neq \emptyset$ since $p \in U$. We next show that U is an open subset of M. If $q \in U$, there is a path α in M from p to q. Since M is a manifold, there is an open subset $W \subseteq M$, with $q \in W$ such that W is either homeomorphic to $\mathbf{R}^k$ or R^k_+; call this homeomorphism h. Let $x \in W$. Since $\mathbf{R}^k$ and R^k_+ are path connected, we can find a path β in $h(W)$ from $h(q)$ to $h(x)$. Let $\alpha' = h^{-1} \circ \beta$; then α' is a path in W from q to x. Finally, $\alpha * \alpha'$ is a path in M from p to x where $*$ denotes concatenation of the paths; recall Definition 8.07. Thus U is an open subset of M since, for every point $p \in U$, there is an open subset $W \subseteq U$.

Next, a variation of the argument above shows that V is open. Suppose $q \in V$. We can find a W, open in M, which is homeomorphic to $\mathbf{R}^k$ or R^k_+. No point $x \in W$ of W can be the endpoint of a path α in M beginning at p. This is because, as we have shown, for any point $x \in W$ there is a path α' in W from x to q. The path $\alpha * \alpha'$ would then be a path in M from p to q, contradicting the definition of V. This means that V is an open subset of M.

We conclude that V must be an empty subset; otherwise, U and V would form a disconnecting partition of M, contradicting the assumption that M is connected. ∎

Proposition 13.07 basically says that, topologically, boundary points do not look like interior points. In particular, this means that a manifold with a non-empty boundary cannot be homogeneous. However, the next proposition states that a connected manifold with an empty boundary *is* homogeneous.

One needs the connectedness hypothesis if the manifold has components which are not homeomorphic. We will later prove that a sphere and a torus are not homeomorphic. If X is a subset with two components, S and T, with S homeomorphic to the sphere and T homeomorphic to a torus, then X cannot be homogeneous since a homeomorphism must take a component to a component; Proposition 7.22.

A basic property of manifolds is homogeneity. The proof is difficult but we will outline it at the end of this section; most easily accessible proofs are involve manifolds with additional structure such as in [31, 37].

Proposition 13.36 *If $M \subseteq \mathbf{R^n}$, and M is a connected manifold with $\partial M = \emptyset$, then M is homogeneous.* ◼

As noted in the remark on page 293, a manifold M with a non-empty boundary is not homogeneous, but we can obtain a version of homogeneity, Proposition 13.37, by treating ∂M specially.

Proposition 13.37 *Suppose $M \subseteq \mathbf{R^n}$, and M is a compact, connected manifold with $\partial M \neq \emptyset$. If p and q are points of $Int(M)$, then there is a homeomorphism $h: M \to M$ such that $h(p) = q$ and $h|_{\partial M} = Id|_{\partial M}$.* ◼

We prove a special case of Proposition 13.37, Proposition 13.36, below. We will use this special case for a discussion in Chapter 14, as well as in our outline for the proof, in the case $n = 2$, of Proposition 13.36. Also, this proof will indicate what is involved in the more general proofs.

Proposition 13.38 *Suppose D^2 is the standard unit disk. If p and q are points of $Int(D^n)$, then there is a homeomorphism H of D^2 such that $H(p) = q$ and $H|_{\partial D^n} = Id|_{\partial D^n}$.*

Proof of Proposition 13.38 for the case of D^2: Here is the strategy. First, we investigate a “lower-dimensional” version of our problem: given two points of $Int(D^1)$ is there a homeomorphism fixed on ∂D^1 that takes one point to the other? Since D^1 is a closed interval $[-1, 1]$, and ∂D^1 is the two-point set $\{-1, 1\}$, this is not hard to do. We then apply that result to obtain a map of a square to itself, considering one coordinate at a time. This map is not quite we want. It sends the boundary of the square to itself, but it is not the identity map there. The next step is to improve this by changing the map near the boundary of the square. Finally since the square is homeomorphic to D^2, we “transfer” this to a map of the square to itself to a map of D^2 to itself. Now for some details.

Let $h: D^2 \to I \times I$ be a homeomorphism, and let $P = h(p)$ and $Q = h(q)$. Write coordinates of these points as $P = (x_P, y_P)$ and

$Q = (x_Q, y_Q)$. Define $f: I \to I$ to be the piecewise linear function such that

$$f(0) = 0, f(x_P) = x_Q, \text{ and } f(1) = 1.$$

Also define $g: I \to I$ to be the piecewise linear function such that

$$g(0) = 0, g(y_P) = y_Q, \text{ and } g(1) = 1,$$

(Problem 13.18); see Figure 13-14. (Note: f and g are solutions of a corresponding one-dimensional problem.)

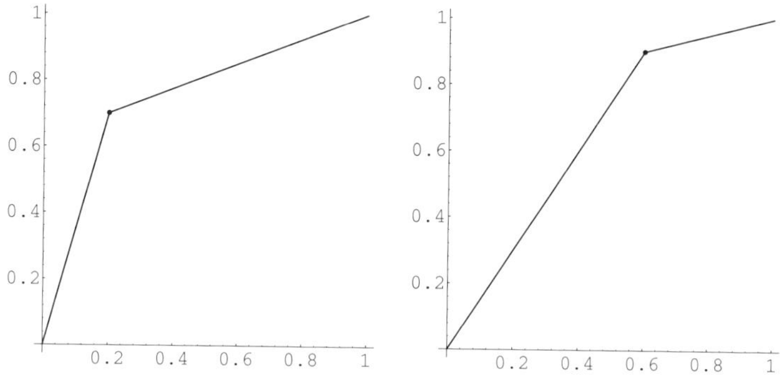

Figure 13-14 Graphs of the functions $f(x)$ and $g(x)$ for the points of Figure 13-15; see proof of Proposition 13.38. The graph of f passes through the points $(0,0), (0.2, 0.7)$, and $(1,1)$, whereas the graph of g passes through the points $(0,0), (0.6, 0.9)$, and $(1,1)$.

Define a map $G: I \times I \to I \times I$ by $G(x, y) = (f(x), g(y))$. Since f and g are homeomorphisms, G is a homeomorphism with $G(P) = Q$. (We will shortly find a need to modify f and g slightly.) We can check that $G(\partial(I \times I)) = \partial(I \times I)$, but, as can be seen in Figure 13-15, $G|_{\partial(I \times I)} \neq Id|_{\partial(I \times I)}$.

We find an ϵ so that the sub-square $X = [\epsilon, 1-\epsilon] \times [\epsilon, 1-\epsilon]$ contains P and Q. Our idea is to use the map G that we have on this sub-square, and use a different map on $I \times I - X$. Figure 13-16 gives an idea of how this is to be done. Conceptually, the map is not difficult to understand, but the details involve a bit of work since we have not developed a set of advanced tools for constructing homeomorphisms.

The first order of business is to make sure that $G(X) = X$. In order to do this, we will alter our definitions for our functions f and g slightly. We want to define these piecewise linear functions so that

$$\begin{aligned} f(x) &= x \text{ if } 0 \le x \le \epsilon \text{ or, if } \epsilon \le x \le 1, \\ f(x_P) &= x_Q, \end{aligned}$$

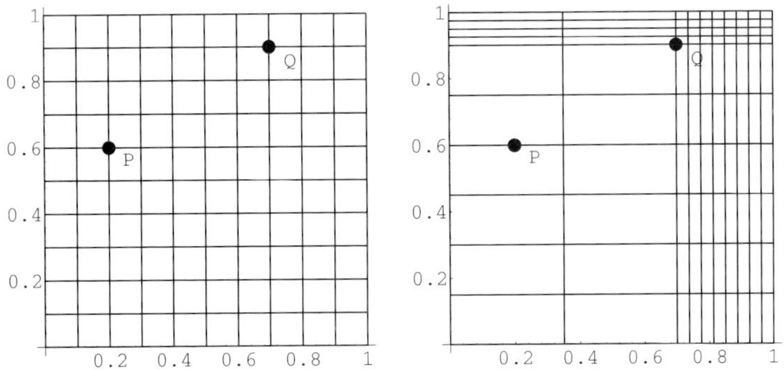

Figure 13-15 Example of a homeomorphism, G, of the square; see proof of Proposition 13.38. At the left is shown a regular grid and two points, $P = (.2, .6)$ and $Q = (.7, .9)$. On the right, we show image of the grid under an isotopy (as defined in the proof) which takes P to Q. We define G, using functions shown in Figure 13-14. Note, in the left figure, that P is the intersection of the third vertical line (beginning on the left) and the seventh horizontal line (beginning at the bottom) and that, in the right figure, Q is the image of the third vertical line and the seventh horizontal line.

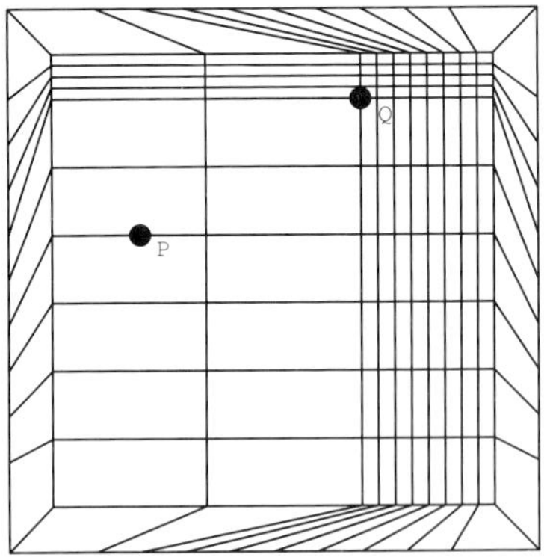

Figure 13-16 The map G, using the modified functions shown in Figure 13-17; see proof of Proposition 13.38. Shown are images of a regular grid under G.

$$\begin{aligned} g(x) &= x \text{ if } 0 \le x \le \epsilon \text{ or, if } \epsilon \le x \le 1, \\ g(y_P) &= y_Q; \end{aligned}$$

see Figure 13-17. We will now use these modified functions to define a map from X to X by $G(x, y) = (f(x), g(y))$.

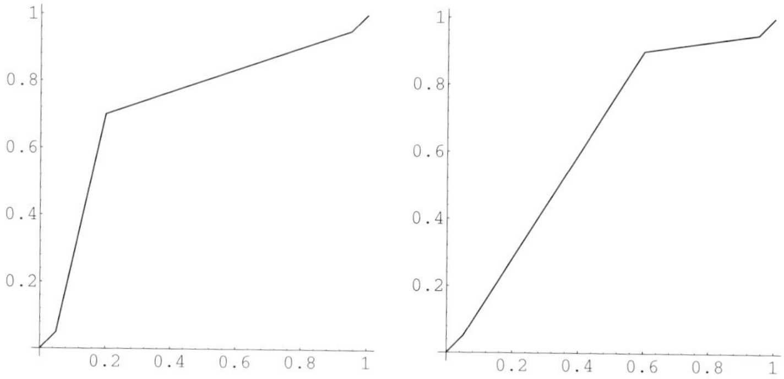

Figure 13-17 New choices for f and g; see proof of Proposition 13.38.

We subdivide $I \times I - X$ into the union of four trapezoids, as showing Figure 13-18. The sub-square is the subset $X = [\epsilon, 1-\epsilon] \times [\epsilon, 1-\epsilon]$. These trapezoids that are above, below, to the left of, and to the right of X, are denoted $T_A, T_B, T_L,$ and T_R respectively.

Our homeomorphism will send X to X and each trapezoid to itself. We define the map on

$$T_B = \{(x, y) \in \mathbf{R}^2 \mid 0 \le y \le \epsilon \text{ and } y \le x \le 1 - y\};$$

the other trapezoids are treated similarly. We define a homeomorphism ψ from $I \times I$ to T_B by

$$\psi(x, y) = (y((1-2\epsilon)x + \epsilon) + (1-y)x, \epsilon y).$$

(To help verify ψ is a homeomorphism, let J_t be the horizontal line segment $J_t = \{(x, y) \in I \times I \mid y = t\}$ where $0 \le t \le 1$. To understand how this has been defined, note that for any $0 \le t \le 1$, $\psi(J_t)$ is the horizontal line segment with endpoints $(\epsilon t, \epsilon t)$ and $(1 - \epsilon t, \epsilon t)$.) Define $t_f : I \times I \to I \times I$ by

$$t_f(x, t) = ((1-t)x + tf(\epsilon + (1-2\epsilon)x), t).$$

Note that $f(x, 0) = (x, 0)$ and $f(x, 1) = (f(x), 1)$. Also note, on the bottom edge of X the map looks like the restriction of f to $[\epsilon, 1-\epsilon]$. The $f(\epsilon + (1-2\epsilon)x)$ is a reparameterization so that at $x = 0$ we get

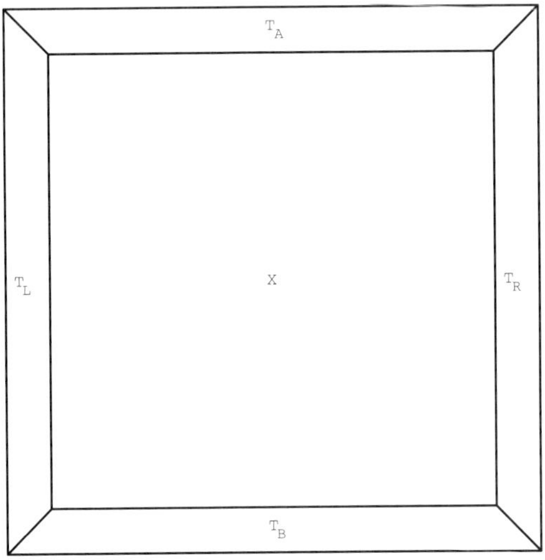

Figure 13-18 The four trapezoids, T_A, T_B, T_L, and T_R, used in the proof of Proposition 13.38.

$f(\epsilon)$, and at $x = 1$ we get $f(1-\epsilon)$. (Note: a definition such as t_f is a standard construction—in the terminology of Chapter 18, Definition 18.10, t_f is the trace of the isotopy between Id_I and f.)

We will define $G|_{T_B} = \psi^{-1} \circ t_f \circ \psi$, and similarly define G on the other three trapezoids. As a final step we transfer this function to a homeomorphism of D^2 via conjugation with h. Let $H = h^{-1} \circ G \circ h$; then H is a homeomorphism of D^2, fixed on ∂D^2, such that $H(p) = q$. ∎

The proof of Proposition 13.36 is difficult and depends on basic results too technical to prove here. However, we sketch a proof. Suppose x and y are points of M.

The idea is to find an open subset, $U \subseteq M$, homeomorphic to $\mathbf{R}^\mathbf{k}$, which contains x and y. This is the "hard part" of the proof. (Briefly the idea of the proof is to first show that there is a simple path—that is non-intersecting path—in M from x to y. Proposition 13.35 gives us a path. In another major step, we improve this and show that, in fact, there is a path from α from x to y such that α is an embedding. We then obtain U as a "thin" open subset containing the image of this path—we choose a small ϵ, and then U is the set of points if M a distance less than ϵ from a point of the path. The final step is to show that, for a small enough choice of ϵ, this subset is homeomorphic to $\mathbf{R}^\mathbf{n}$.)

Once we have such a homeomorphism, h, the remainder of the proof is not difficult. The idea is to use homogeneity of $\mathbf{R}^\mathbf{n}$ to move x to y inside U, while keeping points fixed outside of U.

Using h we obtain another homeomorphism, $h': U \to \mathbf{R^k}$, so that $h'(x)$ and $h'(y)$ are contained in the interior of the standard, open, unit ball $D^k \subseteq \mathbf{R^k}$. For example, if we let $M = \max(|h'(\vec{x})|, |h'(\vec{y})|)$, then we could define $h'(\vec{x}) = (1/(M+1))\vec{x}$. By Proposition 13.38, there is homeomorphism $f: D^k \to D^k$ such that $f(h'(x)) = h'(y)$, which is the identity on ∂D^k. Define $F: M \to M$ by

$$F(p) = \begin{cases} p & \text{if } p \notin U \\ (h')^{-1} \circ f \circ h' & \text{if } p \in U. \end{cases}$$

Continuity of F follows from the gluing lemma. F is a homeomorphism of M with $F(x) = y$. Details for this paragraph is left as an exercise. (Problem 13.17)

13.5 Smooth manifolds

Historically, the study of manifolds arose from analysis where the focus is on subsets of $\mathbf{R^n}$, for which one can consider the notion of a differentiable function. Most definitions of derivative require an open domain, so the definition below is in two parts, first for open subsets of $\mathbf{R^n}$, then for arbitrary subsets. The subsets we are most interested in are manifolds.

Definition 13.39 *Suppose U is an open subset of $\mathbf{R^n}$ and $f: U \to \mathbf{R^m}$ is written as*

$$(x_1, \ldots, x_n) = (f_1(x_1, \ldots, x_n), \ldots, f_m(x_1, \ldots, x_n)).$$

We say f is **smooth** *if each of the functions f_i have derivatives of all orders.*

If $X \subseteq \mathbf{R^n}$ and $f: X \to \mathbf{R^m}$, we say f is **smooth** *if there is an open subset U of $\mathbf{R^n}$, and a smooth map $F: U \to \mathbf{R^m}$ such that $F|_X = f$ and F is smooth.*

A differentiable function is continuous, but there are functions such as $f(x) = |x|$ which are continuous but not differentiable. When applying topology to these problems, the hypothesis that functions have continuous second derivatives is often all that is required. However, it is traditional to assume functions have derivatives of all orders—these are called "smooth functions." For many applications, assuming existence of higher derivatives is more for convenience than necessity.

Definition 13.40 *Suppose $X \subseteq \mathbf{R^n}$, $Y \subseteq \mathbf{R^m}$, and $f: X \to Y$ is a homeomorphism. If f and f^{-1} are also smooth, then we say F is a* **diffeomorphism**. *If a diffeomorphism exists, then we say that X and Y are* **diffeomorphic**.

We obtain a definition for a smooth manifold by replacing "homeomorphic" by "diffeomorphic" in Definition 13.01.

Definition 13.41 *Let $X \subseteq \mathbf{R^n}$, X is a* **smooth manifold** *with boundary, of dimension k if, for every point $x \in X$, there is an open subset, U, of X such that either*

(a) *U is diffeomorphic to $\mathbf{R^k}$.*
(b) *U is diffeomorphic to R^k_+, and x corresponds to a point of $\mathbf{R}^{k-1}$.*

A smooth manifold is sometimes called a "differentiable manifold." The study of smooth manifolds is called "differential topology."

Example 13.42 Clearly, $\mathbf{R^n}$ and R^n_+ are smooth manifolds, for any n. Similarly, any open subset of $\mathbf{R^n}$ or R^n_+ is a smooth manifold.

To show S^n is a smooth manifold, we use stereographic projection (see Proposition 4.06), to get a smooth map from $S^n - \{P\}$ to $\mathbf{R^n}$ where $P = (0, 0, \ldots, 0, -1)$ is the "south pole." Then we use a similar map for $S^n - \{Q\}$ where $Q = (0, 0, \ldots, 0, 1)$ is the "north pole," (Problem 13.42). ◆

The following idea is a familiar one from calculus. A regular value is a value which is not the image of any critical point.

Definition 13.43 *Suppose $f: \mathbf{R^n} \to \mathbf{R^1}$ is a smooth function, $y \in \mathbf{R^1}$. We say that y is a* **regular value** *of f if, for all points $x \in f^{-1}(y)$, there is a j such that $\frac{\partial f}{\partial x_j}(x) \neq 0$.*

The next proposition provides a way of constructing examples of smooth manifolds; for proofs see [7, 31, 22, 21].

Proposition 13.44 *Suppose $f: \mathbf{R^n} \to \mathbf{R^1}$ is a smooth function, $y \in \mathbf{R^1}$, and y is a regular value for f. Then $f^{-1}(y)$ is a smooth $(n-1)$-dimensional manifold with an empty boundary.* ∎

Manifolds obtained by Proposition 13.44 are familiar constructions. For maps of $\mathbf{R^2}$, these are the "level curves"; for $\mathbf{R^3}$, these are "level surfaces."

We can use Proposition 13.44 to get an alternative proof that S^n is a smooth manifold since $S^n = f^{-1}(1)$ where $f: \mathbf{R^{n+1}} \to \mathbf{R^{n+1}}$ is the smooth function $f(\vec{x}) = |\vec{x}|$.

To delve further into differential topology would entail a separate book (such as [21, 22]), but at least we can state one basic problem.

Question 13.45 Suppose X and Y are homeomorphic smooth manifolds. Are they diffeomorphic?

The answer is "no"! A long and interesting story, we mention only a few of the highlights. Basically, in relatively low dimensions, there is not much of a problem. But there are smooth 7-dimensional manifolds that are homeomorphic to S^7 but *not* diffeomorphic to S^7. As a matter of fact, there are exactly 27 different (mutually non-diffeomorphic) such examples.

As a rule, in topology, a situation is simpler in the compact case compared to the non-compact case. Question of 13.45 for $\mathbf{R^n}$ is a notable exception. If X is homeomorphic to $\mathbf{R^n}$, then it is diffeomorphic to $\mathbf{R^n}$, except for $\mathbf{R^4}$, in which case there are infinitely many non-diffeomorphic examples.

*13.6 General Topology and Chapter 13

We have looked at a circle as being a subset of $\mathbf{R^n}$ for some n. For example, as the standard circle in the plane, the boundary of a square in the plane or even a simple closed curve in space. These subsets are all homeomorphic. In other words, they are images of different embeddings A goal of using an abstract viewpoint is to focus on properties of the manifold itself, and not on properties of the particular embedding. In fact, we would really like to suppress the hypothesis that the circle is a subset of some Euclidian space, altogether.

We might propose the following as the "obvious generalization" of an n-dimensional manifold: a manifold is a topological space $\{X, \mathcal{T}\}$ such that, for each $x \in X$, there is an open subset containing x which is homeomorphic to $\mathbf{R^n}$ (or R^n_+). This is almost, but not exactly, what we want, as shown by this next example.

Example 13.46 Let $X = \{(x, y) \in \mathbf{R^2} \mid x = 0\}$ be the x-axis in the plane. Let x_0 be the point $(0, 1)$, and let $M = \{x_0\} \cup X$. Let S denote the open subsets of X derived from the standard topology of $\mathbf{R^2}$.

Define a topology $\mathcal{T}$ for M as follows. The open subsets of $\mathcal{T}$ which do not contain x_0 are the subsets of S. The open subsets of $\mathcal{T}$ which do contain x_0 are sets of the form $(S - (0, 0)) \cup \{x_0\}$ where $S \in S$.

It can be verified that $\{M, \mathcal{T}\}$ is a non-Hausdorff topological space, and that every point of M is contained in an open subset homeomorphic to $\mathbf{R^1}$. (The Hausdorff property fails for the two points $(0, 0)$ and $(0, 1)$.)

◆

Example 13.46 shows that our proposed definition is not entirely successful. What we really have in mind is to define manifold, abstractly, as a topological space $\{M, \mathcal{T}\}$, so that any such M will have an embedding into $\mathbf{R^n}$ for some n. Since it is non-Hausdorff, Example 13.46 cannot be embedded in any metric space, let alone $\mathbf{R^k}$.

There is one other problem.

Example 13.47 Let $\mathcal{T}$ be the standard topology for $\mathbf{R^1}$, let $\mathcal{D}$ be the discrete topology for $\mathbf{R^1}$, and let $X = \{\mathbf{R^1}, \mathcal{T}\} \times \{\mathbf{R^1}, \mathcal{D}\}$. One can, roughly, describe X as uncountably many disjoint copies of $\mathbf{R^1}$. Clearly, X is a 1-dimensional manifold, according to our proposed definition. The problem is that X is not second countable, but any subset of $\mathbf{R^n}$ must be second countable, the remark on page 176. ◆

These are the only problems; thus we are lead to the following definition:

Definition 13.48 *A* **topological** k**-manifold** *is a topological space, $\{M, \mathcal{T}\}$ which is Hausdorff, second countable and such that for each $x \in M$ there is an open subset $U \in \mathcal{T}$ such that either*

(a) *U is homeomorphic to $\mathbf{R}^k$.*

(b) *U is homeomorphic to R^k_+, and x corresponds to a point of $Bd(R^k_+, \mathbf{R^k})$.*

The next theorem shows that this definition satisfies our expectations:

Proposition 13.49 *Let $\{M, \mathcal{T}\}$ be a topological k-manifold. Then there is an embedding of M into $\mathbf{R^n}$, for some n.* ▮

The proof of this theorem is usually found as a corollary of the following more general, and difficult, theorem: Any subset of $\mathbf{R^n}$ with topological dimension n is embeddable in $\mathbf{R^{2n+1}}$; see [17, 25].

If M is a topological k-manifold, it is possible to define a concept of a smooth manifold *without* the assumption that $M \subseteq \mathbf{R^n}$, for some n. To simplify the discussion, we will assume that $\partial M = \emptyset$; the argument in case of non-empty boundary is not difficult, only verbose. The point is to show how one can go about transferring ideas, which seem to be firmly rooted in $\mathbf{R^n}$, to the more abstract world of general topology.

Definition 13.50 *Suppose M is a topological k-manifold and U and V open subsets of M, with homeomorphisms (called* **local coordinate maps***) $\phi: U \to \mathbf{R^k}$ and $\psi: V \to \mathbf{R^k}$, as in Definition 13.48(a). Assume also that $U \cap V \neq \emptyset$.*

Let $W = \phi(U \cap V)$. Since ϕ is a homeomorphism, W is an open subset of $\mathbf{R^k}$. Now consider $\psi \circ \phi^{-1}: W \to \mathbf{R^k}$; such a homeomorphism is called a **transition function**.

We can verify that ϕ is an embedding. For a manifold to be smooth, we require that all transition functions be *smooth* embeddings:

Definition 13.51 *Suppose $M \subseteq \mathbf{R}^n$ is a topological k-manifold written as a union of open subsets, $M = \cup_\alpha U_\alpha$, such that each U_α is homeomorphic to $\mathbf{R}^k$ by coordinate function $f_\alpha : U_\alpha \to \mathbf{R}^k$. If we can find choices for U_α and f_α so that all transition functions are smooth, then we say that M is a* **smooth manifold.**

We can similarly transfer the idea of smooth function:

Definition 13.52 *Suppose M is a smooth k-manifold, and M′ a smooth k′-manifold, in the sense of Definition 13.51. Suppose $f: M \to M'$ is a continuous function. We say that f* **is smooth** *if, for all coordinate functions of M, $\phi : U_\alpha \to \mathbf{R}^k$, and for all coordinate functions of M', $\phi' : U'_{\alpha'} \to \mathbf{R}^{k'}$, with $Im(\phi) \cap U'_{\alpha'} = W \neq \emptyset$, the map $\phi' \circ f \circ \phi^{-1}$ is smooth.*

Note, in the definition above, the domain of definition of $\phi' \circ f \circ \phi^{-1}$ is the set $Z = \phi(f^{-1}(W))$.

We seem to have conflicting definitions for smooth manifold, Definitions 13.41 and 13.51. The following definition, resolves the issue:

Proposition 13.53 *If $X \subseteq \mathbf{R}^n$ is a smooth manifold in the sense of Definition 13.41, it is a smooth manifold in the sense of Definition 13.51.*

If X is a topological space which is a smooth manifold in the sense of Definition 13.51, there exists a smooth embedding, in the sense of Definition 13.52, of X into some $\mathbf{R}^n$, so that its image is a smooth manifold in the sense of Definition 13.41. ■

A proof of the proposition above can be found in any differential topology text such as [6, 7, 21, 22]. Furthermore, in a high enough dimension, all these embeddings are equivalent (see [22]):

Proposition 13.54 *Let $\{M, \mathcal{T}\}$ be a smooth k-manifold. Let $f: M \to \mathbf{R}^n$ and $g: M \to \mathbf{R}^m$, be two embeddings. Then f and g are smoothly stably equivalent.* ■

Proofs of both propositions above may be found in [22].

13.7 Problems for Chapter 13

13.1 Show that if X is a 2-dimensional manifold and if Y is a 1-dimensional manifold then X and Y are not homeomorphic.

13.2 Prove that D^2 is a 2-dimensional manifold and $\partial D^2 = S^1$. (Hint: For points of S^1, you might consider the complex exponential map.)

13.3 Prove Proposition 13.15.

13.4 Prove Proposition 13.33 in the case that either $\partial M = \emptyset$ or $\partial N = \emptyset$.

13.5 Prove Proposition 13.17.

13.6 Let M be a k-dimensional manifold. Show that each component of M is a k-dimensional manifold.

13.7 Suppose $M \subseteq \mathbf{R}^n$ and M is a k-dimensional manifold. Show that $int(M) \subseteq Int(M)$. (Note: Be careful to distinguish int and Int.) Also show that $\partial M \subseteq Bd(M)$.

13.8 Prove that if M is a connected k-manifold with $1 < k$ and $x \in M$, then $M - \{x\}$ is a connected set.

13.9 Prove Proposition 13.15.

13.10 Prove that the standard torus does not have the fixed-point property.

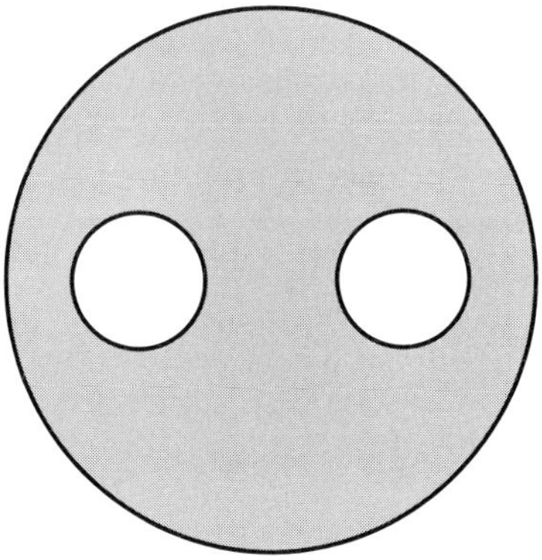

Figure 13-19 A "disk with two holes," P; see Problem 13.11.

13.11 Let M be a torus with two holes. Let $P = D^2 - \left(N_{\frac{1}{4}}((\tfrac{1}{2},0)) \cup N_{\frac{1}{4}}((-\tfrac{1}{2},0))\right)$; P is called a "disk with two holes," see Figure 13-19. Referring to the terminology in the remarks of page 291, show there exists an embedding of P into M so that

(a) each of the three circles in ∂P is a non-trivial circle in M,
(b) no pair of circles is algebraically trivial in M

Note that, because of the embedding of P, the union of all three circles is algebraically trivial in M.

13.12 Verify the statement of the remark on page 285, that for any two rational numbers, r and q, that $L_{(r,q,0)}$ is a circle.

13.13 Verify the proof in Exercise 13.27 for an arbitrary point of T^2.

13.14 Show that $T^2 - L$ and L fail to be locally connected at every point where L is as in Example 13.27.

13.15 Verify that the subsets U and V of Example 13.29 are path connected.

13.16 Verify that, for the subsets defined in Example 13.29, $V \cup T^2 = D_{(2,1)}$, $V = int(D_{(2,1)}) = Int(D_{(2,1)}$ and $Cl(V, \mathbf{R}^3) = D_{(2,1)}$.

13.17 Supply the details for the claims of the final paragraph of the proof of the remark on page 298.

13.18 Write formulas for the piecewise functions f and g, as described in the proof of Proposition 13.38.

13.19 Prove directly that, for any n, S^n is homogeneous.

13.20 Prove, directly that the torus, T^2, is homogeneous. (Hint: Use the map $P: \mathbf{R}^2 \to T^2$ and the homogeneity of $\mathbf{R}^2$.)

14. KNOTS AND KNOTTINGS

OVERVIEW: We introduce some basic concepts of knot theory, with focus on the concepts of isotopy and ambient isotopy.

14.1 Knots and isotopy

Knot theory is inspired by the study of subsets of $\mathbf{R}^3$ that are homeomorphic to circles. We examine some basic definitions and examples. By the end of this chapter, we will not have answered any of the basic questions of knot theory, but we can, at least, motivate and clearly state some of these basic questions.

Example 14.01 In Example 13.26 we considered several curves on the standard torus which give circles in $\mathbf{R}^3$. In order to visually render curves in $\mathbf{R}^3$ clearly, one represents a curve in $\mathbf{R}^3$ as a very thin tube whose center is the given curve. In Figure 14-1 we see the curve shown in Figure 13-5. In Figure 14-2 we see the curve shown in Figure 13-7. ◆

Intuitively, the knotted circles shown in Figures 14-1 and 14-2 seem to be inequivalent subsets of $\mathbf{R}^3$, in the sense of Definition 6.11. We are unable to prove this here. Any substantial study of knot theory involves algebraic techniques which are well beyond the scope of this text but interested readers can find further information in [1, 12, 8, 19, 26, 27, 29]. We will see how the study of knots motivates the need to reexamine and refine our notions of equivalent subsets and equivalent embeddings.

A basic idea of "same" in Euclidian geometry is the concept of congruence. A traditional definition of congruence is: two subsets of the plane are congruent if one can rigidly move one so that it coincides with the other.

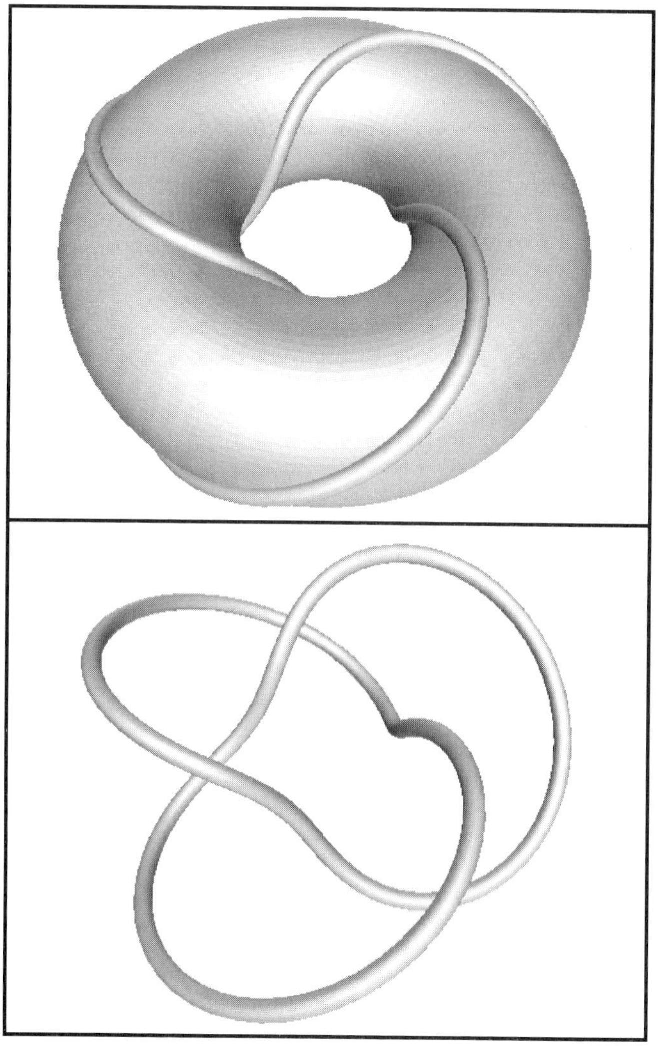

Figure 14-1 The trefoil knot, shown as a thin tube; see Example 14.01. Top, the thin tube and the standard torus in $\mathbf{R}^3$; below, only the knot is shown.

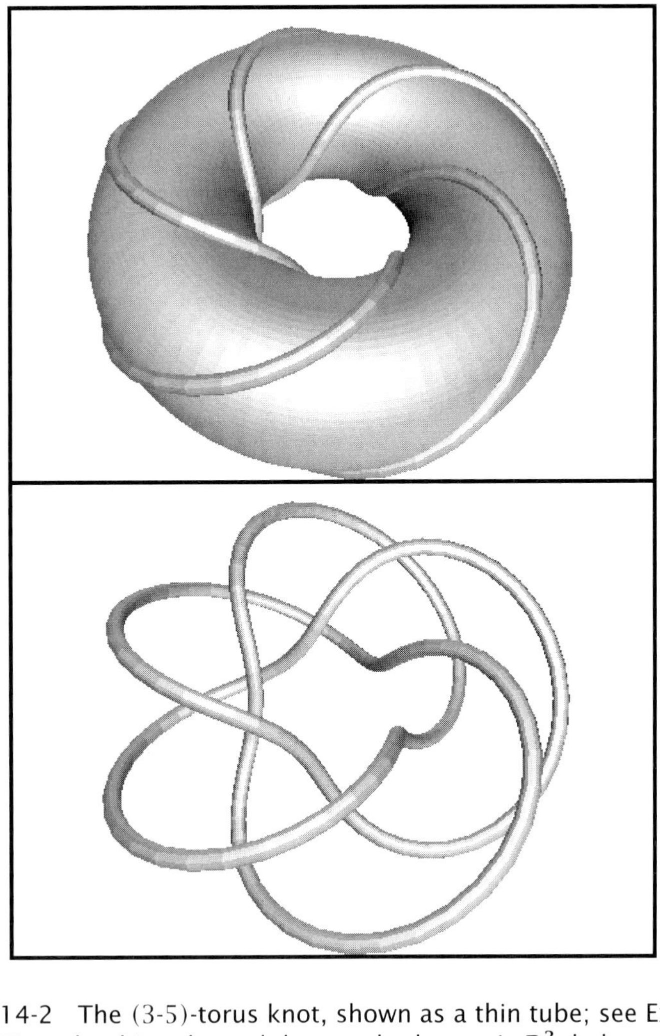

Figure 14-2 The (3-5)-torus knot, shown as a thin tube; see Example 14.01. Top, the thin tube and the standard torus in $\mathbf{R}^3$; below, only the knot is shown.

We previously discussed rigid motion, Definition 2.27. This notion of rigid motion differs in two ways from the description above of congruence. First, rigid motion has been defined to be a map of $\mathbf{R^n}$ to $\mathbf{R^n}$; a definition of congruence involves a map of a *subset* of $\mathbf{R^n}$. Second, the idea of a motion involves a continuous change (in time). This is absent in our definition of rigid motion. Isotopy, which we next define, allows us to articulate the idea that two embeddings differ by a motion.

First, we introduce notation:

Definition 14.02 *Let $X \subseteq \mathbf{R^n}$ and $Y \subseteq \mathbf{R^m}$, and suppose $F: X \times I \to Y$ where I denotes the unit interval. For $t \in I$, we define the* **associated parametric** *function $F_t : X \to Y$ by $F_t(x) = F(x,t)$. Equivalently, $F_t(x) = F \circ i_t$ where $i_t(x) = x \times t$, as in Definition 6.03.*

We have discussed a notion of two *subsets* being equivalent, Definition 6.11, and also two *embeddings* being equivalent, Definition 6.12. As parallels, we introduce two new notions, one for embeddings and one for subsets.

Definition 14.03 *Let $X \subseteq \mathbf{R^n}$ and $Y \subseteq \mathbf{R^m}$, and suppose f and g are embeddings from X to Y. We say f is* **isotopic** *to g, if there is a continuous map $F: X \times I \to Y$ such that, for all $t \in I$, F_t is an embedding of X into Y, $F_0 = f$ and $F_1 = g$.*

Example 14.04 Suppose $X \subseteq \mathbf{R^n}$, and f and g continuous functions from X to $\mathbf{R^m}$. Then, by Proposition 6.09, we have two embeddings of X corresponding to the graphs of these functions: $G_f(x) = (x, f(x))$ and $G_g(x) = (x, g(x))$. For an example of a simple, very useful isotopy, we note that G_f and G_g are isotopic functions. One defines $F: X \times I \to \mathbf{R^m}$ by $F(x,t) = (x, (1-t)f(x) + tg(x))$. Each F_t is an embedding corresponding to the graph of $(1-t)f(x) + tg(x)$. In Figure 14-3 we see an example of this for two simple real-valued functions. Roughly, this isotopy can be described as sliding the graph of one function in the y-direction towards the second function where we think of $y \in \mathbf{R^m}$ and the graphs as subsets of $\mathbf{R^n} \times \mathbf{R^m}$.

Of particular note is the expression $(1-t)f(x) + tg(x)$ which is $f(x)$ if $t = 0$ and $g(x)$ if $t = 1$. We will see similar expressions used in further discussions and have used such an expression previously in the proof of Proposition 13.38. ◆

In Definition 14.03, we defined isotopy as a relation between embeddings; the next definition involves a relation between subsets.

Definition 14.05 *Let $Y \subseteq \mathbf{R^m}$, and suppose X_0 and X_1 are subsets of Y. We say X_0 is* **isotopic** *to X_1, if there is a subset $X \subseteq \mathbf{R^n}$ and a continuous map $F : X \times I \to Y$ such that, for all $t \in I$, each F_t is an embedding of X into Y, and $F_0(X) = X_0$ and $F_1(X) = X_1$.*

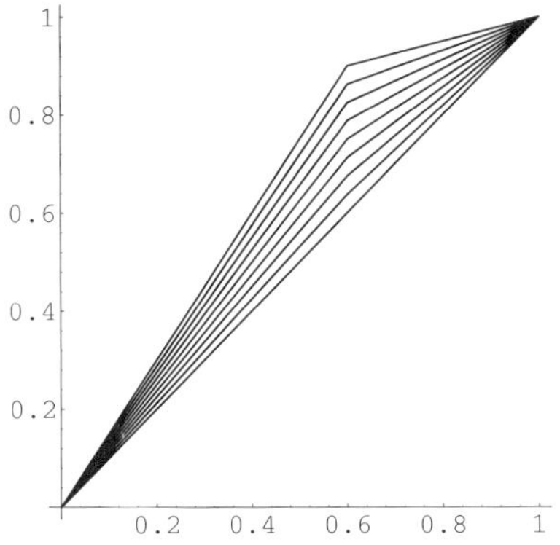

Figure 14-3 Isotopy of the identity map and a simple, piecewise-linear map g; see Example 14.04. We see a sequence of functions that begins with the identity (whose graph is the diagonal) and ends with $g(x)$.

In either of Definitions 14.03 or 14.05, the function F_t is called an "isotopy."

Roughly, an isotopy is motion of a flexible object. The idea is to think of the t variable as time for an interval which begins at time 0 and ends at time 1. The subset $X_t = H_t(X)$ is thought of as the position of X at time t.

Recall our definition of equivalent embeddings, Definition 6.12. In that definition, two embeddings, f_0 and f_1 of X into Y, were equivalent if there was a homeomorphism $h : Y \to Y$ with $f_1 = h \circ f_0$. It is in that spirit that we define ambient isotopy.

Definition 14.06 *Let $X \subseteq \mathbf{R^n}$ and $Y \subseteq \mathbf{R^m}$. Let f and g be embeddings from X to Y. We say f is* **ambiently isotopic** *to g, if there is a continuous map $H : Y \times I \to Y$ such that each H_t is a homeomorphism of Y onto Y, $H_0 = Id|_Y$, and $g = H_1 \circ f$.*

Here is a version of ambient isotopy for subsets:

Definition 14.07 *Let $Y \subseteq \mathbf{R^m}$, X_0 and X_1 subsets of Y. We say X_0 is* **ambiently isotopic** *to X_1, if there is a continuous map $H : Y \times I \to Y$ such that each H_t is a homeomorphism of Y onto Y, $H_0 = Id|_Y$ and $H_1(X_0) = X_1$.*

In either of Definitions 14.06 or 14.07, the map H is called an "ambient isotopy."

Ambient isotopy implies isotopy in an appropriate sense. For example, suppose two embeddings, f and g, of X into Y are ambiently isotopic via an ambient isotopy H_t. We can define an isotopy of f and g

by $F : X \times I \to Y$ where $F(x,t) = H_t \circ f(x)$. If X_0 and X_1 are equivalent subsets via an ambient isotopy H_t, then the map $F : X \times I \to Y$ defined by $F(x,t) = H_t(x)$ provides an isotopy between these two sets.

Now that we have defined our basic notion of ambient isotopy, we can define what it means for a circle to be unknotted. In the definition below we are mostly concerned with subsets of $\mathbf{R^n}$ with $3 \leq n$.

Definition 14.08 *The* **standard circle** *in* $\mathbf{R^n}$, $2 \leq n$, *corresponds to the unit circle in the first two coordinates:* $\Sigma_0 = \{(x_1, \ldots, x_n) \in \mathbf{R^n} \mid x_1^2 + x_2^2 = 1$ *and* $x_i = 0$ *for* $2 < i\}$. *If* $X \subseteq \mathbf{R^n}$, *and* X *is homeomorphic to a circle, we say* X *is an* **unknotted circle** *if* X *is ambiently isotopic to* Σ_0.

Example 14.09 Clearly, ambiently isotopic subsets of $\mathbf{R^n}$ must be equivalent subsets of $\mathbf{R^n}$, but the converse is not true. In the lower portion of Figure 13-11 there are two knots—the trefoil knot and its mirror image. These knots are equivalent but, using algebraic topology, it can be shown that these are *not* ambiently isotopic. That is, there is a right-hand trefoil knot and a left-hand trefoil knot that are equivalent subsets but are not equivalent knots in terms of ambient isotopy. ◆

Example 14.10 In this example we describe a motion which transforms a square into a circle. In Figure 14-4 we see images of an isotopy of a square in the plane.

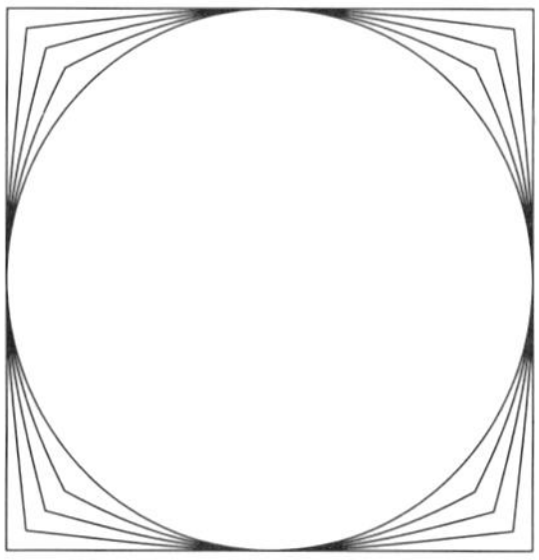

Figure 14-4 Four stages of an isotopy which takes a square to a circle; see Example 14.10.

Let X_0 be the square with corners $(1,1), (-1,1), (1,-1), (-1,-1)$, and let X_1 be the unit circle. An isotopy $F : X_0 \times I \to \mathbf{R}^2$ is given by

$$F(\vec{x}, t) = (1-t)\vec{x} + t\frac{\vec{x}}{|\vec{x}|}.$$

In Figure 14-4 we see the sets

$$X_0 = F_0(X_0), F_{\frac{1}{4}}(X_0), F_{\frac{1}{2}}(X_0), F_{\frac{3}{4}}(X_0), \text{ and } F_1(X_0) = X_1.$$

In fact, X_0 and, X_1 are *ambiently* isotopic. The idea is to view $\mathbf{R}^2$ as the origin together with a family of concentric squares. We continuously deform this family until, at the end, we have a family of concentric circles. To write this ambient isotopy we use a function $mab : \mathbf{R}^2 \to \mathbf{R}^1$. This is defined by $mab(\vec{x})$, the maximum of the absolute value of the coordinates of $\vec{x}$. (In Cartesian coordinates, $mab(x, y) = \max(|x|, |y|)$.) For any number, $0 < a$, $SQ(a) = mab^{-1}(a)$ is the square with corners $(a, a), (-a, a), (a, -a)$, and $(-a, -a)$. In particular, $X_0 = SQ(1)$. Let $C(a)$ denote the circle in $\mathbf{R}^2$ with radius a and center $\vec{0}$.

Define $H : \mathbf{R}^2 \times I \to \mathbf{R}^2$ by

$$H(\vec{0}, t) = \vec{0}; \quad H(\vec{x}, t) = (1 - t)\vec{x} + t\Big(mab(\vec{x})\frac{\vec{x}}{|\vec{x}|}\Big).$$

Roughly speaking, the ambient isotopy H_t takes points on $SQ(a)$ and slowly slides them, along rays from $\vec{0}$, to points on the circle $C(a)$. We leave verification of the continuity of H_t as an exercise—the key is to verify continuity at points of $\{\vec{0}\} \times I$ (Problem 14.1). ◆

Example 14.11 Let us relate congruence, rigid motion, and ambient isotopy. Consider two triangles: triangle X_0 has vertices $(1, 1), (1, 2)$, and $(2, 1)$, and triangle X_1 has vertices $(2, 3), (2, 4)$, and $(3, 3)$; see Figure 14-5.

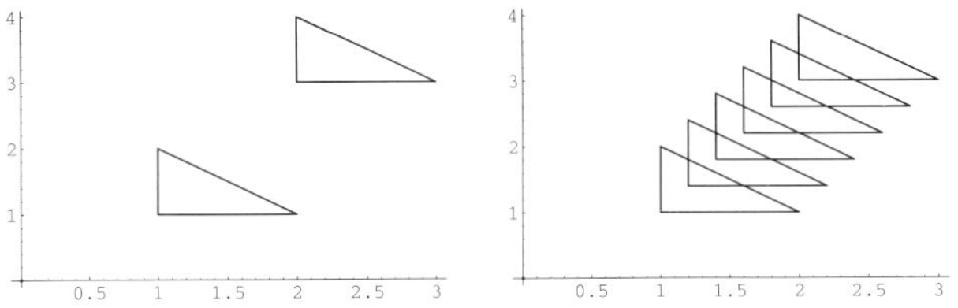

Figure 14-5 On the left are two triangles, X_0 and X_1. On the right, several of images, $H_t(X_0)$, of the triangle under ambient isotopy H_t that takes X_0 to X_1; see Example 14.11.

These triangles are congruent. The affine map T, defined by $T(\vec{x}) = \vec{x} + (\vec{i} + 2\vec{j})$, is a rigid motion such that $T(X_0) = X_1$. There is also an ambient isotopy H_t which takes X_0 to X_1 such that for each t, H_t is the rigid motion $H(\vec{x}, t) = (1 - t)\vec{x} + tT(\vec{x})$. ◆

Not all rigid motions can be expressed as ambient isotopies.

Consider the triangle, X_0 (as in Example 14.11) with vertices $(1, 1)$, $(1, 2)$, and $(2, 1)$; and a triangle, X_2, with vertices $(1, -1)$, $(1, -2)$, and $(2, -1)$. Using a rigid motion, which corresponds to reflection in the x-axis, $R(\vec{x}) = M\vec{x}$ where $M = \left(\begin{smallmatrix} 1 & 0 \\ 0 & -1 \end{smallmatrix}\right)$; see Figure 14-6. Note that $R(X_0) =$

X_2. In this case, we cannot find an ambient isotopy H_t where, for each t, H_t is a rigid motion with $H_1(X_0) = X_2$.

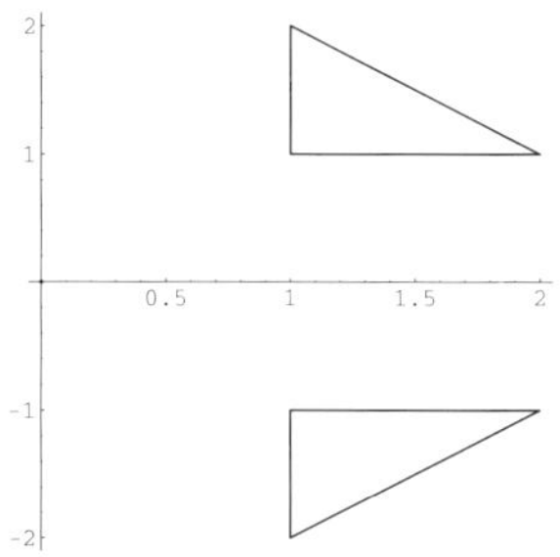

Figure 14-6 A rigid motion which does not correspond to an ambient isotopy using rigid motions. The two triangles shown are related by a rigid motion—reflection in the x-axis. See the remark on page 312.

Suppose there were such an ambient isotopy. It could then be written

$$H_t\vec{x} = M_t\vec{x} + B_t$$

where each M_t is a 2×2 matrix. The four entries of M_t are continuous functions of x and y where $\vec{x} = x\vec{i} + y\vec{j}$. Consider the map $\phi : I \to \mathbf{R}^1$ defined by $\phi(t) = |M_t|$ where $|M_t|$ denotes the determinant. Then ϕ is a continuous function. However, as noted in the remark on page 52, a rigid motion has a matrix of determinant $+1$ or -1. Thus we can view ϕ as a continuous function from I to $\{-1\} \cup \{+1\}$ (Problem 14.7). For any ambient isotopy, H_0 is the identity map; thus $\phi(0) = +1$. Since I is connected, it follows that $\phi(I) = \{+1\}$. In particular $|M_1| = +1$; however, the matrix for the rigid motion R has determinant -1.

We have noted that manifolds are homogeneous. This homogeneity can be expressed by using ambient isotopies:

Proposition 14.12 *Suppose M^n is an n-dimensional manifold with $\partial M = \emptyset$, $p, q \in M$. Then there is an ambient isotopy, H_t of M such that $H_1(p) = q$.* ∎

For manifolds with boundary, we have

Proposition 14.13 *Suppose M^n is an n-dimensional manifold with $\partial M \neq \emptyset$, $p, q \in Int(M)$. Then there is an ambient isotopy H_t of M such that $H_1(p) = q$, and such that for all t, $0 \leq t \leq 1$, $H_t|_{\partial M} = Id|_{\partial M}$.* ∎

We outline the proof for a special case of Proposition 14.13—the case for the manifold D^2. We will use this special case later in this chapter. We will show that if $p, q \in Int(D^2)$, then there is an ambient isotopy H_t of D^2, such that $H_1(p) = q$, and such that for all t, $0 \le t \le 1$, $H_t|_{\partial D^2} = Id|_{\partial D^2}$.

Our definition of H_t is an elaboration of the constructions used in the proof of Proposition 13.38. In that proof, we began with a homeomorphism $h: D^2 \to I \times I$ and an ϵ so that the points $P = h(p)$ and $Q = h(q)$ were inside $I \times I$ with a border of width ϵ removed. We then defined a pair of piecewise-linear homeomorphisms of I, f, and g, and used these to get a homeomorphism G of $I \times I$.

We show that this G is ambiently isotopic to the identity map. Recall the functions f and g, defined in that proof; these are piecewise-linear homeomorphisms. Each of these is isotopic to the identity map.

Specifically, we have two isotopies of I: $f_t(x) = (1-t)x + tf(x)$ and $g_t(x) = (1-t)x + tg(x)$. The idea is to insert these in place of f and g in the formulas presented in the proof of Proposition 13.38. Define an ambient isotopy of $I \times I$ by $G_t(x, y) = (f_t(x), g_t(y))$; see Figure 14-7. One can check that G_t is an ambient isotopy, that $G_1(P) = Q$, and $G_t|_{\partial(I \times I)} = Id|_{\partial(I \times I)}$.

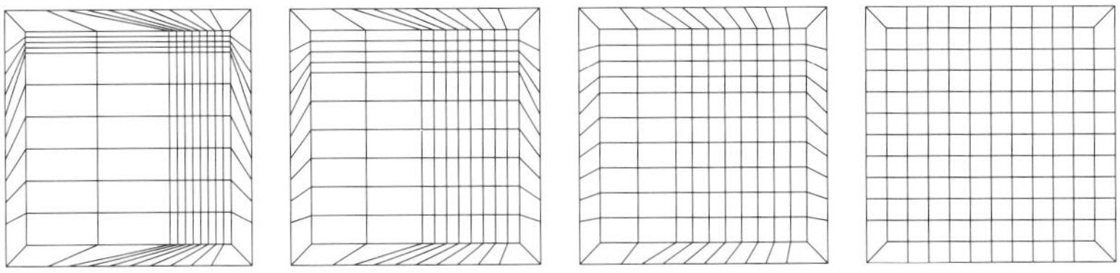

Figure 14-7 An ambient isotopy G_t, fixed on the boundary, between a homeomorphism of the square and the identity map. From left to right: $G_0 = G$ (the homeomorphism of Figure 13-15), two intermediate stages $G_{1/3}$, and $G_{2/3}$, and finally the identity map, denoted G_1. See proof of Proposition 14.13 for D^2.

As a final step, we transfer this to an ambient isotopy of D^2 by conjugation by h. Let $H_t = h^{-1} \circ G_t \circ h$; then H_t is an ambient isotopy of D^2, fixed, for all t, on ∂D^2 such that $H_1(p) = q$. ∎

Ambiently isotopic subsets must be equivalent subsets. If X_0 and X_1 are ambiently isotopic subsets of Y, via ambient isotopy H_t, then X_0 and X_1 are equivalent subsets since H_1 is a homeomorphism of Y with $H_1(X_0) = X_1$.

However, isotopic subsets may not be equivalent subsets. Let X and Y denote the interval $[-1, 1]$ in $\mathbf{R}^1$. Let $X_0 = X$ and let $X_1 = [-\frac{1}{2}, \frac{1}{2}]$. It is clear that X_0 and X_1 are *not* equivalent subsets of Y since $Y - X_0 = \emptyset$ and $Y - X_1 = [-1, -\frac{1}{2}) \cup (\frac{1}{2}, 1]$.

The example above also shows that isotopy does not, in general, imply ambient isotopy.

The homeomorphism of $\mathbf{R}^1$, defined $f(x) = -x$, is not ambiently isotopic to the identity map. This follows from the fact that f is monotone decreasing and the following proposition:

Proposition 14.14 *If $f: \mathbf{R}^1 \to \mathbf{R}^1$ is a homeomorphism which is isotopic to the identity map of $\mathbf{R}^1$, then f is monotone increasing.*

Proof: Suppose h_t is an isotopy of $\mathbf{R}^1$ with $h_0 = Id|_{\mathbf{R}^1}$ and $h_1 = f$. Suppose $p, q \in \mathbf{R}^1$ with $p < q$. For $0 \le t \le 1$ define $g(t) = h_t(q) - h_t(p)$. Since $h_0 = Id|_{\mathbf{R}^1}$, $0 < g(0)$. Since each h_t is one-to-one, $g(t) \neq 0$ for all $0 \le t \le 1$. It follows that $0 < g(t)$ for all $0 \le t \le 1$. In particular, $0 < g(1)$. Since $0 < g(1) = h_t(q) - h_t(p) = f(q) - f(p)$ we have $f(p) < f(q)$. Since p and q were any points of $\mathbf{R}^1$ with $p < q$, we have shown that f is monotone increasing. ∎

Subsets can be equivalent and yet not ambiently isotopic. If we are considering subsets of a set X, the problem stems from the fact that there may be self-homeomorphisms of X that are not isotopic to the identity map on X.

In the example which follows, the underlying problem is that the homeomorphism of $\mathbf{R}^1$, given by $f(x) = -x$, is not isotopic to the identity.

Let $Y = [-3, 3]$ be the closed interval in $\mathbf{R}^1$. Let $X_0 = \{0\} \cup [1, 2]$, and $X_1 = \{0\} \cup [-2, -1]$; see Figure 14-8.

Figure 14-8 Two subsets of the line not ambiently isotopic. On the left, X_0, and on the right, X_1; see the remark on page 315.

Using the homeomorphism h defined by $H(x) = -x$, we see that X_0 and X_1 are equivalent subsets of Y. However there is no ambient isotopy which takes X_0 to X_1. We leave details as an exercise (Problem 14.2), but the general idea is to show that any homeomorphism which gives an equivalence of these subsets must be monotone decreasing.

In fact, one can prove that X_0 and X_1 are not isotopic subsets of Y (Problem 14.3).

The example given in the remark on page 314 has the merit of being simple, but it does not really illustrate an important distinction between the idea of isotopy and ambient isotopy for knotted circles.

A key example by illustrated in Figure 14-9. Here we have a subset which is a "circle with a knot in it." The isotopy shrinks this knot smaller and smaller, until, at $t = 1$, the knot has vanished.

Using algebraic topology, one can show the intuitively obvious fact that these two subsets are *not* equivalent subsets of $\mathbf{R}^3$, and thus not ambiently isotopic.

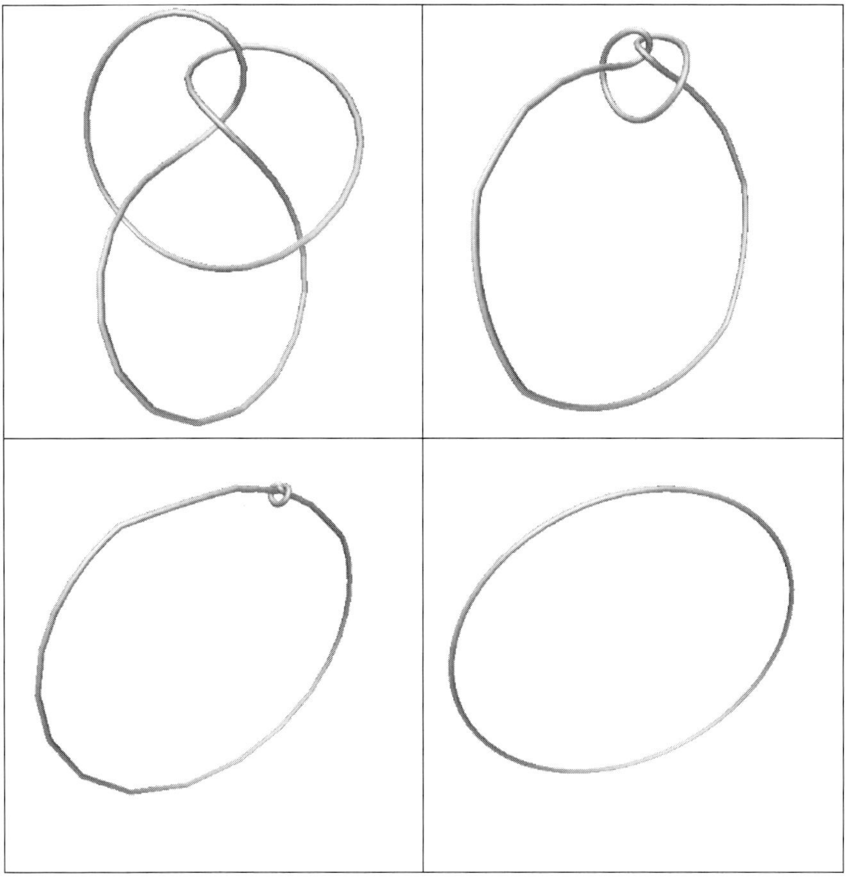

Figure 14-9 An isotopy of a circle which is not an ambient isotopy. Four stages are shown: the "knotted portion" shrinks until, at lower right, we have an unknotted circle; see the remark on page 315.

Thus, when we study knot theory, we consider equivalent subsets, or ambiently isotopic subsets, for our basic equivalence of knots. However, the idea of isotopy, a "motion" of the object of study, is much easier to understand than equivalence of subsets or ambient isotopy, which require attention to the entire surrounding space. In the next sections we consider two ways of dealing with these problems by considering additional conditions leading to study of PL knots and smooth knots.

14.2 Wild knots, smooth knots, PL knots

Despite the fact that isotopy does not, in general, imply ambient isotopy, one often sees these concepts used interchangeably in the literature when discussing manifolds. The reason is that there is a group of theorems, called "ambient isotopy extension theorems," that roughly say that, for *smooth* manifolds, isotopy of sets implies ambient isotopy of sets. Here is one; for proof see [7, 22]

Proposition 14.15 *Suppose $X \subseteq \mathbf{R^n}$ where X is a smooth closed manifold. If there is an smooth isotopy of $h: X \times I \to \mathbf{R^n}$, then there is a smooth ambient isotopy $H: \mathbf{R^n} \times I \to \mathbf{R^n}$ such that $H|_{X\times I} = h$.* ∎

Suppose we shrink a smooth knot to a point as indicated in Figure 14-9. This is an isotopy of a smooth knot in the following sense: it is an isotopy $F: S^1 \times I \to \mathbf{R}^3$ such that, for each $t \in I$, $F|_{S^1\times\{t\}}$ is a smooth knot.

However, this is not a *smooth* isotopy. Suppose p is the location on S^1 towards which the knotted portion seems to be shrinking. The problem is smoothness of the map $F: S^1 \times I \to \mathbf{R}^3$ at the point $(1, p)$. Smoothness would imply that all points $(s, t) \in S^1 \times I$ close to $(1, p)$ have tangent vectors to $F(S^1 \times \{t\})$ close to tangent vector v_0 to $F(S^1 \times \{1\})$ at $F(1, p)$. Intuitively, the problem is that the small, shrinking knot has points close to $F(1, p)$ with tangent vectors very different from v_0; see Figure 14-10.

Example 14.16 There is yet another basic problem in the study of circles in $\mathbf{R}^3$.

Historically, the mathematical theory of knotted circles began by representing a knot as a piecewise-linear closed path in $\mathbf{R}^3$ with no self intersections; see Figure 14-11. Of course since we are taking a topological viewpoint, we want to study subsets of $\mathbf{R}^3$ which are *equivalent* to a polygonal closed path. (For example, the square is a polygonal path and it is equivalent to a round circle.)

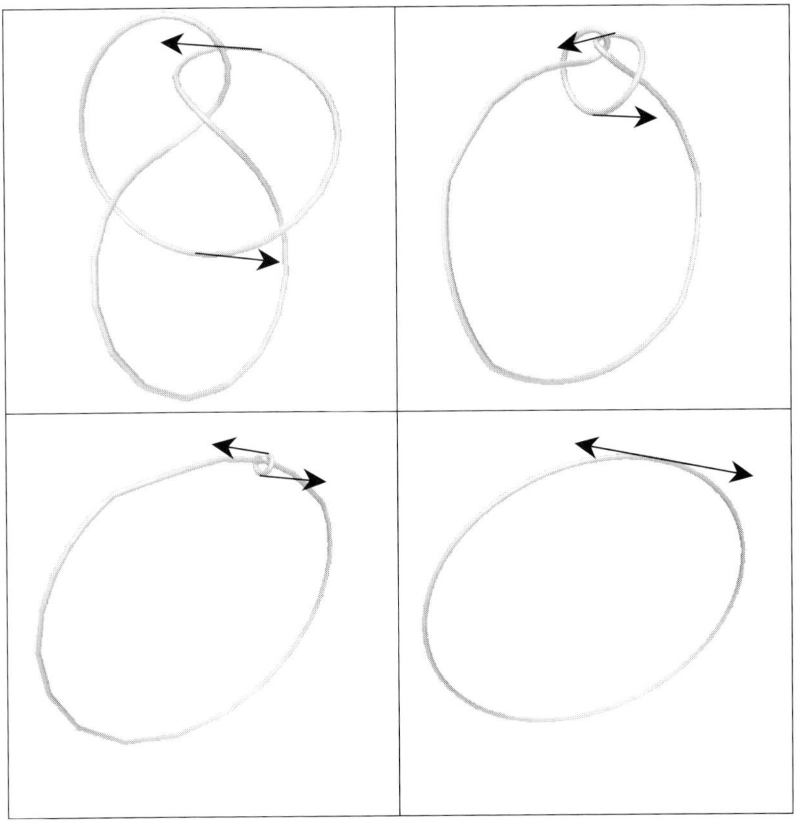

Figure 14-10 A continuous isotopy which is not a smooth isotopy. A pair of tangent vectors are shown for two points of the circle during the isotopy of Figure 14-9. Note that nearby points of the isotopy do not have tangent vectors that are close.

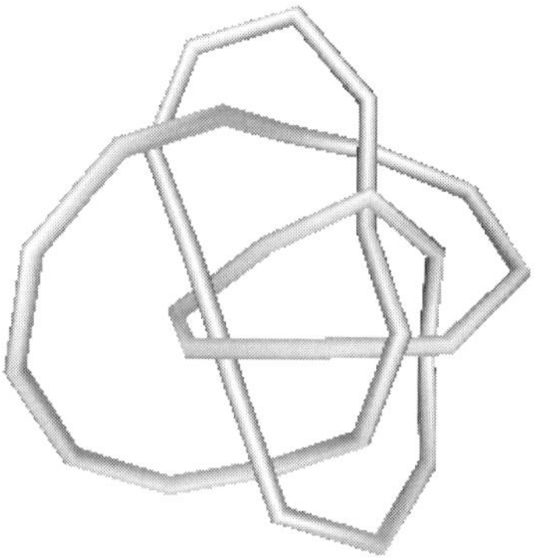

Figure 14-11 A polygonal knot; see Example 14.16.

But consider the subset W of $\mathbf{R}^3$, homeomorphic to the closed interval I, pictured in Figure 14-12.

As in other graphics, we depict this curve as a thin tube. Roughly, we can describe W as obtained by tying an infinite number of smaller and smaller knots in a row and including the limiting point. It can be shown that W is not equivalent to a piecewise-linear path; we will shortly make some definitions to articulate this.

By piecing two of these together, as indicated in Figure 14-13, we get a subset $\mathcal{W}$, of $\mathbf{R}^3$, homeomorphic to S^1. ◆

In order to further our discussion of examples such as Example 14.16, we provide the following definitions. For generality, we define these as subsets of $\mathbf{R}^n$, but in our examples we are concerned with $\mathbf{R}^n = \mathbf{R}^3$. We use the idea of a piecewise-linear function, Definition 3.47.

Definition 14.17 *Suppose $f: I \to \mathbf{R}^n$ is a piecewise-linear map. The image of f is called a* **polygonal path** *in* $\mathbf{R}^n$.

Suppose $\{t_i\}$ is the collection of points, so that $f|_{[t_i,t_{i+1}]}$ is a linear map. The image of the i-th point, t_i, is the i-th **vertex** *of the polygonal path. The image of the i-th interval, $f([t_i, t_{i+1}])$, is the i-th* **edge** *of the polygonal path.*

The polygonal path is **closed** *if $f(0) = f(1)$; the polygonal path is* **simple** *if $f|_{(0,1)}$ is one-to-one, and f is two-to-one at the endpoints.*

A **polygonal knot** *is a simple, closed polygonal path; a* **polygonal arc** *is a simple polygonal path which is not a closed path.*

Clearly, if P is a polygonal knot, then P is homeomorphic to a circle; a polygonal arc is homeomorphic to the interval I (Problems 14.8, 14.9).

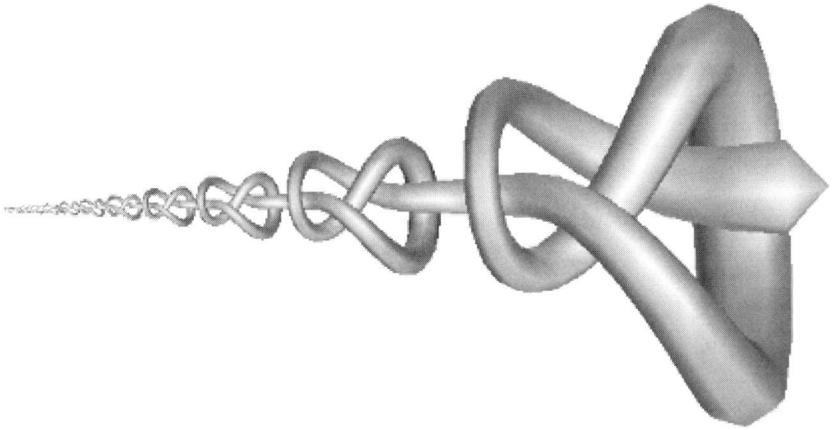

Figure 14-12 A wild embedding, W, of an interval; see Example 14.16. This graphic can be viewed in two ways. We can interpret it as an image of an embedding of I, shown as a tube, thickened so as to get a good three-dimensional picture. Or, we can interpret it as a surface—an image of a wild embedding of a 2-sphere. In either case, the limiting point of the knots on the left is a point of the subset.

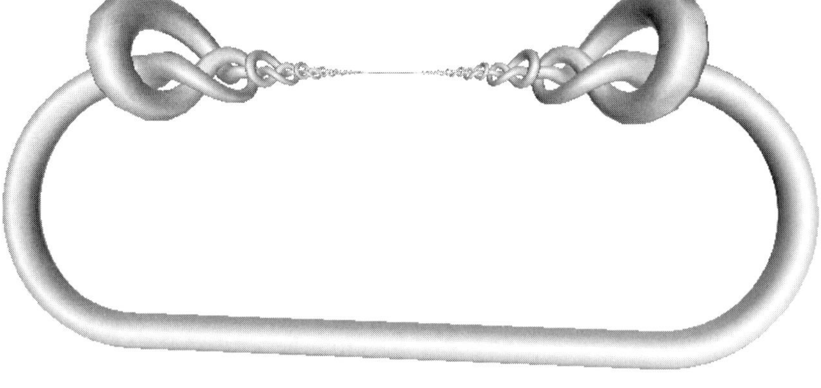

Figure 14-13 A wild embedding, $\mathcal{W}$, of a circle; see Example 14.16.

Definition 14.18 *Let C be a subset of $\mathbf{R^n}$ which is homeomorphic to a circle. We say that C is a* **tame knot** *if C is equivalent (as a subset of $\mathbf{R^n}$) to some polygonal knot. If C is not a tame knot, we say that C is a* **wild knot**.

To go along with the ideas of tame or wild subset, we have companion notions of tame or wild embeddings.

Definition 14.19 *If X is a circle (or union of disjoint circles or an interval) and $f : X \to \mathbf{R^n}$ is an embedding, then we say that f is a* **tame embedding** *if $f(X)$ is a tame knot (or tame link, or tame arc, respectively). An embedding which is not tame is called a* **wild embedding** .

Example 14.20 In Example 14-4 we saw that a square in the plane is equivalent to the unit circle in the plane. So S^1, though clearly not a polygonal path, is equivalent to one, and so is tame. ◆

Definition 14.21 *Let A be a subset of $\mathbf{R^n}$ which is equivalent (as a subset of $\mathbf{R^n}$) to some polygonal arc. We say that A is a* **tame arc**. *If A is homeomorphic to I, and A is not a tame arc, we say that A is a* **wild arc**.

Proposition 14.22 *The subset W described in Example 14.16 is a wild arc; the subset $\mathcal{W}$ is a wild circle.* ▮

Proposition 14.22 is plausible since it seems intuitively clear that one could not construct W (or $\mathcal{W}$) without using an infinite number of vertices. The usual method of proof of Proposition 14.22 is to use algebraic topology to show that the complement of W (or $\mathcal{W}$) is not homeomorphic to the complement of any polygonal knot; see [19].

The definitions of a smooth knotted circle and a polygonal knot are quite different, but in a sense they are equivalent:

Proposition 14.23 *If $X \subseteq \mathbf{R^n}$ is a smooth circle, it is ambiently isotopic to a polygonal knot. If $X \subseteq \mathbf{R^n}$ is a polygonal knot, it is ambiently isotopic to a smooth circle.* ▮

Conceptually, the proof of Proposition 14.23 is simple, but there are some lengthy technical details, found in [12]. The reason a polygonal knot is not smooth is due to the corners at the vertices. To show that a polygonal knot is ambiently isotopic to a smooth knot, the idea is to take a polygonal knot and, near each vertex, "round the corners." That is, replace the corner by a small smooth curve; see Figure 14-14. This needs a bit of care. For example, one cannot simply use circular arcs for the small smooth curves since we would not be able to get the resulting curve to have continuous second derivatives (Problem 14.10). Also, care must be used to avoid self-intersections.

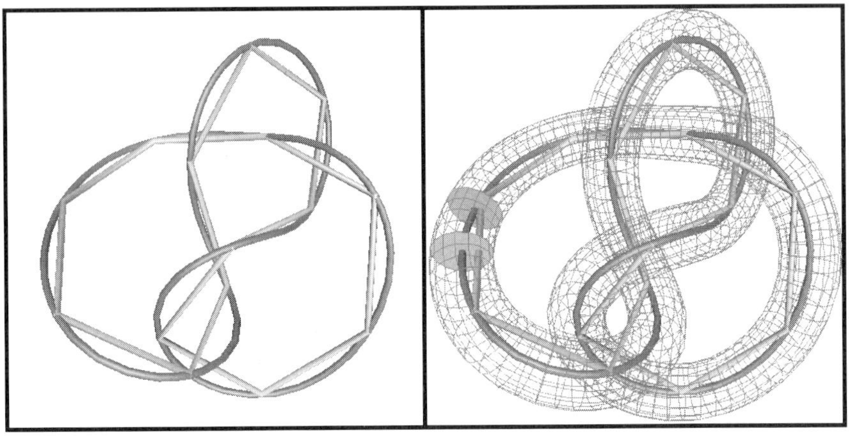

Figure 14-14 On left, a knot (smooth dark curve) and a PL approximation; see the remark on page 322. On right is a tube about the knot, shown in wire-frame, and two examples of transverse disks. Note that the knot and the PL approximation each intersect each transverse disk in a point.

The other part of Proposition 14.23 basically says that smooth knots are tame. The idea here is to select some suitable points on the smooth curve and form a simple closed polygonal path using these points as vertices. We outline a proof that the particular knots defined in Example 13.26, shown in Figures 14-1 and 14-2, are tame knots that will illustrate the general case.

Here, in outline, is our procedure:

1. Find a solid tube W about K (more details on this step later). A "solid tube" is a 3-dimensional manifold in $\mathbf{R}^3$ homeomorphic to $S^1 \times D^2$ with K corresponding to $S^1 \times \{\vec{0}\}$. Let $f: S^1 \times D^2 \to W$ be such a homeomorphism with $f(S^1 \times \{\vec{0}\}) = K$. For each $\vec{x} \in S^1$, the set $F_x = f(\{\vec{x}\} \times D^2)$ will be called a **transverse disk to** K; see Figure 14-14. We also have a corresponding open disk, $\overset{\circ}{F}_x = f(\{\vec{x}\} \times \overset{\circ}{D}{}^2)$.

2. Choose finitely many points, close enough together on K so that, if we join consecutive points, we get polygonal secant approximation K' such that K' meets each open transverse disk $\overset{\circ}{F}_x$ in a single point.

3. For each open disk $\overset{\circ}{F}_x$, we consider two points: $x = K \cap \overset{\circ}{F}_x$ and $x' = K' \cap \overset{\circ}{F}_x$. Using our proof of Proposition 13.38, we know we

can move x to x' by an isotopy, fixed on the boundary (in the sense of boundary of a manifold) of F_x.

Since we have explicit formulas from proof of Proposition 13.38, and using the fact that our tube is a product, we can see that we can do this continuously as we go from one disk to another. This allows us to define an isotopy h_t of W, fixed on ∂W, such that $h_1(K) = K'$.

4. Let $W' = \overline{\mathbf{R}^3 - W}$. We can now define an ambient isotopy H_t of $\mathbf{R}^3$ by $H_t|_W = h_t$ and $H_t|_{W'} = Id_{W'}$

We next indicate how to construct W of step 1, above. To prove the general case, the key step is showing we can find a solid tube for any smooth knot. In general, this step takes a lot of work. However, we can rely on geometry for our specific examples.

Suppose we are concerned with a torus knot $K = L_{(a,b,0)}$ where a and b are relatively prime.

Each page H^2_α in $\mathbf{R}^3$ (recall Definition 2.10) meets the standard torus T in a circle, call it C_α. Also, $K \cap H^2_\alpha$ is a collection of a points $x^\alpha_1, \dots x^\alpha_a$ placed equidistant along this circle. In H^2_α we can find a collection of *disjoint* closed disks, $D^\alpha_1, \dots D^\alpha_a$, of some radius ρ with the center of D^α_i being x^α_i. Let

$$W = \bigcup_{\substack{0 \le \alpha < 2\pi \\ i=1,\dots,a}} D^\alpha_i.$$

To complete our set of elementary definitions for knot theory, we add:

Definition 14.24 *Suppose that L is a subset of* $\mathbf{R}^3$ *which, with finitely many components, each of which is a tame knot, then L is called a* **tame link**

We can now at least say what knot theory is. Knot theory is the study of tame knots and links in $\mathbf{R}^3$ where two are considered equivalent if they are ambiently isotopic. The Figures 5-17 and 5-18, viewed as embedded circles rather than embedded tubes, are examples of links in $\mathbf{R}^3$. Figure 5-17 shows a link with four components; Figure 5-18 shows a link of 16 components.

Here are some of the basic questions:

Question 14.25 Given two tame knots (or tame links) k_0 and k_1 in $\mathbf{R}^3$, how do we determine whether they are equivalent subsets ?

Recalling Example 14.09, we ask:

Question 14.26 Given two tame knots (or tame links) k_0 and k_1 in $\mathbf{R}^3$, how do we determine whether they are ambiently isotopic subsets?

Question 14.27 Given a knot, how can one determine if it is tame?

*14.3 General Topology and Chapter 14

Advanced topics related to knot theory are covered in Section *18.5.

One successful way of looking at knot theory involves function spaces. For knotted circles in $\mathbf{R}^3$, consider the function space, $\mathcal{F}$ all continuous of S^1 into $\mathbf{R}^3$. Since S^1 is compact we can use a the natural metric on such functions or, equivalently, use the compact-open topology.

There is a subset, $\mathcal{E}_0$, in this function space consisting of embeddings. An isotopy is a path in $\mathcal{E}_0$. Furthermore, for $n \in \mathbf{N}$ we consider the subset $\mathcal{E}_n$ of $\mathcal{F}$ consisting of all functions which have exactly n points of self-intersection. Much can be learned by investigating the intersection of a path in $\mathcal{F}$ between two embeddings and examining the intersection of such a path with the subsets $\mathcal{E}_n$.

14.4 Problems for Chapter 14

14.1 Verify the continuity of the map H defined in Example 14.10.

14.2 Verify the claim, made in the remark on page 315, that X_0 and X_1 are not ambiently isotopic subsets.

14.3 Verify the statement of the remark on page 315 that X_0 and X_1 are not isotopic subsets of Y. (Hint: Suppose $F : X_0 \times I \to Y$ were an isotopy. Define $f_0 : I \to Y$ by $f_0(t) = F((0,t)$, and define $f_1 : I \to Y$ by $f_1(t) = F((1,t)$. Obtain a contradiction by showing that, for some t, we must have $f_0(t) = f_1(t)$.) Note: this problem, is about isotopy; Problem 14.2 is about *ambient* isotopy.

14.4 Prove that any two line segments in $\mathbf{R}^3$ are ambiently isotopic subsets.

14.5 Let f_0 and f_1 be embeddings of the unit interval I into $\mathbf{R}^3$ such that $f_0(I)$ and $f_1(I)$ are each line segments. Prove that f_0 and f_1 are ambiently isotopic embeddings.

14.6 Let $X_1 = \{(x,y) \in \mathbf{R}^2 : y = 0 \text{ and } -3 \leq x \leq 3\}$, $X_2 = \{(x,y) \in \mathbf{R}^2 : x = 0 \text{ and } -3 \leq y \leq 3\}$ and let $X = X_1 \cup X_2$. (Then X is homeomorphic to the letter x.) Let $A = \{(x,y) \in \mathbf{R}^2 : y = 0 \text{ and } -1 \leq x \leq 1\}$, and let $B = \{(x,y) \in \mathbf{R}^2 : y = 0 \text{ and } 0 \leq x \leq 2\}$. (See Figure 14-15.) Show that A and B are isotopic subsets but not ambiently isotopic subsets of X.

14.7 Verify the claim, made in the remark on page 312, that the map ϕ defined there is a continuous function from $\mathbf{R}^2$ to $\mathbf{R}^1$.

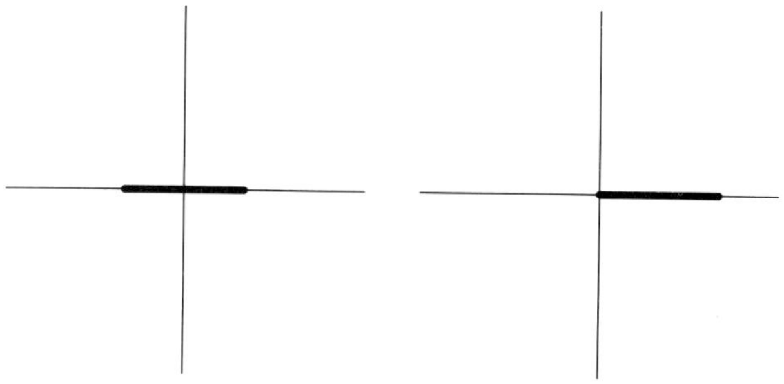

Figure 14-15 On the left is the subset A of Problem 14.6, shown as a thickened line segment; on the right is the subset B.

14.8 Verify that a polygonal knot is homeomorphic to a circle.

14.9 Verify that a polygonal arc is homeomorphic to the interval I.

14.10 Explain the sentence, in the outline of the proof for Proposition 14.23: "... one cannot simply use circular arcs for the small smooth curves since we would not be able to get the resulting curve to have continuous second derivatives."

15. SIMPLE CONNECTIVITY

OVERVIEW: We define simply connectivity, homotopy, and deformation retract. We show that the circle is not simply connected. Important applications include proof that D^2 has the fixed-point property.

15.1 Simple connectivity and homotopy

It certainly seems that the sphere and the torus cannot be homeomorphic. However, this is not easy to verify. None of the topological properties we have established distinguish them. Both are compact, path-connected 2-dimensional manifolds. Cut points do not help—if we remove a point from each, they remain path-connected. The new idea we use is to distinguish these, and many other subsets, is simple connectivity.

Our new strategy is to probe the given subset with continuous maps from the standard circle.

Definition 15.01 *Suppose $X \subseteq \mathbf{R^n}$, and $f: S^1 \to X$ is a continuous map; f is called a* **loop** *in X. We say a loop in X is* **nul-homotopic** *if there is continuous function $F: D^2 \to X$ such that $F|_{S^1} = f$.*

It is easy to give examples of nul-homotopic loops. If $X \subseteq \mathbf{R^n}$, any continuous map $F: D^2 \to X$ is a nul-homotopy of the loop F_{S^1}. It is hard to provide a detailed proof for an example of a loop which is not nul-homotopic—this is the content of Section 15.3. To motivate the utility of this concept, consider the following example.

Example 15.02 In $\mathbf{R}^3$, let C be the standard circle in the first two coordinates: $C = \{(x, y, z) \mid x^2 + y^2 = 1 \text{ and } z = 0\}$. Let L be the line corresponding to the z-axis, and let L' be the line, parallel to L with $x = 0$ and $y = -2$; see Figure 15-1.

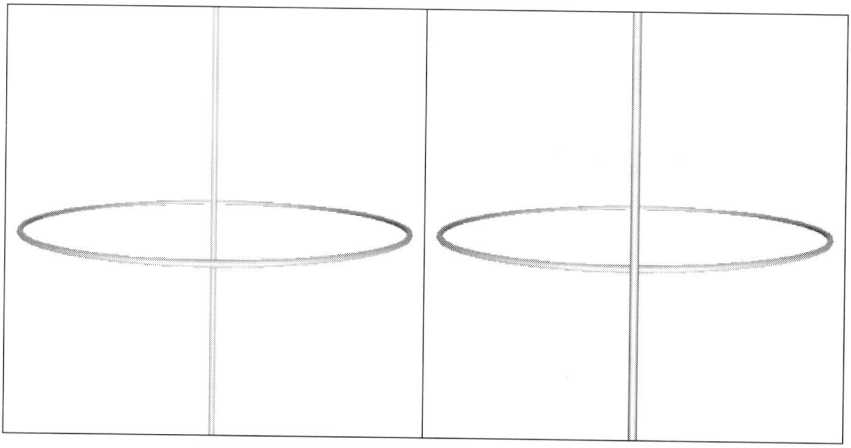

Figure 15-1 Two non-equivalent subsets of $\mathbf{R}^3$. On the left is X, a line linked about a circle. On the right is Y, different line and that same circle. Here the circle does not link the line; see Example 15.02. (In this figure, the direction of the positive y-axis is away from the reader.)

Let $X = L \cup C$ and $Y = L' \cup C$. We would like to show that X and Y are not equivalent subsets of $\mathbf{R}^3$. Intuitively, the explanation is that the circle C is linked about L, whereas C is not linked about L'. However, it take some work to articulate and prove this statement.

The first step is to view the circle C as a loop. Now, C is a *subset* of $\mathbf{R}^3$, and a loop is a *function.* We can associate a loop to C. Let $i: \mathbf{R}^2 \to \mathbf{R}^3$ be the standard inclusion. Since C does not intersect either L or L', it can be considered as a loop in X (in which case we denote it as f), or as a loop in Y (in which case we denote it as g).

We can see that g is a nul-homotopic loop in $\mathbf{R}^3 - L$; define $G: D^2 \to \mathbf{R}^3 - L'$, by $G = H \circ i|_{D^2}$. Let $B = g(D^2)$; then

$$B = \{(x, y, z) \mid (x-2)^2 + y^2 \leq 1 \text{ and } z = 0\}.$$

If h were a homeomorphism of $\mathbf{R}^3$ such that $h(X) = Y$, then (Problem 15.1) $h(L) = L'$ and $h(C) = C$. Using h, we could define a nul-homotopy $F: D^2 \to \mathbf{R}^3 - L$ of f by $F = h \circ G$. It really doesn't seem that this is possible (and it is not). The disk $F(D^2) = h(B)$ would have its boundary sent to C, yet it would not intersect L. However, to prove this is not an simple matter. So, we will not conclude this proof until the end of this chapter; see Example 15.36. ◆

The notion of nul-homotopy gives rise to a topological invariant:

Definition 15.03 *Suppose $X \subseteq \mathbf{R}^n$. We say X is* **simply connected** *if every loop in X is nul-homotopic.*

Proposition 15.04 *Suppose $X \subseteq \mathbf{R^n}$. The assertion "X is simply connected" is a topological property.* ▮ *(Problem 15.2)*

In a certain perspective this is a generalization of the concept of path connectedness. Recall $S^0 = \{-1, 1\} = \partial D^1$. We can reformulate path connectedness as follows:

Definition 15.05 *Suppose $X \subseteq \mathbf{R^n}$. We say X is* 0-**connected** *if, for every map $f: S^0 \to X$, there is a continuous function $F: D^1 \to X$ such that $F|_{S^0} = f$.*

Clearly, X is 0-connected if and only if X is path connected. The next definition combines these two concepts.

Definition 15.06 *Suppose $X \subseteq \mathbf{R^n}$. We say that X is* 1-**connected** *if it is path connected and simply connected.*

Proposition 15.07 *For any n, $\mathbf{R^n}$ is simply connected.*

Proof: We use vector notation for points of $\mathbf{R^n}$, polar coordinates for S^1 and D^2. Let $g: S^1 \to \mathbf{R^n}$ be a loop. We define a map $G: D^2 \to \mathbf{R^n}$ by

$$G(r, \theta) = r\overrightarrow{g(1, \theta)};$$

G is continuous by Proposition 3.49. The image of G is the union of all line segments in $\mathbf{R^n}$ from a point $\overrightarrow{g(1, \theta)}$ to $\vec{0}$. ▮

To more fully understand simple connectivity, we introduce the concept of homotopy, an important basic relationship between functions.

Definition 15.08 *Let $X \subseteq \mathbf{R^n}$ and $Y \subseteq \mathbf{R^m}$. Let f and g be two continuous maps from X to Y. We say f is* **homotopic** *to g, if there is a continuous map $F: X \times I \to Y$ such that $F \circ i_0 = f$ and $F \circ i_1 = g$.*

In terms of coordinates, F is a map of a subset of $\mathbf{R^{n+1}}$, and we have

$$\begin{aligned} F(x_1, x_2, ..., x_n, 0) &= f(x_1, x_2, ..., x_n), \text{ and} \\ F(x_1, x_2, ..., x_n, 1) &= g(x_1, x_2, ..., x_n). \end{aligned}$$

As in Definition 14.02, for a homotopy F we may define continuous functions from X to Y, $F_t(x) = F(x, t)$ or, equivalently, $F_t(x) = F \circ i_t$. Then $F_0 = f$ and $F_1 = g$. Sometimes a homotopy is called a "continuous one-parameter family of functions", where"parameter" refers to the variable t. A homotopy is a one-parameter family of continuous functions; an isotopy (see Definition 14.03) is a one-parameter family of *embeddings*.

The next two propositions provide alternative expressions for simple connectivity.

Definition 15.09 *Suppose $X \subseteq \mathbf{R^n}$, $Y \subseteq \mathbf{R^m}$, and $f: X \to Y$. We say f is a* **constant function** *if, for some point $y_0 \in Y$, and for all $x \in X$, $f(x) = y_0$. If a function is homotopic to a constant function, then we say f is* **nul-homotopic**.

A path homotopy is a homotopy which is fixed at the endpoints:

Definition 15.10 *Suppose $f: I \to X$ and $g: I \to X$ are paths in X with $f(0) = g(0)$ and $f(1) = g(1)$. A* **path homotopy** *from f to g is a homotopy, F_t, from f to g such that, for all $0 \leq t \leq 1$, $F_t(0) = f(0)$ and $F_t(1) = f(1)$.*

Example 15.11 Consider the following two paths in $\mathbf{R}^2$. For $0 \leq t \leq 1$, define $f(t) = (t, \sin(2\pi t))$ and $g(t) = (t, \sin(\pi t))$. These two paths, from $(0,0)$ to $(1,0)$, are path homotopic. In fact, $F(t,s) = (t, (1 - s)\sin(2\pi t)) + s\sin(\pi t)$ is such a path homotopy; see Figure 15-2. ◆

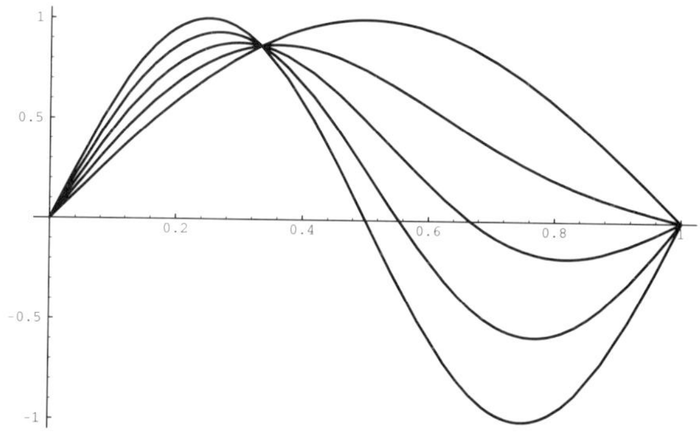

Figure 15-2 Four stages of a path homotopy, from $f(t) = (t, \sin(2\pi t))$ to $g(t) = (t, \sin(\pi t))$; see Example 15.11.

A nul-homotopy of a loop in X is a map $F: S^1 \times I \to X$ where the circle $F(S^1 \times \{1\})$ is sent to a single point. This gives rise, in a natural way, to a map of a disk into X; thus:

Proposition 15.12 *Suppose $X \subseteq \mathbf{R^n}$; X is simply connected if and only if every continuous map from the unit circle into X is nul-homotopic.* ▮ *(Problem 15.7)*

Two paths in X from a to b give rise to a loop obtained by concatenation (Definition 8.07) of the one path with the "reverse" of the other. The reverse of a path $f(t)$ is the path $f(1-t)$.

Proposition 15.13 *Suppose $X \subseteq \mathbf{R^n}$. Then X is simply connected if and only if for any pair of points, $a, b \in X$, any two paths from a to b are path homotopic.* ▮ *(Problem 15.14)*

Here is a basic property of simple connectivity:

Proposition 15.14 *Suppose $X \subseteq \mathbf{R}^n$ and $Y \subseteq \mathbf{R}^m$. If X and Y are simply connected, if and only if $X \times Y$ is simply connected.* ▮ (Problem 15.3)

15.2 Retracts and deformation retracts

The basic classification of subsets of $\mathbf{R}^n$ we have considered is classification by homeomorphism type. Another important classification, deformation type, is the topic of Chapter 16.

The notion of deformation retract is a key concept used in the definition of deformation type. The connection to simple connectivity and deformation retract is stated in Proposition 15.23.

Definition 15.15 *Suppose $A \subseteq X \subseteq \mathbf{R}^n$. We say A is a* **retract** *of X if there is a continuous function $r: X \to A$ such that, for all $a \in A$, $r(a) = a$. In this case we say the map r is a* **retraction** *of X to A.*

Definition 15.16 *Suppose $A \subseteq X \subseteq \mathbf{R}^n$. We say A is a* **deformation retract** *of X, if there is a homotopy $F: X \times I \to X$, fixed on A, such that F_0 is the identity map of X, and F_1 is a retraction of X to A. (In other words, $F(x,0) = x$ for all $x \in X$, $F(x,1) \in A$ for all $x \in A$, and $F(a,t) = a$ for all $a \in A$ and all $0 \leq t \leq 1$.). Such a homotopy, if it exists, is called a* **deformation.**

Example 15.17 Let $X = I$, and let $A = \{0\}$. Then A is a deformation retract of X. Define a deformation by $F(x,t) = x(1-t)$. In particular, A must be a retract of X.

However, let $B = \{0,1\}$. Then B is *not* a retract of X. Any map from X to B must be a constant map, because components of B are points, and X is connected. In particular, we could not have both 0 and 1 sent to themselves. ◆

Example 15.18 In the standard ball D^n, let $A = \{\vec{0}\}$. Then A is a deformation retract of D^n. The deformation is given by $F(\vec{x},t) = t\vec{x}$. ◆

Example 15.19 This example is of particular importance. Let $\vec{0}$ denote the origin of $\mathbf{R}^n$. Then S^{n-1} is a deformation retract of $\mathbf{R}^n - \vec{0}$. The deformation is given by

$$F(\vec{x},t) = (1-t)\vec{x} + t\frac{\vec{x}}{|\vec{x}|}.$$

More generally, since $\mathbf{R^n}$ is homogeneous, for any $\vec{x} \in \mathbf{R^n}$, $\mathbf{R^n} - \{\vec{x}\}$ deforms to a subset of $\mathbf{R^n}$ homeomorphic to an $(n-1)$-sphere (Problem 15.8). ◆

Example 15.20 Let Z be the subset of $\mathbf{R^4}$ corresponding to orthogonal 2×2 matrices, M with $|M| = 1$. Let X_+ correspond to the non-singular 2×2 matrices with positive determinant. Then Z is a deformation retract of X_+.

In terms of the corresponding affine maps, we can say that the set of non-singular linear maps with positive determinant deform to the set of orthogonal maps of determinant +1.

Looking at the argument outlined in Example 8.19, one can see that what is shown there is a description of what happens to a particular point durning the deformation. One needs to check for continuity as we consider a nearby matrix. We leave details to the reader (Problem 15.20). ◆

Example 15.21 Let X be the topologist's $\sin(1/x)$ curve. As in Example 8.09, write $X = L \cup \Gamma$. Then L is a retract of X; define r by $r(x, y) = (0, y)$ for all $(x, y) \in X$. But L is not a deformation retract of X. Suppose there were a deformation, $F: X \times I \to X$. Let $g \in \Gamma$. Then $F|_{\{g\} \times I}$ would be a path in X from a point of Γ to a point of L. This is not possible since Γ and L are path components of X; see Example 8.18. ◆

The example above shows that if A is a retract of X, then it may be possible that X is connected, A is path connected, A is a retract of X, and yet X is not necessarily path connected. However, suppose that A is a *deformation* retract of X. If $a, b \in X$ and $F: X \times I \to X$ is the deformation, one can obtain a path in X from a to b by concatenating paths corresponding to the subsets $F(\{a\} \times I)$ and $F(\{b\} \times I)$ with a path in A from $F(a)$ to $F(b)$. Thus we obtain:

Proposition 15.22 *Suppose $X \subseteq \mathbf{R^n}$; and A is a deformation retract of X. Then A is path connected if and only if X is path connected.* ▮ *(Problem 15.12)*

We have a similar proposition concerning simple connectivity:

Proposition 15.23 *Suppose $X \subseteq \mathbf{R^n}$ and A is a deformation retract of X. Then A is simply connected if and only if X is simply connected.*

Proof: Let $F: X \times I \to X$ denote the deformation of X to A, fixed on A, .

We first show that if X is simply connected, so is A. Let $g: S^1 \to A$ be a loop in A. Since X is simply connected, there is a map $G: D^2 \to X$ with $G|_{S^1} = g$. Then the map $G' = f_1 \circ G$ is a map $G': D^2 \to X$ with $G'|_{S^1} = g$. Thus g is nul-homotopic, and A is simply connected.

We next prove that if A is simply connected, then so is X.

Let $g: S^1 \to X$ be a loop in X. Let D be the disk in $\mathbf{R}^2$ of radius 2 and center $\vec{0}$, and let S denote the circle in $\mathbf{R}^2$ of radius 2 and center $\vec{0}$. It is convenient to define $G: D \to X$. Later, we use scaling $H(\vec{x}) = 2\vec{x}$ to get a homeomorphism $H: D^2 \to D$ which will then give rise to a map from D^2 to X.

Let Y be the annulus $Y = \{\vec{x} \in \mathbf{R}^2 \text{ with } 1 \leq |x| \leq 2\}$; then $D = D^2 \cup Y$, and $D^2 \cap Y = S^1$. The strategy is to define our map on D in two parts. Using the deformation, we define a map of Y that connects a loop in X to a loop in A; we obtain the map on D^2, using the hypothesis that A is simply connected.

As defined, we have $S^1 \times I \subseteq \mathbf{R}^3$ and $\subseteq \mathbf{R}^2$. Define a homeomorphism $\phi: Y \to S^1 \times I$ by

$$\phi(x, y) = \left(\frac{\vec{x}}{|\vec{x}|}, |\vec{x}| - 2\right).$$

We next define a map $\Gamma: D \to X$. Roughly, Γ traces out the image of the loop g under the homotopy on Y and is the given nul-homotopy on D^2. If $\vec{x} \in Y$, define $\Gamma(\vec{x}) = F \circ (g \times Id|_I) \circ \phi$:

$$Y \xrightarrow{\phi} S^1 \times I \xrightarrow{g \times Id|_I} X \times I \xrightarrow{F} X.$$

Note that $\Gamma|_{S^1}$ is a loop in A, which we are assuming is simply connected. By Proposition 15.12 we can find a nul-homotopy of this loop. We use this nul-homotopy to define $\Gamma|_{D^2}$. By the gluing lemma, we obtain a continuous map $\Gamma: D \to X$.

We are almost there—we need a map of D^2, not D. Let $G = \Gamma \circ H$. This is a nul-homotopy in X of the loop g. ∎

15.3 The circle is not simply connected

We show that the circle is not simply connected. Fundamental to the proof is the map $p: \mathbf{R}^1 \to S^1$, defined by $p(x) = (\cos(x), \sin(x))$. The proof hinges on the following two key lemmas.

Proposition 15.24 Path lifting lemma. *Suppose $f: I \to S^1$ is a path; let $x_0 = f(0)$. Let $\tilde{x}_0 \in \mathbf{R}^1$ be a point of $p^{-1}(x_0)$. Then there is a unique path $\tilde{f}: I \to \mathbf{R}^1$ such that $\tilde{f}(0) = \tilde{x}_0$ and $p \circ \tilde{f} = f$. In other words, the following diagram commutes:*

$$\begin{array}{ccc} & & \mathbf{R}^1 \\ & \tilde{f} \nearrow & \downarrow p \\ I & \xrightarrow{f} & S^1 \end{array}$$

Definition 15.25 *If f and $\tilde{f}$ are as in Proposition 15.24, we say that $\tilde{f}$ is a* **lifting** *of f.*

Proposition 15.26 Path homotopy lifting. *Suppose $F: I \times I \to S^1$ is a path homotopy, and $F(0,0) = x_0$. Let $\tilde{x}_0$ be a point of $p^{-1}(x_0)$. Then there is a (unique) path homotopy, $\tilde{F}: I \times I \to \mathbf{R}^1$, such that $\tilde{F}(0,0) = \tilde{x}_0$, and such that $p \circ \tilde{F} = F$. In other words, the following diagram commutes:*

$$\begin{array}{ccc} & & \mathbf{R}^1 \\ & \tilde{F} \nearrow & \downarrow p \\ I \times I & \xrightarrow{F} & S^1. \end{array}$$

∎ *(Problem15.16)*

Definition 15.27 *If $\tilde{F}$ and F are as in Lemma 14.3, we say that $\tilde{F}$ is a* **lifting** *of the homotopy F.*

In Proposition 15.26, the map $\tilde{F}$ is a path homotopy. Suppose F is a path homotopy between paths f_0 and f_1 where the endpoints of these paths are x_0 and x_1. If $F(1,0) = \tilde{x}_1$, then $\tilde{F}$ will be a path homotopy in $\mathbf{R}^1$, between the paths $\tilde{f}_0$ and $\tilde{f}_1$. These paths have endpoints $\tilde{x}_0$ and $\tilde{x}_1$. The particular observation that $\tilde{f}_0(1) = \tilde{f}_1(1)$ will be important to us in the proof that the circle is not simply connected.

Before proving Propositions 15.24 and 15.26, we show their utility by proving the following proposition.

Proposition 15.28 *The circle S^1 is not simply connected.*

Proof (assuming Propositions 15.24 and 15.26): By Proposition 15.13, it is enough to show that there are two paths, f and g, in S^1, with $f(0) = g(0)$, and $f(1) = g(1)$ such that f and g are not path homotopic.

Our two paths will be from $(1,0)$ to $(-1,0)$: $f(t) = (\cos(\pi t), \sin(\pi t))$, and let g be defined by $g(t) = (\cos(-\pi t), \sin(-\pi t))$. Note that the image of f is the upper semicircle; the image of g is the lower semicircle.

We argue by contradiction. Suppose that there is a path homotopy F from f to g. Let $x_0 = f(0) = (1,0)$. Let $\tilde{x}_0 = 0$. Let $\tilde{f}$ and $\tilde{g}$ be the lifting of the paths given by Proposition 15.24. By Proposition 15.26, there is a path homotopy $\tilde{F}$, which is a lifting of F. In particular, $\tilde{f}$ and $\tilde{g}$ are path homotopic. This implies that $\tilde{f}(1) = \tilde{g}(1)$. However, the maps $\tilde{f}$ and $\tilde{g}$ are uniquely determined and therefore must be the maps $\tilde{f}(t) = \pi t$ and $\tilde{g}(t) = \pi t$. Thus $\tilde{f}(1) = \pi$ and $\tilde{g}(1) = -\pi$. ∎

Proof of Proposition 15.24: The idea of the proof is this. If the image of f is small enough (by which we mean the image of f not all of S^1) then we can define $\tilde{f}$, as we next explain. Suppose $q \notin Im(f)$, Then $J =$

$S^1 - \{q\}$ is homeomorphic to an open interval. Furthermore consider $p^{-1}(J)$. If $\tilde{J}$ is a component of $p^{-1}(J)$, then $\tilde{J}$ is homeomorphic to J and, in fact, $p|_{\tilde{J}}$ is a homeomorphism from $\tilde{J}$ to J. Let $\tilde{J}_0$ be the component of $p^{-1}(J)$ which contains $\tilde{x}_0$. Let $p_0 = p|_{\tilde{J}_0}$; this is a homeomorphism, and we can now define our lifting by $\tilde{f} = p_0^{-1} \circ f$ (Problem 15.15).

Here is the outline of the general case. Divide I into pieces which are small enough and do not go completely around the circle. We then define our lifting for these pieces, and then, sequentially, "patch" these functions together. (In outline, the proof of Proposition 15.26 is the same; see hints for Problem 15.16.) Here are the details of this process.

Let $A = S^1 - \{(-1,0)\}$, and $B = S^1 - \{(1,0)\}$. For each n, we divide I into 2^n equal intervals I_j^n where $j = 1, \ldots, 2^n$. (We choose the subscript j so that for a fixed n we have I_j^n to the left of I_i^n if and only if $j < i$. For example, if $n = 2$, we divide I into four intervals, I_1^2, I_2^2, I_3^2, and I_4^2, arranged from left to right in I.)

We claim that we can subdivide I into small enough subintervals such that, for each, its image via f does not go completely around the circle. That is, there is an integer, N, such that if we subdivide I into intervals I_j^N,then for each j, $1 \leq j \leq 2^N$, we have either $I_j^n \subseteq f^{-1}(A)$ or $I_j^n \subseteq f^{-1}(B)$.

We prove this claim in Proposition 15.29, below.

Let C be a component of $f^{-1}(A)$; C is an open interval. Moreover, $p|_C$ is a homeomorphism of C onto A. (This property of p is critical to the proof.) Similarly, if C is a component of $f^{-1}(B)$, then C is an open interval and $p|_C$ is a homeomorphism of C onto B.

We begin by defining the lifting for the first subinterval I_1^N. Since $x_0 = (1,0)$, $x_0 \in A$, and $\tilde{x}_0 = 2q\pi$ for some $q \in \mathbf{N}$. There is a unique lifting of $f|_{I_1^N}$ whose image lies in the interval $\tilde{U}_1 = (2q\pi - \pi, 2q\pi + \pi)$. Let q_1 denote the inverse of the homeomorphism $p|_{\tilde{U}_1}$; so $q_1 : A \to \tilde{U}_1$. Define $\tilde{f}|_{I_1^N}$ by $\tilde{f}|_{I_1^N} = q_1 \circ f$. For $x \in I_1^N$, we see that $p \circ \tilde{f} = f$.

Next, we consider I_2^N. Now $x_1 = f(1/2^N)$ is the image under f of the left-hand endpoint of I_2^N (which is the same as the right-hand endpoint of I_1^N). Let $\tilde{x}_1 = \tilde{f}(1/2^N)$, the image under $\tilde{f}$ of the right-hand endpoint of I_1^N. We know that $I_2^N \subseteq U$ where $U = A$ or $U = B$. Let $\tilde{U}_2$ be the component of $p^{-1}(U)$ which contains $\tilde{x}_1$. Let q_2 denote the inverse of the homeomorphism $p|_{\tilde{U}_2}$; so $q_2 : U \to \tilde{U}_2$.

Define $\tilde{f}|_{I_2^N}$ by $\tilde{f}|_{I_2^N} = q_2 \circ f$. The two functions $\tilde{f}|_{I_1^N}$ and $\tilde{f}|_{I_2^N}$ are each continuous, defined on closed subsets, and they agree on the intersection. Thus they define a continuous function on $I_1^N \cup I_2^N$. Furthermore, on $I_1^N \cup I_2^N$, $p \circ \tilde{f} = f$. Continuing in this way, in a finite number of steps, we define the function $\tilde{f} : I \to \mathbf{R}^1$. The function $\tilde{f}$ is unique since it is uniquely defined at each step. ∎

This next proposition is a consequence of Proposition 10.29 on the Lebesgue number, but it is worthwhile giving a separate proof for this case.

Proposition 15.29 *If W_0 and W_1 are two open subsets of I such that $W_0 \cup W_1 = I$, then there is an integer N such that if we subdivide I into 2^N equal subintervals I_j^N, $j = 1..., 2^N$, then either $I_j^N \subseteq W_0$ or $I_j^N \subseteq W_1$.*

Proof: Suppose this were not true. Then consider $N = 2$ where we divide I into two equal sub-intervals. Let X_2 be the union of those sub-intervals of the form I_j^2 such that neither $I_j^2 \subseteq W_0$ or $I_j^2 \subseteq W_1$. Then X_2 is closed since it is a finite union of closed intervals, and non-empty by our hypothesis. We may similarly define X_3 as the union of all subintervals I_j^3 such that neither $I_j^3 \subseteq W_0$ or $I_j^3 \subseteq W_1$. Note that $X_2 \supseteq X_3$. Continuing in this way, we may obtain an infinite sequence of non-empty, closed, nested sets X_n. Let $X = \bigcap_{ninN} X_n$. Since I is compact, X is non-empty by Proposition 10.45. Let $x \in X$. By our hypotheses, there is an $\epsilon > 0$ such that $N_\epsilon(x)$ is entirely contained in either W_0 or W_1. Now choose N such that $2^{-N} < \epsilon$. Since $x \in X_{N+1}$, then x is contained in a sub-interval of X_{N+1} of length $2^{-(N+1)}$, which is not entirely in W_0 nor entirely in W_1. But since this subinterval must be contained in $N_\epsilon(x)$, it must be in either W_0 or W_1. Thus we have a contradiction. ∎

For other spheres, S^n with $1 < n$, we have the following important result:

Proposition 15.30 *If $1 < n$ then S^n is simply connected.* ∎

A complete proof of Proposition 15.30 is beyond the scope of this text; however a proof can be found in many algebraic topology texts such as [6, 23, 32, 30, 42, 44]. However we give an outline of two arguments that S^2 is simply connected. Let $f: S^1 \to S^2$ be a loop. Suppose there is $x \in S^2$, which is *not* in the image of f. Then, by Proposition 4.06, there is a homeomorphism, $h: (S^2 - \{x\}) \to \mathbf{R}^2$. Consider the loop $f' = h \circ f$, $f': S^1 \to \mathbf{R}^2$. Since $\mathbf{R}^2$ is simply connected, there is a map $F': D^2 \to \mathbf{R}^2$ such that $F|_{S^1} = f'$. Define $F = h^{-1} \circ F'$. Then $F: D^2 \to S^2$ and $F|_{S^1} = f$ since

$$\begin{aligned} F|_{S^1} &= h^{-1} \circ F'|_{S^1} = h^{-1} \circ f' = h^{-1} \circ (h \circ f) \\ &= (h^{-1} \circ h) \circ f) = Id|_{S^2 - \{x\}} \circ f = f. \end{aligned}$$

Thus we see the loop f is nul-homotopic. So, the only problem (but it is a big problem) is that there may not exist an $x \in S^2$ which is not in the image of f. There are loops which go through *every* point of S^2; see Proposition 12.03.

One can show (this is the difficult part of the proof) that we can "slide" the given loop f by a homotopy to a new loop g so that it misses some point x of S^2. The argument above shows that g is nul-homotopic via a map $G: D^2 \to S^2 - \{x\}$. We can combine these homotopies to a nul-homotopy of f.

A second method involves the analysis of S^2 as the union of two hemispheres. Each hemisphere is simply connected since it is homeomorphic to the standard disk D^2. Then we need to prove a theorem telling us when the union of two simply connected sets is simply connected. There is such a result known as the Siefert-Van Kampen Theorem, and a critical condition is that the intersection of the two pieces is connected. (Note: This theorem does not imply that the circle is simply connected. We can write the circle as the union of semicircles, each of which is simply connected. However, the intersection is not connected.)

15.4 Some implications of the non-simple connectivity of S^1

An important application of the Proposition 15.30, in the case of S^2, is

Proposition 15.31 $\mathbf{R}^2$ *and* $\mathbf{R}^3$ *are not homeomorphic.*

Proof: We proceed indirectly. Suppose there is homeomorphism $h: \mathbf{R}^2 \to \mathbf{R}^3$. Letting $\vec{y} = h(\vec{0})$, it follows that $h|_{\mathbf{R}^2 - \{\vec{0}\}}$ is a homeomorphism from $\mathbf{R}^2 - \{\vec{0}\}$ to $\mathbf{R}^3 - \{\vec{\mathbf{y}}\}$.

By Example 15.19, $\mathbf{R}^2 - \{\vec{0}\}$ deforms to a space homeomorphic to S^1, which is not simply connected. By Proposition 15.23, $\mathbf{R}^2 - \{\vec{0}\}$ is not simply connected. But $\mathbf{R}^3 - \{\vec{\mathbf{y}}\}$ deforms to a subset homeomorphic to S^2. By Proposition 15.30, $\mathbf{R}^3 - \{\vec{\mathbf{y}}\}$ is simply connected. Since simple connectivity is a topological invariant, we have a contradiction. ∎

Here is another important consequence of the fact that the circle is not simply connected (a proof in a slightly different form is in the remark on page 341):

Proposition 15.32 S^1 *is not a retract of* D^2.

Proof: We argue by contradiction. Assume there were a retraction $r: D^2 \to S^1$. The strategy is to contradict the fact that S^1 is not simply connected.

Let $f: S^1 \to S^1$ be any loop in S^1. Consider f as a map into D^2 by setting $f' = i \circ f$ where i is the inclusion $i: S^1 \to D^2$. Since D^2 is simply connected, there is a map $F': D^2 \to D^2$ such that $F'|_{S^1} = f'$.

Now consider $F = r \circ F'$. We have $F: D^2 \to S^1$ and

$$F|_{S^1} = r \circ F'|_{S^1} = r \circ f' = r \circ (i \circ f) = (r \circ i) \circ f = id|_{S^1} \circ f = f.$$

So, f is nul-homotopic and S^1 is simply connected, contradicting Proposition 15.28. ∎

An important consequence of Proposition 15.32 is

Proposition 15.33 *Any continuous map from D^2 to D^2 has a fixed point.*

Proof: Let $f: D^2 \to D^2$ be given. We show there is a point $x \in D^2$ such that $f(x) = x$.

The proof is by contradiction. Assume there is a function f that has no fixed point. For each $x \in D^2$, we have $f(x) \neq x$. Let R_x be the ray which begins at $f(x)$ and which passes through x. Then R_x meets S^1 in a well-defined point, call it $r(x)$. With a bit of work it can be verified that $r(x)$ is continuous (Problem 15.18). Note that if $x \in S^1$, then $r(x) = x$. Thus we have found a retract of D^2 onto S^1. But this contradicts Proposition 15.32. ∎

This theorem is true for n-balls; a proof may be found in almost every text on algebraic topology such as [6, 23, 32, 30, 42, 44]. Many view this result as the most useful result of topology.

Proposition 15.34 Fixed point theorem for D^n *Any continuous map from D^n to D^n has a fixed point.* ∎

In Example 15.17 we have shown a version of this proposition for closed intervals.

Next, a variation of Proposition 15.32 which we use to complete the long-postponed proof for Example 15.02:

Proposition 15.35 *There is no continuous map f of the disk D^2 to the circle S^1 which is a homeomorphism on ∂D^2.*

Proof: Suppose there were such a map f. Without loss of generality, we may assume that $f((1,0)) = (1,0)$, (Problem 15.19). Let $\phi: [0,1] \to D^2$ be defined by $\phi(\theta) = (\cos(\theta), \sin(\theta))$. We have a path $\psi = f \circ \phi$ in S^1 which begins and ends at $(1,0)$. Consider the lifting of this path $\tilde{\psi}$.

Since $f|_{\partial D}$ is a homeomorphism, p must be one-to-one on the image of $\tilde{\psi}$, except at the endpoints. It follows that $\tilde{f}(0) = \pm 1$. On the other hand, we must have $\tilde{f}(0) = 0$ since the map $F: D \to S^1$ also lifts. ∎

Example 15.36 We close this chapter by completing the proof begun in Example 15.02.

Consider the continuous map $\rho: \mathbf{R}^3 - L \to C$, which we define by using cylindrical coordinates $\rho(r, \theta, z) = (1, \theta, z)$; $\rho|_C = Id|_C$

Suppose there were an $F : D^2 \to \mathbf{R}^3 - L$, as in Example 15.02, with $F(S^1) = C$. Identifying C with S^1, $\rho \circ h$ would be a map of a disk into the circle which is a homeomorphism on the boundary of the disk, contradicting Proposition 15.35. ◆

*15.5 General Topology and Chapter 15

From the abstract viewpoint, a homotopy is a path in a function space, if we use the compact-open topology. Clearly, a homotopy F_t represents a function from I to the function space $C(X, Y)$, namely, $t \to F_t$. The continuity of this function is a consequence of the following proposition (which also highlights the utility of the compact-open topology):

Proposition 15.37 *Let $\{X, \mathcal{T}\}$ be a locally compact, path-connected Hausdorff space and consider $C(X, Y)$ with the compact-open topology. If $F: X \times I \to Y$ define $\phi: I \to C(X, Y)$ by $\phi(t) = F_t$. Then F is continuous if and only if ϕ is continuous.* ∎

A proof of the proposition above may be found in many standard topology texts such as [32, 42]. If ϕ is continuous, then F will be continuous since F is a composite of the continuous functions:

$$X \times I \overset{Id_X \times \phi}{\longrightarrow} X \times C(X, Y) \overset{e}{\to} Y$$

where en is the evaluation map; see Proposition 11.15.

As a consequence we have

Proposition 15.38 *Suppose $\{X, \mathcal{T}\}$ is a locally compact Hausdorff space. Then X is 1-connected if and only if $C(S^1, X)$, with the compact open topology, is path connected.* ∎

Considering a loop in X as a closed path has an advantage in that it allows use of concatenation to combine loops. The key observation is that we can extend this operation to obtain a method for combining path homotopies.

Proposition 15.39 *Let X be a topological space, f and g two paths in X. The relation "f is path homotopic to g" is an equivalence relation on the set of all paths in X.*

If f is a path in X, then $[f]$ will denote the path homotopy equivalence class of f. Using concatenation, Definition 8.07, we define an operation on these equivalence classes:

Proposition 15.40 *Suppose f and g are two paths in X with $f(1) = g(0)$. Define $[f] * [g] = [f * g]$. This is a well-defined operation. That is, if F is path homotopic to f, and G is path homotopic to g, then $[f * g] = [F * G]$.* ∎

Note that $[f]$ consists of paths fixed on $f(0)$ and $f(1)$ and paths of $[f * g]$ are fixed at $f(0)$ and $g(1)$, but not necessarily fixed on $f(1)$.

Pay close attention to detail in the equation $[f] * [g] = [f * g]$. The symbol $*$ is used in two very different ways. On the right it refers to concatenation of paths; on the left it refers to an operation on homotopy classes of maps. Alertness to context is sometimes required to properly interpret the meaning of this symbol.

If $f, g \in \mathcal{P}(X, x_0)$ are closed paths in X, then $f(1) = g(0) = x_0$, and the concatenation $f * g$ is defined.

Definition 15.41 *Let X be a topological space, $x_0 \in X$. The fundamental group of X with base point x_0, denoted $\pi_1(X, x_0)$, is the set of path homotopy equivalence classes of paths $f: I \to X$ such that $f(0) = f(1) = x_0$.*

In Definition 15.09 we defined nul-homotopy of a loop. There is a related notion of nul-homotopy of an element of $\mathcal{P}(X, x_0)$.

Definition 15.42 *If X is a topological space, and $f: I \to X$ such that $f(0) = f(1) = x_0$ is a path, we say f is a* **nul-homotopic path** *if f is path homotopic to the constant map of I to x_0. We denote this by $[f] = 1$.*

If the only element of $\pi_1(X, x_0)$ is 1, then we say that $\pi_1(X, x_0)$ **is trivial**

It is not hard to show that, for path-connected spaces, the base point has little import. For example,

Proposition 15.43 *If X is a path-connected topological space, $x_0, x_1 \in X$, $\pi_1(X, x_0)$ is trivial if and only if $\pi_1(X, x_1)$ is trivial.* ∎

Extending ,to the abstract setting, our definition of simple connectivity, Definition 15.03, so that we may speak of simple connectivity of a topological space, we have::

Proposition 15.44 *Let X be a topological space. Then X is simply connected if and only if for all $x_0 \in X$, $\pi_1(X, x_0)$ is trivial. Also, X is 1-connected if and only if for some $x_0 \in X$, $\pi_1(X, x_0)$ is trivial and X is path connected.* ∎

Up to this point we have gained little by using paths and path homotopy rather than loops and homotopy of loops.

We begin to see significance of the operation $[f] * [g]$ when we investigate its properties. Recall we have defined $f(1 - t)$ to be the reverse of the path F; see remark on page 329.

Definition 15.45 *Let X be a topological space, $x_0 \in X$. Suppose $[f] \in \pi_1(X, x_0)$. Write $\alpha = [f]$. The* **inverse of** *α is defined to be $[f(1-t)]$ and is denoted α^{-1}.*

Proposition 15.46 *Let X be a topological space, $x_0 \in X$ write $\alpha = [f]$, $\beta = [g]$ and $\gamma = [h]$, where $[f], [g], [h] \in \pi_1(X, x_0)$, then:*

(a) $(\alpha * \beta) * \gamma = \alpha * (\beta * \gamma)$
(b) $1 * \alpha = \alpha * 1 = \alpha$
(c) $\alpha * \alpha^{-1} = 1$. ∎

In the language of algebra, the set $\pi_1(X, x_0)$, together with the binary operation $*$, has the structure of a group.

The basic type of function considered for groups is homomorphism. The basic equivalence relation for groups is isomorphism. Not only do topological spaces correspond to groups, but continuous maps give rise to homomorphisms, and homeomorphisms give rise to isomorphisms.

Definition 15.47 *Suppose X and Y are path-connected topological spaces, $x_0 \in X$, $y_0 \in Y$, and suppose we have a continuous map $\phi: X \to Y$ with $\phi(x_0) = y_0$. Define a map $\phi_* \pi_1(X, x_0) \to \pi_1(Y, y_0)$ by $\phi_*([f]) = [\phi \circ f]$. We say ϕ_* is the map of fundamental groups* **induced** *by ϕ.*

It needs to be shown that Definition 15.47 is actually well-defined (does not depend on choice of representative of the equivalence classes involved). This is not a difficult matter.

Furthermore we have some nice properties:

Proposition 15.48

(a) *Let X be a topological space with $x_0 \in X$. The identity homeomorphism on X induces the identity isomorphism of $\pi_1(X, x_0)$.*
(b) *Suppose X, Y, and Z are path-connected topological spaces with $x_0 \in X$, $y_0 \in Y$, and $z_0 \in Z$. If we have continuous maps $\phi: X \to Y$ and $\psi: Y \to Z$ with $\phi(x_0) = y_0$ and $\psi(y_0) = z_0$. Then:*

$$\psi_* \circ \phi_* = (\psi \circ \phi)_*.$$

(c) *Let X be a topological space with $x_0 \in A \subseteq X$, A path-connected, A a deformation retract of X with deformation $r: X \to A$. Then $r_*: \pi_1(X, x_0) \to \pi_1(A, x_0)$ is an isomorphism.*

A consequence of Proposition 15.48 is that the fundamental group is a topological invariant, by which we mean:

Proposition 15.49 *If X and Y are homeomorphic path-connected topological spaces with $x_0 \in X$, and $y_0 \in Y$, then $\pi_1(X, x_0)$ is isomorphic to $\pi_1(Y, y_0)$.*

Proof: Suppose $h: X \to Y$ is a homeomorphism, thenŁ

$$h^{-1} \circ h = Id|_X \text{ and } h \circ h^{-1} = Id|_Y.$$

By Proposition 15.48 we have

$$h_*^{-1} \circ h_* = Id|_{\pi_1(X,x_0)} \text{ and } h_* \circ h_*^{-1} = Id|_{\pi_1(Y,y_0)}.$$

Elementary facts about homomorphisms imply that h_* is an isomorphism.

Moreover, a similar proof using, in addition, part (c) of Proposition 15.48, will show that the fundamental group is a deformation type invariant:

Proposition 15.50 *If X and Y are path-connected topological spaces of the same deformation type with $x_0 \in X$, and $y_0 \in Y$, then $\pi_1(X, x_0)$ is isomorphic to $\pi_1(Y, y_0)$.* ∎

Recall the map $p: \mathbf{R}^1 \to S^1$, and let $x_0 = (1, 0)$. By the path lifting lemma, Proposition 15.26, any path in S^1, f, beginning at x_0 will lift to a path, in $\mathbf{R}^1$, which starts at 0, and ends at a number of the form $n2\pi$, for some integer n. By examining this further one can show that this correspondence of f to an integer n gives rise to an isomorphism of $\pi_1(S^1, x_0)$ to the integers:

Proposition 15.51 *The fundamental group of the circle $\pi_1(S^1, x_0)$ is isomorphic to the integers.* ∎

To begin to get an understanding of the power of these ideas, consider the brevity of the following alternate proof of Proposition 15.32 that S^1 is not a retract of D^2. If there were a retract $r: D^2 \to S^1$, then we would have $i \circ r = Id|_{S^1}$. Thus $i_* \circ r_* = Id|_{\pi_1(S^1,x_0)}$, and so $i_* \circ r_*$ is an isomorphism of the integers. However, $\pi_1(D^2, x_0) = 1$ since D^2 is simply connected, and so i_* must map $\pi_1(S^1, x_0)$ to the single element 1. But then the composition $i_* \circ r_*$ could not be an isomorphism since it fails to be one-to-one.

15.6 Problems for Chapter 15

15.1 Verify the assertion in Example 15.02 that if h were a homeomorphism of $\mathbf{R}^3$ such that $h(X) = Y$, then $h(L) = L'$ and $h(C) = C$.

15.2 Prove Proposition 15.04.

15.3 Prove Proposition 15.14.

15.4 Let $X = I$, and let $A = \{0, 1\}$. Show that A is not a retract of X.

15.5 Let X be the topologist's comb of Example 2.43 and let T_0 also be as described in that example. Show that T_0 is a retract of X but not a deformation retract.

15.6 Show that "f is homotopic to g" is an equivalence relation in the set of all continuous maps from X to Y.

15.7 Prove Proposition 15.12.

15.8 Verify the assertion of Example 15.19 that, for any $\vec{x} \in \mathbf{R^n}$, $\mathbf{R^n} - \{\vec{x}\}$ deforms to a subset of $\mathbf{R^n}$ homeomorphic to an $(n-1)$-sphere .

15.9 Suppose $X \subseteq \mathbf{R^n}$.

(a) Suppose $C \subseteq A \subseteq X$ where C is a component of A and $r: A \to X$ is a deformation retraction. Show that $r(C) \subseteq C$.
(b) Suppose $C \subseteq A \subseteq X$ where C is a *path-component* of A and $r: A \to X$ is a deformation retraction. Show that $r(C) \subseteq C$.
(c) Suppose X is the piecewise-linear $\sin(1/x)$ curve, Example 8.11. What can you say about a subset $A \subseteq X$ such that A is a deformation retract of X; what subsets A are possible?

15.10 Prove that the set of non-singular linear maps of $\mathbf{R}^2$ is a deformation retract of the set of all non-singular affine maps of $\mathbf{R}^2$. (Hint: See Example 8.22.)

15.11 Assuming one has defined the paths as in Example 8.19, define a corresponding deformation of X_+ to Z, as mentioned in Example 15.20.

15.12 Prove Proposition 15.22; see outline following to the statement.

15.13 Let C be a circle in $\mathbf{R}^2$, with center p and radius r. Show $\mathbf{R}^2 - C$ has two components one of which is simply connected and the other which is not. (Remark: these are the inside and outside of the circle as defined in Example 10.21.)

15.14 Prove Proposition 15.13. ; see comment prior to statement

15.15 Verify the map defined as $\tilde{f} = p_0^{-1} \circ f$ in the proof of Proposition 15.24 is a lifting of the path f as claimed.

15.16 Prove Proposition 15.26. (Hint: Follow the outline for the proof of the path lifting lemma. Now you will be concerned with maps of the square rather than an interval, so you will have to divide the square into smaller squares, and prove a version of Proposition 15.29. There is one place where you will have to be especially careful with the order in which you define the maps of the sub-squares. Suppose you had divided the square into sixteen equal squares and defined the lifting for and suppose that you had ordered the sub-squares in a spiral fashion as shown on the left of Figure 15-3. The twelfth sub-square meets the previous sub-squares in two disjoint line segments: S, the intersection of sub-square 12 and T, the intersection sub-square 12 and the first sub-square. If you define the lifting map in the twelfth sub-square so that it matches up the way you want on S, how do you know that it will match up correctly on T? The best solution is to make sure you order the sub-squares so that this won't occur. If we use ordering such as on the right in Figure 15-3, then the k-th sub-square meets the union of sub-squares of lower order in a connected subset.)

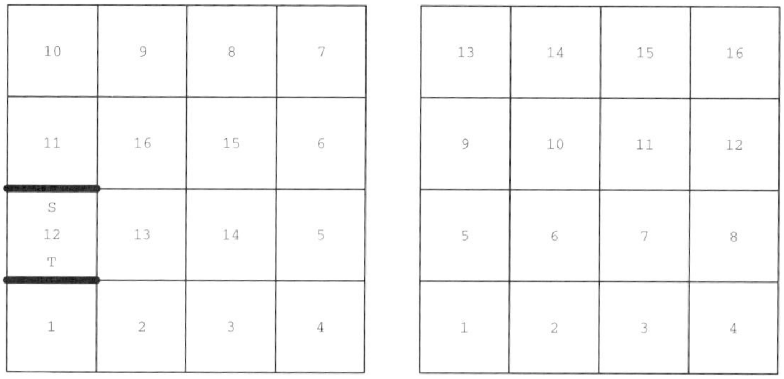

Figure 15-3 In defining a path lifting, one needs to order sub-squares, as on the right, rather than on as the left; see Problem 15.16.

15.17 We consider two tangent circles, C_1 and C_2. Here $C_1 = \{(x, y) \in \mathbf{R}^2 \mid x^2 + y^2 = 1\}$, $C_2 = \{(x, y) \in \mathbf{R}^2 \mid (x - 2)^2 + y^2 = 1\}$. Let $X = C_1 \cup C_2$. Show that X is not simply connected.

15.18 Verify that the function $r(x)$ defined in the proof of Proposition 15.33 is continuous.

15.19 Explain the statement in proof of Proposition 15.35: Without loss of generality, we may assume that $f((1,0)) = (1,0)$.

15.20 Recall the circle C and line L of Example 15.02. Prove that C is a deformation retract of $\mathbf{R}^3 - L$ (Hint: See Example 15.36 for a retraction.)

16. DEFORMATION TYPE

OVERVIEW: The basic equivalence notion we have been using for subsets is homeomorphism type. We introduce the important notion of equivalence, deformation type, which is less discriminating, and thus, often useful. We discuss a basic construction, the mapping cylinder.

16.1 Definitions and examples of deformation type

For many subsets, the next definition, deformation type, is equivalent to an important topological notion called "homotopy type," Definition 16.18. (In Proposition 17.33 we make more explicit subsets for which this is true.)

Recall the definition of deformation retract, Definition 15.16.

Definition 16.01 *Suppose $X \subseteq \mathbf{R^n}$ and $Y \subseteq \mathbf{R^m}$. We say X and Y have the same* **deformation type** *if there is a finite sequence of subsets, X_i, each a subset of some Euclidian space, where $1 \leq i \leq n$, such that $X_1 = X$, $X_n = Y$, and that for all i with $1 \leq i \leq n-1$ we have either:*

(a) *X_i is a deformation retract of X_{i+1}.*
(b) *X_{i+1} is a deformation retract of X_i.*
(c) *X_i and X_{i+1} are homeomorphic.*

The following notations may be helpful. Let $X \swarrow Y$ denote X is a deformation retract of Y, $X \searrow Y$ denote Y is a deformation retract of X, and $X \approx Y$ denote X and Y are homeomorphic. Then, for example, if we have

$$X = X_0 \searrow X_1 \approx X_2 \swarrow X_3 \searrow X_4 = Y,$$

then X and Y have the same deformation type

The next two propositions follow from basic properties of homeomorphism and deformation retract.

Proposition 16.02 *Suppose $X \subseteq \mathbf{R^n}$ and $Y \subseteq \mathbf{R^m}$. "X is the same deformation type as Y" is an equivalence relation.* ∎ *(Problem 16.1)*

Proposition 16.03 *Suppose $X \subseteq \mathbf{R^n}$ and $Y \subseteq \mathbf{R^m}$. If X and Y have the same deformation type, then there is a one-to-one correspondence between the path components of X and the path components of Y.* ∎ *(Problem 16.2)*

Example 16.04 One consequence of Proposition 16.03 is that there are subsets which are not of the same deformation type. In fact, there is a countable collection of subsets X_i (say of the line) which all have distinct deformation type. We just let X_i consist of i distinct points. ◆

Proposition 16.05 *Suppose $X \subseteq \mathbf{R^n}$ and $Y \subseteq \mathbf{R^m}$. If X and Y have the same deformation type then X is simply connected if and only if Y is simply connected.* ∎ *(Problem 16.3)*

Example 16.06 The subsets of the plane corresponding to "the letter Q" and "the letter P" are of the same deformation type, but they are not homeomorphic. If we "shrink the tail of the Q" and "shrink the stem of the P," then each of these sets becomes homeomorphic to a circle.

One simple type of subset is a contractible set.

Definition 16.07 *If $X \subseteq \mathbf{R^n}$, $p \in X$, and $\{p\}$ is a deformation retract of X, then we say that X is* **contractible**.

For example all of the subsets described in Example 7.28 are contractible (Problem 16.5).

Example 16.08 The topologist's comb, Example 2.43 is contractible, however the related subset, shown in Figure 16-1, is not contractible. A detailed proof of this is not an easy exercise, but the basic reasoning is clear. Roughly, the idea is that points on the teeth to the left would have to travel upwards during the deformation; points of the teeth to the right would have to travel down; thus continuity would then be violated in the middle. ◆

From Proposition 15.22 and 15.23, we conclude

Proposition 16.09 *If $X \subseteq \mathbf{R^n}$ and is a contractible subset, then X is 1-connected.* ∎

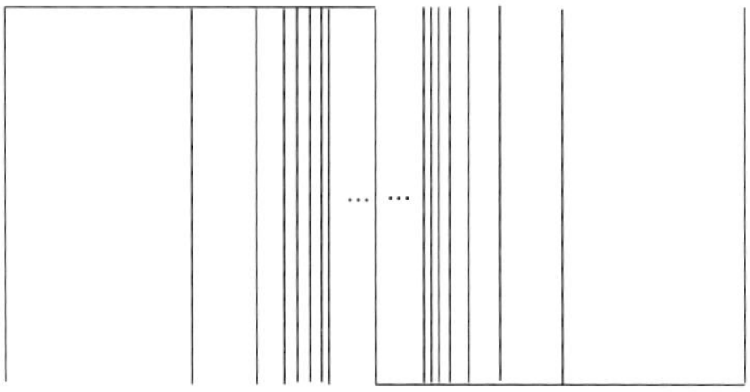

Figure 16-1 A variation of the topologist's comb which is not contractible; see Example 16.08.

The notion of deformation type will be of little usefulness unless we have a plentiful supply of examples of deformation retracts. An excellent source comes from the next construction, the "mapping cylinder."

The idea of a mapping cylinder is to associate a subset to a function $f: X \to Y$. We already have one way to do this, the graph of the function, Definition 3.27. The mapping cylinder is a very different construction. It is assembled from $X \times I$ and Y and is not, in any natural way, a subset of $X \times Y$.

Definition 16.10 *Let $X \subseteq \mathbf{R}^n$, $Y \subseteq \mathbf{R}^m$, and $f: X \to Y$ a continuous map.*

If $Z \subseteq \mathbf{R}^k$ we say that Z is the **mapping cylinder** *of f if there are continuous maps $F: X \times I \to Z$ and $j: Y \to Z$ that satisfy the following conditions:*

(a) *j is an embedding of Y into Z.*
(b) *$F|_{X\times[0,1)}$ is an embedding of $X \times [0,1)$ into Z with $F(x,1) = j(f(x))$, for all $x \in X$.*
(c) *$Z = Im(j) \cup Im(F|_{X\times[0,1)})$ with $Im(j) \cap Im(F|_{X\times[0,1)}) = \emptyset$.*

We use the notation M_f for the mapping cylinder. If $i_0 : X \to X \times I$ is the inclusion map $i_0(x) = (x,0)$, then the map $i = F \circ i_0$ is called the **natural inclusion** *of X into M_f. The map j is called the* **natural inclusion** *of Y into M_f.*

Here is a simple example.

Example 16.11 Let $X \subseteq \mathbf{R}^1$, $X = Y = [-2,2]$, and let $f : X \to Y$ be defined by $f(x) = \frac{x}{2}$. The image of f is $[-1,1]$. In Figure 16-2 we show $M_f \subseteq \mathbf{R}^2$. ◆

The next result shows that, in the mapping cylinder M_f, if we identify Y with its image under the natural inclusion, $M_f \searrow Y$. In particular, M_f and Y have the same deformation type.

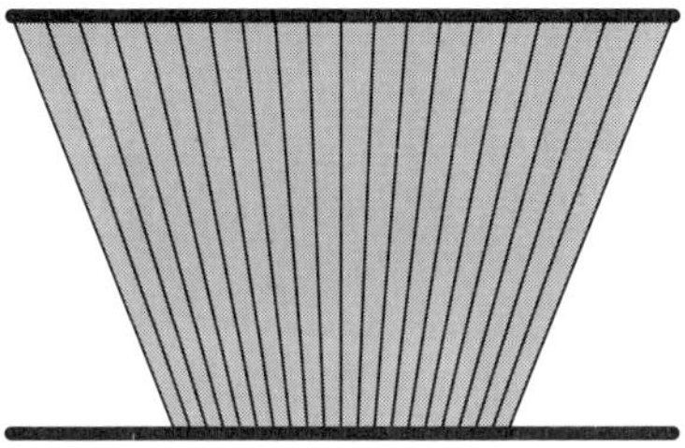

Figure 16-2 An example of a mapping cylinder, M_f; see Example 16.11. The dark line segement at the top is the image of X under the natural inclusion; the dark line segement at the bottom is the image of Y under the natural inclusion. The gray polygon is the image of $X \times I$. Also shown are the intervals $I_x = F(x,t)$ for values of x, $\{-2, -1.8, \ldots 1.8, 2\}$.

Proposition 16.12 *Suppose $X \subseteq \mathbf{R}^n$ and $Y \subseteq \mathbf{R}^m$. If f is a continuous function, $f : X \to Y$, and if M_f is the mapping cylinder of f, then $j(Y)$ is a deformation retract of M_f.*

Proof: For $x \in X$, let $I_x = F(x,t)$. I_x is an embedded interval with one end $F(x,1)$ corresponding to x, and the other $F(x,0) = j(f(x))$ corresponding to $f(x)$.

The idea is this: at time t, shrink each interval I_x towards $j(f(x))$, so that at time t the interval has length $1 - t$. More exactly, define the deformation G_t, $0 \leq t \leq 1$, as follows. If $z \in j(Y)$, define $G_t(z) = z$. If $z \notin j(Y)$, then $z = F((x,s))$ for some $(x,s) \in X \times I$, and we define

$$G_t(z) = F(x, s(1-t) + t).$$

(See Figure 16-3.) Details are left as an exercise (Problem 16.9). ∎

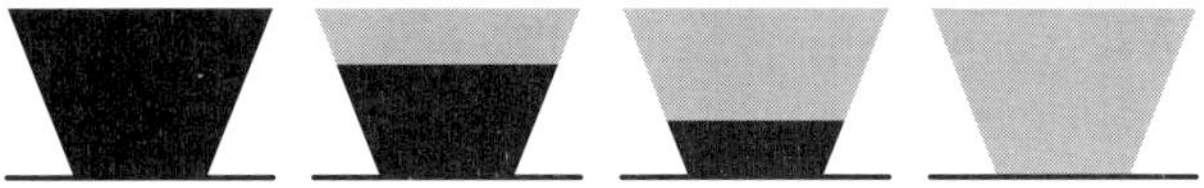

Figure 16-3 A deformation of the mapping cylinder shown in Figure 16-2 to the subset corresponding to Y; see proof of Proposition 16.12. Shown in black are the images of the deformation. On the left, the identity map; on the right, the image is the subset Y.

We remark that the mapping cylinder of a map clearly contains within it all the information of the function in a geometric way. One can locate a point $x \in X$ as a point in M_f, using the natural inclusion and locate $f(x)$ as the other point of the interval I_x.

Example 16.13 We can reformulate our proof that the n-dimensional ball D^n is contractible, Example 15.18. Let S^n be the standard n-sphere in $\mathbf{R}^{n+1}$, with origin $\vec{0}$ and let $f : S^n \to \{\vec{0}\}$ be the constant map $f(x) = \vec{0}$. Then the mapping cylinder M_f is homeomorphic to D^{n+1}. By Proposition 16.12, it is contractible. ◆

Example 16.14 Considered as a subset of the plane, the letter X is homeomorphic to the mapping cylinder M_f where X and Y are subsets of $\mathbf{R}^1$; $X = \{1, 2, 3, 4\}$, $Y = \{0\}$, and $f : X \to Y$. ◆

Example 16.15 The standard Möbius band in $\mathbf{R}^3$, Example 13.28, has the structure of a mapping cylinder.

Consider the map $f: S^1 \to S^1$ which, in polar coordinates, is $f(1, \theta) = (1, 2\theta)$. (In terms of complex numbers, f is the restriction to numbers of norm 1, of $f(z) = z^2$.) The map f is two-to-one. In Figure 16-4 we can see the two-to-one nature of f as two line segments which begin at the boundary of the Möbius band have a common endpoint on the center circle.

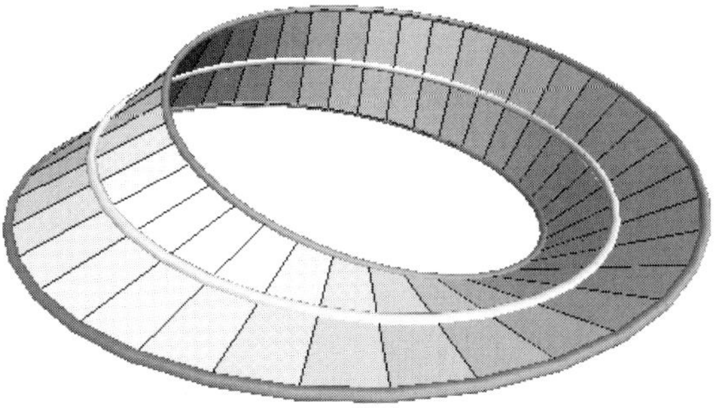

Figure 16-4 The standard Möbius band in $\mathbf{R}^3$ as a mapping cylinder M_f; see Example 16.15. The domain of f corresponds to the boundary circle shown as a gray tube. The domain is the light center circle. Dark line segments join points of the domain of f to image points.

Thus the Möbius band has the deformation type of a circle. This implies the Möbius band is not simply connected, by Proposition 16.05. ◆

Example 16.16 Consider the standard solid torus in $\mathbf{R}^3$; see Definition 13.24. Let $f: S^1 \times S^1 \to S^1$ be projection on the second factor. It can be shown that the standard solid torus in $\mathbf{R}^3$ has the structure of the mapping cylinder M_f.

If $y \in S^1$, the circle $S^1 \times \{y\}$ will correspond to a meridian circle of the standard torus. The map $f|_{\{S^1 \times \{y\}\}}$ will map that circle to the point

y, and that mapping cylinder is a disk, one of the disks $D(\alpha)$ used to define the solid torus; see Example 13.23.

This implies that the solid torus has the deformation type of a circle, and thus is not simply connected, by proposition 16.05. ◆

The uncountable collection of mutually non-homeomorphic subsets of the plane, Example 7.28, since they are contractible, are equivalent in the sense of deformation type. This raises the question

Question 16.17 Is there an uncountable collection of subsets of the plane, no two of which are of the same deformation type?

The answer to this is yes. We sketch one solution and leave details as a challenging exercise. The exercise is either complete description of the subsets and complete the suggested proof, or to construct alternative candidates and alterative proofs.

For $n \in \mathbf{N}$ let $a_n = (-1/n, 0)$. Let L be the line segment with endpoints $(0,0)$ and $(1/3,0)$. Let S be the subset obtained by joining together with an infinite collection of scaled and translated versions of the piecewise-linear $\sin(1/x)$ curve of Example 8.11; see Figure 16-5. As in the proof for Example 7.28, we construct a collection of subsets which encode binary strings.

For a binary s string we define a set X_s to be the union of the points a_n, the line segment L, the set S, together with a some circles, one for each 1 in s, placed as indicated in Figure 16-5.

One of the key steps in the proof is use of statements of Problem 3.

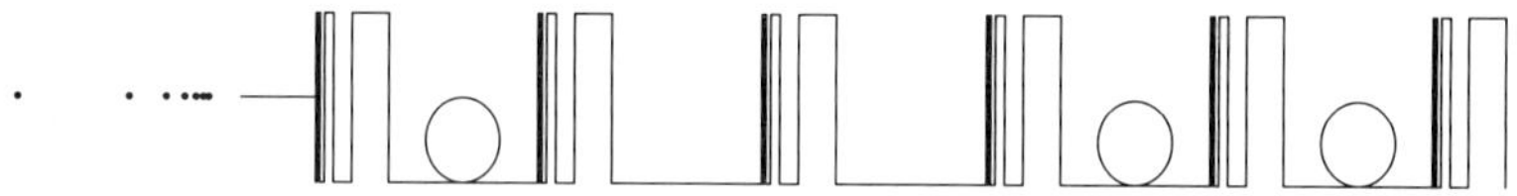

Figure 16-5 A subset of $\mathbf{R}^2$ which encodes the binary string 0.10011...; used to show there are uncountably many distinct deformation types. See the remark on page 349

*16.2 General Topology and Chapter 16

Homotopy type is an important basic definition in topology:

Definition 16.18 *Let X and Y be topological spaces. Suppose there are continuous maps $f: X \to Y$, $g: Y \to X$ such that $g \circ f$ is homotopic to the*

identity map of X, $Id|_X$, and $g \circ f$ is homotopic to $Id|_Y$. Then we say that f and g are **homotopy equivalences** *and that X and Y have the same* **homotopy type.**

We need to verify that, as the language suggests, "X has the homotopy type of Y" *is* an equivalence relation.

If $r: X \to A$ is a deformation retraction, then r is a homotopy equivalence, and the inclusion $i: A \to X$ is a homotopy inverse of r. Thus deformation type implies homotopy type:

Proposition 16.19 *If X is a topological space and $A \subseteq X$, with A a deformation retract of X, then A and X have the same homotopy type.*

If X has the same deformation type as Y, then X has the same homotopy type as Y. ∎

In Chapter 17, Proposition 17.33 provides a converse to Proposition 16.19 which applies to certain common topological spaces such as simplicial complexes, Definition 17.07, or the more general collection called "cell complexes," Definition 17.30.

16.3 Problems for Chapter 16

16.1 Prove Proposition 16.02.

16.2 Prove Proposition 16.03.

16.3 Prove Proposition 16.05.

16.4 Group the letters of the alphabet according to deformation type. Try different styles of letters and even different alphabets, if you know them.

16.5 Show that all the subsets described in Example 7.28 are contractible.

16.6 Let $X = \mathbf{R}^2 - (C_1 \cup C_2)$ where C_1 and C_2 are circles of radius 1 and 2 respectively with center at the origin. Show that one component of X has the deformation type of a point and the other two have the deformation type of a circle.

16.7 Let $C_1 = \{(x, y) \in \mathbf{R}^2 : x^2 + y^2 = 1\}$, $C_2 = \{(x, y) \in \mathbf{R}^2 : x^2 + y^2 = 4\}$, $C_3 = \{(x, y) \in \mathbf{R}^2 : (x - 3)^2 + y^2 = 1\}$. Let $A = C_1 \cup C_2$, and $B = C_1 \cup C_3$. Show that A and B of are not equivalent subsets of the plane. (Hint: Begin with Problem 16.6.)

16.8 Let $X = \{(x, y) \in \mathbf{R}^2 : y = 0$ and $x = 1/n$ where n is a positive integer$\}$. Let $Z = X \cup \{(0, 0)\}$. Are X and Z of the same deformation type ? Justify your answer.

16.9 Verify the claim of Proposition 16.12, that the given map is, in fact the deformation desired.

16.10 Let $f : X \to X$ be the identity map of X. Show that M_f is homeomorphic to $X \times I$. (Note: If $f : S^1 \to S^1$ is the identity map, then $S^1 \times I$ is a cylinder. It is this example and the result above that give the mapping cylinder its name.)

17. COMPLEXES

OVERVIEW: Simplicial complexes provide a rich and varied source of examples of subsets of $\mathbf{R^n}$. *We present a brief introduction to this topic, emphasizing connections to topology.*

17.1 Simplicial complexes

We have seen some very strange subsets of $\mathbf{R^n}$, such as the topologist's comb (Example 2.43), the topologist's $\sin(1/x)$ curve (Example 3.30), the Cantor set (Example 5.01) and the Hawaiian earring (Example 2.42). There are times when a topologist wants to simplify mathematics by restricting attention to a collection of subsets which *excludes* such sets. One such collection are the manifolds, described in Chapter 13. However, there are many simple subsets of interest that are not manifolds (for example, the union of two intersecting lines in the plane). An alternative is to focus on the set of simplicial complexes, Definition 17.07. None of the these "strange" examples are a simplicial complex.

The set of simplicial complexes is useful enough to give rise to a branch of topology called "piecewise-linear topology," [24]. Associated to a simplicial complexes are algebraic structures. In algebraic topology, these ideas are developed and extended to enable certain important calculations (see any standard text in the subject such as [15, 23, 44]). In computer graphics objects are usually represented as a finite collection of line segments and triangles in $\mathbf{R^3}$; the union of these has the structure of a simplicial complex. So the collection of simplicial complexes includes most structures commonly used in computer graphics.

In spirit, a simplicial complex is a generalization of the notion of manifold (Definition 13.01). A closed k-manifold is a subset of $\mathbf{R^n}$ which is the union of pieces each of which is homeomorphic to $\mathbf{R^n}$. Each of the pieces are of the same dimension, namely, dimension k. For a simplicial complex, the pieces are special, simple, closed subsets of some $\mathbf{R^k}$ where different pieces can have different dimensions.

Roughly speaking, a simplicial complex is a subset of $\mathbf{R^n}$ which is the union of sets, called "simplexes." Examples of simplexes are line segments, triangles, solid tetrahedra. The standard simplexes of Definition 2.15 are examples of simplexes.

In order to define a general (non-standard) simplex, we need some terminology. Three points, $x_0, x_1, x_2 \in \mathbf{R}^3$, determine a triangle, if they do not lie on a line. Four points, $x_0, x_1, x_2, x_3 \in \mathbf{R}^3$, determine a tetrahedron, if they do not all lie on a plane. To articulate this idea to higher dimensions, we use the language of linear algebra.

Definition 17.01 *Suppose $x_0, \ldots, x_k$ are points of $\mathbf{R^n}$. We say that $x_0, \ldots, x_k$ are* **affinely independent** *if the only solution of $\sum_0^k a_i \vec{x_i} = \vec{0}$ where $a_0, \ldots, a_k$ are real numbers such that $\sum_0^k a_i = 0$, is the (trivial) solution $a_i = 0$ for $0 \le i \le k$.*

Equivalently, we could say that the $k+1$ points $x_0, \ldots, x_k$ are affinely independent points if and only if the k vectors $\{\vec{x_1} - \vec{x_0}, \ldots, \vec{x_k} - \vec{x_0}\}$ are affinely independent (Problem 17.3). The reason this definition is not preferred is that it seems to give the first point a special role.

Example 17.02 Suppose we have three points, x_0, x_1, x_2, in $\mathbf{R}^3$. How can we describe points on the line segment L with endpoints x_0 and x_1, or a point in the triangle T with vertices x_0, x_1, and x_2?

Let us consider a few examples. The point $m = \frac{1}{2}\vec{x_0} + \frac{1}{2}\vec{x_1}$ is on L; in fact, m is the midpoint of L. The point $p = \frac{1}{3}\vec{x_0} + \frac{2}{3}\vec{x_1}$ is the point one-third the way from $\vec{x_0}$ to $\vec{x_1}$. Let $c = \frac{1}{3}\vec{x_0} + \frac{1}{3}\vec{x_1} + \frac{1}{3}\vec{x_2}$. Then $c \in T$; in fact, it is the centroid of T. The points m and c are vector averages of a set of vertices The point p is a weighted average of the vertices. A few more examples and bit of linear algebra leads us to the next definition. ◆

Definition 17.03 *Suppose $S = \{x_0, \ldots, x_k\}$ is a set of affinely independent points of $\mathbf{R^n}$. The k-***simplex** *with vertices $x_0, \ldots, x_k$ is the set, $\sigma(S)$, of points that satisfy $0 \le a_i$ for $0 \le i \le k$, $\sum_0^k a_i \vec{x_i}$, and $\sum_0^k a_i = 1$. We say the points $x_0, \ldots, x_k$ are* **vertices** *of $\sigma(S)$. The number k is called the* **dimension** *of the simplex.*

To get a feeling for Definition 17.03, look at special cases considered in Problems 17.1 and 17.2. As in Problem 2.27, one can show that any k-simplex in $\mathbf{R^n}$ is a closed subset of $\mathbf{R^n}$.

A "face" of a simplex, Definition 17.05, is a special subset of a simplex. Proposition 17.04 tells us a face is, in fact, a simplex.

Proposition 17.04 *Suppose $S = \{x_0, \ldots, x_k\}$ is a set of affinely independent points of $\mathbf{R^n}$, and $\sigma(S)$ is the k-simplex with vertices $x_0, \ldots, x_k$. Suppose $T \subseteq S$, $T \neq \emptyset$. Then T is a set of affinely independent points of R^n, and $\sigma(T) \subseteq \sigma(S)$.* ∎ *(Problem 17.4)*

In fact, the points of τ are exactly those points of σ which can be written $\sum_0^k a_i \vec{x_i}$ with $0 \le a_i$ where $\sum_0^k a_i = 1$ *and* $a_i = 0$ for all vertices of σ which are not vertices of τ.

Definition 17.05 *Suppose $S = \{x_0, \ldots, x_k\}$ is a set of affinely independent points of $\mathbf{R^n}$, and σ is the k-simplex with vertices $x_0, \ldots, x_k$. Suppose $T \subseteq S$ and that T contains $k' + 1$ points. Let τ be the k'-simplex whose vertices are the points of T; we say that τ is a k'-**face** of σ. If the dimension of the face is not significant, one simply refers to τ as a **face** of σ.*

So, every k-simplex is a face of itself. Sometimes we want to not allow this as "proper":

Definition 17.06 *If σ is a k simplex, any face of σ which is not equal to σ is called a **proper face** of σ.*

Vertices of a k-simplex are 0-faces. A triangle in $\mathbf{R^n}$ is a 2-simplex whose proper faces are: three 0-faces (the vertices) and three 1-faces (the edges).

Here, then, is our definition of simplicial complex:

Definition 17.07 *Suppose $S \subseteq \mathbf{R^n}$, and suppose $\Sigma = \{\sigma_a^{k_a}\}_{a \in A}$ is a collection of simplexes in $\mathbf{R^n}$ where k_a is the dimension of $\sigma_a^{k_a}$. Then S is a* **simplicial complex** *if we can write $S = \bigcup_{a \in A} \sigma_a^{k_a}$ such that*

(a) *(Face condition) If $\sigma_a^{k_a} \in \Sigma$, then Σ contains every face of $\sigma_a^{k_a}$.*
(b) *(Intersection condition) If any two of these simplexes meet, the intersection is a face of each of them.*
(c) *(Local finiteness condition) For any point $x \in \mathbf{R^n}$, there is an ϵ-neighborhood of x which meets only finitely many simplexes of S.*

If S can be written as a simplicial complex, using a finite number of simplexes, then S is a **finite** *simplicial complex.*

The set of simplexes Σ is called a **simplicial structure** *for S.*

The **dimension** *of S is the largest dimension of any simplex in its simplicial structure.*

The motivation for the face condition, Definition 17.07(a), is convenience rather than a necessity. It facilitates certain induction arguments. If a collection of simplexes is missing some faces, we can simply enlarge the collection by adding all such faces. This new collection would satisfy the face condition.

The intersection condition, Definition 17.07(b), says that simplices "fit together in a simple way"; see Figure 17-1.

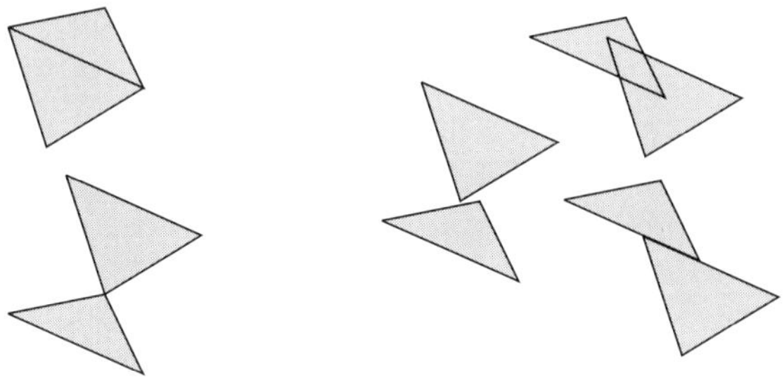

Figure 17-1 Some pairs of intersecting 2-simplexes in $\mathbf{R}^2$. On the left, the pairs satisfy the intersection condition of Definition 17.07. The three pairs on the right fail to satisfy this condition.

If a finite collection of simplexes fails this condition, we can re express the union of these simplexes as the union of another collection that does satisfy this condition; see Example 17.18 and Proposition 17.19.

Clearly, local finiteness, Definition 17.07(c), is of significance only for infinite complexes. It is imposed in order to keep things simple. Basically, we avoid having a point of $\mathbf{R}^n$ be a limit point of vertices of a simplicial complex; see Example 17.08. Another subset which fails the local finiteness condition is the suspension of natural numbers shown in Figure 2-14. This set is a union of line segments, but we do not have local finiteness at the two suspension points.

Example 17.08 The line $\mathbf{R}^1$ has a simplicial structure where the vertices are the integers, and the edges are the closed intervals, $[n, n+1]$, between consecutive integers; see top of Figure 17-2.

On the other hand, $\mathbf{R}^1$ is homeomorphic to the open interval $(-1, 1)$, and this is not a simplicial complex, (Problem 17.08). For example, we cannot use the points $\{\ldots, -\frac{2}{3}, -\frac{1}{2}, 0, \frac{1}{2}, \frac{2}{3}, \ldots\}$ as vertices, with edges the closed intervals between consecutive entries in this list; see bottom of Figure 17-2. This is a union of 1-simplexes but is not a simplicial structure. The problem is that the local finiteness, Definition 17.07(c), fails at 1 and at -1. ◆

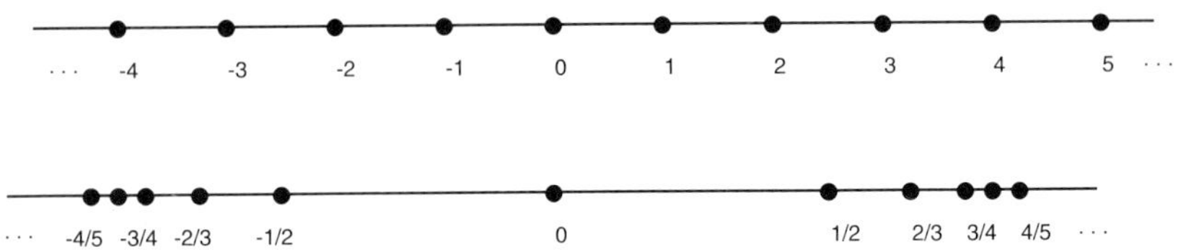

Figure 17-2 Top: The line as a simplicial complex. Bottom: The open interval is not a simplicial complex; see Example 17.08.

The perimeter of a triangle is a simplicial complex; it is homeomorphic to a circle which is not a simplicial complex. In Example 17.08, we saw that $\mathbf{R}^1$ is a simplicial complex, homeomorphic to $(-1, 1)$, which is not a simplicial complex.

So clearly, "X is a simplicial complex" is not a topological property. This is not a big problem, but accounts for the next definition.

Definition 17.09 *Suppose that X is a subset of $\mathbf{R}^n$. We say X is* **triangulable** *if X is homeomorphic to a simplicial complex.*

So, "X is triangulable" *is* a topological property. A circle and an open interval are triangulable sets.

The standard k-simplex Δ^k in $\mathbf{R}^{k+1}$ (see Definition 2.14) together with all of its faces give it the structure of a (finite) simplicial complex. We note that Δ^k is homeomorphic to D^k (Problem 17.16). We have not given a proof of this, but a guide to the proof, for Δ^2, given in Problems 4.33 and 4.34. So, the closed k-ball D^k is a triangulable.

Example 17.10 The proper faces of a $k+1$-simplex form a simplicial complex homeomorphic to S^k. Thus, for all k, the k-dimensional sphere S^k is triangulable. ◆

17.2 Examples and properties of simplicial complexes

Next, we state two topological properties of simplicial complexes.

Proposition 17.11 *If $S \subseteq \mathbf{R}^n$ is a simplicial complex, then S is a finite complex if and only if S is compact.* ▮ *(Problem 17.5)*

Another topological property follows from the local finiteness condition of Definition 17.05.

Proposition 17.12 *If $S \subseteq \mathbf{R}^n$ is a simplicial complex, then S is locally connected.* ▮ *(Problem 17.12)*

Example 17.13 The topologist's comb (Example 2.43), the topologist's $\sin(1/x)$ curve (Example 3.30), and the Cantor set (Example 5.01), are not triangulable since none are locally connected. ◆

Example 17.14 For any n, $\mathbf{R}^n$ is a simplicial complex. We do not prove this, but we will provide an outline for a proof.

We have a simplicial structure for $\mathbf{R}^1$, Example 17.08. We would like to use this and make an inductive argument.

The problem is that a Cartesian product of simplexes is not a simplex. The Cartesian product of two 1-simplexes is a rectangle and not a triangle. That is not a serious problem since we can simply divide the square by adding a diagonal. Using this idea we can obtain a simplicial structure for $\mathbf{R}^2$; see Figure 17-3.

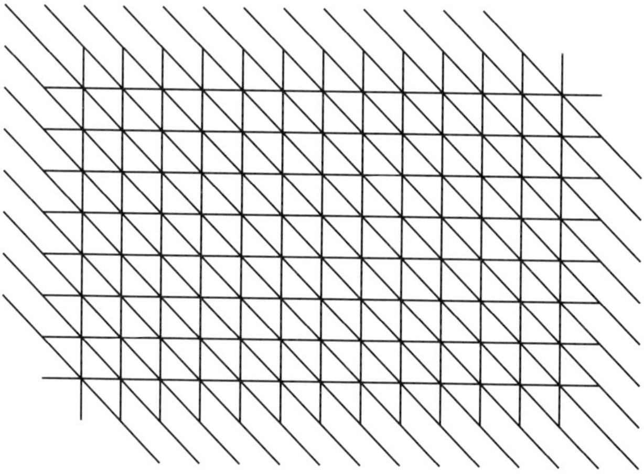

Figure 17-3 Using horizontal, vertical, and diagonal lines to give $\mathbf{R}^2$ the structure of a simplicial complex.

In this process we do not need to introduce new vertices. This gives an idea of how to proceed in general for $\mathbf{R}^n$. We leave the description of simplicial structure of $\mathbf{R}^3$ as an exercise (Problem 17.15). As a start, note that we can express $\Delta^1 \times \Delta^2$ as a union of three 3-simplexes; see Figure 17-4.

This subdivision is not difficult, but needs some care to do correctly. The problem is how to choose the simplexes. This subdivision is not

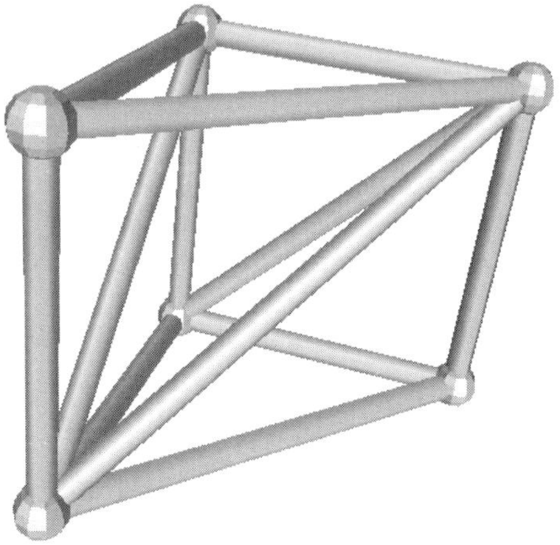

Figure 17-4 The 1-skeleton of a simplicial complex corresponding to the product of a 1-simplex and a 2-simplex; see Example 17.14

unique. (It is not unique for a square, actually, since there are two diagonals.) If we just divide each product of simplexes, individually, the result might not fit together properly to give a simplicial structure. Even describing the simplexes takes some thought. The first really challenging case is $\Delta^2 \times \Delta^2$.

One implication of the local finiteness condition is

Proposition 17.15 *If $S \subseteq \mathbf{R}^n$ and S is a simplicial complex, then there are, at most, countably many simplexes in any simplicial structure.*

Proof: This follows from the remark on page 245, that every uncountable subset of $\mathbf{R}^n$ has a limit point (Problem 17.13). ∎

Example 17.16 A 0-dimensional simplicial complex is just a (countable) collection of isolated points. If the points were not isolated, we would violate the local finiteness condition of Definition 17.07.

A 1-dimensional simplicial complex in $\mathbf{R}^n$ is just a (countable) collection of isolated points, vertices, together with some edges which connect those points so that the edges are disjoint, except possibly at a vertex. ◆

A 1-dimensional complex is frequently called a "graph":

Definition 17.17 *A 1-dimensional simplicial complex in $\mathbf{R}^n$ called a* **graph** *in $\mathbf{R}^n$, or sometimes a* **topological graph** *or a* **spatial graph**.

Example 17.18 Any finite collection of simplexes can be given a simplicial structure, as we see in Proposition 17.19. The proof, which we omit, is technical, but the idea is simple. In Figure 17-5 we see two 2-simplexes (triangles) in $\mathbf{R}^3$. The union can be made into a simplicial complex in many ways. On the right, in Figure 17-5, is shown a simplicial structure with eight 2-simplexes.

The next proposition states that we can do this process in general.

◆

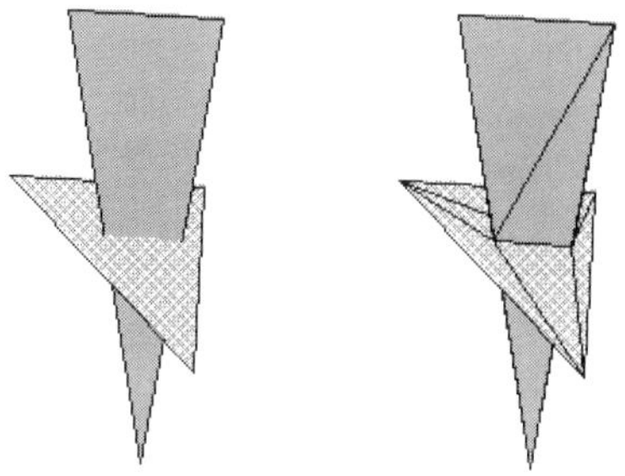

Figure 17-5 Two triangles in space, on left. The union as a simplicial complex, on the right.

Proposition 17.19 *If* $\Sigma = \{\sigma_i^{k_i}\}_{0 \le i \le N}$ *is a finite collection of simplexes of* $\mathbf{R}^\mathbf{n}$*, then the set* $S = \cup_{i=1}^{N} \sigma_i^{k_i}$ *is a simplicial complex.* ∎

In Proposition 17.19 we are not saying that the given Σ is a simplicial structure for S, but only that there exists some simplicial structure for S.

In Example 17.08 we expressed $(-1, 1)$, which is not a simplicial complex, as an infinite union of 1-simplexes; thus we see the need for the need for finiteness in Proposition 17.19.

Definition 17.20 *Suppose S and S' are simplicial complexes in* $\mathbf{R}^\mathbf{n}$*, with simplicial structures Σ and Σ', respectively. We say S' is a* **subcomplex** *of S if $\Sigma' \subseteq \Sigma$*

Of course, it follows from the Definition 17.20 that if S' is a subcomplex of S, then $S' \subseteq S$.

A basic technique in the study of simplicial complexes is induction. As a consequence, the idea of a k-skeleton of a simplicial complex is a useful one:

Definition 17.21 *Suppose S is a simplicial complex. The k-***skeleton** *of S is the simplicial complex associated with the set of all faces of S of dimension less than or equal to k. The associated set is denoted by $S^{(k)}$*

Any k-skeleton of S is a subcomplex of S and, in fact, $S^{(0)} \subseteq S^{(1)} \subseteq S^{(2)} \ldots \subseteq S$. In computer graphics, the 1-skeleton of an object is sometimes called a "wire-frame rendering."

Example 17.22 Suppose σ is a 3-simplex in $\mathbf{R}^3$. Then the 0-skeleton $\sigma^{(0)}$ consists of the four vertices of σ. The 1-skeleton $\sigma^{(1)}$ consists of these together with the six edges. The simplicial structure for the 2-skeleton $\sigma^{(2)}$ consists of all of these vertices and edges, together with the four triangular faces of σ. ◆

When are two simplicial complexes "the same"? The idea of a simplicial complex is to write a subset as a union of special subsets—the simplexes. In topology, we say two subsets are the same if they are homeomorphic. When those subsets are simplicial complexes, we will want to take this additional structure into account somehow. The simplest way to do this would be to say that a simplicial complex X is the same as a simplicial complex X' if there is a homeomorphism $h: X \to X'$, and the restriction of h to a simplex of X maps that simplex to a simplex of X'. The problem with this idea is that it distinguishes the square and the triangle and thus is too restrictive a notion for topology.

Example 17.23 One problem with the notion of a simplicial complex is that it is a very rigid idea, in a geometric sense. In Figure 17-6 we see some homeomorphic subsets of $\mathbf{R}^3$. Four of these are expressed as a union of two triangles joined only along a common edge. If the idea of simplicial complex is to have some topological significance, we would like to say that these examples are (somehow) equivalent.

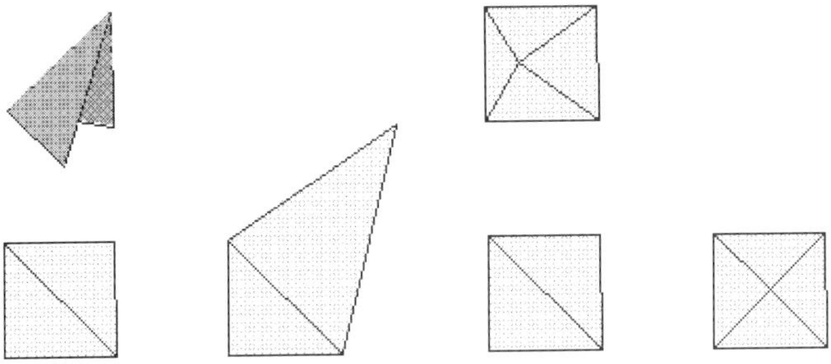

Figure 17-6 Some simplicial complexes in $\mathbf{R}^3$ that we wish to consider as equivalent complexes. All are planar except at top left.

One obvious problem is that these examples do not have the same number of simplexes. Let us deal with this first. The idea is to allow subdivision of simplexes. ◆

Definition 17.24 *Suppose $\sigma \subseteq \mathbf{R}^n$ is a simplicial complex with two simplicial structures Σ and Σ'. We say Σ' is a* **subdivision** *of Σ if every simplex of Σ' is a subset of a simplex of Σ.*

We can show that

Proposition 17.25 *If Σ' is a subdivision of Σ, then every simplex of Σ is a union of simplexes of Σ'.* ▮ *(Problem 17.17)*

Definition 17.26 *Suppose $X \subseteq \mathbf{R}^n$ and $Y \subseteq \mathbf{R}^m$ are simplicial complexes. We say X and Y are* **PL-homeomorphic** *if there is a subdivision X' of X and a subdivision Y' of Y and a homeomorphism $g: X \to Y$ such that g maps each simplex σ of X' homeomorphically onto a simplex of Y'.*

In Definition 17.26, we required $g|_\sigma$ to be a homeomorphism. It is not difficult to show that, if we have such a homeomorphism, we can obtain a "slightly nicer" homeomorphism, g', so that $g|'_\sigma$ is the restriction of an affine map (Problem 17.12). The key observation is that vertices of σ are must be sent to vertices of $g|_\sigma$; this determines an affine map which maps σ homeomorphically onto $g(\sigma)$. Furthermore since this is such a simple procedure, we can check that, if we find a similar affine map for another simplex, say τ of X', the two affine maps will agree on the intersection, if any, of these simplexes.

The map g' is called a "piecewise-linear homeomorphism." Two simplicial complexes are "PL-homeomorphic" if, allowing for subdivisions, we can find a piecewise-linear homeomorphism between them.

Now we can ask an important fundamental question:

Question 17.27 Suppose X and Y are simplicial complexes with X homeomorphic to Y'; are X and Y PL-homeomorphic ?

The answer to Question 17.27, in general, is no. That this seems surprising is due to the fact that it is true for 3-dimensional complexes. This general problem of relating the PL-homeomorphism to topological homeomorphism is called the "Hauptvermutung." (This German word means "principle result" and was the name associated with this problem at a time when many thought the answer would be yes.)

Putting this another way, a given simplicial complex X may correspond to two incompatible simplicial structures. So a simplicial structure associated with X is not determined by X.

How do manifolds relate to simplicial complexes?

Question 17.28 if $X \subseteq \mathbf{R}^n$ and X is an n-manifold, is it triangulable?

The answer here is also, no. But it is not a problem for $n \leq 3$.

*17.3 General Topology and Chapter 17

Suppose $S \subseteq \mathbf{R^n}$ is a simplicial complex, Definition 17.07. To simplify initial discussion, assume S is a finite simplicial complex. Consider the various k skeleta of X, Definition 17.21, $S^{(0)} \subseteq S^{(1)} \subseteq S^{(2)} \subseteq \cdots \subseteq S$.

We can loosely describe S in $\mathbf{R^n}$ as follows. Start with some finite collection of points in $\mathbf{R^n}$; these will be the vertices, of $S^{(0)}$. Next, one at a time, glue some line segments to these points, attaching only endpoints of the segments to the vertices. These line segments do not intersect at except at vertices. We have now "assembled" $S^{(1)}$. Next, one at a time, we glue some (solid) triangles in such a way that vertices of the triangle are sent to vertices of $S^{(1)}$, and the edges of the triangles are glued to the line segments of $S^{(1)}$. The end-result is $S^{(2)}$. We then glue some tetrahedra and obtain $S^{(3)}$, etc.

The idea of a cell complex captures the idea of something constructed by gluing together simple pieces in a simple way. The key is to use the notion of decomposition spaces to capture the gluing idea. A k-simplex is homeomorphic to D^k. So, if we are to abstract, we may as well use k-balls, rather than k-simplexes, as our basic object.

Definition 17.29 *Suppose Y is a topological space, and let f be a continuous function $f: S^{k-1} \to Y$. The topological space Y with an* **attached k-cell** *via f is the quotient space obtained from the disjoint union $Y \underset{\circ}{\bigcup} D^k$, using partition P, whose the non-degenerate elements of the partition are subsets $(f^{-1}(y) \times \{1\}) \underset{\circ}{\bigcup} \{y\}$ where $y \in Im(f)$. This quotient space is denoted $Y \underset{f}{+} D^k$*

The image of D^k under the quotient map is called a **closed k-cell**; *the image of $\overset{\circ}{D^k}$ is called an* **open k-cell**.

The map $D^k \to Y \underset{f}{+} D^k$ is called the **cell map** *for f.*

Briefly, a "cell complex" is a space built up by attaching cells in order of increasing dimension.

Definition 17.30 *Let X be a Hausdorff topological space. We say that X is an* **n-dimensional cell complex** *if there are subsets $X^{(0)} \subseteq X^{(1)} \subseteq X^{(2)} \subseteq \cdots \subseteq X$ as follows:*

(a) *$X^{(0)}$ is a collection of points with the discrete topology.*
(b) *$X^{(1)}$ is obtained by attaching 1-disks to $X^{(0)}$.*
(c) *Inductively, for $k \leq n$, $X^{(k)}$ is obtained by attaching k-disks to $X^{(k-1)}$.*
(d) *If the intersection of K with each closed cell Z is a closed subset of Z, then K is a closed subset of X*

Condition (d) of Definition 17.30 only has significance for infinite complexes. The Hawaiian earring H, Example 2.42, is not a cell complex. We can write H as a single 0-cell at the origin and infinitely many 1-cells. Here each closed 1-cell is a circle. The set $K = \{(1/n, 0)\}_{n \in N}$, is not a closed subset of H. The subset K does not contain the limit point at the origin yet, for each 1-cell, the intersection with K is closed; it is a single point. Similarly, subsets such as the piecewise-linear $\sin(1/x)$ curve; Example 8.11 are not a cell complexes.

There are two basic reasons that cell complexes give a mathematically desirable collection of objects. In the first place, they are more "economical" than piecewise-linear structures in that they generally use fewer pieces. More importantly, they are better-suited to study the notion of homotopy.

Example 17.31 For any $1 \le n$, S^n is a cell complex consisting of two cells: a single 0-cell, and a single n-cell. The most economical way to look at S^2 as a simplicial complex is as four vertices, six edges, and four 2-dimensional faces. ◆

Here is an example of how cell complexes are useful for homotopy. It basically says that sliding a cell by a homotopy of the attaching map does not change homotopy type.

Proposition 17.32 *Suppose X is a topological space, and f_0 and f_1 are two maps of S^{k-1} to X, and f_1 and f_2 are homotopic maps in X. Then $X_1 = X \underset{f_0}{+} D^k$ and $X_2 = X \underset{f_1}{+} D^k$ have the same deformation type.*

We outline the proof. To simplify discussion, we consider the case $k = 2$; however, the proof is really no different for other dimensions.

Let $X_1 = X \underset{f_0}{+} D^2$ and $X_2 = X \underset{f_1}{+} D^2$, and suppose we have a homotopy $F: S^1 \times I \to X$ between $F|_{X \times \{0\}} = f_0$ and $F|_{X \times \{1\}} = f_1$.

Write S^2 as the union of two disks and an annulus, $S^2 = N \cup A \cup S$, where N is the "north polar disk," S is the "south polar disk," A corresponds to the "equatorial and temperate zones." So $C_N = N \cap A$ is a "polar circle;" similarly, with $C_S = S \cap A$. In fact, we can define homeomorphisms $n: N \to D^2$, $s: S \to D^2$, and $a: A \to S^1 \times I$, which agree on the polar circles.

Consider

$$X_3 = X \underset{f_0}{+} D^k \underset{f_1}{+} D^k.$$

Let $e_0: D^k \to X_3$ denote the cell map for f_0, and let $e_1: D^k \to X_3$ denote the cell map for f_1.

We can define a continuous function $\phi: S^k \to X_3$ by

$$\phi(x) = \begin{cases} e_0 \circ n(x) & x \in N \\ F \circ a(x) & x \in A \\ e_1 \circ s(x) & x \in S. \end{cases}$$

Now consider that we use ϕ to attach a k-cell to X_3:

$$X_4 = X \underset{f_0}{+} D^k \underset{f_1}{+} D^k \underset{\phi}{+} D^{k+1}.$$

In Figure 17-7, we show a simple dimensional example. Here we are attaching two 1-cells to a circle. In this dimension N and S are arcs and A is a pair of arcs (homeomorphic to $S^0 \times I$). In this example, X is the unit circle S^1. We attach a 1-cell at to the points $(-1,0)$ and $(1,0)$ to obtain X_1. Both ends of a 1-cell are attached to $(0,-1)$ to obtain X_2. You are invited to describe a homotopy between these two attaching maps.

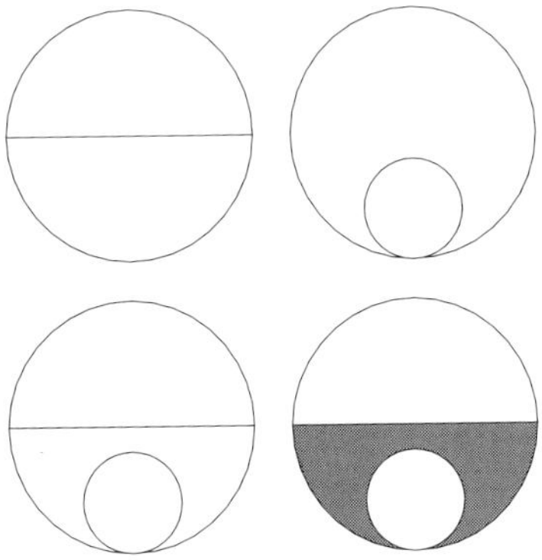

Figure 17-7 Illustration of the proof of Proposition 17.32. In this case, we show X_1, upper left, has the homotopy type of X_2, upper right. We construct X_3, lower left, and attach a 2-cell to it, obtaining X_4, lower right. Both X_1 and X_2 are deformation retracts of X_4.

It is not hard to verify that X_1 is a deformation retract of X_4, and that X_2 is a deformation retract of X_4. Notationally, $X_1 \swarrow X_4 \searrow X_2$. Thus X_1 and X_2 have the same deformation type; thus they have the same homotopy type. ∎

For cell complexes, the idea of deformation type and homotopy type coincide:

Proposition 17.33 *Suppose X and Y are cell complexes of the same homotopy type; then X and Y have the same deformation type.* ∎

This is a difficult theorem, but we can give an idea of how the proof goes. A proof can be found in most texts that discuss homotopy type of cell complexes such as [42, 44]; it often appears as part of a proof of a more general theorem.

Consider the mapping cylinder, Definition 1.80. Using Proposition 6.41, we can consider X and Y as subsets of M_f.

It is not hard to show that, in general, Y is a deformation retract of M_f. The deformation of $X \times I$ to the subset $X \times \{1\}$, gives rise, in a natural way, to a deformation, towards the image of $X \times \{1\}$, of the subset of M_f corresponding to the image of $X \times I$.

The hard part of the proof comes in showing that, if f is a homotopy equivalence and X and Y are cell complexes, then X is a deformation retract of M_f.

This implies that X and Y have the same deformation type, $X \swarrow M_f \searrow Y$. The use of the mapping cylinder in this proof indicates why this construction is so important.

17.4 Problems for Chapter 17

17.1 Suppose that x_0 and x_1 are distinct points of $\mathbf{R}^3$. Show that x is a point on the line segment L, between x_0 and x_1, if and only if there are numbers a_0 and a_1 with $0 \le a_0$ and $0 \le a_1$, $a_0 + a_1 = 1$ and $\vec{x} = a_0\vec{x_0} + a_1\vec{x_1}$. (Hint: You might want to use the fact that $x \in L$ if and only if there is a λ, $0 \le \lambda$, with $\vec{x} = \vec{x_0} + \lambda(\vec{x_1} - \vec{x_0})$.)

17.2 Suppose that x_0, x_1, and x_3 are affinely independent points of $\mathbf{R}^2$. Let T be the triangle with vertices x_0, x_1, and x_3. Verify that T is the 2-simplex with vertices x_0, x_1, and x_3 as defined in Definition 17.03.

(Hint: For any three affinely independent points, p, q, and r of $\mathbf{R}^2$, the line L containing p and q divides $\mathbf{R}^2$ into two pieces only one of which contains the point r. Let $H_{pq}(r)$ be this half-plane which contains r, together with the line L. Note that $T = H_{pq}(r) \cap H_{rp}(q) \cap H_{qr}(p)$, and that $x \in H_{pq}(r)$ if and only if there are numbers λ and μ, $0 \le \mu$, such that $x = \vec{p} + \lambda(\vec{p} - \vec{q}) + \mu(\vec{q} - \vec{r})$. Draw a vector diagram.)

17.3 Prove the assertion of the remark on page 353.

17.4 Supply the details of Proposition 17.04.

17.5 Prove Proposition 17.11.

17.6 Prove Proposition 17.12.

17.7 Let X be the open interval $(-1, 1) \subseteq \mathbf{R}^1$. Show that X is not a simplicial complex. (Hint: One idea is to show that no matter what simplicial structure one might want for X, we would have the points -1 and 1 limit points of the set of vertices of that structure, contradicting the local finiteness condition of Definition 17.07; see Example 17.08.)

17.8 Suppose Σ' is a subdivision of Σ, $\sigma \in \Sigma$. Show the collection of all simplices of $\sigma' \in \Sigma'$ such that $\sigma' \subseteq \sigma$ is a simplicial structure for σ.

17.9 Suppose $X \subseteq \mathbf{R}^\mathbf{n}$ is a simplicial complex with $p, q \in X$. Show that if there is a path in X from p to q, it can pass through only finitely many different vertices of X.

17.10 Suppose $X \subseteq \mathbf{R}^\mathbf{n}$ is a simplicial complex. Show that X is connected if and only if the 1-skeleton, $X^{(1)}$ is connected.

17.11 Suppose $X \subseteq \mathbf{R}^\mathbf{n}$ is a simplicial complex. Show that X is path connected if and only if the 1-skeleton, $X^{(1)}$ is path connected.

17.12 Give details of the definition of homeomorphism g' the remark on page 361. Verify that it is a homeomorphism.

17.13 Prove Proposition 17.15.

17.14 Not all simple maps from a line segment to a line segment are linear. A common mistake, made so often that it is called the "standard mistake," is that radial projection gives a linear map between line segments.

For example, in $\mathbf{R}^2$ let $a = (0,0), b = (1,0), c = (\sqrt{2}/2, \sqrt{2}/2)$ and $p = (0,1)$;' see Figure 17-8. Radial projection from the point p will map the line segment between a and c onto the line segment between a and b. For $q \in L$, the function $r = f(q)$ is defined as follows. Draw the ray from p through q; the intersection of this ray with L is r.

Show that this function f is *not* linear. (Hint: Obtain a formula for f using, the similar triangles Δpsq and Δpar.)

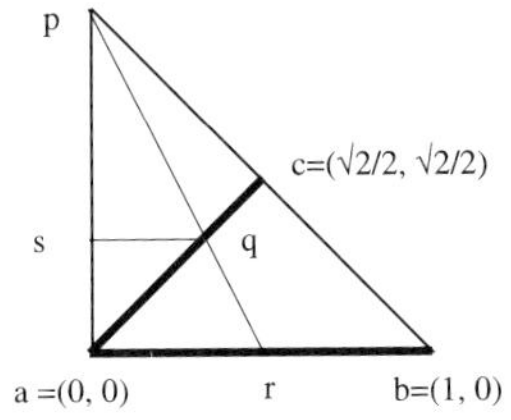

Figure 17-8 Figure for Problem 17.14.

17.15 Describe a simplicial structure for $\mathbf{R}^3$; see Example 17.15.

17.16 Prove that Δ^k is homeomorphic to D^k; see the remark on page 356.

17.17 Prove Proposition 17.25.

18. HIGHER DIMENSIONS

OVERVIEW: Here we discuss some subsets of $\mathbf{R}^4$ *with emphasis on surfaces, such as the Klein bottle, the projective plane, and knotted surfaces. We also consider the* 3*-dimensional sphere.*

18.1 The circle and torus in higher dimensions

In this text, for the most part, we use $\mathbf{R}^n$ as a convenient way of unifying our concepts and notations. Our focus has been on $\mathbf{R}^1$, $\mathbf{R}^2$, and $\mathbf{R}^3$.

However there have been times that dimensions higher than three have been mentioned:

(a) Definition of Cartesian product, Definition A.39.
(b) Graphs of functions, the remark on page 82.
(c) 2×2 matrices; see the remark on page 55.
(d) The standard 3-sphere and 3-simplex, as defined are subsets of $\mathbf{R}^4$.

In this chapter we concentrate on subsets of $\mathbf{R}^4$. To avoid subscripts, we can write points of $\mathbf{R}^4$ as (x, y, z, w). To understand subsets of $X \subseteq \mathbf{R}^4$, there are two general methods, projection and slicing. (Actually, these are the same approaches one uses to understand subsets of $\mathbf{R}^3$.)

In these coordinates, the "standard projection" $\pi\colon \mathbf{R}^4 \to \mathbf{R}^3$ is $\pi(x, y, z, w) = (x, y, z)$. We try to understand $X \subseteq \mathbf{R}^4$ by examining $\pi(X) \subseteq \mathbf{R}^3$. (We could consider other projections, of course, but equivalently we can think of rotating X in $\mathbf{R}^4$ followed by projection by π. For example, projection into yzw-space is equivalent to the composition of a rotation of $\mathbf{R}^4$ that interchanges the x-axis and the w-axis, followed by π.)

The problem with projection is that $\pi|_X$ is most likely not one-to-one. This takes some getting used to. But there is a similar problem, in principle, to drawing a picture on a two-dimensional page of a three-dimensional object.

A second method involves intersections. The most common is the method of slicing by hyperplanes. A "hyperplane" is a subset, homeomorphic to $\mathbf{R}^3$, defined as

$$\{(x, y, z, w) \in \mathbf{R}^4 \mid Ax+By+Cz+Dw = E;\ A, B, C \text{ and } D \text{ not all zero}\}.$$

Usually we consider hyperplanes where one of the coordinates is held constant. Most often we consider $P_{w_0} = \{(x, y, z, w) \in \mathbf{R}^4 \mid w = w_0\}$. We can fill up $\mathbf{R}^4$ with such a family of hyperplanes. Since each P_{w_0} can be identified with $\mathbf{R}^3$ in a canonical way, the intersection $X \cap P_{w_0}$, called a "slice" corresponds to a subset of $\mathbf{R}^3$. A slice of an object in $\mathbf{R}^4$ is analogous to a two-dimensional crossection of a subset or $\mathbf{R}^3$. We consider the set of slices, one for each value of w_0. Unlike the projection method, we do not have two points of X appear as one. The disadvantage is that, in each hyperplane we have only some of the points of X. We now have introduced the problem of relating the points of X from distinct slices. The problem of relating points in distinct slices is analogous to visualizing an entire building from a set of floor plans.

For example, if we slice the unit sphere $S^3 \subseteq \mathbf{R}^4$ by a hyperplane P_{w_0} where $-1 < w_0 < 1$, we obtain a sphere of radius $\sqrt{1 - w_0^2}$.

Another method, mostly useful for objects with a certain symmetry is to use generalized cylindrical coordinates and view $\mathbf{R}^4$ as the union of 3-dimensional pages; see Definition 2.10. This is a variation of the slicing by hyperplanes where we intersect X with each page.

Here is a basic example of a Cartesian product in $\mathbf{R}^4$.

Definition 18.01 *The* **standard torus in $\mathbf{R}^4$** *is the subset $S^1 \times S^1$ where we identify $\mathbf{R}^4$ with $\mathbf{R}^2 \times \mathbf{R}^2$ in the usual way.*

It is not hard to show that the standard torus in $\mathbf{R}^4$ is homeomorphic to the standard torus in $T^2 \subseteq \mathbf{R}^3$. Moreover, we can view T^2 as a subset of $\mathbf{R}^4$:

$$T^2 \subseteq \mathbf{R}^3 = (\mathbf{R}^3 \times 0) \subseteq (\mathbf{R}^3 \times \mathbf{R}^1) \approx \mathbf{R}^4.$$

An interesting exercise is to show that T^2 and $S^1 \times S^1$ are equivalent subsets of $\mathbf{R}^4$ and thus

Proposition 18.02 *The subsets $S^1 \times S^1$ and T^2 are stably equivalent subsets.* ▮ *(Problem 18.4)*

One might think that knot theory in $\mathbf{R}^4$ would consider embeddings of S^1 into $\mathbf{R}^4$. However, all such knots are equivalent; see [24, 37]:

Proposition 18.03 *Any two tame knottings of S^1 into $\mathbf{R}^4$ are ambiently isotopic.* ▮

A reformulation of this is

Proposition 18.04 *Any two tame knottings of S^1 into $\mathbf{R}^3$ are stably equivalent.*

While we can't give the proof of Proposition 18.03 here, the next example indicates why one should believe the statement.

Intuitively, the reason circles can be knotted in $\mathbf{R}^3$ is that one cannot pass a portion of the circle through itself. For example, consider the knots in Figure 18-1.

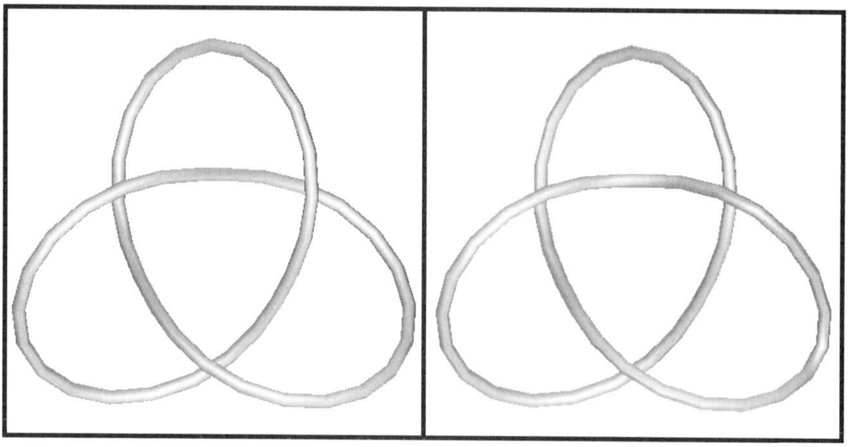

Figure 18-1 By changing the crossing at the upper right of a knot (on left) we can obtain an unknotted circle (on right); see the remark on page 369.

If we could somehow pass the knot through itself near a crossing point we could unknot any circle. A "crossing point" refers to the apparent self-intersection of the curve from a given point of view. (A circle is "unknotted" if it is ambiently isotopic to the standard unit circle in the plane.)

This is basically the same problem that we have in Example 15.02 where we see that we cannot move a circle in $\mathbf{R}^3$ "linked about a line" to one which is not. The next example shows that this linking cannot happen in $\mathbf{R}^4$. In a similar way, the phenomenon of knotting of circles cannot happen in $\mathbf{R}^4$.

Example 18.05 Recall the subsets X and Y of $\mathbf{R}^3$, considered in Example 15.02. Consider X and Y as subsets of $\mathbf{R}^4$ under standard inclusion. We show that these are isotopic subsets of $\mathbf{R}^4$.

We cannot move L to L' in $\mathbf{R}^3 - C$, but we show that we *can* move L to L' in $\mathbf{R}^4 - C$. Not only can it be done, but it is very easy to do.

We can understand the situation by slicing with hyperplanes parallel to xyw-coordinate hyperplane.

For fixed value z_0, let $P_{z_0} = \{(x, y, z, w) \in \mathbf{R}^4 \mid z = z_0\}$. If $z_0 \neq 0$, the intersections we get are single points, l_{z_0} and l'_{z_0} where

$$\begin{aligned} l_{z_0} = P_{z_0} \cap X &= \{(0, 0, z_0, 0)\}, \text{ and} \\ l'_{z_0} = P_{z_0} \cap Y &= \{(0, -2, z_0, 0)\}. \end{aligned}$$

But if $z_0 = 0$, the intersections with X, and with Y, are each the union of a point and a circle in P_0: $P_0 \cap X = \{l_0\} \cup C$ and $P_0 \cap Y = \{l'_{z_0}\} \cup C$; see Figure 18-2.

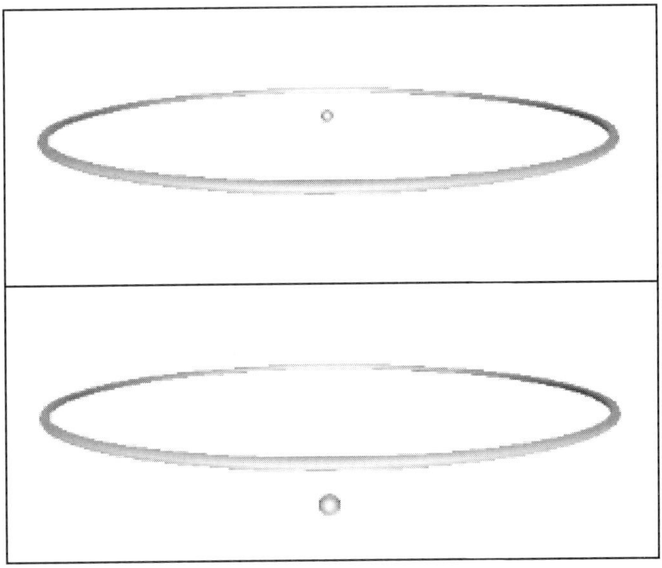

Figure 18-2 Intersections of two subsets of $\mathbf{R}^4$, with a three-dimensional hyperplane, P_0, given by $z = 0$; see Example 18.05. Top shows $X \cap P_0$, a circle and a point, inside the circle in the plane of that circle; bottom shows $Y \cap P_0$, a circle and a point, outside the circle in the plane of that circle.

Clearly, there is an isotopy of the point l_0 in $P_0 - C$ that takes l_0 to l'_{z_0}: lift l_0 up a bit in the w-direction, slide it above l'_{z_0}, then drop it down onto l'_{z_0}. We can use this to define an isotopy, Φ_t, of L in $\mathbf{R}^4 - C$ such that $\Phi_1(L) = L'$. Roughly speaking, what Φ does, for each $l_{z_0} \in L$, is to move the point l_{z_0} to l'_{z_0} in P_{z_0} in the same way that we move l_0 to l'_{z_0} in P_0.

An isotopy of a point is just a path. So let $\phi: I \to P_0 - C$ be a path from l_0 to l'_{z_0}. Viewing P_0 as xyw-space we can write parametric equations:

$\phi(t) = (x(t), y(t), w(t))$. Then the map

$$\Phi_t = (x(t), y(t), z, w(t))$$

is our isotopy of L in $\mathbf{R}^4 - C$ with $\Phi_1(L) = L'$.

In fact, since manifolds are homogeneous, Proposition 13.36, there is an *ambient* isotopy of $P_0 - C$ that takes l_0 to l'_{z_0}. This can then be used to show there is an *ambient* isotopy of L in $\mathbf{R}^4 - C$ such that $\Phi(1)(L) = L'$ (Problem 18.3). ◆

The isotopy described in Example 18.05 shows, that, in effect, it is possible to pass one curve through another by use of a fourth dimension. In a similar sense one can, in effect; seem to pass a portion of a curve through itself by using the fourth dimension. In this way we can prove that there are no (tame) knotted circles in $\mathbf{R}^4$, as asserted in Proposition 18.03.

18.2 The Klein bottle, the projective plane

There are some very simple, important 2-dimensional manifolds that are not subsets of $\mathbf{R}^3$. However, they *are* subsets of $\mathbf{R}^4$, and we describe these next. They are the Klein bottle and the projective plane.

The simpiest way of describing these is to use four-dimensional cylindrical coordinates (see Definition 2.09), writing points of $\mathbf{R}^4$ as (r, θ, z, w) where $0 \le r$ and $0 \le \theta \le 2\pi$. We describe these subsets, using three-dimensional pages—see Definition 2.10.

The Klein bottle is easiest to understand.

Example 18.06 In R^3_+, consider the circle of radius 1 and center $(2, 0, 0)$:

$$C = \{(x, y, z) \in R^3_+ : (x-2)^2 + y^2 = 1 \text{ and } z = 0\}.$$

Next, consider an isotopy F_θ of R^3_+, parameterized by $[0, 2\pi]$ which "flips the circle by a 180° rotation; that is, F_θ is the restriction to R^3_+ of rotation of angle $\theta/2$ about the x-axis. Note that $F_{2\pi}(C) = C$, but $F_{2\pi}$ is *not* the identity map of C since $F_{2\pi}(x, y, 0) = (x, -y, 0)$.

For any θ, there is a standard embedding $p_\theta : R^3_+ \to \mathbf{R}^4$ whose image is the page H_θ, given by

$$p_\theta(x, y, z) = (x, \theta, y, z).$$

Let $C_\theta = p_\theta \circ F_\theta(C)$. The **Klein bottle** K is the union of these circles:

$$K = \bigcup_{0 \le \theta \le 2\pi} C_\theta.$$

In the top part of Figure 18-3 we see the projection $\pi(K)$. To aid in understanding this image, consider images of the circles C_θ under the projection π. If $\theta \neq \phi$, then C_θ and $C\phi$ (except for $C_0 = C_{2\pi}$) lie in different pages, and it follows that $\pi(C_\theta) \cap \pi(C\phi) = \emptyset$. Also, $\pi(C_0)$ is a circle, $\pi(C_\pi)$ is a line segment; otherwise, $\pi(C_\theta)$ is an ellipse which gets increasingly thin as θ gets close to π. ◆

We show that the Klein bottle, K, is homeomorphic to a union of two Möbius bands.

Write the circle $C = L \cup R$ as the union of two semicircular arcs, a 'left arc and a right arc:

$$L = \{(x, y, z) \in R_+^3 : (x-2)^2 + y^2 = 1, x \leq 2, \text{ and } z = 0\},$$
$$R = \{(x, y, z) \in R_+^3 : (x-2)^2 + y^2 = 1, x \geq 2, \text{ and } z = 0\}.$$

The common endpoints of these arcs are

$$e_+ = (2, 1, 0, 0) \text{ and } e_- = (2, -1, 0, 0).$$

Let $E = \{e_+, e_-\}$. Clearly,

$$F_{2\pi}(e_+) = e_- \text{ and } F_{2\pi}(e_-) = e_+.$$

In particular, $F_{2\pi}(E) = E$. Also, that $F_{2\pi}$ maps each arc to itself (in a "reversing manner," as we see from $F_{2\pi}|_E$): $F_{2\pi}(L) = L$ and $F_{2\pi}(R) = R$.

Let $L_\theta = p_\theta \circ F_\theta(L)$, $R_\theta = p_\theta \circ F_\theta(R)$, and $E_\theta = p_\theta \circ F_\theta(E)$. Define

$$M_1 = \bigcup_{0 \leq \theta \leq 2\pi} L_\theta, \quad M_2 = \bigcup_{0 \leq \theta \leq 2\pi} R_\theta, \quad \text{and} \quad G = \bigcup_{0 \leq \theta \leq 2\pi} E_\theta.$$

Since $C = L \cup R$, $E = L \cap R$, we see $K = M_1 \cup M_2$ and $M_1 \cap M_2 = G$; see Figure 18-3. It is easy to check that G is homeomorphic to a circle. Also, M_1 and M_2 are each homeomorphic to a Möbius band. Furthermore, $\partial M_1 = G$ and $\partial M_2 = G$.

Thus we can say that the Klein bottle is homeomorphic to a union of two Möbius bands which meet only along a common boundary.

The next example, the projective plane, can be defined, in a similar manner, as a subset of $\mathbf{R}^4$:

Example 18.07 This time, we begin with a different circle,

$$C = \{(x, y, z) \in R_+^3 : (x-1)^2 + y^2 = 1 \text{ and } z = 0\}.$$

Note that C contains the origin.

As for the Klein bottle, we use the isotopy F_θ of R_+^3 parameterized by $[0, 2\pi]$ given by F_θ is rotation of angle $\theta/2$ about the x-axis.

With p_θ as in Example 18.06, let $C_\theta = p_\theta \circ F_\theta(C)$. The **projective plane** P is the union of these circles:

$$P = \bigcup_{0 \leq \theta \leq 2\pi} C_\theta.$$

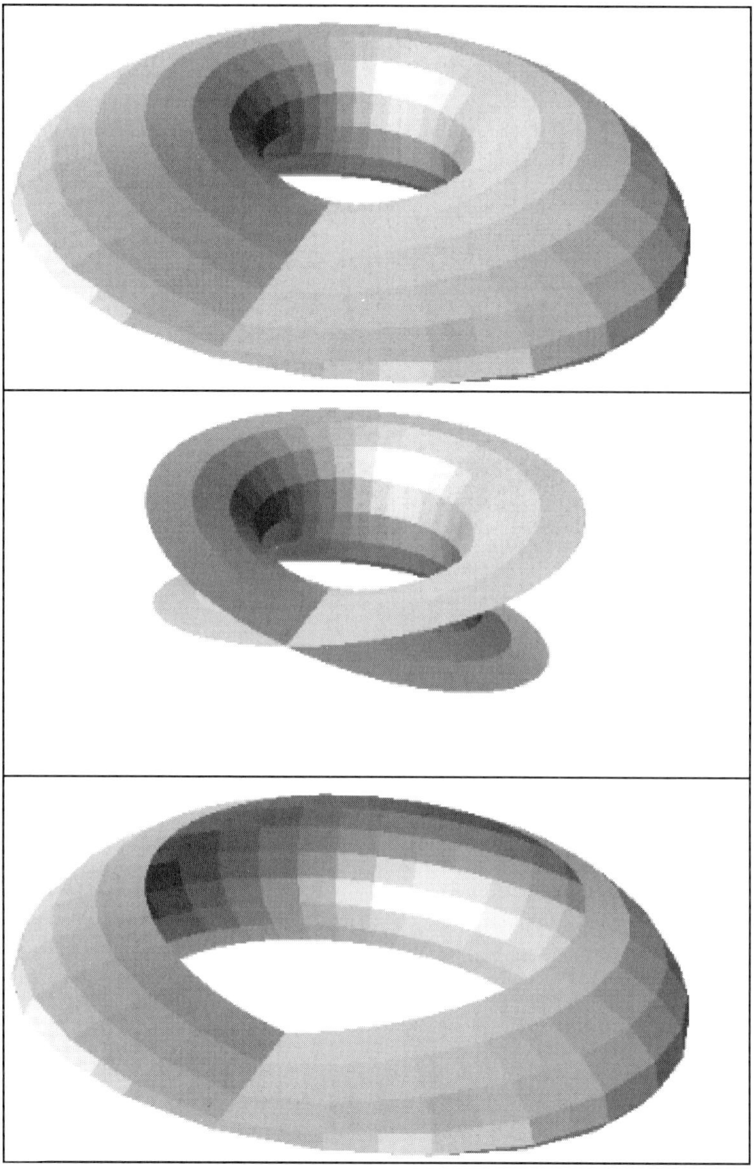

Figure 18-3 The projection of the Klein bottle, K, at top; see Example 18.06. Also shown are projections of the two Möbius bands whose union is K; see the remark on page 372. In the middle is shown $\pi(M_1)$, at the bottom is $\pi(M_2)$.

The projection $\pi(P)$ is shown at the top of Figure 18-4. As with the Klein bottle, the circles C_θ have disjoint projections, and these projections are ellipses, except that $\pi(C_0)$ is a circle and $\pi(C_\pi)$ is a line segment; see Figure 18-4.

Note the distinction between these circles and those used for the Klein bottle. In the construction of the Klein bottle, the circles $C_\theta \cap C_\phi = \emptyset$ if $\theta \neq \phi$ (except for the case $C_0 = C_{2\pi}$). In contrast for the circles used for the projective plane, any two circles C_θ and C_ϕ have the point $(0, 0, 0, 0)$ in common. ◆

The projective plane P is homeomorphic to a union of a disk and a Möbius band. The analysis can be done modeled on discussion in the remark on page 372.

Write the circle $C = L \cup R$ as the union of two semicircular arcs, a left arc and a right arc:

$$L = \{(x, y, z) \in R^3_+ : (x - 2)^2 + y^2 = 1, x \leq 1, \text{ and } z = 0\},$$
$$R = \{(x, y, z) \in R^3_+ : (x - 2)^2 + y^2 = 1, x \geq 1, \text{ and } z = 0\}.$$

The common endpoints of these arcs are: $e_+ = (1, 1, 0, 0)$ and $e_- = (1, -1, 0, 0)$. Let $E = \{e_+, e_-\}$. Clearly,

$$F_{2\pi}(e_+) = e_- \text{ and } F_{2\pi}(e_-) = e_+;$$

in particular, $F_{2\pi}(E) = E$. Also, that $F_{2\pi}$ maps each arc to itself (in a "reversing manner," as we see from $F_{2\pi}|_E$): $F_{2\pi}(L) = L$ and $F_{2\pi}(R) = R$.

Let $L_\theta = p_\theta \circ F_\theta(L)$, $R_\theta = p_\theta \circ F_\theta(R)$, and $E_\theta = p_\theta \circ F_\theta(E)$. Define

$$M_1 = \bigcup_{0 \leq \theta \leq 2\pi} L_\theta, \; M_2 = \bigcup_{0 \leq \theta \leq 2\pi} R_\theta, \text{ and } G = \bigcup_{0 \leq \theta \leq 2\pi} E_\theta.$$

Since $C = L \cup R$, $E = L \cap R$, we see that $K = M_1 \cup M_2$ and $M_1 \cap M_2 = G$. It is easy to check that G is homeomorphic to a circle. As with the Klein bottle, M_2 is homeomorphic to a Möbius band. Furthermore, $\partial M_1 = G$ and $\partial M_2 = G$.

However, M_1 is homeomorphic to a D^2. To see this consider the subset Q of C consisting of the quarter circle:

$$Q = \{(x, y, z) \in R^3_+ : (x - 2)^2 + y^2 = 1, x \leq 1, 0 \leq y \text{ and } z = 0\}.$$

Define $f: Q \to I$ by $f(x, y, z) = x$; f is clearly a homeomorphism. For $0 \leq \theta \leq 4\pi$, define $Q_\theta = p_\theta \circ F_\theta(Q)$. Then, for $\theta, \phi \in [0, 4\pi]$, $Q_\theta \cap Q_\phi = \{(0, 0, 0)\}$, except that $Q_0 = Q_{4\pi}$. Also, one can write

$$M_1 = \bigcup_{0 \leq \theta \leq 4\pi} Q_\theta.$$

Use polar coordinates for D^2, and define $F: D^2 \to M_1$ by $F(r, \theta) = p_{2\theta} \circ F_{2\theta}(f^{-1}(r))$. Note that if D_θ is the diameter of D^2 corresponding

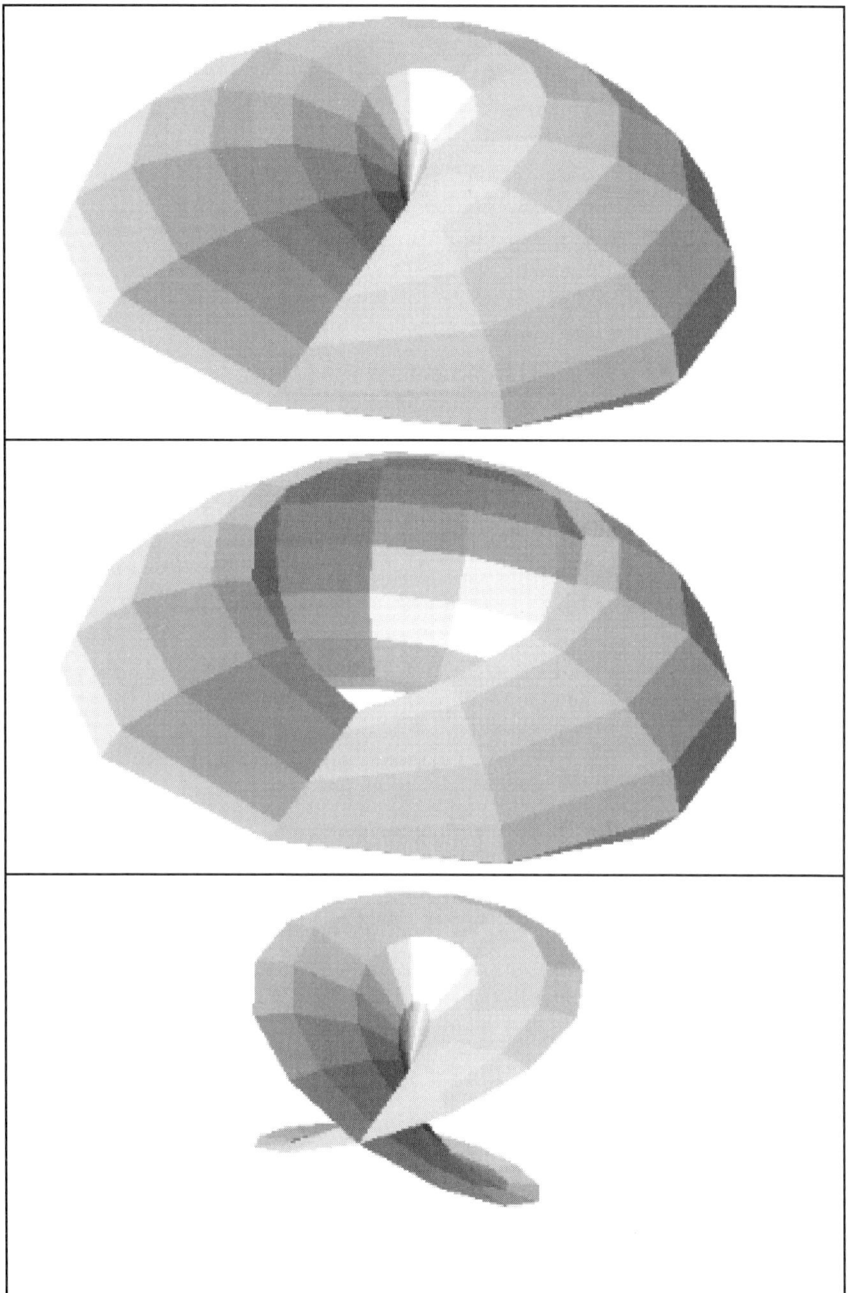

Figure 18-4 The projection of the projective plane and two pieces—a disk and a band; see the remark on page 376.

to points with argument θ or $\theta + \pi$, that $F(D_\theta) = L_\theta$. It can be checked that F is a homeomorphism.

It is possible to describe the projective plane in $\mathbf{R}^4$, using an equation. Define $f(x, y, z) = (x^2 - y^2, xy, xz, yz)$. Then the subset $f|_{S^2}$ is an immersion, and it can be shown that the image of f is homeomorphic to P^2. This does seem like an easier description, but it does not shed light on certain topological aspects. For example, it is not as clear that this subset is the union of a disk and a Möbius band glued together along a common boundary.

18.3 The 3-dimensional sphere, S^3

Let us take a look at the 3-dimensional sphere. It is a subset of $\mathbf{R}^4$ which is a closed 3-dimensional manifold.

Example 18.08 The 3-dimensional sphere is

$$S^3 = \{(x, y, z, w) \in \mathbf{R}^4 : x^2 + y^2 + z^2 + w^2 = 1\}.$$

We will show how to view S^3 as a union of circles and also how to find an interesting family of tori in S^3.

If we project onto the first two coordinates, we get a disk. Define $p: S^3 \to D^2$ by $p(x, y, z, w) = (x, y)$.

Let $p_0 \in D^2$, and write $p_0 = (x_0, y_0)$. If $p_0 \in \partial D^2$, then $x_0^2 + y_0^2 = 1$, and so $p^{-1}(p_0)$ is the single point $\{(x_0, y_0, 0, 0)\}$. Let $S_{\partial D} = p^{-1}(\partial D^2)$; then $S_{\partial D}$ is homeomorphic to the circle S^1 (in fact, the restriction $p|_{S_{\partial D}}$ is a homeomorphism).

If $p_0 \notin \partial D^2$, then $x_0^2 + y_0^2 < 1$ and so $S_{p_0} = p^{-1}(p_0)$ is a circle $z^2 + w^2 = 1 - (x_0^2 + y_0^2)$. We can see that S^3 is the disjoint union of subsets, each homeomorphic to the circle S^1:

$$S^3 = S_{\partial D} \cup \bigcup_{p_0 \notin \partial D^2} S_{p_0}.$$

An interesting exercise in geometry is to use stereographic projection $\sigma: (S^3 - \{(0, 0, 0, -1)\}) \to \mathbf{R}^3$ to see what these circles look like. In general, stereographic projection will take a geometric circle in S^3 to a geometric circle in $\mathbf{R}^3$, except that a circle containing $(0, 0, 0, -1)$ will be sent to a line. As a start, σ will take the circle containing $(0, 0, 0, -1)$ to the line corresponding to the z-axis.

For $0 < r \leq 1$, let C_r be the circle in D^2 with equation $x^2 + y^2 = r^2$. Let $T_r = p^{-1}(C_r)$. Each T_r is homeomorphic to the torus $S^1 \times S^1$.

Thus S^3 is the union of two disjoint circles, $S_{\partial D}$ and $S_{(0,0)}$, together with a collection of disjoint torii T_r, $0 < r < 1$. ◆

Example 18.09 In addition, we can view S^3 as a union of solid torii. Let D_0 be the closed disk in $\mathbf{R}^2$ of radius $1/2$, $D_0 = \overline{N_{1/2}(\vec{0})}$. Let $A = D^2 - N_{1/2}(\vec{0})$; A is an annulus and $D^2 = D_0 \cup A$ with $D_0 \cap A = \partial D_0$. Let

$$M_1 = p^{-1}(A) \text{ and } M_2 = p^{-1}(D_0).$$

Then each of M_1 and M_2 are homeomorphic to a solid torus. This is clear for M_2. In fact, let $s: D^2 \to D_0$ be defined by $s(\vec{x}) = \frac{1}{2}\vec{x}$. Then there is a homeomorphism $t: M_2 \to$ such that the following diagram commutes:

$$\begin{array}{ccc} M_2 & \xrightarrow{t} & S^1 \times D^2 \\ \downarrow p & & \downarrow q \\ D_0 & \xrightarrow{s} & D^2; \end{array}$$

where q is projection onto the second factor. ◆

The next construction is useful for constructing and understanding certain examples in high dimensions.

Definition 18.10 *Suppose $X \subseteq \mathbf{R^n}$, $Y \subseteq \mathbf{R^m}$ and $f: X \times I \to Y$ is a homotopy. The map $F: X \times I \to Y \times I$ is called the* **trace** *of f.*

If the homotopy f is an isotopy, then the trace of F is an embedding.

18.4 Knotted surfaces in $\mathbf{R}^4$

If we wish to pursue knot theory in $\mathbf{R}^4$, Proposition 18.03 shows that we should not look for knotted circles. It turns out that the proper generalization is the study of surfaces in $\mathbf{R}^4$.

The simplest examples of knotting of a surface in $\mathbf{R}^4$, are torii in $\mathbf{R}^4$ obtained by spinning of a knot. Recall the description of the standard torus in $T^2 \subseteq \mathbf{R}^3$, Example 13.23. This subset T^2 was generated by placing a circle in each page H^2_α of $\mathbf{R}^3$. We get an subset of $\mathbf{R}^4$ by doing the same kind of construction, using the three-dimensional pages of $\mathbf{R}^4$.

As a warm-up, we use this to define a simple torus in $\mathbf{R}^4$. For any angle θ, consider the page H^3_θ. In generalized cylindrical coordinates, let $m(\theta)$ be the circle in H^3_θ with center $(2, \theta, 0, 0, 0)$ and radius 1. Let $T'_{(2,1)} = \bigcup_{0 \le \theta < 2\pi} m(\theta)$. This is nothing really new—$T'_{(2,1)}$ is the same as the "standard inclusion" of T^2 into $\mathbf{R}^4$ considered in the remark following Definition 18.01.

We get interesting subsets by replacing the circle $m(\theta) \subseteq H_\theta$ by a fixed knotted circle $k(\theta)$, for each $0 \leq \theta \leq 2\pi$. That is, we have a knotted circle $k \subseteq Int(R^3_+)$, and we let $k(\theta) = p_\theta(k)$; then define

$$\Sigma(k) = \bigcup_{0 \leq \theta \leq 2\pi} m(\theta).$$

Except for $m(0) = m(2\pi)$, we have $m(\theta) \cap m(\phi) = \emptyset$ if $\theta \neq \phi$, and $\Sigma(k)$ is easily seen to be homeomorphic to a torus. It is called the "spun torus" obtained by spinning k.

Let P denote xzw-space in $\mathbf{R}^4$; P is the union of the two pages H_0 and H_π. The intersection of the spun torus will be the disjoint union of two knotted circles, one in each page. The knot in H_0 will be equivalent to the original knot, but the knot in H_π will be the mirror image of this knot. At first it might seem strange that we get the knot and its mirror image rather than two copies of the knot. However, if we consider the situation for a torus in $\mathbf{R}^3$, it becomes clear. Suppose we take a non-isosceles right triangle in the plane and spin his about the z-axis to produce a subset homeomorphic to a torus. If we slice this using the xz-plane, we obtain two triangles in that plane that are mirror images; see Figure 18-5.

We say that a torus in $\mathbf{R}^4$ is an "unknotted torus" if it is ambiently isotopic to $T'(2,1)$; otherwise, we say it is a "knotted torus." Using algebraic topology it is possible to show that if k is knotted, then so is $\Sigma(k)$; in particular, knotted surfaces exist.

We can alter this construction and get knotted spheres in $\mathbf{R}^4$. For this we use an image of a proper embedding of an interval into a page H^3_α. So we begin with a subset k of R^3_+ homeomorphic to an interval, D^1, with the endpoints p and q in ∂R^3_+ such as shown in any of the eight parts of Figure 18-6. Such a subset is called a "proper knotted arc." Let $k(\theta) = p_\theta(k)$, and define

$$\Sigma(k) = \bigcup_{0 \leq \theta \leq 2\pi} k(\theta).$$

Except for $k(0) = k(2\pi)$, we have $k(\theta) \cap k(\phi) = \{p\} \cup \{q\}$ if $\theta \neq \phi$. One can show that $\Sigma(k)$ is homeomorphic to S^2. It is called the **spun sphere obtained by spinning** k.

We say that a sphere in $\mathbf{R}^4$ is an "unknotted sphere" if it is ambiently isotopic to $i(S^2)$ where $i: \mathbf{R}^3 \to \mathbf{R}^4$ is the standard inclusion; otherwise we say it is a "knotted sphere." A proper arc in R^3_+ is an "unknotted proper arc" if it is ambiently isotopic to $Z = \{(x, y, z) \in R^3_+ \mid x^2 + y^2 = 1$ and $z = 0$; otherwise, it is called a "knotted proper arc." Using algebraic topology it is possible to show that if k is a knotted proper arc, then so is $\Sigma(k)$ is a knotted sphere. In particular, knotted spheres exist.

We end with one more set of examples in which an isotopy of a proper arc is used it to construct a knotted surface. The point here is

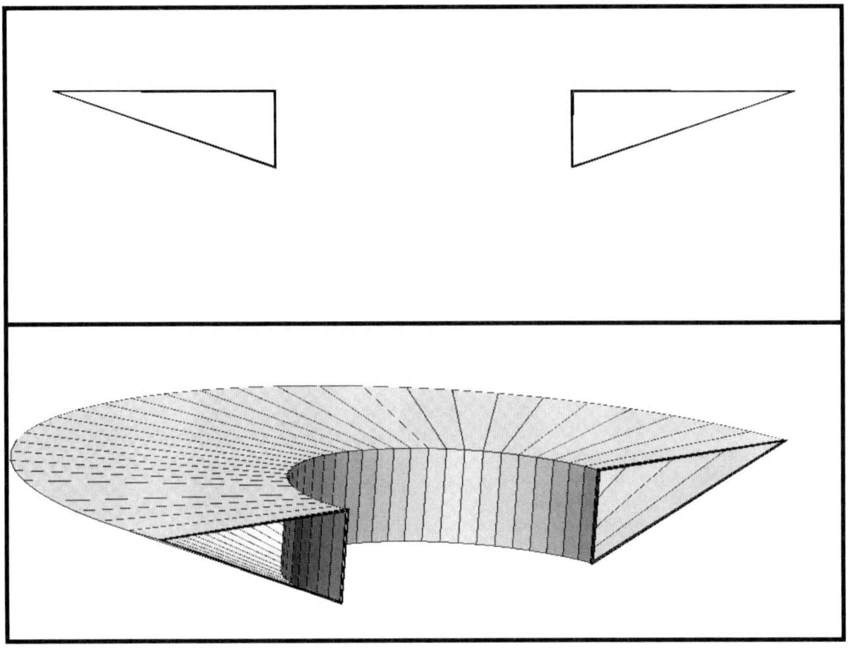

Figure 18-5 Bottom shows spinning a triangle to obtain a torus in $\mathbf{R}^3$ with resulting torus cut in half. Top shows two triangles in the cutting plane. Note these are mirror images; see the remark on page 378.

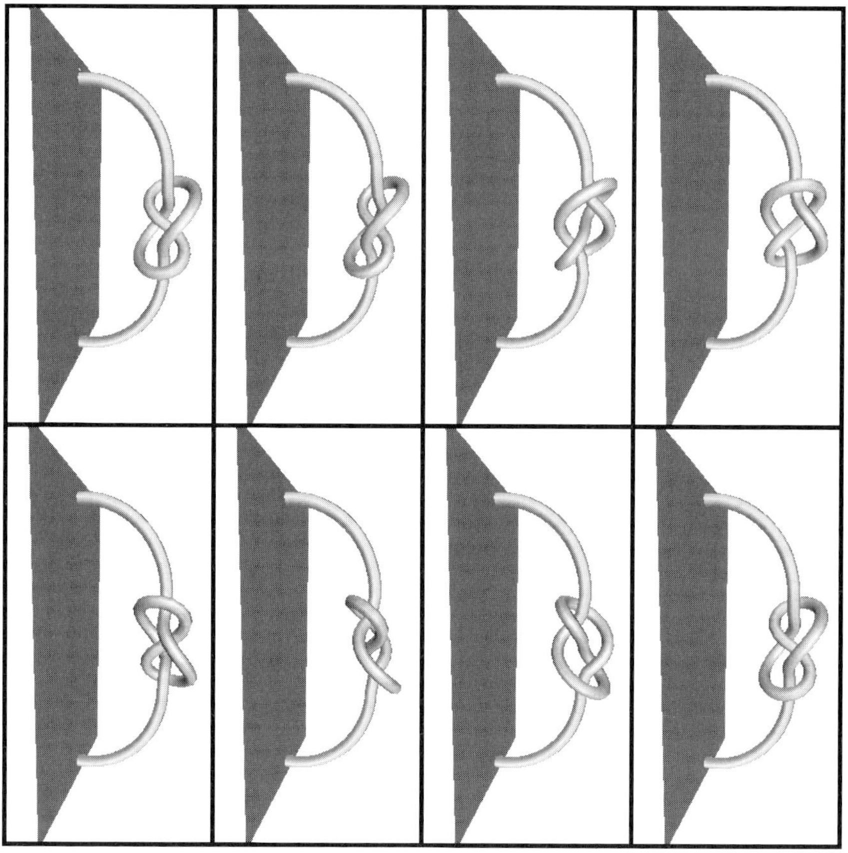

Figure 18-6 Eight knotted arcs. A twisting motion of the arc is shown at upper left; see remark 381. These figures show the knotted arcs in H_0, $H_{\pi/4}$, $H_{\pi/2}$, $H_{3\pi/4}$, H_{π}, $H_{5\pi/4}$, $H_{3\pi/2}$, $H_{7\pi/4}$, and $H_{2\pi} = H_0$. The gray square, seen in perspective, is a portion of ∂H_{α}.

to indicate that there are a lot of interesting surfaces in $\mathbf{R}^4$, defined in terms of topological constructions in $\mathbf{R}^3$.

One can make some simple alterations to the definition of the spinning construction and get additional interesting examples. For example, consider a spun sphere $\Sigma(k) = \bigcup_{0\le\theta\le 2\pi} k(\theta)$. All one needs to be assured that $\Sigma(k)$ is homeomorphic to S^2 are the conditions that each $k(\theta)$ is a proper knotted arc in H_θ and that, except for $m(0) = m(2\pi)$, we have $m(\theta) \cap m(\phi) = \{p\} \cup \{q\}$ if $\theta \neq \phi$ and $k(\theta)$ varies continuously with θ. For example, consider the proper arcs $k(\theta)$ obtained by a twisting motions as shown in Figure 18-6. The subset

$$M = \bigcup_{0\le\theta\le 2\pi} k(\theta),$$

using these $k(\theta)$ as shown is called a "1-twist spinning" of k_0, and the sphere we obtain is called a "twist spun knot." The isotopy shown in Figure 18-6 is called a single twist. By making n twists as we vary θ from 0 to 2π, we obtain other possibilities called n-twist spun knots. It can be shown that this twisting is topologically significant and that twist spun knots are different from spun knots.

*18.5 General Topology and Chapter 18

The natural topic to discuss here, infinite dimensional space, has already been discussed as the topic of infinite products, especially the Hilbert cube.

In the study of high dimensional knotting, most of the interest is in the "co-dimension two" case where one studies embeddings of n-dimensional manifolds in $\mathbf{R}^{n+2}$, and most of what is known relates to the case where the n-dimensional manifold is S^n. So, for example, for knotted three-dimensional spheres in $\mathbf{R}^5$, knotted four-dimensional spheres in $\mathbf{R}^6$, etc. The spinning and twist-spinning constructions have several distinct generalizations in higher dimensions.

18.6 Problems for Chapter 18

18.1 Let $i : \mathbf{R}^3 \to \mathbf{R}^4$ be the standard inclusion. Determine $i(T^2) \cap S^1 \times S^1$.

18.2 Let $S^1 \times S^1$ be standard torus in $\mathbf{R}^4$; what is the image of $S^1 \times S^1$ under the standard projection? What do we get if we slice $S^1 \times S^1$ by hyperplanes P_{w_0}?

18.3 Give details for Example 18.05, for defining an ambient isotopy of L in $\mathbf{R}^4 - C$ such that $\Phi(1)(L) = L'$.

18.4 Prove Proposition 18.02.

18.5 In Example 18.06, suppose we used a different motion of the circle. Suppose, instead, F_θ is rotation of angle $\theta/2$ about the line through $(2,0,0)$ parallel to the z-axis. The circle C rotates about this axis with fixed points $(2,0,\pm 1)$. Note that $F_{2\pi}(C) = C$. Let $K' = \bigcup_{0 \leq \theta \leq 2\pi} C_\theta$. Show that the projection of K', using projection in the w-direction, is a Möbius band.

18.6 Describe in more detail the map $h: D \to D^2$ of Example 18.2, and verify that it is a homeomorphism.

18.7 Referring to Example 18.09, define a maps $q': S^1 \times D^2 \to S^1 \times D^1$ and $s': A \to S^1 \times D^1$, so that the following diagram commutes:

$$\begin{array}{ccc} M_1 & \xrightarrow{t} & S^1 \times D^2 \\ \big\downarrow{\scriptstyle p} & & \big\downarrow{\scriptstyle q'} \\ A & \xrightarrow{s'} & S^1 \times D^1. \end{array}$$

19. THE POINCARÉ CONJECTURE

OVERVIEW: We state and discuss the Poincaré Conjecture, one of the most important unsolved (as yet) problems in mathematics.

19.1 What is S^3?

We return to two of our original questions:

Question 0.1 What is the plane ?

Question 0.3 What is space ?

One way to think about these questions is: given an subset $X \subseteq \mathbf{R^m}$; how could we recognize that X is the plane, or space? We can now reformulate these questions as:

Question 19.01 What topological properties imply that X is homeomorphic to $\mathbf{R}^2$?

Question 19.02 What topological properties imply that X is homeomorphic to $\mathbf{R}^3$?

We could ask a question for any value of n:

Question 19.03 What topological properties imply that X is homeomorphic to $\mathbf{R^n}$?

In many ways the n-sphere S^n is an easier object of study, largely because it is compact. Using generalized stereographic projection, Proposition 4.06, we can view S^n as $\mathbf{R^n}$ with one point added. So, in a sense, there s does not seem to be a lot of difference between S^n and $\mathbf{R^n}$. This leads us to formulate and focus on some similar questions for subsets $X \subseteq S^n$.

Question 19.04 What topological properties imply that X is homeomorphic to $\mathbf{S^n}$?

We discuss some answers to these questions. Although proofs of these statements, except for the first, are beyond the level of this book, we can at least understand the questions and the terms used in the discussion.

Here is an answer for Question 19.04, for $n = 1$:

Proposition 19.05 *If $X \subseteq \mathbf{R^m}$, then X is homeomorphic to $\mathbf{S^1}$ if and only if X is a connected, compact, 1-dimensional manifold with empty boundary.* ∎

Turning our attention to Question 19.04 for $n = 2$, the immediate problem is that there are a lot of closed connected 2-dimensional manifolds. However, it turns out that simple connectedness is the key. We have the following answer; proofs follow for general classification theorems of surfacers as in [42, 43]:

Proposition 19.06 *If $X \subseteq \mathbf{R^m}$, then X is homeomorphic to $\mathbf{S^2}$ if and only if X is a connected, simply-connected, compact, 2-dimensional manifold with empty boundary.* ∎

Based on our answer for $n = 2$, Question 19.04 for $n = 3$, has the following refinement, known as the "Poincaré Conjecture."

Question 19.07 Poincaré Conjecture. Suppose M is a compact connected, simply-connected, 3-dimensional manifold with empty boundary; must it be homeomorphic to S^3?

Question 19.07 was first raised by Henri Poincaré. As of this writing, this question has not been resolved. It is generally considered to be the most important unsolved problem in topology. Consequently, it is one of the major unsolved problems in modern mathematics.

It may seem peculiar to call a question a conjecture, but this is common usage.

Much mathematical effort has been made to try and answer this question. One important observation is the following:

Proposition 19.08 *If M is a compact, connected, 3-dimensional manifold with empty boundary, then M has the deformation type of S^3 if and only if it is simply connected.* ∎

In light of Proposition 19.08, we can restate Question 19.07 as

Question 19.09 The Poincaré Conjecture, version II. Suppose M is a compact connected, 3-dimensional manifold with empty boundary which has the deformation type of S^3; must it be homeomorphic to S^3?

This significance of this reformulation becomes apparent when we consider the Question 19.04 for $3 < n$. We have to abandon hope that simple connectivity is a key hypothesis. For example, consider $M = S^2 \times S^2$. Recall that S^2 is simply connected, Proposition 15.3, and the product of simply-connected subsets is simply connected, Proposition 15.14. Also, the product of connected subsets is connected, Proposition 7.14, the product of compact subsets is compact, Proposition 10.19, and the product of two 2-dimensional manifolds is a 4-dimensional manifold, Proposition 13.33. Thus we see that M is a compact, connected, simply-connected 4-dimensional manifold with empty boundary. However, it can be shown that M is not homeomorphic to S^4 (or even of the same deformation type).

It turns out that the formulation of the Poincaré Conjecture, as stated in Question 19.09, points the way to generalization in higher dimensions.

Question 19.10 The generalized Poincaré Conjecture. If M is a compact, connected, n-dimensional manifold with empty boundary which has the deformation type of S^n; must it be homeomorphic to S^n?

It would seem that, as one increases the dimension of a such a problem, the solutions might become more difficult. However, in this case, the problems are easier in high dimensions. In fact, the hardest case of Proposition 19.11 is for dimension 4.

Here is a solution of the generalized Poincaré Conjecture:

Proposition 19.11 *If M is a compact, connected, n-dimensional manifold, $4 \leq n$, with empty boundary which has the deformation type of S^n, it is homeomorphic to S^n.* ■

*19.2 General Topology and Chapter 19

Although there are topological manifolds that are not smooth manifolds, it turns out that any closed topological manifold that has the homotopy type of S^n is a smooth manifold. The topological solution to the higher-dimensional Poincaré Conjecture then implies that these are homeomorphic to S^n.

There remains the "smooth version" of the Poincaré Conjecture, which asks the question:

Question 19.12 Is a smooth, closed, n-dimensional manifold, which is homeomorphic to S^n, diffeomorphic to S^n?

As mentioned in the remark on page 301, in dimension 7, there are examples of smooth, closed, 7-dimensional manifolds homeomorphic to S^7, but not diffeomorphic to S^7. Other high dimensions are also well understood. The notable exception is dimension 4, in which case the answer is (as of this writing) unknown. That is, it is known that a closed connected 4-dimensional manifold which has the deformation type of S^4 is homeomorphic to S^4, but it is unknown if it must be *diffeomorphic* to S^4.

Part III

APPENDICES

A. SETS AND LOGIC

OVERVIEW: We discuss the relation between sets and logic, and touch on topics that are of particular importance in topology.

We assume some familiarity with the basic concepts of logic and set theory. If you are unfamiliar with these you should seek exposition either from your instructor or from a standard text on logic and sets.

A.1 Logic

One way to view mathematics, especially from the point of view of learning it, is that mathematics is a language. The basics of learning a new language are memorizing the vocabulary and mastering the grammar. So it is with mathematics. From this point of view, what is it that makes mathematics different from other languages?

Grammar is a set of rules that enable us to decide whether a given collection of words makes sense. What makes mathematics different from natural languages is that the grammar includes strict rules of logic, called mathematical logic.

In a mathematical system we study statements, called "elementary propositions." These are statements which, under the assumptions of the mathematical system, are either true or false. This is in contrast to natural languages. For example, the sentence "Beauty is in the eyes of the beholder" is a perfectly grammatical English sentence. However, reasonable people may differ as to whether this sentence is true or false.

In mathematics we encounter complex statements built up from elementary propositions. Mathematical logic gives us formal rules for assigning "true" or "false" to complex statements, based on the truth values of these simpler statements. These rules are based on the *pattern* of the complex proposition and are completely independent of the *content* of the elementary propositions.

Let X be a set, and suppose P and Q are propositions. We list some basic definitions and notations.

(a) $\sim P$ denotes *not P*.
(b) $P \wedge Q$ denotes *P and Q*.
(c) $P \vee Q$ denotes *P or Q*.
(d) $P \rightarrow Q$ denotes *if P then Q*.
(e) $\forall x \in X[P]$ denotes *for all x in X, P (is true).*
(f) $\exists x \in X[P]$ denotes *there exists an x in X such that P (is true).*
(g) $P \leftrightarrow Q$ denotes *P if and only if Q*.

Definition A.01 *Suppose C and K are compound statements. We say C and K are* **logically equivalent** *if, for all possible truth values of the propositions that make up C and K, the truth value of C and the truth value of K are the same. We denote this by* $C \Leftrightarrow K$.

Example A.02 For example,

$$P \leftrightarrow Q \Leftrightarrow [P \rightarrow Q] \wedge [Q \rightarrow P].$$

In words: "P if and only if Q" is logically equivalent to "P implies Q" *and* "Q implies P." ◆

Example A.03 The "double negation" rule of logic, symbolically expressed, becomes $P \Leftrightarrow \sim [\sim P]$. ◆

One logical construction that needs extra care is the "if P, then Q" construction. In natural languages, such a sentence is often taken to mean that "P causes Q." This in turn implies some relation between P and Q. But mathematical logic must deal with the pattern of "if P, then Q," only on the basis of the truth values of P and Q.

The biggest problem for many is: if P is false, then $P \rightarrow Q$ is true, no matter what the truth value is for Q. The consequence of this is that some rather strange sounding sentences are considered to be true. For example: If there is one triangle with four sides, then no triangle has four sides, and the sentence you are now reading is written with purple ink.

There are some cases where is seems reasonable to have the rule that $P \rightarrow Q$ is true, if P is false, no matter what the truth value is for Q. Consider the system of natural numbers. The statement $2 + 2 = 5$ is a proposition about numbers and, of course, is false. Call this statement P. Let Q be the statement $2 + 3 = 6$; this is also false. Now consider the sentence of the form "if P then Q." We wish to decide whether this new sentence is true or false, by regarding the truth values of the propositions P and Q only. We might like to say that

$$\text{if } 2 + 2 = 5 \text{ then } 2 + 3 = 6$$

is true (since we could just add one to each side of the equation P). Next, consider Q to be the statement $2 < 5$; R is true. Consider the sentence

$$\text{if } 2 + 2 = 5 \text{ then } 2 < 5.$$

We would like this to be true (arguing that in general, if a and b are numbers with $a + 2 = b$, then it would follow that $a < b$).

The bottom line is that we must decide once and for all the truth value of $P \to Q$ in the case that P is false. That rule is: "if P is false then $P \to Q$" is true.

The other possibilities for the "if P then Q" are not surprising. If P and Q are both true, then "if P the Q" is true; if P is true and Q is false, then we say that "if P then Q" is false.

Example A.04 From the discussion above about the truth values for $P \to Q \Leftrightarrow$, we can verify $P \to Q \Leftrightarrow (\sim P) \vee Q$. For example,

if x is an even integer then x^2 is an even integer

is equivalent to

an integer x is either odd or has an even square. ◆

The next two formulas are known as "DeMorgan's rules." These are critical in that they describe how negation affects the basic constrictions of *and* and *or.*

Proposition A.05

(a) $\sim [P \vee Q] \Leftrightarrow \sim P \wedge \sim Q$.
(b) $\sim [P \wedge Q] \Leftrightarrow \sim P \wedge \sim Q$. ▮

Example A.06 Using Proposition A.05 and Example A.04, we see that

$$\begin{aligned} \sim [P \to Q] &\Leftrightarrow \sim [\sim P \vee Q] \\ &\Leftrightarrow [\sim\sim P] \wedge \sim Q \\ &\Leftrightarrow P \wedge \sim Q. \end{aligned}$$

This equivalence, $\sim [P \to Q] \Leftrightarrow P \wedge \sim Q$, is important since it is the basis of what is called "argument by contradiction." In this technique one shows that P implies Q by showing that it is false that P is true and Q is false. This implies that $\sim [P \to Q]$ is false; thus $[P \to Q]$ is true.

As a typical example, suppose x is a certain positive number. Consider a proof that if $x < 1$ then $x^2 < 1$. Assume this is false. Using our equivalence, that would mean that $1 < x$ and $x^2 \geq 1$. (We could now finish the argument as follows. If $1 < x$, then $x - 1$ is negative. If $x^2 \geq 1$, then $x^2 - 1 \geq 0$, or by factoring we have $(x - 1)(x + 1) \geq 0$. But this is impossible since $x + 1$ is positive, and a product of a positive number and a negative number must be less than 0.) ◆

The following two equivalences are important and can be thought of as generalizations of DeMorgan's rules, Proposition A.05.

Proposition A.07

(a) $\sim [\forall x \in X[P]] \Leftrightarrow \exists x \in X[\sim P]$.
(b) $\sim [\exists x \in X[P]] \Leftrightarrow \forall x \in X[\sim P]$. ∎

A.2 Sets, some basics

Suppose that X is a set, and A and B are subsets of X. We express the statement "A is a subset of B", using notations of Chapter A.1.

Definition A.08 *We say A is a* **subset** *of B if* $\forall x \in X[x \in A \rightarrow x \in B]$

Example A.09 So

$$A \text{ is not a subset of } B$$

means

$$\exists x \in X[x \in A \wedge x \notin B].$$

We show how to derive this symbolically. Note the technique. We begin with a *complex* statement in (larger) brackets, which is negated. The bracketed statement gets less complex. At the end, the negation only involves negation of *simple* propositions.

$$\begin{aligned} \sim [\forall x \in X[x \in A \rightarrow x \in B]] &\Leftrightarrow \exists x \in X \sim [x \in A \rightarrow x \in B] \\ &\Leftrightarrow \exists x \in X[x \in A \wedge x \notin B] \blacklozenge \end{aligned}$$

The ability to articulate the negation of a complex sentence is important in the construction of mathematical proofs. Some find that a good technique for this is to first express the given sentence symbolically, and then to use some of the rules above. This technique is mostly of use to sort things out in an unfamiliar setting. In practice, one might do this once or twice and then remember the pattern in words, rather than symbols.

Example A.10 We can express "U is an open subset of $\mathbf{R}^n$" as follows:

$$\forall x \in U[\exists \epsilon > 0[N_\epsilon(x) \subseteq U]].$$

(Note: As is common practice, "$\exists \epsilon > 0[P]$" is a shorthand for "there exists a real number ϵ such that $\epsilon > 0$ and P is true"; in symbols, $\exists \epsilon \in R^1[(\epsilon > 0) \wedge P]$. ◆

Example A.11 Using the expression of Example A.10, the sentence "U is *not* an open subset of $\mathbf{R^n}$" can be expressed as

$$\exists x \in U[\forall \epsilon > 0[\exists y \in \mathbf{R^n}[y \in N_\epsilon(x) \wedge y \notin U]]].$$

Here are some details:

$$\begin{array}{ll} \sim [\forall x \in U[\exists \epsilon > 0[N_\epsilon(x) \subseteq U]]] & \Leftrightarrow \\ \exists x \in U \sim [[\exists \epsilon > 0[N_\epsilon(x) \subseteq U]]] & \Leftrightarrow \\ \exists x \in U[\forall \epsilon > 0 \sim [[N_\epsilon(x) \subseteq U]]] & \Leftrightarrow \\ \exists x \in U[\forall \epsilon > 0 \sim [[\forall y \in \mathbf{R^n}[y \in N_\epsilon(x) \longrightarrow y \in U]]] & \Leftrightarrow \\ \exists x \in U[\forall \epsilon > 0[\exists y \in \mathbf{R^n}[y \in N_\epsilon(x) \wedge y \notin U]]]. & \end{array}$$

We translate each of the symbolic statements above into ordinary sentences:

- U is not an open subset of $\mathbf{R^n}$.
- There is an x in U for which it is not true that there is an ϵ so that $N_\epsilon(x)$ is contained in U.
- There is an x in U so that, for all possible ϵ, $N_\epsilon(x)$ is not contained in U.
- There is an x in U so that, for all possible ϵ, it is not true that if y is in $N_\epsilon(x)$ then y must be in U.
- There is an x such that, for all possible ϵ, there is a point in $N_\epsilon(x)$ which is not in U. ◆

Example A.12 Consider the definition of continuity, Definition D.07. Symbolically put, $f: \mathbf{R^1} \to \mathbf{R^1}$ is continuous if

$$\forall a \in R^1[\forall \epsilon > 0[\exists \delta > 0[\forall x \in \mathbf{R^1}[0 < |x-a| < \delta \to |f(x)-f(a)| < \epsilon]]]].$$

Then "f is not a continuous function" can be expressed

$$\exists a \in R^1[\exists \epsilon > 0[\forall \delta > 0[\forall x \in \mathbf{R^1}[\exists x[0 < |x-a| < \delta \wedge |f(x)-f(a)| \geq \epsilon]]]].$$

To see how we use this in an example, consider the function, $f(x)$ defined by

$$f(x) = \begin{cases} 1 & \text{if } x \leq 3 \\ 2 & \text{if } 3 < x. \end{cases} .$$

We show that $f(x)$ is not continuous. Choose $a = 3$ and $\epsilon = 1/2$. Then, for all $\delta > 0$ let $x = a + \frac{\delta}{2}$. Then we have $x \in N_\delta(a)$, yet $|f(x) - f(a)| \geq \epsilon$ since, in fact, we have $f(x) = 3$, $f(a) = 2$, and $|f(x) - f(a)| = 1$. ◆

The empty set $\emptyset$ is the set with no elements. It follows that $\emptyset \subseteq A$ for any set A since the sentence "if $x \in \emptyset$ then $x \in A$" is true, and "$x \in \emptyset$" is, by definition, always false. Recall the discussion in the remark on page 390.

Equality of sets can be defined in terms of the inclusion relation:

Definition A.13 *If A and B are subsets of S, we say A and B are* **equal subsets**, *denoted $A = B$, if and only if $A \subseteq B$ and $B \subseteq A$*

Thus $A = B$ if and only if: for every $x \in X$: (1) if $x \in A$, then $x \in B$, *and* (2) if $x \in B$, then $x \in A$.

The propositions below concerning complements of subsets are used many times in the study of topology.

Definition A.14 *Suppose $B \subseteq A$. Then the* **complement** *of B in A is the set of points which are in A and not in B. Notationally, we write this as $A - B$.*

Here is another useful proposition.

Proposition A.15 *Suppose S is a set and $A \subseteq S$ and $B \subseteq S$, then*

$$A - B = A \cap (S - B). \quad \blacksquare$$

Often, when the subset S is understood in context, one denotes $S - B$ by $-B$. Using this, the equation of Proposition A.15 becomes $A - B = A \cap -B$.

Here is a proposition about sets and complements that is used in the text.

Proposition A.16 *Suppose S is a set and $A \subseteq S$ and $B \subseteq S$, and $C \subseteq S$. If $A \cap C = A - B$, then $A \cap B = A - C$.* $\blacksquare$

In working with complements of sets, the following proposition is basic. It is an immediate consequence of the DeMorgan rules. Often, the equalities below are called the "DeMorgan rules for sets."

Proposition A.17 *Suppose we have two subsets, S_1 and S_2 of S.*

(a) $S - (S_1 \cup S_2) = (S - S_1) \cap (S - S_2)$.
(b) $S - (S_1 \cap S_2) = (S - S_1) \cup (S - S_2)$. $\blacksquare$

In topology we often consider indexed collections of subsets.

Definition A.18 *Let A be a set (to be used as an index set) and let S be a set. If for each $\alpha \in A$, we have a subset S_α, then the collection $\{S_\alpha\}_{\alpha \in A}$ is called an* **indexed collection of sets** *indexed by A.*

If the index set is the integers, $\mathbf{N}$, we say $\{S_\alpha\}_{\alpha \in \mathbf{N}}$ is a **sequence of subsets**.

Example A.19 Here are some typical examples of collections of subsets of $\mathbf{R}^2$ used in topology. The definition neighborhood, $N_\epsilon(x)$, is found in Chapter 1, Definition 1.04.

- Using index set $\mathbf{N}$, define $S_n = N_{\frac{1}{n}}((0,0))$. We might consider the collection $\{S_n\}_{n\in\mathbf{N}}$.
- Using the positive real numbers, R^1_+ as an index set, define $S_\epsilon = N_\epsilon((0,0))$. We might consider the collection $\{S_\epsilon\}_{\epsilon\in R^1_+}$.
- Using $\mathbf{R}^2$ for an index set, define $S_p = N_1(p)$. We might consider the collection $\{S_p\}_{p\in\mathbf{R}^2}$. ◆

You are certainly familiar with the idea of union and intersection of subsets, but perhaps only in the context of a finite number of subsets. Topology (and much of modern mathematics) uses infinite unions and intersections extensively. Pay particular attention to these definitions.

Definition A.20 *Suppose $\{S_\alpha\}_{\alpha\in A}$ is an indexed collection of subsets of S. The* **union** *of the sets S_α is the set $\{s \in S : \exists\alpha \in A[s \in S_\alpha]\}$. Notationally, we express this as: $\bigcup\limits_{\alpha\in A} S_\alpha$.*

Definition A.21 *Suppose $\{S_\alpha\}_{\alpha\in A}$ is an indexed collection of subsets of S. The* **intersection** *of the sets S_α is the set $\{s \in S : \forall\alpha \in A[s \in S_\alpha]\}$. Notationally, this is expressed as $\bigcap\limits_{\alpha\in A} S_\alpha$.*

The following formulas for subsets are generally known as the "distributive laws of sets," and are of fundamental importance.

Proposition A.22 *Suppose A, B, and X are subsets of S, then*

(a) $X \cap (A \cup B) = (X \cap A) \cup (X \cap B)$.
(b) $X \cup (A \cap B) = (X \cup A) \cap (X \cup B)$.

Here are two set equalities that are frequently used when infinite collections of subsets are involved. These are known as the "generalized distributive laws for sets:"

Proposition A.23 *Suppose $\{S_\alpha\}_{\alpha\in A}$ is an indexed collection of subsets of S and $X \subseteq S$ then*

(a) $X \cap (\bigcup\limits_{\alpha\in A} S_\alpha) = \bigcup\limits_{\alpha\in A} (X \cap S_\alpha)$.
(b) $X \cup (\bigcap\limits_{\alpha\in A} S_\alpha) = \bigcap\limits_{\alpha\in A} (X \cup S_\alpha)$.

The generalizations of DeMorgan's rules give the following set equalities:

Proposition A.24

(a) $S - \bigcup_{\alpha \in A} S_\alpha = \bigcap_{\alpha \in A} (S - S_\alpha)$.

(b) $S - \bigcap_{\alpha \in A} S_\alpha = \bigcup_{\alpha \in A} (S - S_\alpha)$. ∎

As an elementary exercise using these definitions, we note that if we have a union of indexed subsets $\bigcup_{\alpha \in A} S_\alpha$ and $A = \emptyset$, then $\bigcup_{\alpha \in A} S_\alpha = \emptyset$. This is because $x \in \bigcup_{\alpha \in A} S_\alpha$ means that $\exists \alpha_0 [\alpha_0 \in A \wedge x \in S_{\alpha_0}]$. But $A = \emptyset$ means $\alpha_0 \in A$ is false. Thus $[\alpha_0 \in A \wedge x \in S_{\alpha_0}]$ must be false. Therefore, $x \in \bigcup_{\alpha \in A} S_\alpha$ must be false, and so $\bigcup_{\alpha \in A} S_\alpha = \emptyset$.

A.3 Cartesian products

Given two sets, a standard way of creating a third is the Cartesian product by taking ordered pairs. (For general topology, as in Section*2.7, one needs a more general treatment of Cartesian product in the case of an infinite product. This is found at the end of this section in Definition A.39 of Section A.4.)

Definition A.25 *Suppose X and Y are sets. The set of all ordered pairs $X \times Y = \{(x, y) : x \in X$ and $y \in Y\}$ is called the* **Cartesian product** *of X and Y. We say that the sets X and Y are* **factors** *of $X \times Y$.*

From Definition A.25 it follows that $\mathbf{R}^2 = \mathbf{R}^1 \times \mathbf{R}^1$.

Consider the equation $\mathbf{R}^3 = \mathbf{R}^1 \times \mathbf{R}^2$. Now $\mathbf{R}^3$ consists of triples of real numbers such as (x, y, z). An element of $\mathbf{R}^1 \times \mathbf{R}^2$, has the form $(x, (y, z))$. In a strict sense, (x, y, z) and $(x, (y, z))$ are not the same—the first is a triple, and the second is a pair. Nevertheless, the association of (x, y, z) with $(x, (y, z))$ is so intuitive that most often these are considered identical.

If $A \subseteq X$ and $B \subseteq Y$, then $A \times B \subseteq X \times Y$. So, in particular, if $A \subseteq \mathbf{R^n}$ and $B \subseteq \mathbf{R^n}$, then $A \times B \subseteq \mathbf{R^{n+m}}$. In terms of coordinates, $A \times B$ is the subset:

$$\{(x_1, \dots, x_{n+m}) \in \mathbf{R}^{n+m} \mid (x_1, \dots, x_n) \in A \text{ and } (x_{n+1}, \dots, x_{n+m}) \in B\}.$$

In this way a point of $A \times B$ is viewed to be an $(n+m)$-tuple of numbers rather than a pair consisting of an n-tuple and an m-tuple.

We have

$$\mathbf{R}^{n+m} = \mathbf{R}^n \times \mathbf{R}^m = \mathbf{R}^m \times \mathbf{R}^n \text{ and}$$

$$\mathbf{R}^n \times (\mathbf{R}^m \times \mathbf{R}^k) = (\mathbf{R}^n \times \mathbf{R}^m) \times \mathbf{R}^k = \mathbf{R}^{n+m+k}.$$

It is not true in general that $A \times B = B \times A$. For example, if $A \subseteq \mathbf{R}^1$, $A = [0, 1]$, and $B \subseteq \mathbf{R}^1$, $B = [0, 2]$, then $(1, 2) \in A \times B$ but $(1, 2) \notin B \times A$. Here is another example: $\mathbf{R}_+^{n+1} = \mathbf{R}_+^1 \times \mathbf{R}^n$, but $\mathbf{R}_+^{n+1} \neq \mathbf{R}^n \times \mathbf{R}_+^1$. On the other hand, there is a sense, homeomorphism, in which $A \times B$ is the same as $B \times A$; see Problem 4.20 of Chapter 4.

We next relate Cartesian product construction with the basic constructions of union, intersection, and complement.

Proposition A.26 *Let $A \subseteq X$ and $B \subseteq Y$ and $C \subseteq Y$.*

(a) $A \times (B \cup C) = (A \times B) \cup (A \times C)$.
(b) $A \times (B \cap C) = (A \times B) \cap (A \times C)$.
(c) $A \times (B - C) = (A \times B) - (A \times C)$.

If $A \subseteq X$, $C \subseteq X$ and $B \subseteq Y$:
(d) $(A \cup C) \times B = (A \times B) \cup (C \times B)$.
(e) $(A \cap C) \times B = (A \times B) \cap (C \times B)$.
(f) $(A - C) \times B = (A \times B) - (C \times B)$. ∎

Finally, the next proposition provides a set equation we will need, which is not a common one and whose proof is a good excercise in the use of several of the propositions above. Make a sketch of these sets, using the closed intervals: $A = B = [0, 1]$ and $A_0 = B_0 = [1/3, 2/3]$.

Proposition A.27 *Suppose A and B are sets, $A_0 \subseteq A$ and $B_0 \subseteq B$; then*

$$A \times B - A_0 \times B_0 = (A \times (B - B_0)) \cup ((A - A_0) \times B). \quad ∎$$

The next definition causes problems for many beginning students of topology.

Definition A.28 *Suppose f is a function, $f : X \to Y$ and $U \subseteq Y$. The* **inverse image** *of U via f is the set $\{x \in X | f(x) \in U\}$.*

The notation for the inverse image of U via f is $f^{-1}(U)$. This notation, as well as the use of the word "inverse" can give rise to considerable confusion. The problem is that f^{-1} is also used as a notation for inverse of a function.

Definition A.29 *A function $f(x)$ is called one-to-one if, for any two points, x and x' of X with $x \neq x'$, $f(x) \neq f(x')$. Suppose $f(x)$ is one-to-one, and let $Z = f(X)$ denote the image of f. The* **inverse function** *from Z to X, denoted by f^{-1}, is defined by $f^{-1}(z) = x$ if and only if $f(x) = z$.*

Look at two functions, $g(x) = x^3$ and $f(x) = x^2$. Now g is a one-to-one function; the inverse of $g(x)$ is $\sqrt[3]{x}$. The inverse function of f does not exist since f is not a one-to-one function. Let U be the interval $U = (-8, 8)$. We can think of $g^{-1}(U)$ either as an inverse image of a set or the image under the inverse function. In either case, we get the same set, $g^{-1}(U) = (-2, 2)$. There is no *function* f^{-1}. However, the inverse, via f, of the set U is defined and $f^{-1}(U) = (-\sqrt{8}, \sqrt{8})$.

In summary, remember—if you see $f^{-1}(U)$ discussed, this does not necessarily mean that there is a *function* f^{-1}.

Here is a list of properties of functions which relate to basic set constructions:

Proposition A.30 *Let $f : X \to Y$ be functions. Suppose that X_1 and X_2 are subsets of X, and Y_1 and Y_2 are subsets of Y; then*

(a) $f(X_1 \cap X_2) \subseteq f(X_1) \cap f(X_2)$.
(b) *If f is a one-to-one map, then $f(X_1 \cap X_2) = f(X_1) \cap f(X_2)$.*
(c) $f^{-1}(Y_1 \cap Y_2) = f^{-1}(Y_1) \cap f^{-1}(Y_2)$.
(d) $f^{-1}(Y_1 - Y_2) = f^{-1}(Y_1) - f^{-1}(Y_2)$.
(e) *If $X_2 \subseteq X_1$, $f(X_1) - f(X_2) \subseteq f(X_1 - X_2)$.*
(f) *If f is a one-to-one map, and $X_2 \subseteq X_1$, then $f(X_1 - X_2) = f(X_1) - f(X_2)$.*
(g) $f(X_1 \cup X_2) = f(X_1) \cup f(X_2)$.
(h) $f^{-1}(Y_1 \cup Y_2) = f^{-1}(Y_1) \cup f^{-1}(Y_2)$.
(i) $X_1 \subseteq f^{-1}(f(X_1))$.
(j) *If f is a one-to-one map, $X_1 = f^{-1}(f(X_1))$.*
(k) $f(f^{-1}(Y_1)) \subseteq Y_1$.
(l) *If f is an onto map, $f(f^{-1}(Y_1)) = Y_1$.*

One can generally remember the above as "sets behave well under f^{-1} but sometimes need additional hypothesis for f."

The following is a useful result concerning one-to-one correspondences. The proof is not an easy one.

Proposition A.31 **The Bernstein-Schroeder Theorem** *Suppose that X and Y are sets. If $f : X \to Y$ and $g : Y \to X$ are functions which are both one-to-one functions, then there is a one-to-one correspondence between X and Y.* ∎

Another important notion is that of an equivalence relation.

Definition A.32 *Let X be a set. An* **equivalence relation** *on X, denoted by $\sim$, is a relation between elements of X which satisfy the following three conditions:*

(a) *For all $x \in X$, $x \sim x$ (the identity property).*
(b) *For all $x, y \in X$, if $x \sim y$, then $y \sim x$ (the symmetry property).*

(c) *For all* $x, y, z \in X$, *if* $x \sim y$ *and* $y \sim z$, *then* $x \sim z$ *(the transitivity property).*

Whenever we have an equivalence relation on X, we can divide X into disjoint non-empty subsets called "equivalence classes" (assuming X is non-empty).

Definition A.33 *If* $x \in X$, *then the* **equivalence class** *containing* x *is defined to be all* $y \in X$ *such that* $x \sim y$. *We use the notation* $[x]$ *for the equivalence class corresponding to* x.

Example A.34 Let L be the set of all lines in the plane. For any two lines, l_1 and l_2, define an equivalence relation $l_1 \sim l_2$ if and only if l_1, and l_2 are parallel lines. This is an equivalence relation.

If l_1 has equation $y = 2x + 1$, then the equivalence class $[l_1]$ would consist of all lines in the plane, with equation $y = 2x + b$ for some number b. ◆

One way of expressing an equivalence relation on X is by use of a partition of X.

Definition A.35 *Suppose* X *is a set; a* **partition** *of* X *is a collection* $\mathcal{P}$ *of disjoint non-empty subsets of* X *such that the union of these subsets is* X. *That is, if* $P \in \mathcal{P}$ *and* $Q \in \mathcal{P}$, *then either* $P = Q$ *or* $P \cap Q = \emptyset$, *and also* $X = \bigcup_{P \in \mathcal{P}} P$.

Proposition A.36 *Suppose* X *is a set with an equivalence relation. If* $x \in X$ *and* $y \in X$, *then either* $[x] = [y]$ *or* $[x] \cap [y] = \emptyset$.

So, if we have an equivalence relation for a set X, the collection of all sets of the form $[x]$ where $x \in X$, is a partition of X.

Conversely, suppose that we have a partition, $\mathcal{P}$ of X. We can then define an equivalence relation on X by $x \sim y$ if and only if there is a $P \in \mathcal{P}$ such that x and y are both in P.

A.4 Sets—topics needed for general topology

For optional, general topology sections we need to discuss a general Cartesian product.

If X and Y are sets, then the Cartesian product, $X \times Y$, is defined to be the set of all ordered pairs $\{(x, y)\}$ where $x \in X$ and $y \in Y$. Similarly the Cartesian product of k sets, say $X_1 \times \cdots \times X_k$, is defined to be all k-tuples, $\{(x_1, \ldots, x_k)\}$ where $x_i \in X_i$. Problems arise, however, in discussing an *infinite* product, especially a product of an uncountable number of sets (see Definition C.05 for a definition of uncountable). If this is your first encounter the topic of infinite products, it may come as a surprise that there are serious mathematical issues to address.

Suppose we have a collection of sets $\{X_\alpha\}_{\alpha \in A}$. Then the Cartesian product of these sets is denoted by $\prod_{\alpha \in A} X_\alpha$. Our first task, and it is not as simple as it might at first seem, is to provide a definition of $\prod_{\alpha \in A} X_\alpha$.

We focus on two particular examples. Let $\mathbf{N}$ be the set of natural numbers, $\mathbf{N} = \{1, 2, \ldots\}$, and $\mathbf{R}^1$ the real numbers. We use $\mathbf{R}^\infty$ to denote $\prod_{n \in \mathbf{N}} R_n^1$, and use $\mathbf{R}^\mathbf{R}$ to denote $\prod_{\alpha \in \mathbf{R}^1} R_\alpha^1$ where R_α^1 is copy of $\mathbf{R}^1$ indexed by α.

Example A.37 If $x \in \mathbf{R}^\infty$, we can write $x = (x_1, x_2, \ldots)$. We might think of x as an "infinite-tuple" of real numbers, but we already have a common term for this—we call x a "sequence of real numbers." Thus $\mathbf{R}^\infty$ is a familiar object—the set of all sequences of real numbers. ◆

The idea of an ordered pair is simple, namely, "one thing, then another," it hardly seems to need explanation. But a key to understanding an *uncountable* product such as $\mathbf{R}^\mathbf{R}$, is to determine: what do we *really* mean by "ordered pair" or "n-tuple"?

Definition A.38 *If X_1 and X_2 are two sets, then an* **ordered pair of points** *of X_1 and X_2 is a function from the set of two elements $\{1, 2\}$ to the disjoint union of X_1 and X_2.*

*More generally, if $X_1, \ldots, X_n$ are n sets, then an n-***tuple of points** *$X_1, \ldots, X_n$ is a function from the set of n elements $\{1, \ldots, n\}$ to the disjoint union of $X_1, \ldots, X_n$.*

Using this as a guide, we are lead to define

Definition A.39 *Suppose we have a collection of sets $\{X_\alpha\}_{\alpha \in A}$. The* **Cartesian product** *of $\{X_\alpha\}_{\alpha \in A}$, denoted, $\prod_{\alpha \in A} X_\alpha$, is the set of all functions from A to the disjoint union of the X_α.*

Example A.40 Using this definition, we see that $\mathbf{R}^\mathbf{R}$ is not that strange. Namely, $\mathbf{R}^\mathbf{R}$ is the set of all functions $f: \mathbf{R}^1 \to \mathbf{R}^1$. Note that this is the set of *all* real-valued functions and not just the continuous ones. ◆

Example A.41 More generally, if X and Y are sets, we see that Y^X is the set of all functions $f: X \to Y$. As a product, we write $Y^X = \prod_{x \in X} Y_x$. Let $|X|$ denote the cardinality of X. Then we can say that Y^X is the Cartesian product of $|X|$ copies of the space Y. ◆

There is a subtle, but important, issue that arises in the discussion of infinite products.

Question A.42 If each X_α is non-empty, then is $\prod_{\alpha \in A} X_\alpha$ non-empty?

Here is why this is not a simple issue. Suppose we are thinking of an uncountable product and we think the answer to Question A.42 should be "yes." Then one is saying "there exists a function from A to the disjoint union of the X_α." Here is the problem. If there *does* exist such a function, how is it defined? The problem is that we are asserting the existence of something, but we may not have a method for actually finding it.

This is a real mathematical problem in that there is no way to prove that such a function exists, based on the set theory we have mentioned to date. Briefly, here is the situation. There is a statement in set theory called the "Axiom of Choice" which roughly says that "given any collection of non-empty sets, there exists an element in the disjoint union, namely, just choose one element from each set."

Definition A.43 The Axiom of Choice: *Suppose we have a collection of non-empty sets* $\{X_\alpha\}_{\alpha \in A}$*; then there is a function* $f: A \to \bigcup_{\alpha \in A} X_\alpha$ *with* $f(\alpha) \in X_\alpha$ *for all* $\alpha \in A$.

It is known that the Axiom of Choice is independent of the other axioms of set theory. The controversy about accepting the Axiom of Choice has to do with some of the implications. Using the Axiom of Choice, it is possible to prove some remarkable results, so-called "paradoxes." On the other hand, if we do not allow the Axiom of Choice, we find ourselves very limited in what we can do. In this text, we accept the Axiom of choice as one of the axioms of set theory.

The following is a well-known example, that illustrates the kind of constructions that can arise where one defines a set with no effective way of explicit construction.

Example A.44

Consider the following equivalence relation for $\mathbf{R}^1$: $x \sim y$ if and only if $x - y$ is a rational number. This gives rise to a partition of $\mathbf{R}^1$ into a collection of subsets $\{X_\alpha\}_{\alpha \in A}$; see Proposition A.36 . (The index set is uncountable since each X_α is a countable set. If A were countable, then $\mathbf{R}^1 = \cup_{\alpha \in A} X_\alpha$ would be countable (see Problem C.2), contradicting the uncountability of $\mathbf{R}^1$ (see Problem C.3)).

Our problem is to construct a set S that contains exactly one point from each X_α. The Axiom of Choice implies that S exists, but no explicit construction of S has ever been done. ◆

B. NUMBERS

OVERVIEW: This appendix provides a reference to basic properties of the real number system. We concentrate on the nature of the greatest lower bound axiom, and how it is used.

For purposes of reference, we list properties of real numbers:

Definition B.01 *The* **real number system** *is a set of objects with two operations (addition and multiplication), one relation (equality), and one additional property (the greatest lower bound property.) The two operations and relations satisfy the following familiar laws of algebra:*

(a) *For all numbers x, y, and z, we have*
$x + y = y + x$ *and* $xy = yx$;
$x + (y + z) = (x + y) + z$ *and* $x(yz) = (xy)z$;
$x(y + z) = xy + xz$.

(b) *There are two special numbers denoted 0 and 1 such that, for every x $x + 0 = x$ and $1x = x$.*

(c) *For every x, there is a w with $x + w = 0$.*

(d) *For every x, with $x \neq 0$, there is a w such that $xw = 1$.*

(e) *There is a subset of the real numbers, called the "positive numbers," denoted R_+ such that*
if $x \in R_+$ *and* $y \in R_+$ *then* $x + y \in R_+$ *and* $xy \in R_+$;
if $x \neq 0$ *then* $x \in R_+$ *or* $-x \in R_+$, *but not both;*
$0 \notin R_+$.

(f) *The set of numbers satisfies the greatest lower bound property.*

We describe the familiar order relation if we define $x < y$ to mean $y - x \in R_+$.

In order to state the least upper bound axiom fully, we need some definitions.

Definition B.02 *Let S be a set of real numbers. We say b is a* **lower bound** *for S if, for all $x \in S$, we have $b \leq x$.*

Definition B.03 *Let S be a set of real numbers. We say b_0 is a greatest lower bound for S, if b_0 is a lower bound for S, and $b \leq b_0$ for any lower bound b of S. The* **greatest lower bound** *of S, is often called the* **infimum** *of S. We use the notations $g.l.b(S)$ or $\inf(S)$.*

Definition B.04 The Greatest Lower Bound Property. *Let S be a non-empty subset of real numbers. If S has a lower bound, then there exists a greatest lower bound for S.*

Symmetric with the greatest lower bound property, we have the least upper bound property.

Definition B.05 *Let S be a set of real numbers. We say b is an* **upper bound** *for S if, for all $x \in S$, we have $x \leq b$.*

Definition B.06 *Let S be a set of real numbers. We say b_0 is a* **least upper bound** *for S, if b_0 is an upper bound for S and $b_0 \leq b$ for any upper bound b of S. The least upper bound of S is often called the* **supremum** *of S. We use notations $l.u.b.(S)$ and $\sup(S)$.*

Definition B.07 The Least Upper Bound Property. *Let S be a non-empty subset of real numbers. If S has an upper bound, then there exists a least upper bound for S.*

Proposition B.08 *Given any real number a, there is a natural number N, such that $a < N$.*

Proof: Our argument is indirect. If the conclusion of our proposition were false; then $\mathbf{N}$ would be a non-empty subset of real numbers, and a would be an upper bound of $\mathbf{N}$. By the least upper bound property, there is a least upper bound for $\mathbf{N}$; call this least upper bound A. For all $n \in \mathbf{N}$, $n < A$, and A is the smallest number with this property. Now, if $n \in \mathbf{N}$, then $(n+1) \in \mathbf{N}$. Thus, for all $n \in \mathbf{N}$, $n + 1 < A$. This means that, for all $n \in \mathbf{N}$, that $n < A - 1$. So, $A - 1$ is an upper bound of $\mathbf{N}$. But since $A - 1 < A$, A was not the least upper bound of $\mathbf{N}$. Thus we have arrived at a contradiction. ∎

Proposition B.09 *Given any positive real number a, there is an integer N with $1/N < a$.*

Proof: Suppose the proposition is false, then we would have that for all $n \in \mathbf{N}$, $a < 1/n$. But then, for all $n \in \mathbf{N}$, we have $n < 1/a$. But this now contradicts Proposition B.09. ∎

The set of real numbers has both the least upper bound property and the greatest lower bound property. However, we only need to assume one of these. If we assume the set of axioms in the list above, together with the greatest lower bound property, then we can deduce the least upper bound property.

These properties are important and powerful because they assert the existence of numbers. For example, how do we know that $\sqrt{2}$ exists? That is, how do we know there is a solution of the equation $x^2 = 2$?

We give the proof as a typical example of the way that the greatest lower bound axiom is used in a proof.

Proposition B.10 *There is a positive real number which is a solution of the equation $x^2 = 2$.*

Proof: Here is the basic outline of the argument.

Let S be the set of all positive real numbers whose square is greater than 2. This set is non-empty since $3 \in S$. The set S has a lower bound; for example, 0 is a lower bound. Thus S has a greatest lower bound, call it B. We then wish to show that $B^2 = 2$. We argue by contradiction. If we assume that $B^2 \neq 2$, then there are two cases: $B^2 < 2$ and $2 < B^2$. In the first case, we contradict the assumption that B is the *greatest* lower bound of S. In the second case, we contradict the assumption that B is a lower bound of S.

In the first case, we assume $B^2 < 2$. We show that if this were true, then there is a natural number n such that $(B + \frac{1}{n})^2 < 2$. If we establish this, then we would have a contradiction since $(B + \frac{1}{n})$ is a lower bound for S, bigger than B which was supposed to be the *greatest* lower bound. We now show how to find our natural number, n.

Note that

$$(B + \frac{1}{n})^2 = B^2 + \frac{2B}{n} + \frac{1}{n};$$

thus we need to find an n such that

$$B^2 + \frac{2B}{n} + \frac{1}{n} < 2.$$

We are assuming that $B^2 < 2$; thus $0 < 2 - B$. Since $(\frac{1}{2})^2 < 2$, then we must have $\frac{1}{2} \leq B$. In particular, $0 < B$, and we may make use of Proposition B.09. The next steps involve making some manipulations of the inequalities at hand to find how large we should choose our n. What we present here is the end result of this. These manipulations can be reconstructed by reading the next portion of the proof backwards.

By Proposition B.09, there is an n such that

$$\frac{1}{n} < \frac{2 - B^2}{4B} \text{ and at the same time have } \frac{1}{n} < \frac{2 - B^2}{2}.$$

Since $1 \leq n$, we would also have $\frac{1}{n^2} < \frac{2-B^2}{2}$. This implies that

$$\frac{2B}{n} < \frac{2 - B^2}{2} \text{ and } \frac{1}{n^2} < \frac{2 - B^2}{2}.$$

Adding these inequalities we get

$$\frac{2B}{n} + \frac{1}{n^2} < 2 - B^2.$$

From this we get

$$B^2 + \frac{2B}{n} + \frac{1}{n^2} = (B + \frac{1}{n})^2 < 2.$$

This completes the proof of our first case.

In the second case, we assume that $2 < B^2$, and thus $0 < B^2 - 2$. We show that there is a natural number, n, such that $2 < (B - \frac{1}{n})^2$. Then we would have a contradiction since $B - \frac{1}{n} \in S$ and $B - \frac{1}{n} < B$, contradicting the assumption that B is a lower bound of S. We now show how to find our natural number, n.

Since $0 < B$, by Proposition B.09, there is an n such that

$$\frac{1}{n} < \frac{B^2 - 2}{2B}.$$

Then $\frac{2B}{n} < B^2 - 2$.

From this we get

$$2 < B^2 - \frac{2B}{n} < B^2 - \frac{2B}{n} + \frac{1}{n^2} = (B - \frac{1}{n})^2.$$

This completes the proof in the second case, which concludes the proof. ∎

By modifying the proof of Proposition B.09, one can show the following propositions.

Proposition B.11

If a and b are any two real numbers with $a < b$, then there is a rational number r such that $a < r < b$. In fact, one can find a number r of the form $r = \frac{q}{2^n}$ where q is an integer and n is a natural number. ∎

If r is any rational number, then $r\sqrt{2}$ is an irrational number. (If $r\sqrt{2} = q$ for some rational number q, then $\sqrt{2} = \frac{q}{r}$, but this would imply that $\sqrt{2}$ is rational.) Similar to Proposition B.09 one can show:

Proposition B.12 *Given any positive real number a, there is an integer, N, with $\frac{\sqrt{2}}{N} < a$.* ∎

This can then be used to prove

Proposition B.13 *If a and b are any two real numbers with $a < b$, then there is an irrational number x such that $a < x < b$. In fact, one can find a number x of the form $x = \frac{q\sqrt{2}}{2^n}$ where q is an integer and n is a natural number.* ∎

It is a familiar fact that any real number can be written in decimal notation. We examine what this means and why it is true, with particular attention to the role of the least upper bound property.

Let x be a real number. We say x is written in decimal notation if we write: $x = n.d_1d_2d_3 \ldots.$ where n is an integer, and, for each i, d_i is an integer with $0 \leq d_i \leq 9$. This notation is a shorthand for

$$x = n + \frac{d_1}{10} + \frac{d_2}{10^2} + \frac{d_3}{10^3} + \frac{d_4}{10^4} + \cdots.$$

That this infinite sum is well-defined follows from the least upper bound property. Let n be an integer, and d_i be digits. Let $P_k = n + \frac{d_1}{10} + \frac{d_2}{10^2} + \frac{d_3}{10^3} + \cdots. \frac{d_k}{10^k}$. ($P_k$ is the k-th partial sum.) We show $\{P_k\}$ is bounded above by $n+1$. Let $w = \frac{d_1}{10} + \frac{d_2}{10^2} + \frac{d_3}{10^3} + \frac{d_4}{10^4} + \cdots$. Let $x = \frac{1}{10} + \frac{1}{10^2} + \frac{1}{10^3} + \frac{1}{10^4} + \cdots$. Since each $d_i \leq 9$, we have $w \leq 9x$. Using the standard summation formula for a geometric series, we see that $x = 1/9$ and thus $w \leq 1$. Then it follows from the least upper bound property that $\{P_k\}$ has a least upper bound. Thus we may make the following definition.

Definition B.14 *Given an integer n and digits $d_1, d_2, \ldots$, the* **real number** $n + \frac{d_1}{10} + \frac{d_2}{10^2} + \frac{d_3}{10^3} + \frac{d_4}{10^4} + \cdots$. *is defined to be the least upper bound of the numbers $\{P_k\}$ where $P_k = n + \frac{d_1}{10} + \frac{d_2}{10^2} + \frac{d_3}{10^3} + \cdots. \frac{d_k}{10^k}$.*

Proposition B.15 *Every real number can be written in decimal notation.*

Proof (Outline): Let x be a real number. Let S be the set of real numbers less than or equal to x, which can be written in decimal notation. We can show that S is non-empty. Also note that x is an upper bound for S. Let B be a least upper bound for S. We then show that $x = B$ and $B \in S$. ∎

In the last step of the proof above, we need to show that two numbers, about which we have little direct information, are equal. The following is useful.

Proposition B.16 *If x and y are two real numbers such that for any positive number ϵ, we have $|x - y| < \epsilon$, then $x = y$.* ∎

For example, one can use this to show:

Proposition B.17 *If $x = n.d_1d_2d_3 \ldots$ where, for some K, we have $d_K \neq 9$ and, for all k, with $k > K$, we have $d_k = 9$. Then we can also write $x = n.d_1d_2 \ldots (d_K + 1)000 \ldots.$* ∎

For example, 1.999999... = 2.000000..., and 1.23299999... = 1.23300000.... Thus we see that decimal notation is not unique. However, this is the only type of ambiguity.

Definition B.18 *If $x = n.d_1d_2d_3\ldots$, and there is a K, such that for all k, with $k > K$, we have $d_k = 9$, we say that the decimal expression $n.d_1d_2d_3\ldots$ has a* **terminal sequence of 9's.**

Every real number can be written uniquely in decimal notation, if we avoid such terminal sequeces of 9's:

Proposition B.19 *If $x = n.d_1d_2d_3\ldots$ and $y = m.c_1c_2c_3\ldots$, where neither of these two decimal expressions has a terminal sequence of 9's, then $x = y$ if and only if $n = m$ and for all i, we have $d_i = c_i$.* ∎

The discussion above has been for the familiar decimal system, that is, the system which uses base 10. However, one can develop notations for numbers using other bases. Perhaps less familiar is the decimal notation in the base 2 and base 3. For example, in the base 2 system, we write the number eleven as 1011 , and using base of 3 we would write eleven as 102.

To obtain the decimal expression of a number x in the base 2 system, we write $x = n.d_1d_2d_3\ldots$ where n is an integer (written in binary form), and each d_i is a binary digit. That is, for each i, d_i is either 0 or 1. This notation is a shorthand for saying that we can write

$$x = n + \frac{d_1}{2} + \frac{d_2}{2^2} + \frac{d_3}{2^3} + \frac{d_4}{2^4} + \cdots.$$

For example, the number five and three fourths, written as a binary expression, becomes 101.11000....

Similarly, we could consider the base three system. To obtain the decimal expression of a number x in the base 3 system, we write $x = n.d_1d_2d_3\ldots$ where n is an integer (written in base 3), and each d_i is a base 3 digit; that is, for each i, d_i is 0, 1, or 2. This notation is a shorthand for

$$x = n + \frac{d_1}{3} + \frac{d_2}{3^2} + \frac{d_3}{3^3} + \frac{d_4}{3^4} + \cdots.$$

For example, the number eleven and one-fourth, written as a decimal expression in the base three system, becomes 102.0202020202....

Every real number can be expressed as a binary decimal expression, and this expression is unique except for the case of a terminal sequence of 1's. For example, in the binary system we have

$$101.1101011111111\ldots = 101.1101100000000\ldots.$$

For decimal expressions in the base 3, the only problems arise with numbers which have a terminal sequence of 2's. For example, in the base 3 system, we have

$$2010.01202102222222\ldots = 2010.01202110000000\ldots.$$

C. CARDINALITY OF SETS

OVERVIEW: Cardinality of a set is a measure of its size. We review, and give examples of cardinality. Of special importance is the construction of one-to-one correspondences for infinite sets. Some of the techniques of the proofs presented here may be useful as guides for constructions needed for exercises in the text.

Definition C.01 *We say a set X is* **finite set** *if there is an integer n and a one-to-one correspondence between X and the set of integers from 1 to n. In this case, we say X* **has exactly n elements**.

Definition C.02 *A set X is* **infinite** *if it is not finite.*

Definition C.03 *If there is a one-to-one correspondence between a set and all of the integers, then we say the set is* **countably infinite**.

Definition C.04 *We say a set is* **countable** *if it is either countably infinite or finite.*

For example, every subset of the integers is countable.

Definition C.05 *A set which is not countable is called* **uncountable**.

Proposition C.06 *A subset of a finite set is a finite set. A subset of a countable set is a countable set.* ∎

Proposition C.07 *If $A \subseteq B$, and A is an uncountable set, then B is an uncountable set.* ∎

The following result shows that there are exactly as many integers as there are natural numbers.

Proposition C.08 *The set of integers is a countable set.*

Proof: We need to define a one-to-one correspondence between the set of natural numbers and the set of integers. We can think of such a correspondence as an infinite list in which we find each integer once and only once. It is not hard to construct such a list. For example

1. 0

2. 1

3. -1

4. 2

5. -2

6. 3

7. ...

This one-to-one correspondence can be described by the function

$$f(x) = \begin{cases} \frac{n}{2} & \text{if } n \text{ is even} \\ \frac{1-n}{2} & \text{if } n \text{ is odd.} \end{cases} \blacksquare$$

Proposition C.09 *The set of positive rational numbers is countably infinite.*

Proof: We construct an infinite list which contains each rational number, exactly one time. We begin by writing all the rational numbers in an array:

1/1	2/1	3/1	4/1	5/1	6/1	7/1	8/1...
1/2	2/2	3/2	4/2	5/2	6/2	7/2	8/2...
1/3	2/3	3/3	4/3	5/3	6/3	7/3	8/3...
1/4	2/4	3/4	4/4	5/4	6/4	7/4	8/4...
1/5	2/5	3/5	4/5	5/5	6/5	7/5	8/5...
1/6	2/6	3/6	4/6	5/6	6/6	7/6	8/6...
1/7	2/7	3/7	4/7	5/7	6/7	7/7	8/7...
1/8	2/8	3/8	4/8	5/8	6/8	7/8	8/8...
⋮	⋮	⋮	⋮	⋮	⋮	⋮	⋮

Now consider the path, indicated by the arrows, which begins in the corner of this array and sweeps back and forth in a diagonal way.

1/1	→	2/1		3/1	→	4/1		5/1	→	6/1		7/1	→	8/1
	↙		↗		↙		↗		↙		↗		↙	
1/2		2/2		3/2		4/2		5/2		6/2		7/2		8/2
↓	↗		↙		↗		↙		↗		↙		↗	
1/3		2/3		3/3		4/3		5/3		6/3		7/3		8/3
	↙		↗		↙		↗		↙		↗		↙	
1/4		2/4		3/4		4/4		5/4		6/4		7/4		8/4
↓	↗		↙		↗		↙		↗		↙		↗	
1/5		2/5		3/5		4/5		5/5		6/5		7/5		8/5
	↙		↗		↙		↗		↙		↗		↙	
1/6		2/6		3/6		4/6		5/6		6/6		7/6		8/6
↓	↗		↙		↗		↙		↗		↙		↗	
1/7		2/7		3/7		4/7		5/7		6/7		7/7		8/7
	↙		↗		↙		↗		↙		↗		↙	
1/8		2/8		3/8		4/8		5/8		6/8		7/8		8/8

Consider the list obtained by writing these numbers in the order that they occur along this path.

1. 1/1
2. 2/1
3. 1/2
4. 1/3
5. 2/2
6. 3/1
7. 4/1
8. 3/2
9. 2/3
10. ...

This list clearly contains all positive rational numbers. However, there are duplications in this list. For example, the number 1 is found in the forms 1/1, 2/2, 3/3, etc. We modify our rule as follows. Follow the path as indicated above, and write the positive rational numbers in the order that they occur along this path; however, do not put the number in the list if it has already been represented previously on the list. Thus our new list begins as follows.

1. 1/1
2. 2/1
3. 1/2

4. 1/3
5. 3/1
6. 4/1
7. 3/2
8. 2/3
9. 1/4
10. ...

Clearly, such a list gives a one-to-one correspondence. ∎

Proposition C.10 *The set of rational numbers between 0 and 1 is countably infinite.*

Proof: We use the idea of the proof of the previous theorem; however, we need to modify the rule. Here is the rule, with the modification shown in italics. Follow the path as above, and write the positive rational numbers in the order that they occur along this path; however, do not put the number in the list if it has already been represented previously, *nor if the number is larger than* 1. With this rule, our new list begins as follows.

1. 1/1
2. 1/2
3. 1/3
4. 2/3
5. 1/4
6. 1/5
7. 3/4
8. 3/5
9. 1/6
10. ...

This gives our desired one-to-one correspondence. ∎

Proposition C.11 *The set of all numbers between* 0 *and* 1 *is uncountable.*

Proof: We argue by contradiction. If the set of all numbers between 0 and 1 were countable, we could construct an infinite list that contains all real numbers between 0 and 1. We show that this cannot happen by giving a method whereby, if we had any such list, there must be some number between 0 and 1 missing from the list.

Suppose we had a list of all real numbers between 0 and 1. Write the numbers in decimal notation, making sure not to use terminal sequences of 9's; see Proposition B.17. If the j-th digit of the i-th number on the list is denoted a_{ij}, the list will look like this:

1. $0.a_{11}a_{12}a_{13}a_{14}a_{15}a_{16}a_{17}a_{18}a_{19}\ldots$
2. $0.a_{21}a_{22}a_{23}a_{24}a_{25}a_{26}a_{27}a_{28}a_{29}\ldots$
3. $0.a_{31}a_{32}a_{33}a_{34}a_{35}a_{36}a_{37}a_{38}a_{39}\ldots$
4. $0.a_{41}a_{42}a_{43}a_{44}a_{45}a_{46}a_{47}a_{48}a_{49}\ldots$
5. $0.a_{51}a_{52}a_{53}a_{54}a_{55}a_{56}a_{57}a_{58}a_{59}\ldots$
6. $0.a_{61}a_{62}a_{63}a_{64}a_{65}a_{66}a_{67}a_{68}a_{69}\ldots$
7. $0.a_{71}a_{72}a_{73}a_{74}a_{75}a_{76}a_{77}a_{78}a_{79}\ldots$
8. $0.a_{81}a_{82}a_{83}a_{84}a_{85}a_{86}a_{87}a_{88}a_{89}\ldots$
9. ...

Consider the number B whose decimal notation is $0.b_1b_2b_3b_4b_5b_6b_7b_8\ldots$ where

$$b_i = \begin{cases} a_{ii} - 1 & \text{if } a_{ii} \neq 0 \\ 1 & \text{if } a_{ii} = 0. \end{cases}$$

Now B is a number which is not on the list, because it differs from the i-th number on the list in the i-th decimal place. Note also that B does not have a terminal sequence of 9's since no b_i can be 9.

For example, if our list happened to be

1. 0.123123123123123 ...
2. 0.314159265358979 ...
3. 0.198819891990199 ...
4. 0.333333333333333 ...
5. 0.000000010661066 ...
6. 0.112123123412345 ...
7. ...

the number B would be 0.007212 .... ∎

There is a one-to-one correspondence between the points of the line and the points of the plane, as the next proposition shows.

Proposition C.12 *There is a one-to-one correspondence between the set of all real numbers and the set of all complex numbers.*

Proof: We use the Bernstein-Schroeder Theorem. The standard inclusion map $f(x) = (x, 0)$ provides a one-to-one map of $\mathbf{R}^1$ into $\mathbf{R}^2$. We next need to define a one-to-one map, g, from $\mathbf{R}^2$ to $\mathbf{R}^1$.

Let (X, Y) be a point of the plane. Write each of X and Y in decimal form, using digits x_i, y_i, X_i, and Y_i, and allowing no terminal sequence of 9's:

$$X = \pm X_n \dots X_1.x_1x_2x_3 \dots \text{ and } Y = \pm Y_m \dots Y_1.y_1y_2y_3 \dots.$$

There are only two possible signs for a real number, but there are four possible sign combinations for the numbers X and Y. Here is a way of encoding this information. If Y is positive, replace the string Y_n ...Y_1 by the string corresponding to twice this number; call this new string Z_k ... Z_1. If Y is negative, replace the string Y_n ...Y_1 by the string corresponding to twice this number plus 1; call this new string Z_k ...Z_1 (do not retain the minus sign).

We now have two numbers, written in decimal form:

$$\pm X_n \dots X_1.x_1x_2x_3 \dots \text{ and } Z_k \dots Z_1.y_1y_2y_3 \dots.$$

The number $g(X, Y)$ is the number formed by alternating the digits of these two numbers (filling in zeros to the left of the string, if necessary, to force n and k to be equal) and retaining the sign of X. We thus obtain the number

$$g(X, Y) = \pm X_n Z_n \dots X_1 Z_1.x_1y_1x_2y_2x_3y_3 \dots.$$

One can check that g is one-to-one.

For example, suppose that

$$\begin{aligned} g(X, Y) &= -12345.67891011121314\dots. \text{ Write this as} \\ & -1\underline{2}3\underline{4}5.\underline{6}7\underline{8}9\underline{1}0\underline{1}1\underline{1}2\underline{1}3\underline{1}4\dots. \end{aligned}$$

Then, looking at the underlined digits, we see that we must have had $x = -24.6811111\dots$. From the non-underlined portion, we first extract the integer 135; this is an odd number, and $135 = 2 \cdot 67 + 1$. So Y must have been $-67.7901234\dots$. ∎

A "binary string" is a sequence of zeros and ones, for example, 00110101.... Use of binary strings for labels is natural in many constructions in topology. It is important that this is an uncountable set. One could prove this by using Proposition C.07; we offer, below, a more direct proof.

Proposition C.13 *There is a one-to-one correspondence between real numbers in* $[0, 1]$ *and infinite binary strings of zeros and ones.*

Proof: We can turn a binary string into a real number in the unit interval by placing a zero and decimal point to the left of the string. For the string 00110101 ...we would get the number 0.00110101 ....

The problem is that two strings may correspond to the same number; for example, $0.0\bar{1}$ and $0.1\bar{0}$; see the remark on page 406. For this proof, agree on the convention that a real number is represented by a binary decimal without use of $\bar{1}$.

We define a one-to-one correspondence f between the set of binary strings and the real numbers in the unit interval. Any binary string which does not have a $\bar{1}$ or a $\bar{0}$ corresponds to the real number as indicated above. We complete our definition of our one-to-one correspondence by finding a one-to-correspondence between the binary strings with terminal zeros or terminal ones, and the real numbers in the unit interval whose decimal expansion has a terminal sequence of zeros. Below is the beginning of this correspondence.

$0\overline{1}$	$0.01\overline{0}$
$1\overline{0}$	$0.11\overline{0}$
$00\overline{1}$	$0.001\overline{0}$
$10\overline{1}$	$0.101\overline{0}$
$01\overline{0}$	$0.011\overline{0}$
$11\overline{0}$	$0.111\overline{0}$
$000\overline{1}$	$0.001\overline{0}$
$100\overline{1}$	$0.101\overline{0}$
$010\overline{0}$	$0.010\overline{0}$
$110\overline{0}$	$0.111\overline{0}$
$0000\overline{1}$	$0.0001\overline{0}$
$0100\overline{1}$	$0.0101\overline{0}$
$1000\overline{1}$	$0.1001\overline{0}$
$1100\overline{1}$	$0.1101\overline{0}$
$0010\overline{0}$	$0.0011\overline{0}$
$0110\overline{0}$	$0.0111\overline{0}$
$1010\overline{0}$	$0.1011\overline{0}$
$1110\overline{0}$	$0.1111\overline{0}$
...	...

To define this correspondence we first note that if a binary string, b, has a terminal sequence of zeros, then, except for $b = \overline{0}$, we can write b as $x0\overline{0}$ where x is a finite binary string ending in a one. (If b has a terminal sequence of zeros, x represents an odd integer.) If b has a terminal sequence of ones, then, except for $b = \overline{1}$, we can write b as $x\overline{1}$ where x is a binary string that ends in a zero. (In this case, the string x represents an even natural number.) Consider the correspondence that takes a binary string of the form $x\overline{0}$ or $x\overline{1}$, and associates to it a real number whose binary decimal notation is $0.x1\overline{0}$. This gives us a one-to-one correspondence. (This correspondence is similar to the function in the proof of Proposition C.08. By taking a string and appending a symbol 1 to the end, we are essentially multiplying by two and adding one since a left shift of a binary integer is multiplication by two.)

This is almost the one-to-one correspondence we wish. The problem is that there are two binary strings not accounted for, namely, $\overline{0}$ and $\overline{1}$. Also there are three real numbers not considered, namely, $0.0\overline{0}$, $1.\overline{0}$, and $0.0\overline{1}$. However, one can take care of this problem by adjusting the start of our correspondence; thus obtaining a one-to-one correspondence as desired. ∎

C.1 Problems for Appendix C

C.1 Prove Proposition C.07.

C.2 Prove that a countable union of countable sets is countable.

C.3 Show that the set of all real numbers is uncountable.

C.4 Show that the set of all points (x, y), in the plane, such that x and y are both rational, is a countable set.

C.5 Show that the set of all polynomial equations, in two variables with integer coefficients, is countable.

C.6 Show that the set of all binary strings is uncountable by using the technique of the proof of Proposition C.11.

C.7 Let F be the set all functions from $\mathbf{R}^1$ to $\mathbf{R}^1$. Show that there does not exist a one-to-one correspondence between F and $\mathbf{R}^1$. (This means that F has more than an uncountable number of elements! Hint: Think about our proof that the real numbers are uncountable. Your proof might begin: Suppose there is a one-to-one correspondence ϕ from the set of functions whose from $\mathbf{R}^1$ to $\mathbf{R}^1$. We define a function $g: \mathbf{R}^1 \to \mathbf{R}^1$ such that for all $x \in \mathbf{R}^1$, $g(x) \neq f(\phi(x))$, ...)

D. SUMMARY FROM CALCULUS

OVERVIEW: For reference purposes, we list here familiar definitions and statements of basic results on limits and continuity. In the first part of this appendix, all functions considered are functions of a single real variable.

The following definitions are found in most calculus texts.

Definition D.01 *Suppose $f(x)$ is a real-valued function. We define the* **limit** *of $f(x)$ at a is L by $\lim_{x \to a} f(x) = L$ if, for every $\epsilon > 0$, there exists a $\delta > 0$ such that if $0 < |x - a| < \delta$, then $|f(x) - L| < \epsilon$.*

Definition D.02 *Suppose $f(x)$ is a real-valued function, defined (at least) for an interval (x_0, a). Then we define the* **left-hand limit** *of $f(x)$ at a is L by $\lim_{x \to a^-} f(x) = L$, if, for every $\epsilon > 0$, there exists a $\delta > 0$ such that if $a - \delta < x < a$, then $|f(x) - L| < \epsilon$.*

Definition D.03 *Suppose $f(x)$ is a real-valued function, defined (at least) for an interval (a, x_0). Then we define the* **right-hand limit** *of $f(x)$ at a is L by $\lim_{x \to a^+} f(x) = L$ if, for every $\epsilon > 0$, there exists a $\delta > 0$, such that if $a < x < a + \delta$, then $|f(x) - L| < \epsilon$.*

The relationship between these types of limits can be expressed:

Proposition D.04 *Suppose $f(x)$ is a real-valued function. Then $\lim_{x \to a} f(x)$ exists if and only if $\lim_{x \to a^+} f(x)$ and $\lim_{x \to a^-} f(x)$ exist and are equal. In this case, $\lim_{x \to a} f(x) = \lim_{x \to a^+} f(x) = \lim_{x \to a^-} f(x)$.* ∎

Example D.05 Let

$$f(x) = \begin{cases} 1 & \text{if } x \leq 3 \\ 2 & \text{if } 3 < x. \end{cases}$$

Then $\lim_{x \to 3^+} f(x) = 2$ and $\lim_{x \to 3^-} f(x) = 1$. Thus f is not continuous at 3. ◆

Definition D.06 *Suppose $f(x)$ is a real-valued function and $a \in \mathbf{R}^1$. Then f is* **continuous at a point** *a if* $\lim_{x \to a} f(x) = f(a)$

Putting the preceding definitions together, we get the following expression of continuity:

Suppose $f(x)$ is a real-valued function, and a a number. Then f is continuous at a if for any number $\epsilon > 0$ there exists a $\delta > 0$ such that if $|x - a| < \delta$, then $|f(x) - f(a)| < \epsilon$.

Definition D.07 *Suppose $f(x)$ is a real-valued function, defined on $\mathbf{R}^1$; f is a* **continuous** *function if f is continuous at every point of $\mathbf{R}^1$.*

Definition D.08 *Suppose $f(x)$ is a real-valued function, defined (at least) for an interval (x_0, a); f is* **continuous from the left** *at a if $\lim_{x \to a^-} f(x) = f(a)$.*

Definition D.09 *Suppose $f(x)$ is a real-valued function, defined (at least) for an interval (a, x_0); f is* **continuous from the right** *at a if $\lim_{x \to a^+} f(x) = f(a)$.*

Definition D.10 *Suppose $f(x)$ is a real-valued function, defined (at least) for an closed interval $[a, b]$. Then f is* **continuous on the interval** *$[a, b]$, if all of the following hold:*

(a) *f is continuous from the right at a.*
(b) *f is continuous from the left at b.*
(c) *f is continuous at all other points of $[a, b]$.*

The following is a collection of general results about continuity of functions. We state these in terms of continuity at a point; similar results are true for right-handed and left-handed continuity, etc.

Proposition D.11 *Suppose f and g are functions continuous at x_0; then*

(a) *$f + g$ is continuous at x_0.*
(b) *$f - g$ is continuous at x_0.*
(c) *$f \cdot g$ is continuous at x_0.*
(d) *If $g(x_0) \neq 0$ then $\frac{f}{g}$ is continuous at x_0.*
(e) *If k is a real number kf is continuous at x_0.*
(f) *If h is continuous at $f(x_0)$, then $h \circ f$ is continuous at x_0.* ∎

Here, for reference, is a list of the more common continuous functions. One can obtain more by using Proposition D.11

Proposition D.12 *The following functions are continuous (where defined):*

(a) *All polynomials.*
(b) $x^{\frac{p}{q}}$ *where* $\frac{p}{q}$ *is a rational number.*
(c) $\sin(x), \cos(x), \tan(x), \cot(x), \csc(x), \sec(x)$.
(d) $\arcsin(x), \arccos(x), \arctan(x)$.
(e) $\sinh(x), \cosh(x), \tanh(x), \coth(x)$.
(f) $\exp(x), \ln(x), \log(x)$. ▮

Proposition D.13 *Suppose f is a continuous function defined, on an interval $[a,b]$. Then f has a maximum value, M, and a minimum value, m. That is, there is a number C with $a \le C \le b$ with $f(C) = M$, and for all $x \in [a,b]$, $f(x) \le M$. Also, there is a number c with $a \le c \le b$ with $f(c) = m$ and for all $x \in [a,b]$, $m \le f(x)$.* ▮

Proposition D.14 Intermediate Value Theorem. *Suppose f is a continuous function defined on an interval $[a,b]$, and suppose $f(a) \neq f(b)$. If c is between $f(a)$ and $f(b)$ (that is, either $f(a) < c < f(b)$ or $f(b) < c < f(a)$), then there is an x_0 with $a < x_0 < b$ such that $f(x_0) = c$.*

Proposition D.15 Uniform Continuity. *Suppose $f\colon I \to \mathbf{R}^1$ is continuous. For any $\epsilon > 0$, we can find a number δ such that for all x and $y \in I$, if $|x - y| < \delta$, then $|f(x) - f(y)| < \epsilon$.*

We remark that Proposition D.15 is not an immediate consequence of continuity since the claim here is that same the number δ can apply to *any* pair of numbers in I.

Example D.16 Also, it is important, in Proposition D.15 that the domain of the function is a closed interval.

For example consider $f(x) = x^2$, defined for all $x \in \mathbf{R}^1$. It is false that for any $\epsilon > 0$, we can find a number δ such that for all x and $y \in \mathbf{R}^1$, if $|x - y| < \delta$, then $|f(x) - f(y)| < \epsilon$.

That is, there is an $\epsilon > 0$ such that for any $\delta > 0$, we can find x and y with $|x - y| < \delta$ yet $|x^2 - y^2| \ge \epsilon$. We will chose $\epsilon = 2$, and suppose that $\delta > 0$ is given. Choose a number x such that $\delta + 1/\delta < x$. Suppose y is a number with $|x - y| < \delta$; then $x - \delta < y < x + \delta$ and, in particular, $1/\delta < y$. But now

$$|x^2 - y^2| = |(x+y)(x-y)| = |(x+y)||(x-y)| > (1/\delta + 1/\delta)\delta = 2 = \epsilon. \blacklozenge$$

Now we consider functions of several variables.

Suppose f is a function $f : \mathbf{R^n} \to \mathbf{R}^1$. Writing points of $\mathbf{R^n}$ as vectors, with $|\vec{x}|$ denoting the length of the vector $\vec{x}$, we generalize Definition D.01:

Definition D.17 *Suppose $f(\vec{x})$ is a real-valued function. We define the* **limit** *of $f(\vec{x})$ at $\vec{a}$ is L by $\lim_{\vec{x}\to\vec{a}} f(\vec{x}) = L$ if, for every $\epsilon > 0$, there exists a $\delta > 0$ such that if $0 < |\vec{x} - \vec{a}| < \delta$, then $|f(\vec{x}) - L| < \epsilon$.*

At a point of $\mathbf{R^1}$, there are two directions "right" (the positive direction), and "left," (the negative direction). But from a point in the plane there are infinitely many directions. Replacing our right-handed limits and left-handed limits, we have the notion of a directional limit. Here we use the notion of open ball in $\mathbf{R^n}$, Definition 1.04. We note that this definition does not have a standard notation, and many calculus books mention this type of limit only in discussing the directional derivative. Note in the definition below that we restrict $\vec{x}$ to be of the form $\vec{a} + \delta\vec{u}$. Geometrically, this means that $\vec{x}$ lies on a line through $\vec{a}$, with direction $\vec{u}$.

Definition D.18 *Suppose $f(\vec{x})$ is a real-valued function, $f : \mathbf{R^n} \to \mathbf{R^1}$, defined (at least) for an open ball containing $\vec{x}$. Let $\vec{u}$ be a unit vector in $\mathbf{R^n}$. Let $\vec{x} = \vec{a} + \delta\vec{u}$. Then we define the* **directional limit** *of $f(\vec{x})$ at $\vec{a}$ in direction $\vec{u}$ is L $\lim_{\vec{x}\to\vec{a}} f(\vec{x}) = L$ if for every $\epsilon > 0$ there exists a $\delta > 0$ such that if $0 < |\vec{x} - \vec{a}| < \delta$, then $|f(\vec{x}) - L| < \epsilon$.*

Example D.19 Limits in different directions may be different.

Let $f(x,y) = \frac{xy}{x^2+y^2}$. If $\vec{u} = \vec{i}$ and we let $\vec{x} = x\vec{i}$, then $\lim_{\vec{x}\to\vec{0}} f(\vec{x}) = \lim_{x\to 0} \frac{0}{x^2+0^2} = 0$. Similarly, if $\vec{u} = \vec{j}$, then $\vec{x} = y\vec{j}$, then $\lim_{\vec{x}\to\vec{0}} f(\vec{x}) = \lim_{y\to 0} \frac{0}{0^2+y^2} = 0$. However, if $u = \frac{1}{\sqrt{2}}(\vec{i} + \vec{j})$, then we consider points where the x and y coordinates are equal, so $\lim_{\vec{x}\to\vec{0}} f(\vec{x}) = \lim_{y\to 0} \frac{yy}{y^2+y^2} = \frac{1}{2}$ ◆

The definitions of continuity, below, are modeled on Definitions D.06 and D.07

Definition D.20 *Suppose $f(\vec{x})$ is a real-valued function $f : \mathbf{R^n} \to \mathbf{R^1}$, and $\vec{a} \in \mathbf{R^n}$. Then f is* **continuous at point** *$\vec{a}$ if $\lim_{\vec{x}\to\vec{a}} f(\vec{x}) = f(\vec{a})$*

Definition D.21 *Suppose $f(\vec{x})$ is a real-valued function $f : \mathbf{R^n} \to \mathbf{R^1}$. We say f is a* **continuous** *function if f is continuous at every point.*

As in Proposition D.11, we have

Proposition D.22 *Suppose f and g are functions from $\mathbf{R^n}$ to $\mathbf{R^1}$, continuous at x_0, then*

(a) *$f + g$ is continuous at x_0.*
(b) *$f - g$ is continuous at x_0.*
(c) *$f \cdot g$ is continuous at x_0.*
(d) *If $g(x_0) \neq 0$ then $\frac{f}{g}$ is continuous at x_0.*
(e) *If k is a real number kf is continuous at x_0.*

(f) *If h is continuous at $f(x_0)$, then $h \circ f$ is continuous at x_0.*

If $f: \mathbf{R}^1 \to \mathbf{R}^1$ is a continuous function, and we define $F: \mathbf{R}^2 \to \mathbf{R}^1$ by $F(x, y) = f(x)$ we see that F is a continuous function. For example, $F(x, y) = x^2$. Similarly, $G(x, y) = y^2$ is continuous. By Proposition D.22, $x^2 + y^2$ and x^2y^2 are continuous. In this way we see that polynomials in several variables are continuous, as well as functions such as $\sin(x^2 + y^2)$.

The "Inverse Function Theorem" is found in many calculus texts. Recall that if $U \subseteq \mathbf{R}^\mathbf{n}$, with $x_0 \in U$ and $f: U \to \mathbf{R}^\mathbf{n}$ written

$$f(x_1, \ldots, x_n) = (f_1(x_1, \ldots, x_n), \ldots, f_n(x_1, \ldots, x_n)),$$

the "Jacobian determinant" of f at x_0 is the determinant of the $n \times n$ matrix $\frac{\partial f_i}{\partial x_j}(x_0)$.

Proposition D.23 Inverse Function Theorem. *Suppose $U \subseteq \mathbf{R}^\mathbf{n}$, with $x_0 \in U$ and $f: U \to \mathbf{R}^\mathbf{n}$ where the Jacobian determinant of f at x_0 is non-zero; then there is an open subset $V \subseteq U$ such that $f|_V$ is one-to-one.* ∎

The reason that this is called the Inverse Function Theorem is that we can locally define an inverse function $f^{-1}: f(V) \to V$.

If $U \subseteq \mathbf{R}^\mathbf{k}$, with $x_0 \in U$ and $f: U \to \mathbf{R}^\mathbf{n}$ written

$$f(x_1, \ldots, x_k) = (f_1(x_1, \ldots, x_k)), \ldots, f_n(x_1, \ldots, x_k))$$

the differential of f a x_0, is the matrix, $\frac{\partial f_i}{\partial x_j}(x_0)$. This is an $n \times k$ matrix, and, if $n \neq k$, it will not be square. Using the Inverse Function Theorem we can show:

Proposition D.24 *Suppose $U \subseteq \mathbf{R}^\mathbf{k}$, with $x_0 \in U$ and $f: U \to \mathbf{R}^\mathbf{n}$, with $k \leq n$, written*

$$f(x_1, \ldots, x_k) = (f_1(x_1, \ldots, x_k)), \ldots, f_n(x_1, \ldots, x_k))$$

and the differential of f at x_0 has rank k. Then there is an open subset, $V \subseteq \mathbf{R}^\mathbf{k} \times \mathbf{R}^{\mathbf{n}-\mathbf{k}} = \mathbf{R}^\mathbf{n}$, and a function $F: V \to \mathbf{R}^\mathbf{n}$ such that $F|_V$ is one-to-one and such that $F|_{V \cap (\mathbf{R}^\mathbf{k} \times \{0\})} = f$. ∎

Although we do not prove Proposition D.24 here, we outline a proof of a special case that gives the basic idea of the general proof.

Here is a proof of Proposition D.24 for the case $k = 1$, $V = \mathbf{R}^1$, and $n = 2$. We are considering a function $f(x) = (f_1(x), f_2(x))$ and $x_0 \in V$. The differential of f is the tangent vector

$$\left(\frac{df_1}{dx}, \frac{df_2}{dx}\right).$$

The requirement of the theorem is that

$$\vec{v_0} = (\frac{df_1}{dx}(x_0), \frac{df_2}{dx}(x_0))$$

is a non-zero vector. Writing points of $\mathbf{R}^2$ in vector notation, let $\vec{y_0} = f(x_0)$

Let Φ be an non-singular affine transformation of $\mathbf{R}^2$ such that $\Phi(\vec{y_0}) = (0,0)$ and $\Phi(\vec{v_0}) = \vec{i}$. Now we can write $\Phi \circ f = (\phi_1(x), \phi_2(x))$. Define $G: (V \times \mathbf{R}^1) \to \mathbf{R}^2$ by $G(x,y) = (\phi_1(x), \phi_2(x) + y)$. Note that $G(x_0, 0) = (0,0)$ and the differential of G at $(x_0, 0)$ is the 2×2 identity matrix which, of course, has non-zero determinant. By the Inverse Function Theorem, there is an open subset V on which G is one-to-one. Now define $F = \Phi^{-1} \circ G$. One can now verify that F has the desired properties. Note that non-singular affine transformations are one-to-one. ∎

Example D.25 It is helpful to have an example in hand to follow the proof above. Suppose that $f(x) = (\cos(x), \sin(x))$ and that $x_0 = 0$. Then $f(x_0) = (1,0)$ and the tangent vector at $(1,0)$ is $\vec{j}$. In this case, we can take Φ to be the affine transformation: rotation of angle $-\frac{\pi}{2}$ followed by translation by $\vec{j}$. Then $\Phi \circ f(x) = (\cos(x - \frac{\pi}{2}), \sin(x - \frac{\pi}{2}) + 1)$. And $G(x,y) = (\cos(x - \frac{\pi}{2}), \sin(x - \frac{\pi}{2}) + 1 + y)$. Clearly, G is not a one-to-one function—for example, $G(x,y) = G(x + 2\pi, y)$. However, one can check that G is one-to-one on $V = (-\frac{\pi}{2}, \frac{\pi}{2}) \times \mathbf{R}^1$. ◆

Finally, we provide a proof of the Cauchy-Schwarz inequality:

Proposition D.26 Cauchy-Schwarz Inequality *Suppose* $\vec{u}, \vec{v} \in \mathbf{R^n}$; *then*

$$|\vec{u} \cdot \vec{v}| \leq |\vec{u}||\vec{v}|.$$

(Here $||\vec{u} \cdot \vec{v}||$ *denotes the absolute value of the dot product, and* $|\vec{u}|$ *denotes length of the vector.)*

Proof: If either $\vec{u} = \vec{0}$ or $\vec{v} = \vec{0}$, the inequality is true (in fact, is an equality with both sides 0). So we assume that $\vec{u} \neq \vec{0}$ and $\vec{v} \neq \vec{0}$.

First, consider the special case: $|\vec{u}| = 1$ and $|\vec{v}| = 1$. In this case we want to show that $|\vec{u} \cdot \vec{v}| \leq 1$.

For any two real numbers, x and y,

$$0 \leq (x - y)^2 = x^2 - 2xy + y^2.$$

This give us our key algebraic fact: for any two real numbers, x and y, $2xy \leq x^2 + y^2$. Now we have

$$|\vec{u} \cdot \vec{v}| = |\sum_1^n u_i v_i| \leq \sum_1^n |u_i v_i| \leq \sum_1^n \frac{1}{2}(u_i^2 + v_i^2) =$$

$$\frac{1}{2}\left(\sum_1^n u_i^2 + \sum_1^n v_i^2\right) = \frac{1}{2}(1+1) = 1.$$

Having established this, we can take any two non-zero vectors, $\vec{u}, \vec{v}$ in $\mathbf{R^n}$, and apply our inequality to the unit vectors $\frac{\vec{u}}{|\vec{u}|}$ and $\frac{\vec{v}}{|\vec{v}|}$, concluding that

$$\frac{\vec{u}}{|\vec{u}|} \cdot \frac{\vec{v}}{|\vec{v}|} \leq 1.$$

Multiplying both sides of this inequality by $|\vec{u}||\vec{v}|$ gives the Cauchy-Schwarz Inequality. ▮

E. STRATEGY IN PROOF

OVERVIEW: We discuss strategy for generating and articulating mathematical arguments.

E.1 Proofs in general

What is a proof? How much detail is needed for a proof to be complete?

These are not just philosophical questions, but important practical issues. Students are frequently surprised to find that some answers begin: "It depends ...." Here is a frequently used criterion: "A proof is a (correct) argument that will convince whoever you are trying to convince."

Mathematics is a language, and a proof is a communication. As with all communications, the level of detail one needs to communicate clearly depends very much on the people involved. It is dependent on the context of the communication; in particular, on what other bonds of communication have been established between the two parties. For example, the level of detail expected of a student from the instructor at the end of a course may be less than at the beginning, if the instructor becomes confident that the student can correctly "fill in the details."

The best way to learn how to construct proofs is by reading, understanding, and remembering proofs of others. The understanding of proofs is the hard part. It is not enough to be able to verify that each step follows from previous ones. The difficult part is to understand how the author arrived at the proof. Sometimes the author of a proof will explicitly provide the reader with a guide to the construction of the proof. But often the author will take great pains in covering up all evidence of this construction—not generally with malicious intent, but rather to achieve an elegantly brief presentation. Such elegance may bring tears of joy to a mature mathematician but tears of frustration to the beginner.

Too often, from reading standard textbooks a student gets the false impression that mathematics develops in an orderly and logical progression and that one's basic strategy is attaining insight into "the big picture" through proper abstraction, so as to avoid mistakes and blind alleys. Nothing could be further from the truth. Most progress in mathematics is based primarily on trial and error investigation of particular examples and flawed, but hopefully improving, attempts to describe relationships that apply to a large collection of these examples. In short, mathematics is not much different than any other intellectual pursuit.

There are no firm rules for how to prove things in mathematics, only guidelines. This really should be no surprise since a proof is a communication. The nature of communication is that there are many ways to say the same thing, but no ideal best way.

Strategy in mathematics is very similar to the strategy of a coach of a sports team. One never knows, in advance, what plan will be successful. One improves strategy by developing a collection of plans that have been successful in the past. Sometimes there are clues for which plan to try first, and sometimes one just picks one at random and sees what happens.

In most mathematics books, the order of presentation is definition, theorem, proof, example. This order is considered to be most logical and is certainly an efficient and concise exposition.

The student is frequently misled into thinking that this is how mathematics is done. But it is important to keep in mind that what is usually presented is the end product of a long series of refinements of presentation.

In fact, frequently the order in which mathematics is done is the reverse of the definition, theorem, proof, example order:

1. One has examples one is trying to understand.

2. One manages to do some mathematical analysis, find some connections, and establish some argument. One then tries to understand the nature of the argument better and summarize the essential ideas. It is often necessary to go through this cycle several times.

3. To summarize the process, one states carefully and completely what one has demonstrated.

4. In the process of doing this one frequently finds it necessary to re-phrase and refine the concepts involved so that the statement arrived at in the previous step is correct and complete.

These steps, in order, can be roughly described as

1. Example
2. Proof
3. Theorem
4. Definition

These four categories give a good list to organize possible strategies for a proof.

1. Example: Is there an example that would guide us to the proof? We hope that the truth of the statement for this example might serve as a model for a general argument.
2. Proof: Perhaps there is remembered argument that resolves a similar statement; perhaps one can reuse this argument, altered slightly to obtain the desired proof
3. Theorem: Perhaps the problem in question is a consequence of a previously established result. Sometimes one might need a combination of such results.
4. Definition: Some proofs are based primarily on the definitions. By necessity most of the first things one proves relating to a particular topic will be of this kind.

E.2 Specifics for topology proofs

In the previous section, we discussed generalities about how to prove things.

The text has be written so that, in most chapters, there is a key proof in which the strategy of proof is discussed in detail. Also, we provide a large number of examples.

We have tried to devise a special index, to be helpful for searching through the examples, propositions and remarks.

For example, suppose one wants to show that certain subset is a closed subset. In the index under "closed subset," one has a list of examples of closed subsets, and a list of propositions about closed subsets, as well as a list of remarks. In addition, there is a list of examples of subsets which are not closed.

Bibliography

[1] Adams, Colin, *The Knot Book*, W. H. Freeman (1994).

[2] Alexandrov, P. S. *Combinatorial Topololgy*, Dover Publications (1998).

[3] Banchoff, T. and Wermer, J., *Linear Algebra through Geometry* (2nd Ed.) Springer-Verlag (1992)

[4] Banchoff, T., *Beyond the Third Dimension: Geometry, Computer Graphics and Higher Dimensions*, Scientific American Library (1990).

[5] Barnsley, Michael, *Fractals Everywhere*, Academic Press (1988).

[6] Bredon, G., *Topology and Geometry*, Springer-Verlag (1997).

[7] Bröcker, H. and Jñich, K., *Introduction to Differential Toplogy*, Cambridge Univ. Press, (1982).

[8] Burde, G. and Zieschang, H., *Knots* , Walter de Gruyter (1985).

[9] Carleson, L. and Gamelin, T. W., *Complex Dynamics*, Springer-Verlag (1993).

[10] Chinn, W. G. and Steenrod N., *First Concepts of Topology*, Random House (1966).

[11] Croom, F. H., *Principles of Topology*, Saunders, (1989).

[12] Crowell, R. and Fox, R. H., *Introduction to Knot Theory*, Ginn and Co., (1963).

[13] Devaney, R. L., *An Introduction to Chaotic Dynamical Systems*, Benjamin/Cummings (1986).

[14] Dugundji, J, *Topology*, Allyn and Bacon (1965).

[15] Eilenberg, S. and Steenrod, N., *Foundations of Algebraic Topology*, Princeton Univ. Press (1952).

[16] Edgar, G. A., *Measure, Topology, and Fractal Geometry*, Undergrad. Text in Math., Springer-Verlag, (1990).

[17] Engelking, R., *Theory of Dimensions, Finite and Infinite*, Helderman Verlag (1995).

[18] Francis, George K., *A Topological Picturebook*, Springer-Verlag (1987).

[19] Fox, R. H., "A quick trip through knot theory," *Topology of 3-Manifolds and Related Topics* , Prentice Hall , 120–167 (1961).

[20] Gamelin, T. W. and Greene, R. E., *Introduction to Topology*, Saunders (1983).

[21] Guillemin, V. and Pollack, A., *Differential Topology*, Prentice Hall (1974).

[22] Hirsch, M. , *Differential Topology*, Grad. Text Math No. 33, Springer-Verlag (1976).

[23] Hocking, J. and Young, G. *Topology*, Addison-Wesley. (1961), Currentl publisher, Dover Publications (1988).

[24] Hudson, J. F. P, *Piecewise Linear Topology* , W. A. Benjamin (1969).

[25] Hurewicz, W. and Wallman, H., *Dimension Theory*, Princeton Univ. Press (1948).

[26] Kauffman, Louis H. "New Invariants in Knot Theory," Amer. Math. Monthly **(95)** (1988) 195-242.

[27] Kauffman, Louis H., *Knots and Physics*, (Second Ed.), World Scientific, (1995).

[28] Kelley, J. L., *General Topology*, Springer-Verlag (1975).

[29] Livingston, C, *Knot Theory*, Carus Math. Monographs, Amer. Math. Soc. (1993).

[30] Massey, W., *Algebraic Topology: An Introduction*, Harcourt, Brace & World. (1967).

[31] Milnor, J., *Topology from the Differentiable Viewpoint*, Univ. Press of Virginia (1965).

[32] Munkres, J., *Topology: A First Course*, Prentice Hall (1975).

[33] Munkres, J., *Elements of Algebraic Topology*, Benjamin/Cummings Hall (1984).

[34] Nadler, S., *Continuum Theory: An Introduction*, Marcel Dekker (1992).

[35] Newman, M. H. A., *Elements of the Topology of Plane Sets of Points*, Cambridge Univ. Press (1961).

[36] Rolfsen, Dale, *Knots and Links*, Publish or Perish Inc. (1976).

[37] Rourke, C. and Sanderson, B. *Introduction of Piecewise-Linear Topology*, Springer-Verlag (1972).

[38] Roseman, D. (with Mayer, D. and Holt, O.) "Twisting and Turning in 4 Dimensions," video (19 min.), produced at the Geometry Center, August 1993. Distributed: Great Media, Nicassio, CA.

[39] Roseman, D. "Unraveling in 4 Dimensions," video (18 min.), produced at the Geometry Center, July 1994. Distributed: Great Media, Nicassio, CA.

[40] Sagan, H, *Space-Filling Curves*, Springer-Verlag (1994).

[41] Schurle, A., *Topics in Topology*, Elsevier North Holland (1979).

[42] Sieradski, A., *An Introduction to Topology and Homotopy*, PWS-KENT (1992).

[43] Stillwell, J., *Classical Topology and Combinatorial Group Theory*, Springer-Verlag, Grad. Text Math. (1995).

[44] Spanier, E. H., *Algebraic Topology*, McGraw-Hill (1996).

[45] Steen, L. A. and Seebach, J. A., *Counterexamples in Topology*, Holt, Rinehart and Winston (1970).

[46] Sumners, D.W., "New Scientific Applications of Geometry and Topology," *Proc. Symp. App. Math.* **45**, Amer. Math. Soc. (1992).

[47] Wagon, Stan, *Mathematica in Action*, W. H. Freedman.

[48] Willard, S., *General Topology*, Addison-Wesley (1970).

Index of Examples, Remarks, and Propositions

Subject Index